PROBLEM BOOK IN RELATIVITY AND GRAVITATION

PROBLEM BOOK IN RELATIVITY AND GRAVITATION

ALAN P. LIGHTMAN
WILLIAM H. PRESS
RICHARD H. PRICE
SAUL A. TEUKOLSKY

PRINCETON UNIVERSITY PRESS
PRINCETON AND OXFORD

Published by Princeton University Press,
41 William Street, Princeton, New Jersey 08540
In the United Kingdom: Princeton University Press,
6 Oxford Street, Woodstock, Oxfordshire OX20 1TR

press.princeton.edu

Cloth: 978-0-691-17777-9
Paper: 978-0-691-17778-6

The Library of Congress has cataloged the
first printing of this book as follows:

Main entry under title:
Problem book in relativity and gravitation.
Includes bibliographical references and index.
1. Relativity (Physics)—Problems, exercises, etc. 2. Gravitation—
Problems, exercises, etc. 3. Astrophysics—problems, exercises, etc.
I. Lightman, Alan P., 1948-
QC173.55P76 530.1'1 74-25630

British Library Cataloging-in-Publication Data is available

Printed on acid-free paper. ∞

Printed in the United States of America

1 3 5 7 9 10 8 6 4 2

CONTENTS

PREFACE

This book contains almost 500 problems and solutions in the fields of special relativity, general relativity, gravitation, relativistic astrophysics and cosmology. The collection is motivated by a simple premise: that the most important content of this field does not lie in its rigorous axiomatic development, nor, necessarily, in its intrinsic aesthetic beauty, but rather does lie in computable results, predictions, and models for phenomena in the real universe. Accordingly, we have aimed for problems whose statement is broadly understandable in physical terms and have tried to make their statement independent of notational conventions. We hope to awaken the reader's curiosity. ("Now how *would* one show that...?") We have steered clear of purely technical problems, found in texts, of the form "prove equation 17.4.38." In our solutions we also try to show the reader "good" ways to compute things, methods and tricks which can vastly reduce the labor of a plug-in and grind-away approach, but we also try to avoid the opposite pitfall of introducing too much confusing but powerful formalism for an easy problem. There is often a lot of leeway in this balance, and the reader should not be surprised if his solutions use a rather smaller (or larger) set of calculational tools.

The first five chapters of this book deal only with special relativity, and are designed for advanced undergraduates and graduate students in any course in modern physics, classical mechanics or electromagnetism. They are arranged roughly in order of increasing sophistication, beginning at about the easy level of *Spacetime Physics* by E. F. Taylor and J. A. Wheeler (Freeman, 1963); there are, however, both easy and difficult problems in each chapter.

The remainder of the book is aimed at the student in a course in general relativity and/or cosmology. The chapters cover aspects of metric geometry, the equations of Einstein's gravitation theory (and some competing theories), the effect of gravitation on other physical phenomena, and applications to a variety of experimental and astrophysical situations. A final chapter deals with some more formal topics whose applications are less direct.

Each chapter begins with an introductory note whose purpose is largely to define the notation used. These by no means constitute a complete or orderly presentation of the material covered in the chapter, but are intended to aid the student familiar with a notation different from ours. We assume that the reader has the benefit of one or more of the following texts (which we have used heavily):

C. W. Misner, K. S. Thorne, and J. A. Wheeler, *Gravitation* (Freeman, 1973) [cited in this book as "MTW"].

S. Weinberg, *Gravitation and Cosmology* (Wiley, 1972) [cited in this book as "Weinberg"].

R. Adler, M. Bazin, and M. Schiffer, *Introduction to General Relativity* 2nd ed. (McGraw-Hill, 1975).

We have also been influenced by the following texts or monographs:

Anderson, J. L., *Principles of Relativity Physics* (Academic Press, 1967).

Batygin, V. V., and Toptygin, I. N., *Problems in Electrodynamics* (Academic Press - Infosearch, 1964).

Hawking, S. W., and Ellis, G. F. R., *The Large-Scale Structure of Space-Time* (Cambridge University Press, 1973).

Landau, L. D., and Lifschitz, E. M., *The Classical Theory of Fields*, 3rd ed., (Addison-Wesley, 1971).

Peebles, P. J. E., *Physical Cosmology* (Princeton University Press, 1971).

Robertson, H. P., and Noonan, T. W., *Relativity and Cosmology* (Saunders, 1968).

Sexl, R. U., and Urbantke, H. K., *Gravitation and Kosmologie*, (Wiener Berichte über Gravitationstheorie, 1973).

We have cited the primary literature where appropriate.

We are pleased to express our appreciation to colleagues who have contributed original problems to this collection: Douglas Eardley, Charles W. Misner, Don Page, Bernard F. Schutz, and our friend and teacher, Kip S. Thorne.

We are also grateful to C. R. Alcock, B. C. Barrois, J. Conwell, H. B. French, K. S. Jancaitis, C. Jayaprakash, S. J. Kovacs and W. A. Russell for valuable help in improving the problems and solutions. Our thanks go to Steve Wilson for preparing most of the illustrations in this book. We acknowledge support from the Department of Physics at the California Institute of Technology while we were there. Of course, we are responsible for the errors which inevitably must be present in a book of this sort. We have tried particularly hard for problems and solutions which are *conceptually* free from error, but we also apologize in advance for the algebraic slips that the diligent reader will certainly find; we invite his corrections.

A. P. LIGHTMAN
W. H. PRESS
R. H. PRICE
S. A. TEUKOLSKY

PASADENA, MAY 1974

NOTATION

It is intended that this book be compatible with several different textbooks, each with its own system of notational conventions. Thus, no single notational system will be used exclusively in this book. In almost all instances, meanings will be clear from the context. The following is a list of the *usual* meanings of some frequently used symbols and conventions.

$\alpha, \beta, \mu, \nu \cdots$	Greek indices range over 0, 1, 2, 3 and represent spacetime coordinates, components, etc.
$i, j, k \cdots$	Latin indices range over 1, 2, 3 and represent coordinates etc. in 3-dimensional space
$e_\alpha, e_j \cdots$	Basis vectors
$\mathbf{A}$	(Any boldface symbol) a spacetime vector, tensor, or form
$\underset{\sim}{A}$	A 3-dimensional vector
$A^\mu, B^\alpha{}_\beta \cdots$	Tensor components
(A^0, A^1, A^2, A^3)	A vector represented by its components
$(A^0, \underset{\sim}{A})$	A vector represented by its time component and spatial part
$\hat{}$	(Caret) indicates unit vector, components in orthonormal basis
$d/d\lambda$	Occasionally used to represent a vector (see Introduction to Chapter 7)
$A(f)$	A vector operating on a function $= A^\alpha f_{,\alpha}$

$\tilde{\boldsymbol{\omega}}$	A one-form
$\otimes$	Outer product, tensor product e.g. $\mathbf{A}\otimes\mathbf{B}$ has components $A^{\mu}B^{\nu}$
$\wedge$	Wedge product (see Introduction to Chapter 8)
∇	Covariant derivative operator (see Introduction to Chapter 7). Also used as in ordinary physics $\nabla\times$ = curl, ∇^2 = Laplacian, etc.
$\nabla_{\mathbf{A}}$	Directional derivative (see Introduction to Chapter 7)
$D/d\lambda$	Covariant derivative along a curve (see Introduction to Chapter 7)
d	Gradient operator as in e.g. the one-form $\widetilde{df}$ (see introduction to Chapter 8)
$\mathfrak{L}$	Lie derivative (see Problem 8.13)
$\Gamma^{\alpha}{}_{\beta\gamma}$	Christoffel symbol (see Introduction to Chapter 7)
$\Box$	d'Alembertian operator $\equiv \nabla^2 - \partial^2/\partial t^2$ in Special Relativity
,	Partial derivative
;	Covariant derivative (see Introduction to Chapter 7)
$R_{\alpha\beta\gamma\delta}$	Riemann tensor (see Introduction to Chapter 9)
$R_{\alpha\beta}$	Ricci tensor $\equiv R^{\gamma}{}_{\alpha\gamma\beta}$
R	Ricci scalar $\equiv R^{\alpha}{}_{\alpha}$. Also scale factor in Robertson-Walker metric.
$G_{\alpha\beta}$	Einstein tensor (see Introduction to Chapter 9)
$C_{\alpha\beta\gamma\delta}$	Weyl (conformal) tensor (see Introduction to Chapter 9)
K_{ij}	Extrinsic curvature tensor (see Introduction to Chapter 9)
τ	Proper time
c	Speed of light (usually taken as unity in the problems)
G	Gravitational constant (usually taken as unity in the problems)
$\mathbf{u}$	4-velocity
$\mathbf{a}$	4-acceleration $\equiv d\mathbf{u}/d\tau$
$\mathbf{p}$ or $\mathbf{P}$	4-momentum

Symbol	Meaning
p or P	Pressure
$T^{\mu\nu}$	Stress-energy tensor (see Introduction to Chapter 5)
$F^{\mu\nu}$	Electromagnetic field tensor (see Introduction to Chapter 4)
J^{μ}	Current density (see Introduction to Chapter 4)
$J^{\mu\nu}$	Angular momentum tensor (see Problems 11.1, 11.2)
$\eta_{\mu\nu}$	Minkowski metric (see Introduction to Chapter 1)
$h_{\mu\nu}$	Metric perturbations (see Introduction to Chapter 13)
C.M.	Center of momentum frame, center of mass
ν, ω	Frequency in cycles per unit time, radians per unit time
γ	Lorentz factor $\equiv (1-v^2/c^2)^{-\frac{1}{2}}$, or photon symbol
$\Lambda^{\alpha}{}_{\beta}$	Lorentz transformation matrix
det	Determinant
Tr	Trace
$\langle\ \rangle$	Average (as in $\langle E \rangle$ = average energy)
$\langle\ ,\ \rangle$	Scalar combination of vector and one-form, as in $\langle \tilde{\omega}, \mathbf{A} \rangle$ (see Introduction to Chapter 8)
[ ]	Antisymmetrization (see Problem 3.17) or commutator (see Introduction to Chapter 8) or discontinuity (as in Problem 21.9)
( )	Symmetrization (see Problem 3.17)
$\varepsilon^{\alpha\beta\gamma\delta}$	The totally antisymmetric tensor (see Problem 3.20)
*	Duality symbol (see Problem 3.25)
Re	Real part
Ω	Solid angle (as in $\int d\Omega$), angular velocity
$P^{\alpha\beta}$	Projection tensor (see Problems 5.18, 6.6)
θ	Expansion (see Problem 5.18)
$\sigma_{\alpha\beta}$	Shear (see Problem 5.18)
$\omega_{\alpha\beta}$	Rotation (see Problem 5.18)

$\mathcal{I}_{jk}$	Reduced quadrupole tensor (see Introduction to Chapter 18)
H_0	Hubble constant
q_0	Deceleration parameter
$M_\odot, R_\odot, \cdots$	Mass, radius, $\cdots$ of sun
z	Redshift factor (see Problem 8.28, Introduction to Chapter 19)
$\mathcal{O}$	Order of magnitude
$\propto$	Proportional to (e.g., $r^3 \propto t^2$) or parallel vector to (e.g., $\mathbf{A} \propto \mathbf{B}$)

PROBLEMS

CHAPTER 1

SPECIAL-RELATIVISTIC KINEMATICS

The path of an observer through spacetime is called the worldline of that observer. The time measured by the observer's own clocks, called his proper time τ, is given by

$$-d\tau^2 \equiv ds^2 = -dt^2 + dx^2 + dy^2 + dz^2 \quad ,$$

where t, x, y, z are the observer's (Minkowski) coordinates along his path. Here, and unless noted otherwise throughout this book, we use units in which c, the speed of light, is unity.

The 4-velocity $\mathbf{u}$, with components $(dt/d\tau, dx/d\tau, dy/d\tau, dz/d\tau)$, and 4-acceleration $\mathbf{a} \equiv d\mathbf{u}/d\tau$, components $(d^2t/d\tau^2, d^2x/d\tau^2, d^2y/d\tau^2, d^2z/d\tau^2)$, are defined on the worldline. The (contravariant) components of these or other 4-vectors are denoted u^α, a^β, A^γ, B^δ, etc., where a Greek index indicates any of the 4 components $t, x, y, z \equiv 0, 1, 2, 3$. Latin indices $i, j, k\cdots$ are used to indicate only the spatial components $x, y, z \equiv 1, 2, 3$.

The Einstein summation convention is used, that is, any repeated literal index is assumed to be summed over its range. For example,

$$\mathbf{V} = V^\mu e_\mu$$

expresses a vector as a sum of contravariant components multiplied by basis vectors, $e_0 \equiv (1,0,0,0)$, $e_1 \equiv (0,1,0,0)$, etc.

The invariant dot product of two 4-vectors is, in Minkowski coordinates,

$$\mathbf{A}\cdot\mathbf{B} = -A^0B^0 + A^1B^1 + A^2B^2 + A^3B^3 \quad .$$

This can be written as $\mathbf{A}\cdot\mathbf{B} = A_\mu B^\mu$, where the numbers A_μ, called covariant components of $\mathbf{A}$, are defined by $A_\mu \equiv \eta_{\mu\nu}A^\nu$, or $A^\mu = \eta^{\mu\nu}A_\nu$

$$\eta_{\mu\nu} \equiv \begin{bmatrix} -1 & 0 & 0 & 0 \\ 0 & 1 & 0 & 0 \\ 0 & 0 & 1 & 0 \\ 0 & 0 & 0 & 1 \end{bmatrix} \qquad (\text{also} \equiv \eta^{\mu\nu}) \ .$$

Vectors are called spacelike, timelike, or null, according to whether their square $\mathbf{v}\cdot\mathbf{v}$ is positive, negative, or zero. 4-velocities are always timelike.

Two Lorentz frames may differ by a relative 3-velocity $\underset{\sim}{v}$ or by a spatial rotation, or by a combination of relative velocity and rotation. If t, x, y, z are the coordinates of one frame, then the coordinates in a different frame are usually written t', x', y', z'. Similarly, vector components in the primed frame are written $A^{\mu'}$, $B_{\nu'}$, etc., and its basis vectors are $\mathbf{e}_{\mu'}$. The basis vectors and the components of vectors in Lorentz frames are related by

$$\mathbf{e}_{\mu'} = \Lambda^{\alpha}_{\ \mu'}\mathbf{e}_{\alpha}\ , \qquad V_{\mu'} = \Lambda^{\alpha}_{\ \mu'} V_{\alpha}$$

$$V^{\mu'} = \Lambda^{\mu'}_{\ \alpha} V^{\alpha} \qquad (\Lambda^{\mu'}_{\ \alpha} \equiv \text{matrix inverse of } \Lambda^{\alpha}_{\ \mu'})$$

where the Λ's are Lorentz transformation matrices. Of special interest are the "boost" transformations involving changes in velocity with no rotation. For a primed frame with velocity β in the x-direction,

$$\Lambda^{\mu'}_{\ \nu} = \begin{bmatrix} \gamma & -\gamma\beta & 0 & 0 \\ -\gamma\beta & \gamma & 0 & 0 \\ 0 & 0 & 1 & 0 \\ 0 & 0 & 0 & 1 \end{bmatrix}, \qquad \gamma \equiv (1-\beta^2)^{-\frac{1}{2}} \ .$$

The velocity between two frames is sometimes parameterized by $\theta \equiv \tanh^{-1}\beta$ ("the rapidity parameter").

A particle of rest mass m and 4-velocity $\mathbf{u}$ has 4-momentum $\mathbf{p} \equiv m\mathbf{u}$. If $m = 0$ (photons), $\mathbf{p}$ is defined by its components in the frame of any observer $p^0 \equiv$ photon energy, $p^i \equiv \underset{\sim}{p} =$ photon 3-momentum.

Problem 1.1. The 4-velocity $\mathbf{u}$ corresponds to 3-velocity $\underset{\sim}{v}$. Express:

(a) u^0 in terms of $|\underset{\sim}{v}|$

(b) $u^j (j=1,2,3)$ in terms of $\underset{\sim}{v}$

(c) u^0 in terms of u^j

(d) $d/d\tau$ in terms of d/dt and $\underset{\sim}{v}$

(e) v^j in terms of u^j

(f) $|\underset{\sim}{v}|$ in terms of u^0.

Problem 1.2. Find the matrix for the Lorentz transformation consisting of a boost v_x in the x-direction followed by a boost v_y in the y-direction. Show that the boosts performed in the reverse order would give a different transformation.

Problem 1.3. If two frames move with 3-velocities $\underset{\sim}{v}_1$ and $\underset{\sim}{v}_2$, show that their relative velocity is given by

$$v^2 = \frac{(\underset{\sim}{v}_1 - \underset{\sim}{v}_2)^2 - (\underset{\sim}{v}_1 \times \underset{\sim}{v}_2)^2}{(1 - \underset{\sim}{v}_1 \cdot \underset{\sim}{v}_2)^2} \ .$$

Problem 1.4. A cart rolls on a long table with velocity β. A smaller cart rolls on the first cart in the same direction with velocity β relative to the first cart. A third cart rolls on the second cart in the same direction with relative velocity β, and so on up to n carts. What is the velocity v_n of the nth cart in the frame of the table? What does v_n tend to as $n \to \infty$?

Problem 1.5. A distant camera snaps a photograph of a speeding bullet (velocity v) with length b in its rest frame. Behind the bullet and parallel to its path is a meter stick, at rest with respect to the camera. The direction to the camera is an angle α from the direction of the bullet's velocity. What will be the *apparent length* of the bullet as seen in the photo? (i.e. How much of the meter stick is hidden?).

Problem 1.6. Tachyons are hypothetical particles whose velocity is faster than light. Suppose that a tachyon transmitter emits particles of a constant velocity $u > c$ in its rest frame. If a tachyonic message is sent to an observer at rest at a distance L, how much time will elapse before a tachyonic reply can be received? How much time will elapse if the distant observer is moving directly away at velocity v, and is at a distance L at the instant he receives the message and replies? (Show that for $u > [1+(1-v^2)^{\frac{1}{2}}]/v$ the reply can be received before the signal is sent!)

Problem 1.7. Frame S′ moves with velocity $\underset{\sim}{v}$ relative to frame S. A rod in frame S′ makes an angle θ' with respect to the forward direction of motion. What is this angle θ as measured in S?

Problem 1.8. Frame S′ moves with velocity $\underset{\sim}{\beta}$ relative to frame S. A bullet in frame S′ is fired with velocity $\underset{\sim}{v}'$ at an angle θ' with respect to the forward direction of motion. What is this angle θ as measured in S? What if the bullet is a photon?

Problem 1.9. Suppose that an observer at rest with respect to the fixed distant stars sees an isotropic distribution of stars. That is, in any solid angle $d\Omega$ he sees $dN = N(d\Omega/4\pi)$ stars, where N is the total number of stars he can see.

Suppose now that another observer (whose rest frame is S′) is moving at a relativistic velocity β in the e_x direction. What is the distribution of stars seen by this observer? Specifically, what is the distribution function $P(\theta',\phi')$ such that the number of stars seen by this observer in his solid angle $d\Omega'$ is $P(\theta',\phi')\,d\Omega'$? Check to see that $\int_{\text{sphere}} P(\theta',\phi')d\Omega' = N$, and check that $P(\theta',\phi') \to \frac{N}{4\pi}$ as $\beta \to 0$. Where will the observer see the stars "bunch up"?

Problem 1.10. Show that $A = 3^{\frac{1}{2}} e_t + 2^{\frac{1}{2}} e_x$ is a unit timelike vector in special relativity. Show that the angle between A and e_t is not real.

Problem 1.11. Two rings rotate with equal and opposite angular velocity ω about a common center. Suppose Adam rides on one ring and Eve on the other, and that at some moment they pass each other and their clocks agree. At the moment they pass, Eve sees Adam's clock running more slowly, so she expects to be ahead the next time they meet. But Adam expects just the reverse. What really happens? Can you reconcile this with Adam's (or Eve's) observations?

Problem 1.12. Define an imaginary coordinate $w = it$. Show that a rotation of angle θ in the x_i, w plane $(i=1,2,3)$, where θ is a pure imaginary number, corresponds to a pure Lorentz boost in t, x, y, z coordinates. How is the boost velocity v related to the angle θ?

Problem 1.13. Show that the curve

$$\begin{aligned} x &= \int r \cos\theta \cos\phi \, d\lambda \\ y &= \int r \cos\theta \sin\phi \, d\lambda \\ z &= \int r \sin\theta \, d\lambda \\ t &= \int r \, d\lambda \ , \end{aligned}$$

where r, θ and ϕ are arbitrary functions of λ, is a null curve in special relativity. Under what conditions is it a null geodesic?

Problem 1.14. Show that an observer's 4-acceleration $du^a/d\tau$ has only 3 independent components, and give the relation of these to the 3 components of ordinary acceleration that he would measure with a Newtonian accelerometer in his local frame.

Problem 1.15. Write the magnitude of the acceleration measured in the observer's frame as an invariant.

Problem 1.16. A particle moves with 3-velocity $\underset{\sim}{u}$ and 3-acceleration $\underset{\sim}{a}$ as seen by an inertial observer $\mathcal{O}$. Another inertial observer $\mathcal{O}'$ has 3-velocity $\underset{\sim}{v}$ relative to $\mathcal{O}$. Show that the components of acceleration of the particle parallel and perpendicular to $\underset{\sim}{v}$ as measured by $\mathcal{O}'$ are

$$\underset{\sim}{a}'_{\parallel} = \frac{(1-v^2)^{3/2}}{(1-\underset{\sim}{v}\cdot\underset{\sim}{u})^3}\, \underset{\sim}{a}_{\parallel}$$

$$\underset{\sim}{a}'_{\perp} = \frac{(1-v^2)}{(1-\underset{\sim}{v}\cdot\underset{\sim}{u})^3}\, [\underset{\sim}{a}_{\perp} - \underset{\sim}{v}\times(\underset{\sim}{a}\times\underset{\sim}{u})] \ .$$

Problem 1.17. An observer experiences a uniform acceleration in the x direction, of magnitude g. Define a coordinate system $(\bar{t}, \bar{x}, \bar{y}, \bar{z})$ for him in the following way: (i) Let the observer be at $\bar{x} = \bar{y} = \bar{z} = 0$ and let $\bar{t}$ be his proper time. (ii) Let his hyperplanes of simultaneity agree with the hyperplanes of simultaneity of an instantaneously comoving inertial frame. (iii) Let the other "coordinate stationary observers" (for whom $\bar{x}, \bar{y}, \bar{z}$ are constant) move in such a way that they are always at rest with respect to the observer on the hyperplanes of simultaneity. At $t = 0$ label all spatial points with the same labels as the momentarily comoving inertial system $t = 0$, x, y, z.

Give the coordinate transformation between t, x, y, z and $\bar{t}, \bar{x}, \bar{y}, \bar{z}$. Show that coordinate stationary clocks cannot remain synchronized.

Problem 1.18. A mirror moves perpendicular to its plane with a velocity v. With what angle to the normal is a ray of light reflected, if it is incident at an angle θ? What is the change in the frequency of the light?

Problem 1.19. A mirror is moving parallel to its plane. Show that the angle of incidence of a photon equals the angle of reflection.

Problem 1.20. A particle of rest mass m and 4-momentum p is examined by an observer with 4-velocity $\mathbf{u}$. Show that:

(a) the energy he measures is $E = -\mathbf{p} \cdot \mathbf{u}$;

(b) the rest mass he attributes to the particle is $m^2 = -\mathbf{p} \cdot \mathbf{p}$;

(c) the momentum he measures has magnitude $|\underset{\sim}{p}| = [(\mathbf{p} \cdot \mathbf{u})^2 + \mathbf{p} \cdot \mathbf{p}]^{\frac{1}{2}}$;

(d) the ordinary velocity v he measures has magnitude

$$|\underset{\sim}{v}| = \left[1 + \frac{\mathbf{p} \cdot \mathbf{p}}{(\mathbf{p} \cdot \mathbf{u})^2}\right]^{\frac{1}{2}}$$

(e) the 4-vector v, whose components in the observer's Lorentz frame are

$$v^0 = 0, \quad v^j = (dx^j/dt)_{\text{particle}} = \text{ordinary velocity} ,$$

is given by $\mathbf{v} = -\mathbf{u} - \dfrac{\mathbf{p}}{\mathbf{p} \cdot \mathbf{u}}$.

Problem 1.21. An iron nucleus emits a Mössbauer gamma ray with frequency ν_0 as measured in its own rest frame. The nucleus is traveling with velocity $\underset{\sim}{\beta}$ with respect to some inertial observer. What frequency does the observer measure when the gamma ray reaches him? Express the answer in terms of $\underset{\sim}{\beta}$, ν_0, and the unit vector $\underset{\sim}{n}$ pointing towards the nucleus at the time it emitted the γ-ray, as measured by the observer.

Problem 1.22. An observer receives light from a source of light which is moving with a velocity $\underset{\sim}{v}$; the angle between $\underset{\sim}{v}$ and the line between observer and source is θ at the time the light is emitted. If the observer sees no net redshift or blueshift, what is θ in terms of $|\underset{\sim}{v}|$?

Problem 1.23. Suppose in some inertial frame S a photon has 4-momentum components

$$p^0 = p^x = E, \quad p^y = p^z = 0 \ .$$

There is a special class of Lorentz transformations – called the "little group of p" – which leave the components of p unchanged, e.g. a pure rotation through an angle α in the y-z plane

$$\begin{bmatrix} 1 & 0 & 0 & 0 \\ 0 & 1 & 0 & 0 \\ 0 & 0 & \cos\alpha & -\sin\alpha \\ 0 & 0 & \sin\alpha & \cos\alpha \end{bmatrix} \begin{bmatrix} E \\ E \\ 0 \\ 0 \end{bmatrix} = \begin{bmatrix} E \\ E \\ 0 \\ 0 \end{bmatrix}$$

is such a transformation. Find a sequence of pure boosts and pure rotations whose product is *not* a pure rotation in the y-z plane, but *is* in the little group of p.

Problem 1.24. Two giant frogs are captured, imprisoned in a large metal cylinder, and placed on an airplane. While in flight, the storage doors accidentally open and the cylinder containing the frogs falls out. Sensing something amiss, the frogs decide to try to break out. Centering themselves in the cylinder, they push off from each other and slam simultaneously into the ends of the cylinder. They instantly push off from the ends and shoot across the cylinder past each other into the opposite ends. This continues until the cylinder hits the ground. Consider how this looks from some other inertial frame, falling at another speed. In this frame, the frogs do *not* hit the ends of the cylinder simultaneously, so the cylinder jerks back and forth about its mean speed β. The cylinder, however, was at rest in one inertial frame. Does this mean that one inertial frame can jerk back and forth with respect to another?

Problem 1.25. Let J_x, J_y, J_z be infinitesimal rotation operators defined so that $1 + iJ_j\theta/2$ is a rotation by a small angle θ around the j-axis. Let K_x, K_y, K_z be infinitesimal boost operators defined so that $1 + iK_j v/2$ is a boost by a small velocity v in the j-direction. Show that the following relationships, and all their cyclic permutations, are true:

$$[J_x, J_y] = 2iJ_z$$
$$[J_x, K_y] = 2iK_z$$
$$[K_x, K_y] = -2iJ_z \ .$$

Find a representation of the Lorentz group in terms of Pauli spin matrices σ_x, σ_y, σ_z, and the unit matrix.

Problem 1.26. Two successive, arbitrary pure Lorentz boosts $\underset{\sim}{v}_1$ and $\underset{\sim}{v}_2$ are equivalent to a pure boost $\underset{\sim}{v}_3$ followed by a pure rotation $\theta\underset{\sim}{n}$, where n is a unit vector. Find the magnitude of θ in terms of $\underset{\sim}{v}_1$ and $\underset{\sim}{v}_2$ and show that $\underset{\sim}{n}\cdot\underset{\sim}{v}_3 = 0$.

Problem 1.27. Show that any proper (non time-reversing, non parity-reversing) homogeneous Lorentz transformation leaves fixed at least one null direction.

Problem 1.28. What is the least number of pure boosts which generate an arbitrary Lorentz transformation? Note: This is a difficult problem!

CHAPTER 2

SPECIAL-RELATIVISTIC DYNAMICS

In our laboratory frame, a particle with 4-momentum $\mathbf{p}$ has total energy $E = p^0$ and 3-momentum $\underset{\sim}{p} = p^i$. If the particle has a nonzero rest mass m, the 4-momentum, 4-velocity $\mathbf{u}$ and 3-velocity $\underset{\sim}{v}$ are related by

$$\mathbf{p} = m\mathbf{u} = m(\gamma, \gamma\underset{\sim}{v}), \qquad \gamma \equiv (1-v^2)^{-\frac{1}{2}} ,$$

so $E = \gamma m$, $\underset{\sim}{p} = \gamma m \underset{\sim}{v}$. The square of a particle's 4-momentum, an invariant in all frames, is

$$\mathbf{p}\cdot\mathbf{p} = -E^2 + \underset{\sim}{p}^2 = -m^2 .$$

The kinetic energy of a particle is $T \equiv E - m$.

The fundamental dynamical law for particle interactions is that in any frame the vector sum of the 4-momenta of all particles is a conserved constant in time.

——— o ——— o ——— o ——— o ——— o ——— o ——— o ———

Problem 2.1. (Compton scattering.) A photon of wavelength λ hits a stationary electron (mass m_e) and comes off with wavelength λ' at an angle θ. Derive the expression

$$\lambda' - \lambda = (h/m_e)(1-\cos\theta) .$$

Problem 2.2.

(a) When a photon scatters off a charged particle which is moving with a speed very nearly that of light, the photon is said to have undergone an inverse Compton scattering. Consider an inverse Compton scattering in which a charged particle of rest mass m and total mass-energy (as seen in the lab frame) $E \gg m$, collides head-on with a photon of frequency $\nu\,(h\nu \ll m)$. What is the maximum energy the particle can transfer to the photon?

(b) If space is filled with black-body radiation of temperature 3^0K and contains cosmic ray protons of energies up to 10^{20} eV, how much energy can a proton of energy 10^{20} eV transfer to a 3^0K photon?

Problem 2.3. Show that it is impossible for an isolated free electron to absorb or emit a photon.

Problem 2.4. A particle of rest mass m_1 and velocity $\underset{\sim}{v}_1$ collides with a stationary particle of rest mass m_2 and is absorbed by it. Find the rest mass m and velocity $\underset{\sim}{v}$ of the resultant compound system.

Problem 2.5. The beta-decay of a neutron is isotropic in the rest frame of the neutron, with the velocity of the emitted electron $v_e = 0.77$. If the neutron is moving with velocity β through the laboratory, what values of the electron's laboratory momentum vector $\underset{\sim}{P}$ are possible?

Problem 2.6. Evaluate the "available energy" of two different proton-proton scattering experiments. The first is of the conventional type, where a beam of protons is accelerated to 30 GeV and allowed to strike a target (liquid hydrogen, for example). In the second, two separate beams of protons are accelerated to 15 GeV each, then directed toward each other and allowed to collide. Evaluate the total energy of two colliding protons in the center of momentum frame for each experiment. To what energy would a beam in the first type of experiment have to be accelerated to match the CM energy of the 15 GeV protons in the second experiment?

Problem 2.7. A particle of rest mass m collides elastically with a stationary particle of equal mass. The incident particle has kinetic energy T_0. What is its kinetic energy after the collision, if the scattering angle is θ?

Problem 2.8. Calculate the threshold energy of a nucleon N for it to undergo the reaction

$$\gamma + N \rightarrow N + \pi$$

where γ represents a photon of temperature 3^0K. Assume the collision is head-on; take the photon energy to be $\sim kT$; $m_N = 940$ MeV; $m_\pi = 140$ MeV. (This effect probably produces a cut-off in the cosmic-ray spectrum at this threshold energy.)

Problem 2.9. Consider the reaction $\pi^+ + n \rightarrow K^+ + \Lambda^0$. The rest masses of the particles are $m_\pi = 140$ MeV, $m_n = 940$ MeV, $m_K = 494$ MeV, $m_\Lambda = 1115$ MeV. What is the threshold kinetic energy of the π to create a K at an angle of 90^0 in the lab in which the n is at rest?

Problem 2.10. Consider the reaction $A \rightarrow B + C$ (with particle masses m_A, m_B, m_C).

(a) If A is at rest in the lab frame, show that in the lab frame particle B has energy $E_B = (m_A^2 + m_B^2 - m_C^2)/2m_A$.

(b) An atom of mass M at rest decays to a state of rest energy $M-\delta$ by emitting a photon of energy $h\nu$. Show that $h\nu < \delta$. In the Mössbauer effect, why is $h\nu = \delta$?

(c) If A decays while moving in the lab frame, find the relation between the angle at which B comes off, and the energies of A and B.

Problem 2.11. Consider the reaction $1+2 \rightarrow 3+4$. The lab frame is defined to be the one in which $\underset{\sim}{P}_2 = 0$. The C.M. frame is defined to be the one in which $\underset{\sim}{P}_1^{\text{C.M.}} + \underset{\sim}{P}_2^{\text{C.M.}} = 0$. Show that:

(a) $E_{\text{total}}^{\text{C.M.}} = (m_1{}^2 + m_2{}^2 + 2m_2E_1)^{\frac{1}{2}}$

(b) $E_1{}^{\text{C.M.}} = [(E_{\text{total}}^{\text{C.M.}})^2 + m_1{}^2 - m_2{}^2]/2E_{\text{total}}^{\text{C.M.}}$

(c) $P_1^{\text{C.M.}} = m_2P_1/E_{\text{total}}^{\text{C.M.}}$

(d) $\gamma_{\text{C.M.}} = (E_1 + m_2)/E_{\text{total}}^{\text{C.M.}}$ ($v_{\text{C.M.}} \equiv$ velocity of C.M. in lab frame, and $\gamma_{\text{C.M.}} \equiv (1 - v^2{}_{\text{C.M.}})^{-\frac{1}{2}}$.)

(e) $v_{\text{C.M.}} = P_1/(E_1 + m_2)$.

Problem 2.12. Consider the elastic collision of a particle of mass m_1 with a stationary particle of mass $m_2 < m_1$. Let θ_{max} be the maximum scattering angle of m_1. In nonrelativistic calculations, $\sin\theta_{max} = m_2/m_1$. Prove that this result also holds relativistically.

Problem 2.13.

(a) If a rocket has engines that give it a constant acceleration of 1 g (relative to its instantaneous inertial frame, of course), and the rocket starts from rest near the earth, how far from the earth (as measured in the earth's frame) will the rocket be in 40 years as measured on the earth? How far after 40 years as measured in the rocket?

(b) Compute the proper time for the occupants of a rocket ship to travel the 30,000 light years from the Earth to the center of the galaxy. Assume they maintain an acceleration of 1 g for half the trip and decelerate at 1 g for the remaining half.

(c) What fraction of the initial mass of the rocket can be payload in part (b)? Assume an ideal rocket that converts rest mass into radiation and ejects all of the radiation out of the back with 100% efficiency and perfect collimation.

Problem 2.14. What is the maximum energy one could get out of a fixed frequency electron cyclotron with accelerating potential V.

Problem 2.15. A new force field $F^{\mu}(x^{\nu})$ is discovered which induces a 4-acceleration $a^{\mu} \equiv du^{\mu}/d\tau = m^{-1}F^{\mu}(x^{\nu})$ on a particle of mass m, at position x^{ν}. Notice that F^{μ} does not depend on u^{ν}. Show that this force is not consistent with special relativity.

CHAPTER 3

SPECIAL-RELATIVISTIC COORDINATE TRANSFORMATIONS, INVARIANTS AND TENSORS

Spacetime in special relativity can be described by more general (curvilinear) coordinates than "inertial" or Minkowski coordinates, e.g. coordinates $x^{\mu'}$,

$$x^{\mu'} = f^{\mu}(x^{\nu})$$

where x^{ν} are Minkowski coordinates, and f^{μ} four arbitrary functions. One then shows that the basis vectors and components of vectors in the new coordinates are related to the old by

$$e_{\alpha'} = \frac{\partial x^{\mu}}{\partial x^{\alpha'}} e_{\mu} \ , \qquad e_{\mu} = \frac{\partial x^{\alpha'}}{\partial x^{\mu}} e_{\alpha'}$$

$$V^{\alpha'} = \frac{\partial x^{\alpha'}}{\partial x^{\mu}} V^{\mu} \ , \qquad V^{\mu} = \frac{\partial x^{\mu}}{\partial x^{\alpha'}} V^{\alpha'}$$

$$V_{\alpha'} = \frac{\partial x^{\mu}}{\partial x^{\alpha'}} V_{\mu} \ , \qquad V_{\mu} = \frac{\partial x^{\alpha'}}{\partial x^{\mu}} V_{\alpha'} \ .$$

In other words, the transformation matrix $\Lambda^{\mu'}_{\ \alpha} \equiv \partial x^{\mu'}/\partial x^{\alpha}$ replaces the less general Lorentz matrices (which are applicable only for transformations between two systems of Minkowski coordinates).

In general coordinates, the relation $\mathbf{A} \cdot \mathbf{B} = A_{\mu} B^{\mu}$ still holds, but we no longer have $A_{\mu} = \eta_{\mu\nu} A^{\mu}$. Rather, corresponding to every coordinate system there is a metric tensor with components $g_{\alpha\beta}$, such that $ds^2 = g_{\alpha\beta} dx^{\alpha} dx^{\beta}$, which leads to $A_{\mu} = g_{\mu\nu} A^{\nu}$ and therefore $\mathbf{A} \cdot \mathbf{B} = g_{\mu\nu} A^{\mu} B^{\nu}$. Note also $A^{\mu} = g^{\mu\nu} A_{\nu}$ where $g^{\mu\nu}$ is the matrix inverse of $g_{\mu\nu}$.

Various formal definitions of a *tensor* are possible. Here, it suffices to say that it is a geometrical object which, like a vector, has components whose numerical values are different in different coordinate systems. A tensor has 4^n components, where n is its rank (number of "slots" or indices for components). Slots may be contravariant or covariant; examples: $T^{\mu\nu}$, $F_{\mu\nu}$, $R^{\alpha}{}_{\beta\gamma\delta}$, $G_{\mu}{}^{\nu}$. Tensors transform with one matrix for each slot, e.g.

$$G_{\mu'}{}^{\nu'} = \Lambda^{\alpha}{}_{\mu'}\Lambda^{\nu'}{}_{\beta}\, G_{\alpha}{}^{\beta} .$$

Tensors may be "contracted" (a covariant and a contravariant index summed) or multiplied as a "direct product" with other tensors or with themselves to form new tensors, e.g.

$$Q_{\mu\nu} = R^{\alpha}{}_{\mu\alpha\nu}, \quad A^{\mu} = G^{\mu}{}_{\nu}B^{\nu}, \quad F_{\mu\nu} = A_{\mu}B_{\nu} .$$

A special case is contraction with the metric tensor, where the same symbol is usually used for the result as in the relation of covariant and contravariant vectors, $F^{\mu}{}_{\nu} = g_{\nu\alpha}F^{\mu\alpha}$. A tensor expression with no free indices, e.g. $F_{\mu\nu}A^{\mu}A^{\nu}$ or $R_{\beta\gamma}F^{\beta\gamma}$ or $A^{\alpha}B^{\beta}g_{\alpha\beta}$, is a scalar and is an invariant number in all frames. The analog of the index free notation $\mathbf{A}$ for a vector A^{μ} is to write, e.g. $\mathbf{T}$ for a tensor $T^{\mu\nu}$. In both cases the existence of covariant or contravariant slots must be deduced from the context.

In index free notation, $\otimes$ represents the direct product, e.g. $F^{\mu\nu}A^{\rho}$ is written $\mathbf{F}\otimes\mathbf{A}$; the contracted product is written with a dot, e.g. $\mathbf{F}\cdot\mathbf{A}$ for $F^{\mu\alpha}A_{\alpha}$.

We denote partial derivatives by a comma, e.g. $f_{,\alpha} \equiv \partial f/\partial x^{\alpha}$.

—— o —— o —— o —— o —— o —— o —— o ——

Problem 3.1.

(I) If 2 events are separated by a spacelike interval, show that

(a) there exists a Lorentz frame in which they are simultaneous, and

(b) in no Lorentz frame do they occur at the same point.

(II) If 2 events are separated by a timelike interval, show that

(a) there exists a frame in which they happen at the same point, and

(b) in no Lorentz frame are they simultaneous.

Problem 3.2. Find 4 linearly independent null vectors in Minkowski space. Can you find 4 which are orthogonal?

Problem 3.3. Show that the only non-spacelike vectors orthogonal to a given nonzero null vector are multiples of it.

Problem 3.4. Show that the sum of two vectors can be spacelike, null, or timelike, independently of whether the two vectors are spacelike, null, or timelike.

Problem 3.5. Show that the cross-sectional area of a parallel beam of light is invariant under Lorentz transformations.

Problem 3.6. Show that $\sum_\mu D^{\mu\mu}$ and $\sum_\mu D_{\mu\mu}$ are not invariant under coordinate transformations, but that $\sum_\mu D_\mu{}^\mu$ is invariant. (Take $\mathbf{D}$ to be a tensor defined by its components $D^{\mu\nu}$.)

Problem 3.7. $F^{\alpha\beta}$ is antisymmetric on its two indices. Show that

$$F_\mu{}^\alpha{}_{,\nu} F^\nu{}_\alpha = -F_{\mu\alpha,\beta} F^{\alpha\beta} .$$

Problem 3.8. In a coordinate system with coordinates x^μ, the invariant line element is $ds^2 = \eta_{\alpha\beta} dx^\alpha dx^\beta$. If the coordinates are transformed $x^\mu \to \bar{x}^\mu$, show that the line element is $ds^2 = g_{\bar{\mu}\bar{\nu}} d\bar{x}^\mu d\bar{x}^\nu$, and express $g_{\bar{\mu}\bar{\nu}}$ in terms of the partial derivatives $\partial x^\mu / \partial \bar{x}^\nu$. For two arbitrary 4-vectors $\mathbf{U}$ and $\mathbf{V}$, show that

$$\mathbf{U} \cdot \mathbf{V} = U^\alpha V^\beta \eta_{\alpha\beta} = U^{\bar{\alpha}} V^{\bar{\beta}} g_{\bar{\alpha}\bar{\beta}} .$$

Problem 3.9. Show that the determinant of the metric tensor $g \equiv \det(g_{\mu\nu})$ is not a scalar.

Problem 3.10. If Λ^α_β and $\tilde{\Lambda}^\alpha_\beta$ are two matrices which transform the components of a tensor from one coordinate basis to another, show that the matrix $\Lambda^\alpha_\gamma \tilde{\Lambda}^\gamma_\beta$ is also a coordinate transformation.

Problem 3.11. You are given a tensor $K^{\alpha\beta}$. How can you test whether it is a direct product of two vectors $K^{\alpha\beta} = A^{\alpha}B^{\beta}$? Can you express the test in coordinate-free language?

Problem 3.12. Prove that the general second-rank tensor in n-dimensions cannot be represented as a simple direct product of two vectors, but *can* be expressed as a sum over many such products.

Problem 3.13. A two index "object" $X^{\mu\nu}$ is defined by the "direct sum" of two vectors $X^{\mu\nu} = A^{\mu} + B^{\nu}$. Is $X^{\mu\nu}$ a tensor? Is there a transformation law to take X to a new coordinate system, i.e. to obtain $X^{\hat{\mu}\hat{\nu}}$ from $X^{\mu\nu}$?

Problem 3.14. Show that a second rank tensor F which is antisymmetric in one coordinate frame $(F_{\mu\nu} = -F_{\nu\mu})$ is antisymmetric in all frames. Show that the contravariant components are also antisymmetric $(F^{\mu\nu} = -F^{\nu\mu})$. Show that symmetry is also coordinate invariant.

Problem 3.15. Let $A_{\mu\nu}$ be an antisymmetric tensor so that $A_{\mu\nu} = -A_{\nu\mu}$; and let $S^{\mu\nu}$ be a symmetric tensor so that $S^{\mu\nu} = S^{\nu\mu}$. Show that $A_{\mu\nu}S^{\mu\nu} = 0$. Establish the following two identities for any arbitrary tensor $V_{\mu\nu}$:

$$V^{\mu\nu}A_{\mu\nu} = \frac{1}{2}(V^{\mu\nu} - V^{\nu\mu})A_{\mu\nu}, \qquad V^{\mu\nu}S_{\mu\nu} = \frac{1}{2}(V^{\mu\nu} + V^{\nu\mu})S_{\mu\nu} .$$

Problem 3.16.

(a) In an n-dimensional metric space, how many independent components are there for an r-rank tensor $T^{\alpha\beta\cdots}$ with no symmetries?

(b) If the tensor is symmetric on s of its indices, how many independent components are there?

(c) If the tensor is antisymmetric on a of its indices, how many independent components are there?

Problem 3.17. We define the meaning of square and round brackets enclosing a set of indices as follows:

$$V_{(\alpha_1,\cdots,\alpha_p)} \equiv \frac{1}{p!}\Sigma V_{\alpha_{\pi_1}\cdots\alpha_{\pi_p}}; \qquad V_{[\alpha_1,\cdots\alpha_p]} \equiv \frac{1}{p!}\Sigma(-1)^\pi V_{\alpha_{\pi_1}\cdots\alpha_{\pi_p}} .$$

Here the sum is taken over all permutations π of the numbers $1,2,\cdots,p$ and $(-1)^\pi$ is $+1$ or -1 depending on whether the permutation is even or odd. The quantity V may have other indices, not shown here, besides the set of p indices $\alpha_1,\alpha_2,\cdots,\alpha_p$, but only this set of indices is affected by the operations described here. The numbers $\pi_1,\pi_2,\cdots,\pi_p$ are the numbers $1,2,\cdots,p$ rearranged according to the permutation π. Thus for example $V_{(\alpha_1\alpha_2)} \equiv \frac{1}{2}(V_{\alpha_1\alpha_2}+V_{\alpha_2\alpha_1})$ or equivalently $V_{(\mu\nu)} = \frac{1}{2}(V_{\mu\nu}+V_{\nu\mu})$.

(a) If F is antisymmetric and T is symmetric, apply these definitions to give explicit formulas for the following: $V_{[\mu\nu]}$, $F_{[\mu\nu]}$, $F_{(\mu\nu)}$, $T_{[\mu\nu]}$, $T_{(\mu\nu)}$, $V_{[\alpha\beta\gamma]}$, $T_{(\alpha\beta,\gamma)}$, $F_{[\alpha\beta,\gamma]}$.

(b) Establish the following formulae: $V_{((\alpha_1\cdots\alpha_p))} = V_{(\alpha_1\cdots\alpha_p)}$; $V_{[[\alpha_1\cdots\alpha_p]]} = V_{[\alpha_1\cdots\alpha_p]}$; $V_{(\alpha_1\cdots[\alpha_\ell\alpha_m]\cdots\alpha_p)} = 0$; $V_{[\alpha_1\cdots[\alpha_\ell\alpha_m]\cdots\alpha_p]} = V_{[\alpha_1\cdots\alpha_\ell\alpha_m\cdots\alpha_p]}$.

(c) Use these notations to show that $F_{\mu\nu} = A_{\nu,\mu} - A_{\mu,\nu}$ implies $F_{\alpha\beta,\nu} + F_{\beta\nu,\alpha} + F_{\nu\alpha,\beta} = 0$. (Half of Maxwell's equations!)

Problem 3.18. Show for any two-index tensor X, that $X_{\alpha\beta} = X_{(\alpha\beta)} + X_{[\alpha\beta]}$ where ( ) and [ ] denote symmetrization and antisymmetrization, respectively. Show that in general

$$Y_{\alpha\beta\gamma} \neq Y_{(\alpha\beta\gamma)} + Y_{[\alpha\beta\gamma]} .$$

Problem 3.19. Prove that the Kronecker delta, $\delta^\mu{}_\nu$, is a tensor.

Problem 3.20. Prove that, except for scaling by a constant, there is a unique tensor $\varepsilon_{\alpha\beta\gamma\delta}$ which is totally antisymmetric on all its 4 indices. The usual choice is to take $\varepsilon_{0123} = 1$ in Minkowski coordinates. What are the components of ε in a general coordinate frame, with metric $g_{\mu\nu}$?

Problem 3.21. In an orthonormal frame, show that

$$\varepsilon_{\alpha\beta\gamma\delta} = -\varepsilon^{\alpha\beta\gamma\delta} .$$

What is the analogous relation in a general coordinate frame with metric $g_{\mu\nu}$?

Problem 3.22. Evaluate $\varepsilon_{\mu\nu\rho\sigma}\,\varepsilon^{\mu\nu\rho\sigma}$.

Problem 3.23. Show that for any tensor $A^{\alpha}{}_{\beta}$

$$\varepsilon_{\alpha\beta\gamma\delta}\, A^{\alpha}{}_{\mu}\, A^{\beta}{}_{\nu}\, A^{\gamma}{}_{\lambda}\, A^{\delta}{}_{\sigma} = \varepsilon_{\mu\nu\lambda\sigma} \det \|A^{\alpha}{}_{\beta}\|$$

where $\|A^{\alpha}{}_{\beta}\|$ is the matrix of the components $A^{\alpha}{}_{\beta}$.

Problem 3.24. Show that four vectors $\mathbf{u}$, $\mathbf{v}$, $\mathbf{w}$, $\mathbf{x}$, are linearly independent if and only if $\mathbf{u}\wedge\mathbf{v}\wedge\mathbf{w}\wedge\mathbf{x} \neq 0$. Show that in this case $\mathbf{u}\wedge\mathbf{v}\wedge\mathbf{w}\wedge\mathbf{x}$ is proportional to the totally antisymmetric tensor ε. (The "wedge" product is defined as the antisymmetrized direct product, e.g. $\mathbf{u}\wedge\mathbf{v} = \mathbf{u}\otimes\mathbf{v} - \mathbf{v}\otimes\mathbf{u}$.)

Problem 3.25. Let $\mathbf{F}$ be an antisymmetric second-rank tensor with components $F^{\mu\nu}$. From $\mathbf{F}$ construct another second-rank, anti-symmetric tensor, $*\mathbf{F}$, called the dual of $\mathbf{F}$, as follows

$$*\mathbf{F} = \frac{1}{2}\,\varepsilon^{\mu\nu\alpha\beta} F_{\alpha\beta}\, \mathbf{e}_{\mu}\otimes\mathbf{e}_{\nu} .$$

Show that $*(*\mathbf{F}) = -\mathbf{F}$.

Problem 3.26. Show that

$$V_{\sigma}V^{\sigma} = -\frac{1}{3!}\,(*V)_{\alpha\beta\gamma}\,(*V)^{\alpha\beta\gamma} .$$

Problem 3.27. The tensor $\delta^{\mu\cdots\lambda}_{\rho\cdots\sigma}$ is defined by

$$\delta^{\mu\cdots\lambda}_{\rho\cdots\sigma} \equiv \det\begin{bmatrix} \delta^{\mu}_{\rho} & \cdots & \delta^{\lambda}_{\rho} \\ \vdots & & \vdots \\ \delta^{\mu}_{\sigma} & \cdots & \delta^{\lambda}_{\sigma} \end{bmatrix} .$$

Show that if there are more than 4 upper (or lower) indices, the tensor identically vanishes.

Problem 3.28. Show that $\delta^{\mu\nu}_{\lambda\kappa} = -\frac{1}{2}\,\varepsilon^{\mu\nu\rho\sigma}\varepsilon_{\lambda\kappa\rho\sigma}$, and generalize to $\delta^{\mu\cdots\nu}_{\lambda\cdots\kappa}$ of other ranks.

Problem 3.29. Show that if the antisymmetric tensor $p^{\alpha\beta}$ is a bivector (i.e. $p^{\alpha\beta} = A^{[\alpha}B^{\beta]}$) then

$$p^{\alpha\beta}\,p^{\gamma\delta} + p^{\alpha\gamma}\,p^{\delta\beta} + p^{\alpha\delta}\,p^{\beta\gamma} = 0$$

(the Plücker relations).

Problem 3.30. In 4-space define the 3-dimensional volume element in a hypersurface $x^\alpha = x^\alpha(a,b,c)$ by $d^3\Sigma_\mu = (1/3!)\,\varepsilon_{\mu\alpha\beta\gamma}\,da\,db\,dc\,[\partial(x^\alpha,x^\beta,x^\gamma)/\partial(a,b,c)]$, where the last factor is a 3×3 Jacobian determinant. Compute the components of $d^3\Sigma_\mu$ for a space-like hypersurface $x^0 = \text{constant}$, parameterized by $x^1 = a$, $x^2 = b$, $x^3 = c$.

Problem 3.31. Show that the invariant proper volume element in 4-dimensional space is given by

$$dV = (-g)^{\frac{1}{2}}\,d^4x$$

where $d^4x = dxdydzdt$ in the coordinate system of the metric $g_{\mu\nu}$.

Problem 3.32. Show that the proper 3-volume element of an observer with 4-velocity $\mathbf{u}$ is $d^3V = (-g)^{\frac{1}{2}}\,u^0 d^3x$, and show that this is a scalar invariant.

Problem 3.33. What is the invariant volume element of contravariant momentum d^4P for 4-dimensional momentum space? What is the invariant 3-volume "on the mass shell", i.e. when the constraint $(-\mathbf{P}\cdot\mathbf{P})^{\frac{1}{2}} = m$ is imposed?

Problem 3.34. A group of N particles is seen to occupy a volume $dx\,dy\,dz\,dP^x\,dP^y\,dP^z$ in 6-dimensional phase space, so that the number density of particles in phase space $\mathfrak{N}$ is given by

$$N = \mathfrak{N}\, dx\, dy\, dz\, dP^x\, dP^y\, dP^z \ .$$

Show that $\mathfrak{N}$ is a Lorentz invariant, i.e. that all observers will compute the same numerical value for $\mathfrak{N}$.

Problem 3.35. A vector field $J^\alpha(x^\mu)$ satisfies $J^\alpha{}_{,\alpha} = 0$ and J^α falls off faster than r^{-2} at large distances from the origin. (a) Show that $\int J^0 d^3x$ is constant in time. (b) Show that the integral is a scalar, i.e. $\int J^0 d^3x = \int J^{0'} d^3x'$.

CHAPTER 4
ELECTROMAGNETISM

The electromagnetic field is described relativistically by the antisymmetric electromagnetic field tensor (Maxwell tensor) $F^{\mu\nu}$. In any Lorentz frame the components of $F^{\mu\nu}$ are related to the electric and magnetic field strengths, $\underset{\sim}{E}$ and $\underset{\sim}{B}$, in that frame by

$$F^{\mu\nu} = \begin{bmatrix} 0 & E^x & E^y & E^z \\ -E^x & 0 & B^z & -B^y \\ -E^y & -B^z & 0 & B^x \\ -E^z & B^y & -B^x & 0 \end{bmatrix} .$$

Here μ is the row index and ν the column index. Maxwell's equations can be written

$$F^{\mu\nu}{}_{,\nu} = 4\pi J^{\mu}$$

$$F_{\alpha\beta,\gamma} + F_{\gamma\alpha,\beta} + F_{\beta\gamma,\alpha} = 0 ,$$

where $J^{\mu} = (\rho, \underset{\sim}{J})$ is the 4-current density. The Lorentz force law is

$$dp^{\mu}/d\tau = eF^{\mu\nu}u_{\nu}$$

for a particle of charge e, 4-momentum $\mathbf{p}$ and 4-velocity $\mathbf{u}$.

The energy density $\mathcal{E} = (E^2 + B^2)/8\pi$, the Poynting energy flux $\underset{\sim}{S} = (\underset{\sim}{E} \times \underset{\sim}{B})/4\pi$, and the 3-dimensional stress-tensor

$$T^{ij} = [-(E^iE^j + B^iB^j) + \frac{1}{2}\delta^{ij}(E^2 + B^2)]/4\pi$$

are combined to form the electromagnetic stress-energy tensor

$$T^{\mu\nu} = (F^{\mu\alpha}F^{\nu}{}_{\alpha} - \frac{1}{4}\eta^{\mu\nu}F^{\alpha\beta}F_{\alpha\beta})/4\pi .$$

Problem 4.1. Find the magnetic field $\underset{\sim}{B}$ from a current I in an infinitely long straight wire, by appropriate Lorentz transformations and superpositions of the electric field of an infinitely long straight charge distribution.

Problem 4.2. For electric and magnetic fields, show that $B^2 - E^2$ and $\underset{\sim}{E} \cdot \underset{\sim}{B}$ are invariant under changes of coordinates and Lorentz transformations. Are there any invariants which are not merely algebraic combinations of these two?

Problem 4.3. A particular electromagnetic field has its E field at an angle θ_0 to its $\underset{\sim}{B}$ field, and θ_0 is invariant to all observers. What is the value of θ_0?

Problem 4.4. Show that $\mathcal{E}^2 - |\underset{\sim}{S}|^2$ is a Lorentz invariant of the electromagnetic field, where $\mathcal{E}$ is the energy density and $\underset{\sim}{S}$ the Poynting flux.

Problem 4.5. Prove that except when $(\underset{\sim}{B}\cdot\underset{\sim}{E})^2 + (B^2 - E^2)^2 = 0$, there is a Lorentz transformation which will make $\underset{\sim}{E}$ and $\underset{\sim}{B}$ parallel $(\underset{\sim}{E}' \times \underset{\sim}{B}' = 0)$. [Hint: Try $\underset{\sim}{v} = \alpha(\underset{\sim}{E} \times \underset{\sim}{B})$ for some α.]

Problem 4.6. Suppose that $\underset{\sim}{E} \cdot \underset{\sim}{B} = 0$. Show that there is a Lorentz transformation which makes $\underset{\sim}{E} = 0$ if $B^2 - E^2 > 0$, or one that makes $\underset{\sim}{B} = 0$ if $B^2 - E^2 < 0$. What if $B^2 - E^2 = 0$ in addition to $\underset{\sim}{E} \cdot \underset{\sim}{B} = 0$?

Problem 4.7. A collection of charged particles of charges e_i has 3-velocities $\underset{\sim}{v}_i$ and trajectories $\underset{\sim}{x} = \underset{\sim}{z}_i(t)$. The 4-current has components $J^0 = \sum_i e_i \delta^3[\underset{\sim}{x} - \underset{\sim}{z}_i(t)]$; $J^i = \sum_k e_k v^i \delta^3[\underset{\sim}{x} - \underset{\sim}{z}_k(t)]$. Show that this can be written $J^\mu = \sum_k \int e_k \delta^4[x^\alpha - z^\alpha{}_k(\tau)]\, u^\mu_k\, d\tau$ where u^μ_k is the 4-velocity of particle k.

Problem 4.8. Show by explicit examination of components that the equations

$$F_{\alpha\beta,\gamma} + F_{\beta\gamma,\alpha} + F_{\gamma\alpha,\beta} = 0 \qquad F^{\alpha\beta}{}_{,\beta} = 4\pi J^\alpha$$

reduce to Maxwell's equations:

$$\underset{\sim}{\nabla} \cdot \underset{\sim}{B} = 0, \ \ \dot{\underset{\sim}{B}} + \underset{\sim}{\nabla} \times \underset{\sim}{E} = 0, \ \ \underset{\sim}{\nabla} \cdot \underset{\sim}{E} = 4\pi\rho ,$$

$$\dot{\underset{\sim}{E}} - \underset{\sim}{\nabla} \times \underset{\sim}{B} = -4\pi \underset{\sim}{J} .$$

Problem 4.9. If $F^{\mu\nu}$ is the electromagnetic tensor, show that Maxwell's equations in vacuum can be written as $F^{\mu\nu}{}_{,\nu} = 0$ and $*F^{\mu\nu}{}_{,\nu} = 0$. [Here, $*F^{\mu\nu}$ is the dual of $F^{\mu\nu}$; see Problem 3.25.]

Problem 4.10. Write out the $\mu = 0$ component of the Lorentz force equation $du^\mu/d\tau = (e/m)F^{\mu\beta}u_\beta$ expressing $F^{\mu\nu}$ in terms of E_i and B_i, to obtain

$$dP^0/dt = e\underset{\sim}{v} \cdot \underset{\sim}{E} .$$

Problem 4.11. From the spatial components of the Lorentz 4-force equation, find an equation for $d\underset{\sim}{P}/dt$ in terms of $\underset{\sim}{E}$ and $\underset{\sim}{B}$. (Here $\underset{\sim}{P}$ is the spatial part of $\mathbf{P}$).

Problem 4.12. A particle of charge q and mass m is coasting through the lab with velocity $v\underset{\sim}{e}_x$ when it encounters a constant $\underset{\sim}{E}$ field in the y-direction. Find y(x), the shape of the particle's subsequent motion.

Problem 4.13. A particle of charge q, mass m, moves in a circular orbit of radius R in a uniform B field $B\underset{\sim}{e}_z$. (a) Find B in terms of R, q, m and ω, the angular frequency. (b) The speed of the particle is constant since the B field can do no work on the particle. An observer moving at velocity $\beta\underset{\sim}{e}_x$, however, does not see the speed as constant. What is $u^{0'}$ measured by this observer? (c) Calculate $du^{0'}/d\tau$ and thus $dP^{0'}/d\tau$. Explain how the energy of the particle can change since the B field does no work on it.

Problem 4.14. A small test particle (mass m, positive charge q) makes circular orbits around a "fixed" (i.e. very massive) body of positive charge Q. A uniform magnetic field $\underset{\sim}{B}$ perpendicular to the orbital plane serves to keep the particle in orbit. In the inertial frame in which the central body is at rest, the test charge is seen to circle in the plane perpendicular to the B field with an angular frequency ω. What is the charge to mass ratio of the test particle in terms of ω, R, B, Q?

Problem 4.15. Show that the stress-energy tensor for the electromagnetic field is divergenceless (i.e. $T^{\mu\nu}{}_{,\nu} = 0$) in the absence of charge sources.

Problem 4.16. Show that the stress-energy tensor for the electromagnetic field has zero trace.

Problem 4.17. If $T^{\mu\nu}$ is the stress-energy tensor of the electromagnetic field, show that

$$T^{\mu}{}_{\alpha}T^{\alpha}{}_{\nu} = \delta^{\mu}{}_{\nu}[(E^2-B^2)^2 + (2\underset{\sim}{E}\cdot\underset{\sim}{B})^2]/(8\pi)^2 \ .$$

Problem 4.18. Write Ohm's law $\underset{\sim}{J} = \sigma\underset{\sim}{E}$ invariantly in terms of J^{μ}, $F^{\mu\nu}$, σ and u^{μ} (the 4-velocity of the conducting element).

Problem 4.19. Derive the Lorentz force law for a charged particle from the action $\int J^{\mu}A_{\mu}d^4x - m\int d\tau$, where J^{μ} is the 4-current, A_{μ} the vector potential, and $d\tau^2 \equiv -\eta_{\alpha\beta}\,dx^{\alpha}\,dx^{\beta}$.

Problem 4.20.

(a) Show that $\underset{\sim}{E} \to -\underset{\sim}{B}$ and $\underset{\sim}{B} \to \underset{\sim}{E}$ under the "duality transformation" $F \to {*F}$.

(b) Show that if F is a solution of the free-space Maxwell equations, so is $*F$ and also $e^{*\alpha}F \equiv F\cos\alpha + {*F}\sin\alpha$ for arbitrary α. ($F \to e^{*\alpha}F$ is called a "duality rotation".)

Problem 4.21. If one believes that esthetics should be an important consideration in physical laws, then by symmetry Maxwell's laws should read

$$F^{\mu\nu}{}_{,\nu} = 4\pi J^{\mu}$$

$$*F^{\mu\nu}{}_{,\nu} = 4\pi K^{\mu} \ .$$

What would the significance of K be?

Problem 4.22. In Minkowski spacetime, there is an electromagnetic current $J^{\mu}(x^{\nu})$. Show that the solution to Maxwell's equation is

$$F^{\mu\nu}(x^{\alpha}) = \frac{4}{\pi i}\int \frac{r^{[\mu}J^{\nu]}d^4\tilde{x}}{(r_\sigma r^\sigma)^2}$$

where $r^\beta \equiv \tilde{x}^\beta - x^\beta$. (Start by finding a Green's function of $\Box A^\mu = -4\pi J^\mu$.) How are the retarded boundary conditions specified?

Problem 4.23. Find the equation for the convective time rate of change of a magnetic field which is "frozen in" a perfectly conducting fluid, in terms of the expansion, shear, and rotation of the fluid. (See Problem 5.18 for definitions of these quantities.)

CHAPTER 5
MATTER AND RADIATION

A proper description of the energy, momentum and stress of a relativistic fluid or field uses the symmetric tensor T, the stress-energy tensor (also called the energy-momentum tensor). The components of this tensor in the Lorentz frame of an observer are related to the measurements made by that observer in the following way:

$T^{00} \equiv$ density of mass-energy (often denoted ρ).

$T^{0j} = T^{j0} \equiv$ j-component of momentum-density

$= $ j-component of energy-flux

$T^{ij} \equiv$ components of the ordinary stress tensor (e.g. $T^{xx} =$ x-component of pressure).

If $T^{\mu\nu}$ describes *all* fields, fluids, particles etc. present in a system, the interrelation of momentum flow and energy change is summarized by the *equations of motion*:

$$T^{\mu\nu}{}_{,\nu} = 0 \ .$$

The basic concepts of relativistic thermodynamics and hydrodynamics which follow from this are developed in the problems.

With a view to developments later in the book, several problems in this chapter use covariant differentiation, denoted by a semicolon. The reader not yet familiar with this may replace all semicolons by commas (partial differentiation in Minkowski coordinates). Also, the ∇ notation is introduced; e.g. ∇S for $S^{\alpha}{}_{;\beta}$, ∇f for $f_{,\alpha}$, $\nabla \cdot T$ for $T^{\mu\nu}{}_{;\nu}$, etc.

Problem 5.1. Calculate the nonzero components in an inertial frame S of the stress-energy tensor for the following systems:

(a) A group of particles all moving with the same velocity $\underset{\sim}{\beta}=\beta\underset{\sim}{e}_x$ as seen in S. Let the rest-mass density of these particles be ρ_0 as measured in their comoving frame. Assume a high density of particles and treat them in the continuum approximation.

(b) A ring of N similar particles of mass m rotating counter-clockwise in the x-y plane about some point fixed in S at a radius a and angular velocity ω. (The width of the ring is much less than a.) Do not include the stress-energy of whatever forces keep them in orbit. Assume N is large enough that one can treat the particles as being continuously distributed.

(c) Two such rings of particles, one rotating clockwise, the other counter-clockwise, at the same radius a. The particles do not collide or interact with each other in any way.

Problem 5.2. What is the stress energy of a gas with a proper number density (i.e. number density as measured in the local rest frame of the gas) N of noninteracting particles of mass m, if the particles all have the same speed v but move isotropically? (Do *not* assume $v << c$.)

Problem 5.3. In the rest frame of a perfect fluid its stress energy tensor, in terms of mass-energy density ρ and pressure p, is the diagonal tensor

$$T^{\mu\nu} = \begin{bmatrix} \rho & & & 0 \\ & p & & \\ & & p & \\ 0 & & & p \end{bmatrix}.$$

If a fluid element of proper density and pressure, ρ and p is moving with 4-velocity **u**, what is its stress-energy?

Problem 5.4. Find the stress-energy tensor for a uniform magnetic field. What is the average stress-energy if the B field is static but "chaotic" i.e. the direction of the B field varies, and is isotropic on the average?

Problem 5.5. A rod has cross sectional area A and mass per unit length μ. Write down the stress-energy tensor inside the rod when the rod is under a tension F. (Assume that the tension is uniformly distributed over the cross section.)

Problem 5.6. A rope of mass per unit length μ has a static breaking strength F. What is the maximum F can be without violating the "weak" energy condition that T^{00} should be positive to all observers? How close is steel cable to this theoretical maximum strength?

Problem 5.7. An infinitesimally thin rod of length 2a has a point mass m at each of its ends. The center of the rod is fixed in the laboratory and the rod rotates about this point with a relativistic angular velocity ω. (i.e. $\omega\ell$ is comparable with c). Assume the rod is massless. What is $T^{\mu\nu}$ for the rod and particle system?

Problem 5.8. A parallel plate capacitor consists of two large plates of area A, perpendicular to the x-direction, separated by a small distance d. The capacitor is charged so that a uniform electric field of magnitude E is present between the plates; fringe effects at the edge of the plate can be neglected. The "electrostatic mass" of this capacitor is $E^2 Ad/8\pi$ in the rest frame of the capacitor. Show that the electrostatic energy is *smaller* if the capacitor is moving in the x-direction! Consider now that the plates must be held apart. Let the plates be held apart by an ideal gas of proper density ρ_0. Show that the *total* energy (electrostatic + gas) of the capacitor increases with velocity in the x-direction in precisely the same manner that the energy of a point mass does.

Problem 5.9. Consider a system of discrete particles of charge q_i and mass m_i interacting through electromagnetic forces. From the explicit expression for $T^{\mu\nu}$ of the particles show that the total $T^{\mu\nu}$ (particles plus field) is conserved, i.e. that $T^{\mu\nu}{}_{,\nu} = 0$.

Problem 5.10. The specific intensity I_ν of radiation measures the intensity of radiation at a particular frequency ν in a particular direction. It is defined as the flux per unit frequency interval, per unit solid angle. Show that I_ν/ν^3 is a Lorentz invariant.

Problem 5.11. A star emits radiation isotropically in its own rest frame, with luminosity L (energy per unit time). At a particular instant, as measured from the earth, the star is at a distance R, and is moving with a velocity v which makes an angle θ with respect to the direction from the earth to the star. What is the flux of radiation (energy per unit time per unit area) seen by an observer on the earth in terms of R, v and θ evaluated at the instant the radiation was emitted?

Problem 5.12. Consider a spherical particle of mass m which scatters all electromagnetic radiation incident on it, isotropically in its rest frame. Let A be the effective cross sectional area of the particle. Find the equation of motion of the particle in a constant radiation field of intensity S (energy per time per area), and solve it for the case of a particle initially at rest. (Poynting-Robertson effect).

Problem 5.13. A thermally-conducting black sphere with a thermometer attached moves with velocity v through a black body radiation field of temperature T_0. What does the thermometer read?

Problem 5.14. In an electron gas of temperature $T << m_ec^2/k$ a photon of energy $E << m_ec^2$ undergoes collisions and is Compton scattered. Show that, to lowest order in E and T the average energy lost by a photon in a collision is

$$<\Delta E> = (E/m_ec^2)(E-4kT) \ .$$

Problem 5.15. Show that in special relativity the stress-energy of an isolated physical system of finite extent obeys the tensor virial theorem

$$\int T^{ij}\, d^3x = \frac{1}{2}\frac{d^2}{dt^2}\int T^{00}x^ix^jd^3x \ .$$

Problem 5.16. Show that the stress-energy tensor $T^{\mu\nu}$ has a timelike eigenvector if and only if there is a physical observer who sees no net energy flux in any direction. What is the significance of the eigenvalue?

Problem 5.17.

(a) Consider a stressed medium which moves through a particular inertial frame with velocity $|\underset{\sim}{v}| << 1$. Show that to first order in the velocity, the spatial components of the momentum density are

$$g^j = m^{jk} v^k ,$$

where m^{jk}, the "inertial mass per unit volume" is

$$m^{jk} = T^{0'0'}\delta^{jk} + T^{j'k'}$$

in terms of $T^{\mu'\nu'}$, the components of the stress-energy in the rest frame of the medium. What is m^{jk} for a perfect fluid?

(b) Consider an isolated, stressed body at rest and in equilibrium ($T^{\alpha\beta}{}_{,0} = 0$) in the laboratory frame. Show that its total inertial mass, defined by

$$M^{ij} \equiv \int_{\substack{\text{stressed}\\ \text{body}}} m^{ij}\,dx\,dy\,dz$$

is isotropic and equals the rest mass of the body, i.e. show that

$$M^{ij} = \delta^{ij} \int T^{00}\,dx\,dy\,dz \;.$$

Problem 5.18. If **u** is the 4-velocity of a fluid show that $\nabla\mathbf{u}$ can be decomposed as

$$u_{\alpha;\beta} = \omega_{\alpha\beta} + \sigma_{\alpha\beta} + \frac{1}{3}\,\theta P_{\alpha\beta} - a_\alpha u_\beta ,$$

where **a** is the "4-acceleration" of the fluid

$$a_\alpha \equiv u_{\alpha;\beta}\, u^\beta ,$$

θ is the "expansion" of the fluid world lines

$$\theta \equiv \nabla \cdot \mathbf{u} = u^{\alpha}{}_{;\alpha} \,,$$

$\omega_{\alpha\beta}$ is the "rotation 2-form" of the fluid, and $\sigma_{\alpha\beta}$ is the "shear tensor"

$$\omega_{\alpha\beta} \equiv \frac{1}{2}\,(u_{\alpha;\mu}P^{\mu}{}_{\beta} - u_{\beta;\mu}P^{\mu}{}_{\alpha}) \,,$$

$$\sigma_{\alpha\beta} \equiv \frac{1}{2}\,(u_{\alpha;\mu}P^{\mu}{}_{\beta} + u_{\beta;\mu}P^{\mu}{}_{\alpha}) - \frac{1}{3}\,\theta P_{\alpha\beta} \,.$$

Here $\mathbf{P}$ is the projection tensor

$$P_{\alpha\beta} \equiv g_{\alpha\beta} + u_{\alpha} u_{\beta}$$

which projects a vector onto the 3-surface perpendicular to $\mathbf{u}$.

Problem 5.19. Write the first law of thermodynamics for a relativistic fluid. (i.e. Write the law of conservation of mass-energy for a fluid element.)

Problem 5.20. Use the equations of motion $(T^{\mu\nu}{}_{;\nu} = 0)$ to show that the flow of a perfect fluid is isentropic.

Problem 5.21. For a perfect fluid with equation of state $\rho = \rho(n)$ (where n = baryon density) show that $T^{\mu}{}_{\mu}$, the trace of the stress-energy tensor is negative if and only if

$$d \log \rho / d \log n < 4/3 \,.$$

Problem 5.22. Show that the velocity of sound v_s in a relativistic perfect fluid is given by
$$v_s^2 = \partial p/\partial \rho|_{s=\text{constant}} \,.$$

For a high temperature relativistic gas with an equation of state $\rho \approx 3P$ (essentially that for a photon gas) show that $v_s \approx 1/\sqrt{3}$.

Problem 5.23. The velocity of sound in a fluid is $v_s^2 = \partial p/\partial \rho|_{s=\text{constant}}$. Show that $v_s^2 = \Gamma_1 p/(\rho + p)$ where Γ_1 is the adiabatic index

$$\Gamma_1 = \partial \log p / \partial \log n|_{s=\text{constant}} \,.$$

Problem 5.24. What is the speed of sound in an ideal Fermi gas at zero temperature?

Problem 5.25. A relativistic wind tunnel is to be fed from a tank of perfect adiabatic compressed gas. Suppose the gas has an equation of state $p \propto n^{\gamma}$ with γ constant, and the speed of sound in the tank is a. What is the largest wind velocity v_{max} which can be obtained? (No gravitational forces; isentropic flow.)

Problem 5.26. An idealized description of heat flow in a fluid uses the heat flux 4-vector $\mathbf{q}$ with components in the fluid rest frame $q^0 = 0$, q^j = (energy per unit time crossing a unit surface perpendicular to $\mathbf{e}_j$, in the positive j direction). What is the stress-energy tensor associated with the heat flow?

Problem 5.27. Let s, n and $\mathbf{q}$ be respectively entropy per baryon, number density of baryons, and heat flux, all measured in the proper frame of the fluid. In this proper frame $\mathbf{q}$ is purely spatial. Let $\mathbf{S}$ be the entropy density-flux 4-vector. Show that

$$\mathbf{S} = ns\mathbf{u} + \mathbf{q}/T ,$$

where $\mathbf{u}$ is the 4-velocity of the fluid rest frame.

Problem 5.28. A fluid is "perfect" except for admitting some heat conduction, described by a heat flow 4-vector $\mathbf{q}$. Calculate the local rate of entropy generation $\nabla \cdot \mathbf{S}$.

Problem 5.29. In a uniformly accelerating system, show that the condition for thermal equilibrium is not $T = \text{constant} = T_0$ but rather is

$$T = T_0 \exp(-\underset{\sim}{a} \cdot \underset{\sim}{x})$$

where $\underset{\sim}{x}$ is coordinate position in the accelerating frame.

Problem 5.30. The stress energy tensor of a viscous fluid is

$$T^{\alpha\beta} = \rho u^{\alpha} u^{\beta} + p P^{\alpha\beta} - 2\eta\sigma^{\alpha\beta} - \zeta\theta P^{\alpha\beta} .$$

Here η and ζ are respectively the coefficients of shear and bulk viscosity. The definitions of $\sigma^{\alpha\beta}$, θ, $P^{\alpha\beta}$ are those of Problem 5.18. The pressure and density are p and ρ. Show that the viscous terms lead to the production of entropy at a rate

$$S^{\alpha}{}_{;\alpha} = (\zeta\theta^2 + 2\eta\sigma_{\alpha\beta}\sigma^{\alpha\beta})/T$$

where T is the temperature of the fluid. (Hint: First show that $S^{\alpha}{}_{;\alpha} = [d\rho/d\tau + \theta(\rho+p)]/T$ for a fluid without heat flow, then differentiate $\rho u^{\beta} = -T^{\alpha\beta}u_{\alpha}$ to get $d\rho/d\tau$.)

Problem 5.31. From the stress-energy

$$T^{\alpha\beta} = \rho u^{\alpha}u^{\beta} + pP^{\alpha\beta} - 2\eta\sigma^{\alpha\beta} - \zeta\theta P^{\alpha\beta} ,$$

show that the equations of motion derived from $T^{\alpha\beta}{}_{,\beta} = 0$ reduce to the Navier-Stokes equations in the nonrelativistic limit.

Problem 5.32. As in non-relativistic thermodynamics, one defines the specific heat of a gas at constant volume and constant pressure by

$$c_v = T\left.\frac{ds}{dT}\right|_n \qquad c_p = T\left.\frac{ds}{dT}\right|_p .$$

For a perfect Maxwell-Boltzmann gas, show that $c_p = c_v + k$. (Here k = Boltzmann's constant.) Show that the adiabatic index

$$\Gamma_1 \equiv \left.\frac{\partial \log p}{\partial \log n}\right|_s$$

is equal to the ratio of specific heats,

$$\gamma \equiv c_p/c_v .$$

Problem 5.33. For a perfect Maxwell-Boltzmann gas, show that if γ is approximately constant in some regime of interest, then $p = Kn^{\gamma}$ and $\rho = mn + Kn^{\gamma}/(\gamma-1)$ under adiabatic conditions. (K = constant, m = mass of particles.)

Problem 5.34. The invariant equilibrium distribution function of a relativistic gas is

$$\mathcal{N}(p^\alpha, x^\alpha) \equiv \frac{dN}{d^3x\, d^3P} = \frac{(2J+1)/h^3}{\exp[-\frac{P\cdot u}{kT} - \theta] - \varepsilon} .$$

Here J = spin of particles, h = Planck's constant, $\mathbf{u}$ = mean 4-velocity of gas, $\varepsilon = 1, 0$ or -1 for Bose-Einstein, Maxwell-Boltzmann or Fermi-Dirac statistics respectively. The parameter θ is independent of $\mathbf{P}$. The first two moments of $\mathcal{N}$ are

$$J^\mu \equiv \int \mathcal{N} P^\mu \frac{d^3P}{(-\mathbf{P}\cdot\mathbf{u})}, \qquad T^{\mu\nu} \equiv \int \mathcal{N} P^\mu P^\nu \frac{d^3P}{(-\mathbf{P}\cdot\mathbf{u})} .$$

Since $\mathbf{u}$ is the only free vector, these integrals must have the form

$$J^\mu = n u^\mu , \qquad T^{\mu\nu} = (\rho + p) u^\mu u^\nu + p g^{\mu\nu} .$$

(This is the kinetic-theory *definition* of n, ρ, p.)

(a) Obtain 1-dimensional integrals for n, ρ and p.

(b) Show that $dp = (\rho + p)/T\, dT + nkT\, d\theta$.

(c) Use the first law of thermodynamics to identify $kT\theta$ as the chemical potential $\mu = (\rho + p)/n - Ts$.

(d) Show that for a Maxwell-Boltzmann gas, $p = nkT$ for all T.

(e) Show that for a Maxwell-Boltzmann gas, $\rho = n(m + \frac{3}{2} kT)$ is an approximation valid only for $kT << m$. Find the exact expression for ρ/n. What is ρ/n for $kT >> m$? (Here m is the mass of a gas particle.)

Problem 5.35. For a perfect Maxwell-Boltzmann gas, find $\gamma(T)$, the ratio of specific heats, as a function of temperature.

CHAPTER 6
METRICS

Metric geometry, geometry specified by a distance formula $ds^2 = g_{\alpha\beta}\,dx^\alpha dx^\beta$, is the foundation for general relativity and for most of the remaining chapters in this book. The most important metric is of course a spacetime metric, formally a metric which can locally be transformed to the Minkowski metric, i.e. for every point P in the spacetime, there is some coordinate transformation which makes $g_{\alpha\beta} = \eta_{\alpha\beta}$ at P.

—— o —— o —— o —— o —— o —— o —— o ——

Problem 6.1.

(a) Prove that the 2-dimensional metric space described by

$$ds^2 = dv^2 - v^2\,du^2 \tag{1}$$

is just the flat 2-dimensional Minkowski space usually described by

$$ds^2 = dx^2 - dt^2 \ . \tag{2}$$

Do this by finding the coordinate transformations $x(v,u)$ and $t(v,u)$ which take the metric (2) into the form (1).

(b) For an unaccelerated particle, show that the component of the 4-momentum P_u is constant, but that P_v is not.

Problem 6.2. Show that the line element

$$ds^2 = R^2[d\alpha^2 + \sin^2\alpha\,(d\theta^2 + \sin^2\theta\,d\phi^2)]$$

represents a hypersphere of radius R in Euclidean 4-space, i.e. a locus of points a distance R from a given point.

Problem 6.3. The metric for the surface of a globe of the Earth is

$$ds^2 = a^2(d\lambda^2 + \cos^2\lambda \, d\phi^2)$$

where λ is the latitude and ϕ is the longitude. The metric of a flat map of the world, with Cartesian coordinates x and y is $ds^2 = dx^2 + dy^2$; however we are not interested in *this* geometry, but in that of the globe it represents. What is the metric of the globe expressed in x and y coordinates for (a) a cylindrical projection, and (b) a stereographic projection map of the world?

Problem 6.4. Mercator's projection is defined as follows: The map coordinates are rectangular Cartesian coordinates (x, y) such that a straight line on the map is a line of constant compass bearing on the globe.

(a) Show that the map is defined by $x = \phi$, $y = \log \cot \frac{1}{2}\theta$, where (θ, ϕ) are the polar coordinates of a point on the globe.

(b) What is the metric of the globe in (x, y) coordinates?

(c) Show that the great circles are given by $\sinh y = \alpha \sin(x + \beta)$ (except for the special cases $y = 0$ or $x = \text{constant}$).

Problem 6.5. A space purports to be 3-dimensional, with coordinates x, y, z and the metric

$$ds^2 = dx^2 + dy^2 + dz^2 - \left(\frac{3}{13}\,dx + \frac{4}{13}\,dy + \frac{12}{13}\,dz\right)^2 .$$

Show that it is really a two-dimensional space, and find two new coordinates ζ and η for which the line element takes the form

$$ds^2 = d\zeta^2 + d\eta^2 .$$

Problem 6.6. Show that a contraction of a vector $\mathbf{V}$ with the "projection tensor" $\mathbf{P} \equiv \mathbf{g} + \mathbf{u} \otimes \mathbf{u}$ projects $\mathbf{V}$ into the 3-surface orthogonal to the 4-velocity vector $\mathbf{u}$. If $\mathbf{n}$ is a unit spacelike vector show that

$$\mathbf{P} \equiv \mathbf{g} - \mathbf{n} \otimes \mathbf{n}$$

is the corresponding projection operator. Show that there is no unique projection operator orthogonal to a null vector.

Problem 6.7. Show that a conformal transformation of a metric, i.e. $g_{\alpha\beta} \rightarrow f(x^{\mu}) g_{\alpha\beta}$ for an arbitrary function f, preserves all angles. (Figure out how to define angles!) Show that all null curves remain null curves.

Problem 6.8. One can put a metric on the velocity space of a particle by defining the distance between two nearby velocities as their relative velocity. Show that the metric can be written in the form

$$ds^2 = d\chi^2 + \sinh^2\chi\,(d\theta^2 + \sin^2\theta d\phi^2) \ .$$

where the magnitude of the velocity is $v = \tanh \chi$.

Problem 6.9. A manifold which has the topology of a 2-sphere has – in a neighborhood of $\theta = \frac{1}{2}$, $\chi = 0$ – the metric

$$ds^2 = d\theta^2 + (\theta - \theta^3)^2 d\chi^2 \ .$$

The manifold has precisely *one* point which is not locally flat, and that point is a "conical" singularity. Show that there are two different maximal analytic extensions of the metric, i.e. that there are two different ways of extending the metric, and satisfying the condition that there is only one conical singularity. Note that this shows that a metric in a local coordinate patch does not always pin down the global nature of the manifold. (Hint: consider the periodicity of the χ coordinate.)

Problem 6.10. Find the most general form for a spacetime metric that is (spatially) spherically symmetric.

CHAPTER 7

COVARIANT DIFFERENTIATION AND GEODESIC CURVES

The partial derivatives of a vector or tensor with respect to the coordinates of a space (e.g. $A^{\mu}{}_{,\nu}$ or $Q^{\alpha\beta\cdots}{}_{\gamma\delta\cdots,\nu}$) are not themselves components of a tensor. Rather, the curvilinearity of the coordinates (optional in flat space, but inevitable in a curved space) must be taken into account, leading to the idea of covariant differentiation.

The tensor formed by differentiating a tensor $\mathbf{Q}$ with components $Q^{\alpha\beta\cdots}{}_{\gamma\delta\cdots}$ is denoted $\nabla\mathbf{Q}$ and has components denoted

$$
\begin{aligned}
Q^{\alpha\beta\cdots}{}_{\gamma\delta\cdots;\sigma} \equiv\ & Q^{\alpha\beta\cdots}{}_{\gamma\delta\cdots,\sigma} + \Gamma^{\alpha}{}_{\nu\sigma} Q^{\nu\beta\cdots}{}_{\gamma\delta\cdots} \\
& + \Gamma^{\beta}{}_{\nu\sigma} Q^{\alpha\nu\cdots}{}_{\gamma\delta\cdots} + \ldots - \Gamma^{\nu}{}_{\gamma\sigma} Q^{\alpha\beta\cdots}{}_{\nu\delta\cdots} \\
& - \Gamma^{\nu}{}_{\delta\sigma} Q^{\alpha\beta\cdots}{}_{\gamma\nu\cdots}
\end{aligned}
$$

where there is one "correction" term for every index of $\mathbf{Q}$. The Γ's are called Christoffel symbols or [affine] connection coefficients. In a coordibasis they are related to the partial derivatives of the metric by

$$
\Gamma^{\alpha}{}_{\beta\gamma} = g^{\alpha\mu}\Gamma_{\mu\beta\gamma} = \frac{1}{2} g^{\alpha\mu}(g_{\mu\beta,\gamma} + g_{\mu\gamma,\beta} - g_{\beta\gamma,\mu})
$$

(the first equality *defines* $\Gamma_{\mu\beta\gamma}$). The Γ's are sets of numbers, but they are not components of a tensor. (They do not transform like a tensor.)

A covariant derivative which is dotted into a vector is called a directional derivative:

$$
(\nabla\mathbf{Q})\cdot\mathbf{u} \equiv \nabla_{\mathbf{u}}\mathbf{Q} \equiv Q^{\alpha\beta\cdots}{}_{\gamma\delta\cdots;\nu}\, u^{\nu} .
$$

If the vector $\mathbf{u}$ is tangent to a curve parameterized by λ, one sometimes writes $\mathbf{u} = d/d\lambda$ for $u^{\alpha} = dx^{\alpha}/d\lambda$ and

$$\nabla_{\mathbf{u}} \mathbf{Q} \equiv \frac{D\mathbf{Q}}{d\lambda} .$$

If the vector happens to be a basis vector, one writes

$$\nabla_{\mathbf{e}_\alpha} \mathbf{Q} \equiv \nabla_\alpha \mathbf{Q} .$$

In terms of the basis vectors, the connection coefficients can be written

$$\nabla_\beta \mathbf{e}_\alpha = \Gamma^\mu{}_{\alpha\beta} \mathbf{e}_\mu \quad \text{or} \quad \Gamma_{\mu\alpha\beta} = \mathbf{e}_\mu \cdot \nabla_\beta \mathbf{e}_\alpha .$$

The covariant derivative operator ∇ obeys all of the nice rules expected of a derivative operator, except that in curved space $\nabla_{\mathbf{u}}\nabla_{\mathbf{v}} \neq \nabla_{\mathbf{v}}\nabla_{\mathbf{u}}$ (see Chapter 9).

If $\mathbf{u}$ is the tangent vector to a curve, a tensor $\mathbf{Q}$ is said to be parallel propagated along the curve if

$$\nabla_{\mathbf{u}} \mathbf{Q} = 0 .$$

If the tangent vector is itself parallel propagated,

$$\nabla_{\mathbf{u}} \mathbf{u} = 0$$

(tangent vector "covariantly constant") the curve is a geodesic, the generalization of a straight line in flat space. If $x^\alpha(\lambda)$ is the geodesic (with $u^\alpha = dx^\alpha/d\lambda$) then the components of this *geodesic equation* are

$$0 = (\nabla_{\mathbf{u}} \mathbf{u})^\mu = \frac{du^\mu}{d\lambda} + u^\alpha u^\beta \Gamma^\mu{}_{\alpha\beta} .$$

Here λ must be an affine parameter along the curve; for non-null curves this means λ must be proportional to proper length.

If a curve is timelike, $\mathbf{u}$ is its tangent vector, and $\mathbf{a} \equiv \nabla_{\mathbf{u}} \mathbf{u} = D\mathbf{u}/d\tau$, then a vector $\mathbf{V}$ is said to be Fermi-Walker transported along $\mathbf{u}$ if

$$\nabla_{\mathbf{u}} \mathbf{V} = (\mathbf{u} \otimes \mathbf{a} - \mathbf{a} \otimes \mathbf{u}) \cdot \mathbf{V} .$$

Problem 7.1. Show that the connection coefficients $\Gamma^{\alpha}{}_{\beta\gamma}$ do *not* obey the tensor transformation law.

Problem 7.2. For a 2-dimensional flat, Euclidean space described by polar coordinates r, θ, assume that the geodesics are the usual straight lines.

(a) Find the connection coefficients $\Gamma^{\alpha}{}_{\beta\gamma}$, using your knowledge of these geodesics, and the geodesic equation

$$\frac{d^2x^\mu}{ds^2} + \frac{dx^\alpha}{ds}\frac{dx^\beta}{ds}\Gamma^{\mu}{}_{\alpha\beta} = 0 .$$

(b) Next, in the coordinates x, y which are related to r, θ in the usual way, take the covariant structure to be given by $\Gamma^{x}{}_{xx} = \Gamma^{x}{}_{xy} = \cdots = 0$. Using the transformation law for connection coefficients find the connection coefficients in the r, θ coordinates.

(c) Finally, from the line element $ds^2 = dr^2 + r^2 d\theta^2$ find the Christoffel symbols, in the usual way, as derivatives of the metric coefficients $g_{\mu\nu}$. (All three methods, of course, must give the same Christoffel symbols.)

Problem 7.3. Consider the familiar metric space

$$ds^2 = dr^2 + r^2 d\theta^2 .$$

(a) Write the 2 equations that result from the geodesic equation, and show that the following are first integrals of these equations:

$$r^2 \frac{d\theta}{ds} = R_0 = \text{constant}$$

$$\left(\frac{dr}{ds}\right)^2 + r^2\left(\frac{d\theta}{ds}\right)^2 = 1 .$$

(b) Use the results in (a) to get a first-order differential equation for $r(\theta)$. [That is: Eliminate s as a parameter and replace by θ].

(c) Using the fact that the metric space is just flat 2-dimensional

Euclidean space, write down the general equation for a straight line in r, θ coordinates, and show that the straight line satisfies the equation in (b).

Problem 7.4. For the 2-dimensional metric $ds^2 = (dx^2 - dt^2)/t^2$, find all connection coefficients $\Gamma_{\alpha\beta\gamma}$, and find all timelike geodesic curves.

Problem 7.5. Show that the metric tensor is covariantly constant.

Problem 7.6. For a diagonal metric, prove that (in a coordinate frame) the Christoffel symbols are given by

$$\Gamma^{\mu}{}_{\nu\lambda} = 0 , \qquad \Gamma^{\mu}{}_{\lambda\lambda} = -\frac{1}{2g_{\mu\mu}} \frac{\partial g_{\lambda\lambda}}{\partial x^{\mu}}$$

$$\Gamma^{\mu}{}_{\mu\lambda} = \frac{\partial}{\partial x^{\lambda}} (\log(|g_{\mu\mu}|)^{\frac{1}{2}}), \qquad \Gamma^{\mu}{}_{\mu\mu} = \frac{\partial}{\partial x^{\mu}} (\log(|g_{\mu\mu}|)^{\frac{1}{2}}) .$$

Here $\mu \neq \nu \neq \lambda$ and there is no summation over repeated indices.

Problem 7.7. Prove the following indentities:

(a) $g_{\alpha\beta,\gamma} = \Gamma_{\alpha\beta\gamma} + \Gamma_{\beta\alpha\gamma}$.

(b) $g_{\alpha\mu} g^{\mu\beta}{}_{,\gamma} = - g^{\mu\beta} g_{\alpha\mu,\gamma}$.

(c) $g^{\alpha\beta}{}_{,\gamma} = - \Gamma^{\alpha}{}_{\mu\gamma} g^{\mu\beta} - \Gamma^{\beta}{}_{\mu\gamma} g^{\mu\alpha}$.

(d) $g_{,\alpha} = - g g_{\beta\gamma} g^{\beta\gamma}{}_{,\alpha} = g g^{\beta\gamma} g_{\beta\gamma,\alpha}$.

(e) $\Gamma^{\alpha}{}_{\alpha\beta} = (\log |g|^{\frac{1}{2}})_{,\beta}$ in a coordinate frame.

(f) $g^{\mu\nu}\Gamma^{\alpha}{}_{\mu\nu} = - \frac{1}{|g|^{\frac{1}{2}}} (g^{\alpha\nu}|g|^{\frac{1}{2}})_{,\nu}$ in a coordinate frame .

(g) $A^{\alpha}{}_{;\alpha} = \frac{1}{|g|^{\frac{1}{2}}} (|g|^{\frac{1}{2}} A^{\alpha})_{,\alpha}$ in a coordinate frame.

(h) $A_{\alpha}{}^{\beta}{}_{;\beta} = \frac{1}{|g|^{\frac{1}{2}}} (|g|^{\frac{1}{2}} A_{\alpha}{}^{\beta})_{,\beta} - \Gamma^{\lambda}{}_{\alpha\mu} A_{\lambda}{}^{\mu}$ in a coordinate frame .

(i) $A^{\alpha\beta}{}_{;\beta} = \frac{1}{|g|^{\frac{1}{2}}} (|g|^{\frac{1}{2}} A^{\alpha\beta})_{,\beta}$ in a coordinate frame, if $A^{\alpha\beta}$ is antisymmetric.

(j) $\Box S = S_{;\alpha}{}^{;\alpha} = \frac{1}{|g|^{\frac{1}{2}}} (|g|^{\frac{1}{2}} g^{\alpha\beta} S_{,\beta})_{,\alpha}$ in a coordinate frame.

Problem 7.8. Let $A \equiv \det(A_{\mu\nu})$ where $A_{\mu\nu}$ is a second rank tensor. Show that A is not a scalar. (i.e. Show that its value changes under coordinate transformations.) Since A is not a scalar one cannot define $A_{;\alpha} = A_{,\alpha}$. How *should* $A_{;\alpha}$ be defined (in terms of $A_{,\alpha}$ and A)?

Problem 7.9. If a geodesic is timelike at a given point P, show that it is timelike everywhere along its length, and similarly for spacelike or null, geodesics.

Problem 7.10. Derive the geodesic equation from the definition of a geodesic as a curve of extremal length.

Problem 7.11. An affine parameter λ is one for which the equation of geodesic motion has the form

$$\frac{dx^\alpha}{d\lambda^2} + \Gamma^\alpha{}_{\beta\gamma}\frac{dx^\beta}{d\lambda}\frac{dx^\gamma}{d\lambda} = 0 .$$

Show that all affine parameters are related by linear transformations with constant coefficients.

Problem 7.12. Show that in flat spacetime, the conservation law for the 4-momentum of a freely moving particle can be written $\nabla_p p = 0$. Show that particles with nonzero rest mass move along timelike geodesics.

Problem 7.13. Suppose the coordinate x^1 is a cyclic coordinate, i.e. the metric functions $g_{\alpha\beta}$ are independent of x^1. If p is the momentum of an unaccelerated particle, show that the component p_1 is constant along the particle world line.

Problem 7.14. Prove the general relativity version of Fermat's principle: In any static metric ($g_{0j} = g_{\alpha\beta,0} = 0$), consider all null curves between two points in space, $x^j = a^j$ and $x^j = b^j$. Each such null curve $x^j(t)$ requires a particular coordinate time Δt to get from a^j to b^j. Show that the curves of extremal time Δt are null geodesics of the spacetime.

Problem 7.15.

(a) Show that the *geodesics* of the velocity space metric defined in Problem 6.8 are paths of minimum fuel use for a rocket ship changing its velocity.

(b) A rocket ship in interstellar space with velocity $\underset{\sim}{V}_1$ (with respect to the earth) changes its velocity to a new velocity $\underset{\sim}{V}_2$, in a manner that uses up the least fuel. What is the ship's smallest velocity relative to earth during the change?

Problem 7.16. On the surface of a two-sphere, $ds^2 = d\theta^2 + \sin^2\theta\, d\phi^2$, the vector $\mathbf{A}$ is equal to $\mathbf{e}_\theta$ at $\theta = \theta_0$, $\phi = 0$. What is $\mathbf{A}$ after it is parallel transported around the circle $\theta = \theta_0$? What is the magnitude of $\mathbf{A}$?

Problem 7.17. Consider an observer with 4-velocity $\mathbf{u}$ who transports his four basis vectors $\mathbf{e}_\alpha$ along with him according to the transport law $\nabla_{\mathbf{u}}\mathbf{e}_\alpha = A_\alpha{}^\beta \mathbf{e}_\beta$. What is the most general form of $A_\alpha{}^\beta$ if:

(i) the basis vectors are to be orthonormal?

(ii) in addition $\mathbf{e}_{\hat{0}} = \mathbf{u}$? (i.e. the frame is his rest-frame).

(iii) in addition the spatial vectors are to be nonrotating? (i.e. He sees a freely-falling particle move with no Coriolis forces.)

Problem 7.18. Show that the scalar product of two vectors is not altered as they are both Fermi-Walker transported along a curve $\mathcal{C}$.

Problem 7.19. Show that Fermi-Walker transport along a *geodesic* curve is the same as parallel transport.

Problem 7.20. Write the following expressions in index-free notation:

(a) $U_{\alpha;\beta}U^\beta U^\alpha$ (b) $V^\alpha{}_{;\beta}U^\beta - U^\alpha{}_{;\beta}V^\beta$

(c) $T_{\alpha\beta;\gamma}V^\alpha W^\beta U^\gamma$ (d) $W^{\alpha;\beta}V_{\beta;\gamma}U^\gamma$

(e) $W^\alpha{}_{;\gamma\beta}U^\gamma U^\beta + W^\alpha{}_{;\gamma}U^\gamma{}_{;\beta}U^\beta - U^\alpha{}_{;\beta}W^\beta{}_{;\gamma}U^\gamma$.

Problem 7.21. Show that the paths of light rays in a static, isotropic spacetime can be described by taking the space to have a certain spatially varying "index of refraction" $n(x^j)$. What is n in terms of $g_{\alpha\beta}$? Assume $g_{\alpha\beta}$ has the form

$$ds^2 = g_{00}dt^2 - f(dx_1^2 + dx_2^2 + dx_3^2).$$

Problem 7.22. An inebriated astronaut pulses his rocket, firing in a random direction for each pulse. As measured in a momentarily comoving frame, each pulse corresponds to a velocity boost of $\Delta v << c$. Find the probability distribution of his resultant velocity after n boosts, where n is a very large number. Show that the drunk astronaut achieves highly relativistic velocities less efficiently than a sober astronaut (who fires his rocket always in the same direction), and takes on the average $3c/\Delta v$ times as many pulses to achieve the same velocity.

Problem 7.23.

(a) Suppose a vector field **k** is orthogonal to a family of hypersurfaces ("hypersurface-orthogonal"). Show that this implies $k_{[\mu;\nu}k_{\lambda]} = 0$.

(b) What is the geometric interpretation if $k_{[\mu;\nu]}$ also vanishes?

Problem 7.24. Prove that any congruence of null curves that is hypersurface-orthogonal consists of null geodesics.

Problem 7.25. Show that the variational principle

$$\delta \int (g_{\alpha\beta}\dot{x}^\alpha \dot{x}^\beta)\, ds = 0$$

gives the same geodesics as the defining property for geodesics

$$\delta \int (g_{\alpha\beta}\dot{x}^\alpha \dot{x}^\beta)^{\frac{1}{2}}\, ds = 0$$

when s is proper length (*not* an arbitrary parameterization) and $\dot{x} \equiv dx^\alpha/ds$. If $y \equiv (g_{\alpha\beta}\dot{x}^\alpha \dot{x}^\beta)^{\frac{1}{2}}$, show that

$$\delta \int F(y)\, ds = 0$$

gives the same geodesics for *any* monotonic function $F(y)$.

CHAPTER 8

DIFFERENTIAL GEOMETRY: FURTHER CONCEPTS

A vector $\mathbf{B}$ is related to its contravariant components B^μ by

$$\mathbf{B} = B^\mu e_\mu \ ,$$

where the e_μ are basis vectors. The *covariant* components B_μ represent the same vector, but represent it as a different type of "vector", called a one-form. (Loosely, one-forms are often called "covariant vectors".) For one-forms the analog of the above equation is

$$\tilde{\mathbf{B}} = B_\mu \tilde{\boldsymbol{\omega}}^\mu \ ,$$

where $\sim$ indicates a one-form and $\tilde{\boldsymbol{\omega}}^\mu$ are basis one-forms with covariant components $(1, 0, 0, 0)$, $(0, 1, 0, 0)$, etc. For an arbitrary tensor $\mathbf{T}$, with components $T_{\alpha\beta\cdots}{}^{\gamma\delta\cdots}$,

$$\mathbf{T} = T_{\alpha\beta\cdots}{}^{\gamma\delta\cdots}\, \tilde{\boldsymbol{\omega}}^\alpha \otimes \tilde{\boldsymbol{\omega}}^\beta \otimes \cdots \otimes e_\gamma \otimes e_\delta \otimes \cdots \ .$$

The scalar product of two vectors, or of two one-forms, involves the metric tensor, and is denoted by a "dot":

$$\mathbf{A}\cdot\mathbf{B} = g_{\mu\nu} A^\mu B^\nu$$

$$\tilde{\mathbf{A}}\cdot\tilde{\mathbf{B}} = g^{\mu\nu} A_\mu B_\nu \ .$$

(Here $g^{\mu\nu}$ is the matrix inverse of $g_{\mu\nu}$.) The scalar product of a vector with a one-form does not involve the metric, only a summation over an index. This is sometimes distinguished notationally as

$$\tilde{\mathbf{B}}\cdot\mathbf{A} \equiv \langle\tilde{\mathbf{B}}, \mathbf{A}\rangle \equiv B_\mu A^\mu$$

Since $\langle \tilde{\boldsymbol{\omega}}^\mu, e_\nu \rangle = \delta^\mu{}_\nu$, the basis $\tilde{\boldsymbol{\omega}}^\mu$ is said to be "dual" to the basis e_μ. If $\tilde{\mathbf{B}}$ and $\tilde{\mathbf{A}}$ are one-forms corresponding to vectors $\mathbf{B}$ and $\mathbf{A}$ we have, of course, $\mathbf{A}\cdot\mathbf{B} = \tilde{\mathbf{A}}\cdot\tilde{\mathbf{B}} = \langle \tilde{\mathbf{B}}, \mathbf{A} \rangle = \langle \tilde{\mathbf{A}}, \mathbf{B} \rangle$.

A one-form of particular usefulness is $\widetilde{df}$, the *gradient* of any scalar function f. When combined with a vector v, it gives the directional derivative of f along v.

$$\langle \widetilde{df}, v \rangle = \nabla_v f = f_{,\alpha} v^\alpha \ .$$

The basis vectors e_α corresponding to a coordinate system are tangent to the coordinate lines. This motivates the notation for a coordinate basis vector

$$e_\alpha = \frac{\partial}{\partial x^\alpha} \ .$$

Similarly, the coordinate basis one-forms are gradients of the coordinate surfaces,

$$\tilde{\boldsymbol{\omega}}^\alpha \equiv \widetilde{dx}^\alpha \ .$$

Alternatively, since a spacetime metric can locally be transformed to the Minkowski metric, it is always possible to find a set of *orthonormal* basis vectors and one-forms at every point. These are *not* necessarily tangents to or gradients of the coordinates; they are denoted by $e_{\hat{\mu}}$, $\tilde{\boldsymbol{\omega}}^{\hat{\mu}}$, where $\hat{}$ indicates orthonormality. Note that the relations $\langle \tilde{\boldsymbol{\omega}}^\alpha, e_\beta \rangle = \delta^\alpha{}_\beta$, $e_\alpha \cdot e_\beta = g_{\alpha\beta}$, and $\tilde{\boldsymbol{\omega}}^\alpha \cdot \tilde{\boldsymbol{\omega}}^\beta = g^{\alpha\beta}$ always hold. When the basis is orthonormal, then in addition $g_{\alpha\beta} = \eta_{\alpha\beta}$ and $g^{\alpha\beta} = \eta^{\alpha\beta}$. If a local orthonormal frame in spacetime is freely falling (i.e. the basis is that of a freely falling observer), then all the $\Gamma^\alpha{}_{\beta\gamma}$ vanish at the center of the frame.

The commutator of two basis-vector fields

$$\nabla_{e_\alpha} e_\beta - \nabla_{e_\beta} e_\alpha = [e_\alpha, e_\beta] \equiv c_{\alpha\beta}{}^\gamma e_\gamma$$

vanishes identically, $c_{\alpha\beta}{}^\gamma = 0$, if and only if e_α and e_β are the tangent vectors to come coordinate system (a coordinate basis). In a general (not necessarily a coordinate) basis,

$$\Gamma_{\mu\beta\gamma} \equiv e_\mu \cdot (\nabla_\gamma e_\beta) = \frac{1}{2}(g_{\mu\beta,\gamma} + g_{\mu\gamma,\beta} - g_{\beta\gamma,\mu} + c_{\mu\beta\gamma} + c_{\mu\gamma\beta} - c_{\beta\gamma\mu}) .$$

A tensor of rank p with all of its indices covariant, and which is totally antisymmetric on all indices, is called a p-form. A totally antisymmetrized direct product of forms is denoted by a wedge "$\wedge$".

The concepts of Lie differentiation, Lie transport, and exterior differentiation of forms are developed in problems.

——o——o——o——o——o——o——o——

Problem 8.1.

(a) A spacetime has coordinates x^α with basis vectors $\partial/\partial x^\alpha$ and basis one-forms $\widetilde{dx}^\alpha$. What are the values of:

$$<\widetilde{dx}^0, \partial/\partial x^0>, <\widetilde{dx}^2, \partial/\partial x^3>, (\partial/\partial x^0)\cdot(\partial/\partial x^1), \widetilde{dx}^0 \cdot \widetilde{dx}^1, \widetilde{dx}^0 \cdot \widetilde{dx}^0 ?$$

(b) To what vector does the one-form $\widetilde{dx}^1$ correspond?

Problem 8.2. The usual basis for polar coordinates, $e_{\hat{r}} = e_r$, $e_{\hat{\theta}} = r^{-1} e_\theta$ is not a coordinate basis. Consider the one-form basis $\tilde{\boldsymbol{\omega}}^{\hat{i}}$ dual to this basis

$$<\tilde{\boldsymbol{\omega}}^{\hat{i}}, e_{\hat{j}}> = \delta^i{}_j .$$

Find the function f such that $\tilde{\boldsymbol{\omega}}^{\hat{r}} = \widetilde{df}$ and prove that there does not exist a function g such that $\tilde{\boldsymbol{\omega}}^{\hat{\theta}} = \widetilde{dg}$. Do this without imposing any metric on the polar coordinates.

Problem 8.3. In 3-dimensional Euclidean space, what is a necessary and sufficient condition on a field of one-forms $\tilde{\boldsymbol{\sigma}}$ for there to exist a function f such that $\tilde{\sigma} = \widetilde{df}$?

Problem 8.4. If $\boldsymbol{\Omega}_1$ is a p-form and $\boldsymbol{\Omega}_2$ a q-form, show that

$$\boldsymbol{\Omega}_1 \wedge \boldsymbol{\Omega}_2 = (-1)^{pq} \boldsymbol{\Omega}_2 \wedge \boldsymbol{\Omega}_1 .$$

Problem 8.5. The exterior derivative of a differential form $\boldsymbol{\Omega}$ can be defined axiomatically by the following properties:

(i) If $\boldsymbol{\Omega}$ is a p-form, $d\boldsymbol{\Omega}$ is a (p+1)-form;

(ii) $d(\boldsymbol{\Omega}_1 + \boldsymbol{\Omega}_2) = d\boldsymbol{\Omega}_1 + d\boldsymbol{\Omega}_2$;

(iii) For a zero form f (scalar), $\widetilde{df}$ is defined by $\langle \widetilde{df}, v \rangle = \nabla_v f$ for any v;

(iv) $d(\boldsymbol{\Omega}_1 \wedge \boldsymbol{\Omega}_2) = d\boldsymbol{\Omega}_1 \wedge \boldsymbol{\Omega}_2 + (-1)^p \boldsymbol{\Omega}_1 \wedge d\boldsymbol{\Omega}_2$ where $\boldsymbol{\Omega}_1$ is a p-form;

[Note: If p = 0 (i.e. f a scalar) this reads: $\widetilde{d(f\boldsymbol{\Omega})} = \widetilde{df} \wedge \boldsymbol{\Omega} + f\widetilde{d\boldsymbol{\Omega}}$.]

(v) $dd\boldsymbol{\Omega} = 0$ for any $\boldsymbol{\Omega}$.

An alternative definition notes that a p-form is a completely antisymmetric covariant tensor of rank p, and defines the exterior derivative as the completely antisymmetrized covariant derivative. Show that these definitions are equivalent.

Problem 8.6. Consider a 2-form in n-dimensional space:

$$\boldsymbol{\alpha} = f(x^1, x^2, \cdots, x^n) \widetilde{dx}^1 \wedge \widetilde{dx}^2 .$$

Suppose for some region of space including $x^1 = 0$, that

$$d\boldsymbol{\alpha} = 0 .$$

Construct the 1-form

$$\tilde{\boldsymbol{\beta}} = \left[x^1 \int_0^1 f(\xi x^1, x^2, \cdots, x^n)\, d\xi \right] \widetilde{dx}^2$$

and show that $\boldsymbol{\alpha} = d\tilde{\boldsymbol{\beta}}$.

Problem 8.7. The components of the Maxwell tensor $F_{\alpha\beta}$ can be regarded as the components of a 2-form $\mathbf{F}$. Show that Maxwell's equations in vacuum can be written $d\mathbf{F} = 0$, $d{*}\mathbf{F} = 0$.

Problem 8.8. A 3-surface in spacetime is spacelike, timelike, or null, if its normal vector is timelike, spacelike, or null respectively. It is desired to find three orthogonal linearly independent vectors in a 3-surface. Show that for a spacelike surface, these are all spacelike; for a timelike surface, two spacelike and one timelike; for a null surface, one null and two spacelike.

Problem 8.9. Show that the integral $\int_S F^\mu d^3\Sigma_\mu$, where F^μ is some vector field and S is some oriented 3-dimensional hypersurface in spacetime, is independent of the parametrization $x^\mu = x^\mu(a,b,c)$ used to describe S. [See Problem 3.30 for the definition of $d^3\Sigma_\mu$.]

Problem 8.10. [Note: This problem assumes a familiarity with the Cartan calculus, beyond the scope of most relativity texts. Succeeding problems do not depend on it.] In the language of differential forms the generalized Stokes' theorem is

$$\int_\Omega d\boldsymbol{\theta} = \int_{\partial\Omega} \boldsymbol{\theta} \ .$$

To what does this theorem reduce, in the following cases

(a) Ω is 3-dimensional, $\boldsymbol{\theta} = f^k d^2 S_k$

(b) Ω is 4-dimensional, $\boldsymbol{\theta} = f^\mu d^3 \boldsymbol{\Sigma}_\mu$.

(c) Ω is 3-dimensional, $\boldsymbol{\theta} = F^{\mu\nu} d^2 \boldsymbol{\Sigma}_{\mu\nu}$ where $F^{\mu\nu}$ is antisymmetric.

(d) Use the generalized Stokes theorem to derive the familiar relation

$$\oint \underset{\sim}{A} \cdot d\underset{\sim}{\ell} = \int (\underset{\sim}{\nabla} \times \underset{\sim}{A}) \cdot d\underset{\sim}{S} \ .$$

Problem 8.11. Show that there exists *no* tensor with components constructed from the 10 metric coefficients $g_{\alpha\beta}$ and their 40 first derivatives $g_{\alpha\beta,\mu}$ – except g itself and products of it with itself, e.g. $g \otimes g$.

Problem 8.12.

(a) Show that, in a coordinate frame, $\Gamma_{\alpha\beta\gamma}$ is symmetric on the last two indices.

(b) Show that in an orthonormal frame, $\Gamma_{\alpha\beta\gamma}$ is antisymmetric on the first two indices.

Problem 8.13. The Lie derivative of a scalar function is defined to be the directional derivative: $\mathcal{L}_x f = \nabla_x f$. For a vector field y, we define

Lie differentiation as $\mathcal{L}_{\mathbf{x}}\mathbf{y} \equiv [\mathbf{x},\mathbf{y}] = \nabla_{\mathbf{x}}\mathbf{y} - \nabla_{\mathbf{y}}\mathbf{x}$. The Lie derivative obeys all the usual rules for a derivative operator and always gives a tensor of the same rank as the tensor differentiated.

(a) What is the Lie derivative of a 1-form?

(b) What is the Lie derivative of a tensor whose components are $T^{\alpha}{}_{\beta}$?

Problem 8.14. Suppose $\mathbf{A} = d/d\lambda$ is the tangent vector field to a congruence (set of curves) $x^{\alpha} = x^{\alpha}(\lambda)$ and $\mathbf{B}$ is a vector field. Show that the geometrical interpretation of the transport law $\mathcal{L}_{\mathbf{A}}\mathbf{B} = 0$ is that $\mathbf{B}$ connects points of equal λ on neighboring curves of the congruence.

Problem 8.15. Show that Lie differentiation commutes with the operation of contraction.

Problem 8.16. Show that

$$\mathcal{L}_{\mathbf{u}}\mathcal{L}_{\mathbf{v}} - \mathcal{L}_{\mathbf{v}}\mathcal{L}_{\mathbf{u}} = \mathcal{L}_{[\mathbf{u},\mathbf{v}]} \ .$$

Problem 8.17. Another definition of the Lie derivative of a geometric object $\Phi^{A}[x^{\mu}(P)]$ (A represents all the tensor indices, $x^{\mu}(P)$ are the coordinates of a point P) is as follows: Make an infinitesimal point transformation $P_0 \to P_N$ by $x^{\mu}(P_0) = x^{\mu}(P_N) + \xi^{\mu}(P_N)$. (Since ξ^{μ} is infinitesimal, it can in fact be evaluated at either P_0 or P_N.) Also make an infinitesimal coordinate transformation that makes the numerical values of the coordinates of P_N the same as those of P_0 in the original coordinates:

$$\bar{x}^{\mu}(P_N) = x^{\mu}(P_0) \ .$$

Then define

$$\mathcal{L}_{\boldsymbol{\xi}}\Phi^{A}(P_0) = \lim_{\boldsymbol{\xi}\to 0} [\Phi^{A}(P_0) - \bar{\Phi}^{A}(P_N)] \ .$$

Show that this definition is equivalent to that of Problem 8.13 by examining the cases (i) Φ^{A} = a scalar field (ii) $\Phi^{A} = A_{\mu}$ (iii) $\Phi^{A} = T_{\mu}{}^{\nu}$.

Problem 8.18. It is desired to transport a vector v along a curve with tangent vector **u**. Which of the following are necessary for parallel transport? – for Fermi-Walker transport? – for Lie transport?: metric; affine connection; **u**(x) defined off the curve.

Problem 8.19. You are given a vector field $v^i = (-y, x, z^\alpha)$, α = constant, in a 3-space with metric $ds^2 = dx^2 + dy^2 + dz^2$. A certain vector $\underset{\sim}{u}$ is Lie transported along $\underset{\sim}{v}$ from a point A to a point B and is then parallel transported back to A by the reverse route. For what value of α is there a $\underset{\sim}{u}$ which is always left unchanged by this process?

Problem 8.20. Find the most general vector field which is everywhere parallel-propagated along itself. Fermi-Walker transported along itself. Lie transported along itself.

Problem 8.21. If $\boldsymbol{\Omega}$ is a p-form, show that $\mathcal{L}_{\mathbf{x}}(d\boldsymbol{\Omega}) = d(\mathcal{L}_{\mathbf{x}}\boldsymbol{\Omega})$.

Problem 8.22. Vector analysis in 3-dimensional orthogonal curvilinear coordinates is a special case of tensor analysis where $g_{ij} = h_i^2\,\delta_{ij}$ (not summed). The h_i's are functions of the coordinates called "scale factors". Vector components are often referred to the ("physical") orthonormal basis $\tilde{\boldsymbol{\omega}}^{\hat{i}} = h_i \widetilde{dx}^i$ (not summed). Derive expressions for (i) ∇S, (ii) $\nabla \times \underset{\sim}{V}$, (iii) $\nabla \cdot \underset{\sim}{V}$ and (iv) $\nabla^2 S$, where S is a scalar field and $\underset{\sim}{V}$ a vector field.

Problem 8.23. Derive expressions for $\underset{\sim}{\nabla} \cdot \underset{\sim}{A}$ and $\underset{\sim}{\nabla} \times \underset{\sim}{A}$ in spherical polar coordinates.

Problem 8.24. If $F_{\mu\nu} = A_{\nu;\mu} - A_{\mu;\nu}$, prove that $F_{[\mu\nu;\lambda]} = 0$.

Problem 8.25. In an arbitrary spacetime manifold (not necessarily homogeneous or isotropic), pick an initial spacelike hypersurface S_I, place an arbitrary coordinate grid (x^1, x^2, x^3) on it, eject geodesic world lines orthogonal to it, and give these world lines the coordinates (x^1, x^2, x^3) = constant, $x^0 \equiv t = t_I + \tau$ where τ is proper time along the world line, beginning with $\tau = 0$ on S_I. Show that in this coordinate system ("Gaussian normal coordinates") the metric takes on the synchronous form

$$ds^2 = -dt^2 + g_{ij}\,dx^i dx^j .$$

Problem 8.26.

(a) If $g_{\mu\nu}$ and $\bar{g}_{\mu\nu}$ are the components of two symmetric tensors, show that

$$S^\lambda{}_{\mu\nu} = \bar{\Gamma}^\lambda{}_{\mu\nu} - \Gamma^\lambda{}_{\mu\nu}$$

are the components of a tensor. Here the Γ's and $\bar{\Gamma}$'s are Christoffel symbols formed from the tensors g and $\bar{g}$ in the usual way.

(b) Suppose $g_{\mu\nu}$ and $\bar{g}_{\mu\nu}$ have the same geodesics. Then show that

$$S^\lambda{}_{\mu\nu} = \delta^\lambda{}_\mu \Psi_\nu + \delta^\lambda{}_\nu \Psi_\mu$$

where Ψ_μ are the components of a vector.

Problem 8.27. Compute the connection coefficients of the following metric in an orthonormal frame

$$ds^2 = -e^{2\alpha}dt^2 + e^{2\beta}dr^2 + e^{2\gamma}(d\theta^2 + \sin^2\theta\, d\phi^2)$$

$$\alpha, \beta, \gamma = \text{functions of } r \text{ and } t .$$

Problem 8.28. The redshift between two observers (with 4-velocities $\mathbf{u}_A$ and $\mathbf{u}_B$) can be defined in two ways: (i) by the energies of a photon (4-momentum $\mathbf{p}$) travelling along a null geodesic between them: $1 + z \equiv \mathbf{u}_A \cdot \mathbf{p}/\mathbf{u}_B \cdot \mathbf{p}$, or (ii) by the proper time between two null geodesics, emitted $\Delta\tau_A$ apart and received $\Delta\tau_B$ apart, $1 + z \equiv \Delta\tau_A/\Delta\tau_B$. Show that these definitions are equivalent.

CHAPTER 9
CURVATURE

The study of curvature is based on the Riemann curvature tensor:

$$R^{\mu}{}_{\nu\alpha\beta} \equiv \frac{\partial\Gamma^{\mu}{}_{\nu\beta}}{\partial x^{\alpha}} - \frac{\partial\Gamma^{\mu}{}_{\nu\alpha}}{\partial x^{\beta}} + \Gamma^{\mu}{}_{\rho\alpha}\Gamma^{\rho}{}_{\nu\beta} - \Gamma^{\mu}{}_{\rho\beta}\Gamma^{\rho}{}_{\nu\alpha}$$

in a coordinate frame. The covariant components of the Riemann tensor are connected by several symmetries

$$R_{\alpha\beta\gamma\delta} = R_{\gamma\delta\alpha\beta}, \qquad R_{\alpha\beta\gamma\delta} = -R_{\beta\alpha\gamma\delta},$$

$$R_{\alpha\beta\gamma\delta} = -R_{\alpha\beta\delta\gamma}, \qquad R_{\alpha[\beta\gamma\delta]} = 0 .$$

The (symmetric) Ricci tensor and the Ricci scalar are formed from the Riemann tensor:

$$R_{\alpha\beta} \equiv R^{\mu}{}_{\alpha\mu\beta}$$

$$R \equiv R^{\alpha}{}_{\alpha} .$$

The Weyl tensor,

$$C_{\lambda\mu\nu\kappa} \equiv R_{\lambda\mu\nu\kappa} - \frac{1}{2}(g_{\lambda\nu}R_{\mu\kappa} - g_{\lambda\kappa}R_{\mu\nu} - g_{\mu\nu}R_{\lambda\kappa} + g_{\mu\kappa}R_{\lambda\nu}) + \frac{1}{6}(g_{\lambda\nu}g_{\mu\kappa} - g_{\lambda\kappa}g_{\mu\nu})R$$

is also called the conformal tensor due to its invariance under conformal transformations. It vanishes if and only if the metric is conformally flat (i.e. reducible to Minkowski space by a conformal transformation).

The extrinsic curvature tensor of a hypersurface which has unit normal **n** and which is spanned by basis vectors $e_i, e_j \cdots$ is denoted **K**, with components

$$K_{ij} = -e_j \cdot \nabla_i \mathbf{n} .$$

Problem 9.1. On a sphere of radius a try to construct a local Cartesian system in two ways (a) from geodesics and (b) from the (orthogonal) lines of longitude and latitude. Either way there will be deviations from a good Cartesian system (e.g. the sum of the angles in a coordinate box will differ from 2π, or the fractional difference in length between "parallel" sides of a coordinate box will not vanish). Show that such deviations are of order [Area of coordinate patch/a^2].

Problem 9.2. How many independent components does the Riemann tensor have in n dimensions?

Problem 9.3. Mathematical manipulations with the Riemann tensor are often done with computers. Rather than calculate and store the $4^4 = 256$ components of the tensor as R(I,J,K,L) with I,J,K,L = 0,1,2,3, the symmetries of the Riemann tensor can be used to reduce the size of the stored array. Design a subprogram which stores or recalls all components of R(I,J,K,L) in a linear array of dimensions ≤ 21.

Problem 9.4. Compute all the nonvanishing components of the Riemann tensor $R_{ijkl}(i,j,k,l = \theta,\phi)$ for the 2-sphere metric

$$ds^2 = r^2(d\theta^2 + \sin^2\theta\, d\phi^2) .$$

Problem 9.5. Find the Christoffel symbols and Riemann curvature components for the two dimensional spacetime:

$$ds^2 = dv^2 - v^2\, du^2 .$$

Problem 9.6. Set up a coordinate system on a torus (the 2-dimensional surface of a doughnut in Euclidean 3-space). Calculate all components of $g_{\mu\nu}$, $\Gamma^\mu{}_{\alpha\beta}$ and $R_{\alpha\beta\gamma\delta}$.

Problem 9.7. In a space of fewer than 4 dimensions simple expressions can be given for the Riemann tensor.

(a) What is the Riemann tensor in a 1-dimensional space?

(b) Express the Riemann tensor for a 2-dimensional space in terms of the metric and the Ricci scalar.

(c) Express the Riemann tensor for a 3-dimensional space in terms of the metric and the Ricci tensor.

Problem 9.8. Prove the relation

$$2V_{\alpha;[\nu\kappa]} \equiv V_{\alpha;\nu\kappa} - V_{\alpha;\kappa\nu} = V_\sigma R^\sigma{}_{\alpha\nu\kappa}$$

and find its generalization to the commutator of second derivatives for a tensor of arbitrary rank $T_{\alpha\cdots}{}^{\beta\cdots}$.

Problem 9.9. Show that the second derivatives of a scalar field commute (i.e. show that $S_{;\alpha\beta} = S_{;\beta\alpha}$). For third derivatives $S_{;\alpha\beta\gamma}$, compute $S_{;(\alpha\beta)\gamma}$ and $S_{;\alpha[\beta\gamma]}$.

Problem 9.10. Prove that for any second rank tensor

$$A^{\mu\nu}{}_{;\mu\nu} = A^{\mu\nu}{}_{;\nu\mu} \ .$$

Problem 9.11. An infinitesimal circuit in the shape of a parallelogram can be specified by the differential displacements **u**, v representing the sides of the parallelogram. Let a vector **A** be parallel transported around this circuit. (i.e. Displace it successively by **u**, v, −**u**, −v.) Show that the change in **A** due to the transport around this circuit is

$$\delta A^\alpha = -R^\alpha{}_{\beta\gamma\delta} A^\beta u^\gamma v^\delta \ .$$

Problem 9.12. Riemann curvature can also be computed with the Riemann operator **R**

$$R(A,B)C = (\nabla_A \nabla_B - \nabla_B \nabla_A - \nabla_{[A,B]})C$$

(a) Show that the value of **R** at a point P is linear in the arguments **A**, **B**, **C** and depends only on their values *at* P and not on the way in which they vary around P.

(b) Show that

$$(\mathrm{R(A,B)\,C})^\alpha = R^\alpha{}_{\mu\lambda\sigma} C^\mu A^\lambda B^\sigma \ .$$

Problem 9.13. Two nearby geodesics have affine parameters such that nearby points on the two geodesics have very close values of affine parameter λ. Let $u^\alpha \equiv dx^\alpha/d\lambda$ be the tangent to one of the geodesics, and let $\mathbf{n}$ be the differential vector connecting points of equal affine parameter on the two geodesics. Prove the equation of geodesic deviation

$$\frac{D^2 n^\alpha}{d\lambda^2} + R^\alpha{}_{\beta\gamma\delta} u^\beta n^\gamma u^\delta = 0 \ .$$

Problem 9.14. In a suitable coordinate system the gravitational field of the Earth is approximately (to lowest nontrivial order in M/r)

$$\begin{aligned} ds^2 &= -(1-2M/r)dt^2 + (1+2M/r)(dx^2+dy^2+dz^2) \\ r &\equiv (x^2+y^2+z^2)^{\frac{1}{2}} \\ M &= \text{mass of Earth} \quad (c = G = 1) \ . \end{aligned}$$

Suppose a Skylab satellite orbits the Earth in a circular equatorial orbit. What is the orbital period? An astronaut jettisons a bag of garbage into a nearby orbit and watches it move relative to the satellite. At a given time the separation of the Skylab and its garbage is described by the vector

$$\xi^i \equiv x^i(\text{garbage}) - x^i(\text{skylab}) \ .$$

Using the equation of geodesic deviation, find the components of the relative motion ξ^i as a function of time.

Problem 9.15. Prove the cyclic identity

$$R_{\alpha\beta\gamma\delta} + R_{\alpha\delta\beta\gamma} + R_{\alpha\gamma\delta\beta} = 0$$

and the Bianchi identities

$$R_{\alpha\delta\beta\gamma;\nu} + R_{\alpha\delta\nu\beta;\gamma} + R_{\alpha\delta\gamma\nu;\beta} = 0 \ .$$

Problem 9.16. Show that Bianchi identities imply that the Einstein tensor

$$G_{\mu\nu} \equiv R_{\mu\nu} - \frac{1}{2} g_{\mu\nu} R$$

has vanishing divergence (i.e. $G^{\mu}{}_{\nu;\mu} = 0$).

Problem 9.17. Show that the vanishing of the Riemann tensor is a sufficient condition for a spacetime to be Minkowskian, i.e. a coordinate transformation will bring $g_{\mu\nu}$ into the form $\eta_{\mu\nu}$.

Problem 9.18. A beam of light has a circular cross section at some point along its path. Show that the beam experiences no shear (i.e. the cross-section is not deformed into an ellipse) when the Weyl tensor is zero.

Problem 9.19. Compute the Riemann tensor, Ricci tensor, and scalar curvature of the conformally-flat metric $g_{\mu\nu} = e^{2\phi}\eta_{\mu\nu}$ where $\phi = \phi(x^{\mu})$ is an arbitrary function.

Problem 9.20. Compute the Riemann tensor of the following metric in an orthonormal frame:

$$ds^2 = -e^{2\alpha}dt^2 + e^{2\beta}dr^2 + e^{2\gamma}(d\theta^2 + \sin^2\theta\, d\phi^2)$$

$$\alpha, \beta, \gamma = \text{functions of } r, t .$$

What is the Ricci tensor for this metric? The scalar curvature? The Einstein tensor?

Problem 9.21. Consider a Riemann tensor representing a plane gravitational wave, i.e. $R_{\alpha\beta\gamma\delta} = R_{\alpha\beta\gamma\delta}(u)$, where u is "retarded time" $(\nabla u \cdot \nabla u = 0)$. Find the number of independent components of such a Riemann tensor. Do not assume that $R_{\alpha\beta\gamma\delta}$ satisfies the Einstein field equations.

Problem 9.22. At a given instant, the coordinate accelerations of n nearby test particles are measured. What is the smallest n required to measure all components of $F^{\mu\nu}$? of $R^{\mu}{}_{\nu\rho\sigma}$?

Problem 9.23. Let A and B be two linearly independent vectors tangent at a point to a two dimensional surface, in a space of dimension ≥ 2. The Riemannian curvature of the 2-surface at that point is defined as

$$K = \frac{R_{\alpha\gamma\beta\delta}A^{\alpha}A^{\beta}B^{\gamma}B^{\delta}}{(g_{\alpha\beta}\,g_{\gamma\delta} - g_{\alpha\delta}\,g_{\beta\gamma})A^{\alpha}A^{\beta}B^{\gamma}B^{\delta}} \ .$$

Show that K is unchanged if A and B are replaced by linear combinations of A and B.

Problem 9.24. Suppose K is the curvature at a point in a 2-dimensional surface, as defined in Problem 9.23. If A and B are two vectors tangent at a point to the two surface, and A is parallel transported around a small circuit lying in the 2-surface, show that the change in the angle between A and B is of magnitude

$$\Delta\theta = |K\Delta\Sigma|$$

where $\Delta\Sigma$ is the area enclosed by the circuit.

Problem 9.25. Suppose that the curvature K at a point P, as defined in Problem 9.23, does not depend on the 2-surface which is chosen through that point. Show then that

$$R_{\alpha\beta\gamma\delta} = K(g_{\alpha\gamma}\,g_{\beta\delta} - g_{\alpha\delta}\,g_{\beta\gamma}) \ .$$

Problem 9.26. If the Riemann curvature is isotropic, the Riemann curvature tensor can be written as

$$R_{\alpha\beta\gamma\delta} = K(g_{\alpha\gamma}\,g_{\beta\delta} - g_{\alpha\delta}\,g_{\beta\gamma}) \ .$$

Use the Bianchi identities to show (Schur's theorem) that K must be a constant.

Problem 9.27. Show that a space is conformally flat if the Riemann tensor can be written as

$$R_{\lambda\mu\nu\kappa} = K(g_{\lambda\nu}\,g_{\mu\kappa} - g_{\lambda\kappa}\,g_{\mu\nu}) \ .$$

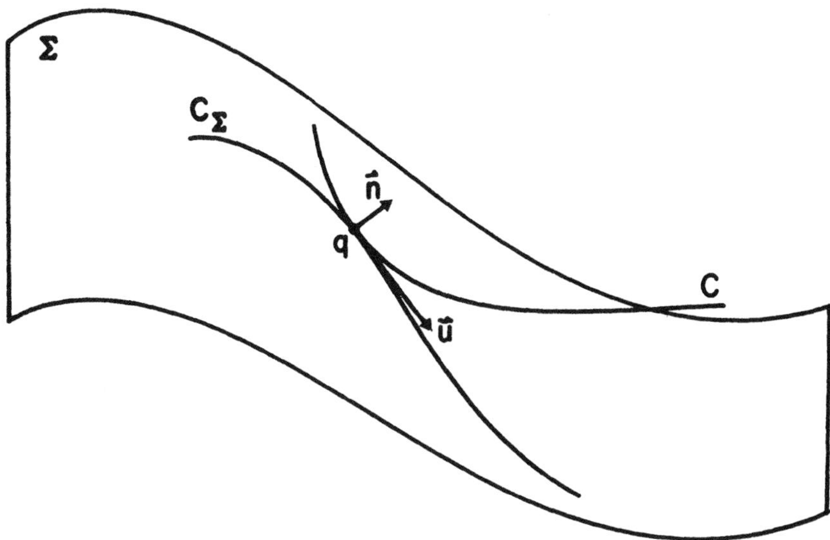

Problem 9.28. Suppose at a point q in a 3-surface Σ two curves are tangent (i) C_Σ a curve in the 3-surface and (ii) C a geodesic of the 4-dimensional space in which Σ is embedded. Let **n** be the unit normal to Σ. The vector $\xi^a = \frac{1}{2} u^a{}_{;\beta} u^\beta$ (where **u** is the tangent vector to C_Σ) measures the rate at which C and C_Σ separate. Show that the rate of separation $\mathbf{n} \cdot \boldsymbol{\xi}$ is

$$\mathbf{n} \cdot \boldsymbol{\xi} = \frac{1}{2} K_{\alpha\beta} u^\alpha u^\beta$$

where $K_{\alpha\beta}$ is the extrinsic curvature tensor for Σ.

Problem 9.29. What is the extrinsic curvature of the τ = constant slice of the metric, $ds^2 = -d\tau^2 + a^2(\tau)\,[\gamma_{ij}(x^k)\,dx^i\,dx^j]$?

Problem 9.30. Prove that the extrinsic curvature of a timelike hypersurface with unit normal vector **n** is $-\frac{1}{2}\mathcal{L}_{\mathbf{n}} P_{\alpha\beta}$, where $P_{\alpha\beta} = g_{\alpha\beta} - n_\alpha n_\beta$ is the projection tensor into the hypersurface.

Problem 9.31. If we neglect gravity, the potential energy due to surface tension of a soap film is proportional to its area. Thus in equilibrium a soap film spanning a fixed closed wire loop will assume a shape of minimum area. Show that this implies that the surface is one whose "mean curvature" $K \equiv K^i{}_i$ is zero.

Problem 9.32. Let $\mathbf{n}$ be the unit normal to a hypersurface Σ, with $\mathbf{n}\cdot\mathbf{n} \equiv \varepsilon = +1$ or -1 if Σ is timelike or spacelike respectively. In Gaussian normal coordinates (see Problem 8.25) based on Σ, the metric is

$$ds^2 = \varepsilon\, dn^2 + {}^{(3)}g_{ij}\, dx^i dx^j \ .$$

Derive the Gauss-Codazzi equations

$${}^{(4)}R^m{}_{ijk} = {}^{(3)}R^m{}_{ijk} + \varepsilon\,(K_{ij} K_k{}^m - K_{ik} K_j{}^m)$$

$${}^{(4)}R^n{}_{ijk} = \varepsilon\,(K_{ik|j} - K_{ij|k}) \ .$$

Here 4 and 3 refer respectively to the spacetime geometry and to the geometry of Σ; a slash denotes covariant differentiation with respect to ${}^{(3)}g_{ij}$; the n index denotes the component on the $\mathbf{n}$ basis vector. Also derive the equation for the remaining component of the Riemann tensor:

$${}^{(4)}R^n{}_{ink} = \varepsilon\,(K_{ik,n} + K_{im} K^m{}_k) \ .$$

Problem 9.33. Using the results of Problem 9.32, derive expressions for ${}^{(4)}G^\alpha{}_\beta$, the components of the Einstein tensor, in Gaussian normal coordinates.

Problem 9.34. The eigenvalues and eigenvectors of the extrinsic curvature tensor are called the principal curvatures and principal directions. Find the principal curvatures and directions for the following surfaces embedded in a 3-dimensional Euclidean space.

(i) sphere: $x^2 + y^2 + z^2 = a^2$.

(ii) cylinder: $x^2 + y^2 = a^2$.

(iii) quadratic surface (compute at origin only): $z = \frac{1}{2}(ax^2 + 2bxy + cy^2)$.

Problem 9.35. Show that if Σ is a 2-dimensional surface in a flat 3-space, then the scalar curvature of Σ is

$$^{(2)}R = \frac{2}{\rho_1 \rho_2}$$

where ρ_1 and ρ_2 are the principal radii of curvature of Σ. What is the analogous formula for a 3-surface embedded in a flat 4-space?

CHAPTER 10
KILLING VECTORS AND SYMMETRIES

Suppose that a geometry has a symmetry such that a vector field $\boldsymbol{\xi}$ exists with the following property: If any set of points is displaced by $\boldsymbol{\xi}\, d\lambda$ ($d\lambda$ a small number) then all distance relationships are unchanged. The vector field is then called a Killing vector for the geometry, and it satisfies Killing's equation

$$\xi_{(\alpha;\beta)} \equiv \frac{1}{2}\left[\xi_{\alpha;\beta} + \xi_{\beta;\alpha}\right] = 0 \ .$$

——— o ——— o ——— o ——— o ——— o ——— o ——— o ———

Problem 10.1. Solve Killing's equations to find the Killing vector fields of the 2-sphere:

$$ds^2 = d\theta^2 + \sin^2\theta\, d\phi^2 \ .$$

Problem 10.2. Show that Killing's equation $\xi_{\alpha;\beta} + \xi_{\beta;\alpha} = 0$ is equivalent to $\mathcal{L}_{\boldsymbol{\xi}}\, \mathbf{g} = 0$, where $\mathbf{g}$ is the metric tensor. Interpret this result geometrically.

Problem 10.3.

(a) Show that the commutator of two Killing vector fields is a Killing vector field.

(b) Show that a linear combination of Killing vectors with constant coefficients is a Killing vector.

Problem 10.4. In Euclidean 3-space show that the 3 Killing vectors describing rotations around the x, y, and z axes are linearly dependent at any given point but that no *constant coefficient* combination of them is zero. Show that the generators of the rotation group O(3) are thus 2-surface forming, even though the group is 3-dimensional. Explain.

Problem 10.5. The metric for an axially symmetric rotating star admits two Killing vectors, $\boldsymbol{\xi}_{(t)}$ and $\boldsymbol{\xi}_{(\phi)}$. Assume there are no other independent Killing vectors. Prove that $\boldsymbol{\xi}_{(t)}$ and $\boldsymbol{\xi}_{(\phi)}$ commute.

Problem 10.6. Show that any Killing vector is a solution of the equation

$$\xi^{\nu;\lambda}{}_{;\lambda} + R^{\nu}{}_{\sigma}\,\xi^{\sigma} = 0 \ .$$

Find a Lagrangian-type variational principle from which this equation can be derived.

Problem 10.7. If $\boldsymbol{\xi}$ is a Killing vector, prove that

$$\xi_{\mu;\alpha\beta} = R_{\gamma\beta\alpha\mu}\,\xi^{\gamma} \ .$$

Problem 10.8. A metric is "stationary" if and only if it has a Killing vector field ξ which is timelike at infinity (the "time" direction is $\partial/\partial t$). There are two ways to define a "static" metric:

(i) stationary and invariant under time reversal, $\partial/\partial t \to -\partial/\partial t$, or

(ii) stationary and $\partial/\partial t$ is hypersurface orthogonal (see Problem 7.23). Show that the two definitions are equivalent.

Problem 10.9. In flat Minkowski spacetime find ten Killing vectors that are linearly independent.

Problem 10.10. If $\boldsymbol{\xi}(x^{\mu})$ is a Killing vector field and $\mathbf{u}$ is the tangent vector to a geodesic, show that $\boldsymbol{\xi}\cdot\mathbf{u}$ is constant along the geodesic.

Problem 10.11. If $\boldsymbol{\xi}$ is a Killing vector and $\mathbf{T}$ is the stress-energy tensor, show that $J^{\mu} \equiv T^{\mu\nu}\xi_{\nu}$ is a conserved quantity i.e. $J^{\mu}{}_{;\mu} = 0$. Interpret $\mathbf{J}$ when $\boldsymbol{\xi}$ is a timelike Killing vector.

Problem 10.12. If $\mathbf{T}$ is the energy-momentum tensor, and $\boldsymbol{\xi}$ is a time Killing vector, show that the integral over a whole spacelike hypersurface

$$\int_{F} T^{\alpha}{}_{\beta}\,\xi^{\beta} d^3\Sigma_{\alpha}$$

is independent of the choice of spacelike hypersurface F.

Problem 10.13. Given a divergenceless stress energy tensor in flat spacetime, i.e.

$$T^{\mu\nu}{}_{;\nu} = 0 \qquad R_{\alpha\beta\gamma\delta} = 0$$

show that one may construct ten global conservation laws and hence ten conserved quantities.

Problem 10.14. If $\boldsymbol{\xi}$ is a timelike Killing vector and $\mathbf{u} = \boldsymbol{\xi}/|\boldsymbol{\xi}\cdot\boldsymbol{\xi}|^{\frac{1}{2}}$ is a 4-velocity, prove that $\mathbf{a} \equiv \nabla_{\mathbf{u}}\mathbf{u} = \frac{1}{2}\nabla \log |\boldsymbol{\xi}\cdot\boldsymbol{\xi}|$.

Problem 10.15. In a stationary metric with time Killing vector $\boldsymbol{\xi}$ the "energy at infinity" $E = -\mathbf{p}\cdot\boldsymbol{\xi}$ of a test particle with 4-momentum $\mathbf{p}$ is conserved. Find the minimum value of E/μ (where μ = particle mass) that the particle can have at a given point in the spacetime, in terms of the norm of $\boldsymbol{\xi}$.

Problem 10.16. Show that a Killing vector is an admissible solution for the vector potential of Maxwell's equations for a test field in a vacuum spacetime. What electromagnetic field corresponds to the Killing vector $\partial/\partial\phi$ in Minkowski space?

CHAPTER 11
ANGULAR MOMENTUM

This chapter contains problems dealing with rotation, angular momentum, spin, etc. in general relativity. Definitions are developed in the problems.

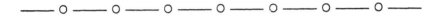

Problem 11.1. In special relativity, when a particle is at an event B and has 4-momentum $\mathbf{p}$, its angular momentum about event A is

$$\mathbf{J} = \Delta\mathbf{x} \otimes \mathbf{p} - \mathbf{p} \otimes \Delta\mathbf{x}$$

where $\Delta\mathbf{x}$ is the 4-vector from event A to event B

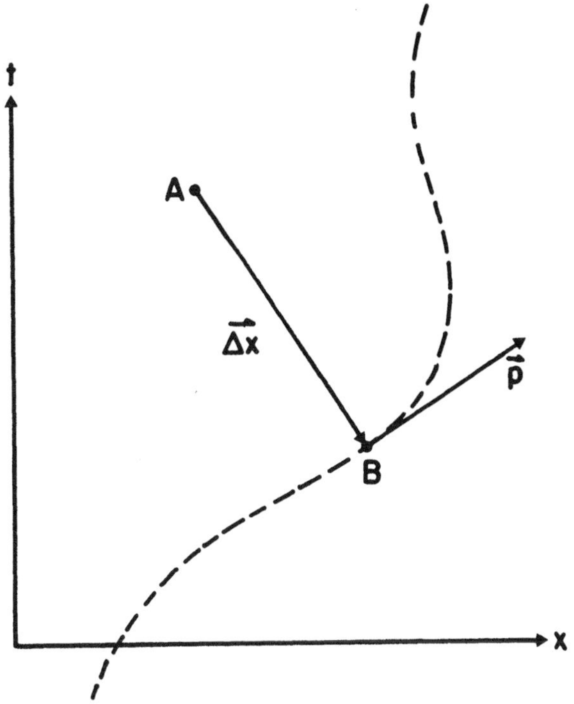

(i) Show that for a freely moving (i.e. unaccelerated) particle $\mathbf{J}$ is conserved – i.e. $d\mathbf{J}/d\tau = 0$.

(ii) Suppose that several particles collide at an event B, thereby producing several other particles. Show that the sum of the angular momenta of the particles about an event A is the same after the collision as before:

$$\sum_{(k)} J_{(k)}\Big|_{\text{after}} = \sum_{(k)} J_{(k)}\Big|_{\text{before}} .$$

Problem 11.2. Show

(a) that the total angular momentum of an isolated system in flat space

$$J^{\alpha\beta} \equiv \int d^3x\,(x^\alpha T^{\beta 0} - x^\beta T^{\alpha 0})$$

is a conserved tensor (when $T^{\alpha\beta}{}_{,\beta} = 0$), but

(b) that it is not invariant under the coordinate translation $x^\alpha \to x^\alpha + a^\alpha$. Show also that

(c) the spin 4-vector defined by

$$S_\alpha \equiv -\frac{1}{2}\,\varepsilon_{\alpha\beta\gamma\delta}\,J^{\beta\gamma}u^\delta$$

is both conserved *and*

(d) invariant under translations. Here u^α is the "center of mass 4-velocity" $u^\alpha \equiv P^\alpha/(-P^\beta P_\beta)^{\frac{1}{2}}$ and P^α is the total momentum $P^\alpha \equiv \int d^3x\, T^{\alpha 0}$.

Problem 11.3. Show that a system's intrinsic spin 4-vector S_α is orthogonal to its 4-velocity u^α.

Problem 11.4. Show that a gyroscope with no applied torques Fermi-Walker transports its spin vector.

Problem 11.5.

(a) If angular momentum is computed about the center of mass of a system show that $J^{\alpha\beta}u_\beta = 0$.

(b) In this case show that the angular momentum ("intrinsic angular momentum") can be found from the spin vector as

$$J^{\alpha\beta}{}_{(\text{C.M.})} \equiv S^{\alpha\beta} = -\varepsilon^{\alpha\beta\gamma\delta}\,S_\gamma u_\delta .$$

Problem 11.6. Two bodies A and B have momenta P_A and P_B and spin S_A and S_B; their centers of mass are on a collision course. After colliding they stick together to form a composite body C (spin S_C). Calculate S_C in terms of P_A, P_B, S_A and S_B.

Problem 11.7. Thomas Precession: Consider a ("classical") spinning electron which Fermi-Walker transports its spin angular momentum, S, as it moves in a circular orbit around an atomic nucleus. As seen in the laboratory frame, the electron moves in a circular orbit of radius r in the x-y plane with constant angular velocity, ω. Calculate S(t), the spin as a function of laboratory time.

Problem 11.8. A nonspherical spinning body in an inhomogeneous gravitational field experiences a torque which causes its intrinsic spin 4-vector S to change with time. If **u** is the 4-velocity of the center of mass of the object, freely moving along a geodesic, show that

$$\frac{DS^\kappa}{d\tau} = \varepsilon^{\kappa\beta\alpha\mu} u_\mu u^\sigma u^\lambda t_{\beta\eta} R^\eta{}_{\sigma\alpha\lambda} .$$

Here $t_{\beta\eta}$ is the "reduced quadrupole moment tensor" $t^{ij} = \int \rho(x^i x^j - \frac{1}{3} r^2 \delta^{ij}) d^3x$ in the rest frame of the center of mass, $t^{\alpha\beta} u_\alpha = 0$ and the Riemann tensor is generated externally to the body in question and is assumed to be approximately constant over the body.

Problem 11.9. Calculate the period of precession of the Earth's axis due to the coupling of tidal forces from the sun and moon with the quadrupole moment of the (slightly nonspherical) Earth.

Problem 11.10. Consider a family of stationary observers in a stationary spacetime, i.e. their 4-velocity is proportional to the time Killing vector $\boldsymbol{\xi}$. Each observer arranges his spatial basis vectors so that they connect to the same neighboring observers for all time t, where $\boldsymbol{\xi} = \frac{\partial}{\partial t}$.

(1) Show that $\mathcal{L}_{\boldsymbol{\xi}} e_{\hat{\alpha}} = 0$, where e_α is a basis vector of the stationary observer.

(2) Show that the rate of change of the components of any tensor quantity Q as measured by the stationary observer in units of t, is

$\frac{dQ^{\hat{\alpha}\cdots\hat{\beta}}}{dt} = (\mathcal{L}_{\boldsymbol{\xi}} Q)^{\hat{\alpha}\cdots\hat{\beta}}$. What is $\frac{dQ^{\hat{\alpha}\cdots\hat{\beta}}}{d\hat{t}}$?

(3) The stationary observer carries a gyroscope with him applying no torques to it. Show that the gyroscope's spin vector precesses (Lens-Thirring effect) with an angular velocity relative to the stationary observer, as measured in units of proper time,

$$\omega^{\alpha} = \frac{\varepsilon^{\alpha\nu\sigma\lambda}\xi_{\nu}\xi_{\sigma;\lambda}}{2\xi^{\gamma}\xi_{\gamma}} .$$

(4) Show that $\omega = 0$ if the spacetime is static, and not merely stationary.

Problem 11.11. A gyroscope is placed in a circular orbit about the Earth and no torques are applied to it. What is the angular velocity of precession for its spin vector relative to a reference frame fixed with respect to the distant stars? ("geodetic precession" and "Lens-Thirring effect".)

CHAPTER 12
GRAVITATION GENERALLY

This chapter contains problems dealing with the physical consequences of gravitational interactions. Most of the problems use the Newtonian limit, in which gravity is represented by a scalar potential U satisfying

$$\nabla^2 U = -4\pi\rho$$

and generating a gravitational acceleration

$$\underset{\sim}{g} = \nabla U \ .$$

(Alternatively, one sometimes uses a potential $\Phi \equiv -U$.) Tidal forces depend on $\partial^2\Phi/\partial x^i \partial x^k$ in Newtonian theory; these forces appear as the R_{j0k0} terms of the Riemann tensor in the appropriate limit of the equation of geodesic deviation in general relativity. Some of the problems explore consequences of gravity's spin-2 nature, and of its weakness in comparison to other fields.

——— o ——— o ——— o ——— o ——— o ——— o ——— o ———

Problem 12.1. A small satellite has a circular frequency ω in an orbit of radius r about a central object of mass m. From the known value of ω show that it is possible to determine neither r nor m individually, but only the effective "Kepler density" $3m/4\pi r^3$ of the object as averaged over a sphere of the same radius as the orbit. Give the formula for ω^2 in terms of this Kepler density.

Problem 12.2. Estimate the height of spring tides and neap tides.

Problem 12.3. If the amplitude of solid-earth tides as a function of time is fourier transformed, there are peaks at certain frequencies. What are the frequencies (or periods) of the 10 strongest peaks?

Problem 12.4. The position of the sun in the sky can in principle be measured by a sensitive tidal gravimeter. What is the angular difference between this position and its position as measured optically? If the actual position of the sun were at its optical position, there would be a force in the direction of the earth's motion. (Why?) If this were the case find the radius of the earth's orbit as a function of time.

Problem 12.5. The "Eddington limit" for the luminosity of a star of mass M is defined as the luminosity at which outward light pressure just balances inward gravitational force everywhere within the star. Calculate the luminosity. (You may assume that all matter is fully ionized hydrogen.)

Problem 12.6. Show that an electron does not fall down when released in the center of an evacuated perfectly conducting closed container in a uniform gravitational field. ("Perfect conductor" ≡ perfectly rigid positive lattice with perfectly mobile conduction electrons.)

Problem 12.7. A tall, cylindrical, insulated bottle of height h is filled with air at 300^0K. It is then sealed and set on a scale at sea level. The scale reads a weight W. For what range of h does the weight W *decrease* as the contents of the bottle are slowly heated.

Problem 12.8. Define a stress tensor for the Newtonian gravitational potential U as follows:

$$T_{jk} \equiv \frac{1}{4\pi}(U_{,j}U_{,k} - \frac{1}{2}\delta_{jk}U_{,n}U^{,n}) .$$

Show that the Newtonian equations of motion for stressed matter with proper density ρ_0 and velocity $\underset{\sim}{v}$ can be written in the form

$$\rho_0 \frac{dv_j}{dt} = -\frac{\partial}{\partial x^k}(T_{jk} + t_{jk})$$

$$(\rho_0 v_j)_{,t} + (T_{jk} + t_{jk} + \rho_0 v_j v_k)_{,k} = 0 \ ,$$

where t_{jk} is the ordinary 3-dimensional stress tensor.

Problem 12.9. Consider an extended body of mass M with several forces $\underset{\sim}{F}_i$ acting on it. Using the equivalence of gravitational mass and energy show that the condition for equilibrium of the body is

$$\sum \underset{\sim}{F}_i (1 - \underset{\sim}{g} \cdot \underset{\sim}{x}_i / c^2) = -M\underset{\sim}{g}$$

where $\underset{\sim}{g}$ is the acceleration of gravity and $\underset{\sim}{x}_i$ denotes the point of application of each of the forces, measured in a local Lorentz frame.

Problem 12.10. From the result of Problem 12.9 show that the equation of hydrostatic equilibrium in a star is

$$\frac{dp}{dr} = -\frac{GM(r)}{r^2}(\rho + p/c^2)$$

where M(r) is defined as the "active" mass interior to the fluid shell at radius r. This shows that in a fluid the "effective inertial mass" density is $\rho + p/c^2$. Notice that this result does not depend on the field equations of general relativity.

Problem 12.11. Show that the Newtonian equation of motion of a test particle in a Newtonian gravitational potential Φ can be written as a geodesic equation in 4-dimensional spacetime. Compute the Christoffel symbols and the Riemann tensor, and show that they are not derivable from a metric.

Problem 12.12. By examining the relative acceleration of a family of test-particle trajectories in Newtonian gravity and comparing with the Newtonian limit of the equation of geodesic deviation, derive the correspondence

$$R_{j0k0} = \frac{\partial^2 \Phi}{\partial x^j \partial x^k}$$

between the Newtonian potential and the Riemann tensor. (A Newtonian test particle is acted on only by gravity; a test particle in a relativistic theory of gravity follows a geodesic.)

Problem 12.13. Write a Newtonian gravitational force law in covariant 4-dimensional language, using a scalar field as the universal Newtonian time function. Show that the resulting theory is consistent with special relativity. Show that signals can be sent faster than the speed of light. Is the theory acausal, i.e. can an observer send signals into his own past?

Problem 12.14. Consider two particles of equal mass in a freely falling elevator. One carries a charge q and the other is neutral. There is a vertical electrical field E in the elevator. Write an equation for the separation of the particles as a function of time, including tidal and electrical effects. Reconcile the existence of both terms with the equivalence principle.

Problem 12.15. New particles of mass m_0 are created which carry a new kind of charge, evidenced by a classical inverse-square force law. A container holds a perfect monatomic gas of these particles in thermal equilibirum. The total charge Q in the container is measured by the average force on test particles outside, and it is observed to vary with temperature as $Q \propto Q_0(1+6kT/m_0)$. What is the spin of the new force field?

Problem 12.16. Show (nonrigorously) that gravity is the only classical, infinite range (massless quanta) pure spin-2 field; i.e. that any other field would couple identically to bulk matter and would thus be indistinguishable from gravity.

Problem 12.17. In "geometrized units of length" (units in which the gravitational constant G, the speed of light c and the Boltzmann constant k are all taken to be unity) give the values of the following, expressed in terms of centimeters: $\hbar$; the charge of an electron; e/m for an electron; the mass of the sun; the luminosity of the sun; 300^0K; one year; one volt.

Problem 12.18. Form "natural" units of mass, length, and time out of the physical constants $\hbar$, G, and c.

Problem 12.19. Estimate the Bohr radius of a "gravitational atom", e.g. two neutrons bound by their gravitational attraction in their lowest energy state.

CHAPTER 13

GRAVITATIONAL FIELD EQUATIONS AND LINEARIZED THEORY

The gravitational field, described by the metric of spacetime $g_{\mu\nu}$, is generated by the stress-energy $T^{\mu\nu}$ of matter. Various field equations relating $g_{\mu\nu}$ to $T^{\mu\nu}$ have been proposed. The most successful to date are the Einstein equations which are the foundation of general relativity:

$$G_{\mu\nu} \equiv R_{\mu\nu} - \frac{1}{2} g_{\mu\nu} R = 8\pi T_{\mu\nu} \tag{1}$$

where $R_{\mu\nu}$ and R are the Ricci tensor and scalar curvature derived from the metric $g_{\mu\nu}$, and $G_{\mu\nu}$ is the Einstein tensor. The equations are non-linear, since the left hand side is not a linear function of the metric.

Some other (less successful) field equations are discussed in the problems, but unless specifically stated, the Einstein field equations are to be assumed.

The equations of motion $T^{\mu\nu}{}_{;\nu} = 0$ are a consequence of Equation (1). Other desirable properties of $T^{\mu\nu}$, called "energy conditions", must be independently postulated on physical grounds.

When the gravitational field is weak, the geometry of spacetime is nearly flat and one writes

$$g_{\mu\nu} = \eta_{\mu\nu} + h_{\mu\nu}$$

where all $|h_{\mu\nu}|$ are $<< 1$. In this case Equation (1) can be solved approximately, by keeping only first-order terms in $h_{\mu\nu}$. A number of problems make use of this "linearized theory."

Problem 13.1. A somewhat generalized form of the Einstein field equations is

$$R_{\mu\nu} - a g_{\mu\nu} R = 8\pi T_{\mu\nu}$$

where a is some dimensionless constant. Show that if a is not $\frac{1}{2}$ the field equations disagree with experiment, even in the Newtonian limit.

Problem 13.2. A metric theory (devised by Nordström in 1913) relates $g_{\mu\nu}$ to $T_{\mu\nu}$ by the equations

$$C_{\mu\nu\rho\sigma} = 0$$

$$R = \kappa g_{\mu\nu} T^{\mu\nu} ,$$

where C is the Weyl tensor. Show that this theory, in the Newtonian limit and with the proper choice of κ, agrees with Newtonian gravitation theory, but that this theory predicts no deflection of starlight passing near the Sun. Does this theory agree with the Pound-Rebka experiments, i.e. are photons redshifted as they rise against the gravitational pull near the surface of the Earth?

Problem 13.3. In the Brans-Dicke theory of gravity (see MTW p. 1070 or Weinberg p. 160 for field equations), the locally measured Newtonian gravitational constant G varies with position and time. Its value at infinity is G_∞. Show that G is a constant inside a self-gravitating spherical shell of mass M and circumference $2\pi R$. If $R >> G_\infty M/c^2$, express G inside the shell in terms of G_∞, M, and R, to lowest order in $(G_\infty M/Rc^2)$.

Problem 13.4. In relativistic quantum mechanics empty space contains virtual particles. It is speculated that the vacuum therefore, has a non-zero stress-energy.

(1) What form must the vacuum stress-energy tensor take, if there is to be no preferred vacuum frame? Show that there is a resulting term in the field equations which can be interpreted as an effective cosmological constant.

(2) Suppose that the vacuum energy is due to the rest mass of virtual

protons or electrons, produced with an average spacing of their Compton wavelength. Is such vacuum stress-energy ruled out by observations?

(3) Ya. B. Zel'dovich has suggested that the mass-energy density should be associated only with the *gravitational* interaction energy of nearby virtual particles (separated by their Compton wavelength). What is the predicted magnitude of the vacuum stress energy here? Is it ruled out by observation?

Problem 13.5. In a local region of spacetime, an observer finds that the Ricci curvature scalar is nearly constant, $R \approx +1/a^2$. Why will the sign to be "+"? If the region of spacetime is filled only with electromagnetic energy what is R?

Problem 13.6. Normally it is assumed that a physically possible $T^{\mu\nu}$ must satisfy the (weak) energy condition, $T^{00} \geq 0$ for *all* physical observers. Assume that $T^{\mu\nu}$ has a timelike eigenvector; how can a given single observer determine whether the $T^{\mu\nu}$ he measures satisfies the condition?

Problem 13.7. The "dominant energy condition" on $T^{\mu\nu}$ requires that the weak energy condition be satisfied (all observers see a nonnegative energy density) and furthermore, that all observers see energy density greater or equal to the magnitude of the energy-flux 3-vector. Show that the statement

$$\mathbf{u} \cdot (-\mathrm{T})^n \cdot \mathbf{u} \leq 0$$

for all nonspacelike vectors $\mathbf{u}$ reduces to the weak energy condition for $n = 1$ and to the dominant energy condition for $n = 2$. What about $n > 2$? [Here $(T^2)_{\mu\nu} \equiv T_\mu{}^\sigma T_{\sigma\nu}$ and so forth.]

Problem 13.8. Is it possible to have a solution of the Einstein field equations in which space is empty to the past of some surface of constant time $t = 0$, but in which there is a nonvanishing $T_{\mu\nu}$ to the future of this surface?

Problem 13.9. A static metric is generated by a perfect fluid. Show that the fluid 4-velocity is parallel to the time Killing vector.

Problem 13.10. At each point on an initial Cauchy hypersurface how many numbers must be specified, to determine uniquely the evolution of the metric field above that hypersurface. [Hint: First show that only spatial components of the Einstein tensor contain second time derivatives of the metric.]

Problem 13.11. For the "nearly Newtonian" metric

$$ds^2 = -(1+2\Phi)dt^2 + (1-2\Phi)\delta_{jk}dx^j dx^k$$

calculate, to lowest nonvanishing order in Φ, the components of the Landau-Lifschitz stress-energy pseudotensor $t^{\alpha\beta}_{L\cdot L}$ (Landau and Lifschitz p. 306). Assume that the field is changing so slowly in time that time derivatives of Φ can be neglected compared to spatial derivatives.

Problem 13.12. A gauge transformation is an infinitesimal coordinate transformation which relabels the coordinates of a point P according to

$$x^{\mu}_{new}(P) = x^{\mu}_{old}(P) + \xi^{\mu}(P) \ .$$

Such transformations induce changes, to first order in ξ, of the functional forms of tensors. Find the gauge transformation laws for scalars, and components of vectors and second rank tensors. For the linearized metric perturbations $g_{\mu\nu} = \eta_{\mu\nu} + h_{\mu\nu}$ show in particular that

$$h^{new}_{\mu\nu}(x) = h^{old}_{\mu\nu}(x) - 2\,\xi_{(\mu,\nu)} \ .$$

Problem 13.13. Show that in linearized theory the components of the Riemann tensor are

$$R_{\alpha\mu\beta\nu} = \frac{1}{2}\,(h_{\alpha\nu,\mu\beta} + h_{\mu\beta,\nu\alpha} - h_{\mu\nu,\alpha\beta} - h_{\alpha\beta,\mu\nu}) \ .$$

Also show explicitly that this Riemann tensor is invariant under a gauge transformation.

Problem 13.14. In linearized theory one often uses the "trace reversed" form of the metric perturbations $\bar{h}_{\alpha\beta} \equiv h_{\alpha\beta} - \frac{1}{2}\eta_{\alpha\beta}h_\sigma{}^\sigma$. Show that there always exists a gauge transformation to the "Lorentz gauge", in which the divergence of $\bar{h}_{\alpha\beta}$ vanishes. Is this gauge transformation unique?

Problem 13.15. Show that in the Lorentz gauge (see Problem 13.14) the linearized field equations reduce to

$$\Box \bar{h}_{\mu\nu} \equiv \bar{h}_{\mu\nu,\alpha}{}^{,\alpha} = -16\pi T_{\mu\nu} ,$$

where $\bar{h}_{\mu\nu}$ is trace-reversed $h_{\mu\nu}$.

Problem 13.16. In linearized theory a plane gravitational wave propagating in empty spacetime can be represented as the real part of a complex expression

$$\bar{h}_{\mu\nu} = \mathrm{Re}\,[A_{\mu\nu} e^{ik_\alpha x^\alpha}]$$

where $A_{\mu\nu}$ is a constant tensor. Show that $\mathbf{k}$ must be a null vector, and that $\mathbf{A}$ is orthogonal to $\mathbf{k}$.

For a particular observer with 4-velocity $\mathbf{u}$ the "transverse-traceless" gauge (a further specialization of the Lorentz gauge) is defined in the observer's unperturbed rest frame $(u^0 = 1,\ u^i = 0)$ such that

$$\bar{h}_{\mu 0} = 0 , \qquad \bar{h}_\mu{}^\mu = 0 .$$

Find the gauge transformation which accomplishes this. Does $\mathbf{A}$ remain orthogonal to $\mathbf{k}$?

Problem 13.17. Show that in linearized theory there is no attractive gravitational force between two thin parallel beams of light.

Problem 13.18. A rigid spherical shell of radius R and total mass M (distributed uniformly on the shell), with negligible thickness, rotates slowly with constant angular velocity Ω with respect to inertial frames far away. Use the linearized equations of gravity to determine ω, the angular velocity of dragging of inertial frames *inside* the shell, to first order in $\Omega R << 1$. Show that

$$\omega \equiv -\frac{g_{0\phi}}{g_{\phi\phi}} = \frac{4}{3}\frac{M\Omega}{R} + \mathcal{O}(\Omega^2 R^2) ,$$

which is constant everywhere inside the shell. [The constancy of ω in the cavity has been interpreted by some to mean that Einstein's equations satisfy Mach's principle to some degree.]

Problem 13.19. In the linearized gravitational theory, show that the equations of motion for matter $T^{\mu\nu}{}_{;\nu} = 0$ are inconsistent with the field equations for the metric perturbation. Show that this inconsistency is of second order in the metric perturbation, hence negligible to first order.

Problem 13.20. A hypothetical particle of negative gravitational mass $-M$ is released from rest at a distance $\ell >> M$ from another fixed particle of equal positive mass $+M$. As seen by a static observer, what is the magnitude and direction of the acceleration of each particle? Calculate the motion of the particles after they are released, making any reasonable approximations necessary.

CHAPTER 14

PHYSICS IN CURVED SPACETIME

This chapter deals with the generalization of the laws of special-relativistic physics (e.g. hydrodynamics, electrodynamics) to curved spacetime. Often this generalization involves only the replacement of partial differentiation by covariant differentiation ("comma-goes-to-semicolon rule"); for example the generalization of the *equations of motion* is from $T^{\mu\nu}{}_{,\nu} = 0$ to $T^{\mu\nu}{}_{;\nu} = 0$; this latter equation, with semicolons, includes the effects of gravity.

—— o —— o —— o —— o —— o —— o —— o ——

Problem 14.1. Write the stress-energy tensor for a single free particle, and show that the equation of geodesic motion follows from $T^{\mu\nu}{}_{;\nu} = 0$.

Problem 14.2. Show that the condition for thermal equilibrium of a static system in general relativity is

$$T(-g_{00})^{\frac{1}{2}} = \text{constant}$$

(where T is the temperature measured by a local static observer) rather than the Newtonian relation $T = \text{constant}$.

Problem 14.3. Derive the Euler equation $(\rho + p)\nabla_{\mathbf{u}}\mathbf{u} = -[\nabla p + (\nabla_{\mathbf{u}}p)\,\mathbf{u}]$ from $T^{\mu\nu}{}_{;\nu} = 0$, and show that this equation has the correct Newtonian limit.

Problem 14.4. Derive the general relativistic equation of hydrostatic equilibrium,

$$\frac{-\partial p}{\partial x^{\nu}} = (\rho + p)\,\frac{\partial}{\partial x^{\nu}} \log\,(-g_{00})^{\frac{1}{2}}$$

and compare it to the Newtonian equation.

Problem 14.5. Show that a highly relativistic fluid $(p = \frac{1}{3}\rho)$ in hydrostatic equilibrium in a gravitational field can never have a free surface (i.e. a surface where $\rho \to 0$).

Problem 14.6. Two identical containers in a static uniform gravitational field contain different substances. The containers and contents have identical time-independent densities of mass-energy, but different (and possibly anisotropic) stresses. Do the containers weigh the same if weighed on the same scale?

Problem 14.7. If $\mathbf{u}$ is the 4-velocity of a perfect gas undergoing adiabatic stationary flow in a stationary gravitational field, prove (the relativistic Bernoulli equation) that along the flow lines

$$u_0 = \text{constant} \times n/(\rho + p) ,$$

where n = baryon number density.

Problem 14.8. Show that the relativistic Bernoulli equation of Problem 14.7 reduces to the correct Newtonian limit for slow velocities and weak gravitational fields.

Problem 14.9. Consider a moving medium with four-velocity $\mathbf{u}(x)$. Choose two arbitary neighboring particles A and B. At each event along the world line of A define the four-vector separation of B from A, $\boldsymbol{\xi}$, as follows: (i) $\boldsymbol{\xi}$ is an *infinitesimal* vector from a given event on A's world line to B's world line; (ii) $\boldsymbol{\xi}$ has vanishing time component, $\xi^{\hat{0}} = 0$, relative to the orthonormal tetrad carried by A.

(a) Define the motion of the medium to be a rigid-body motion if and only if the distance, $(\boldsymbol{\xi}\cdot\boldsymbol{\xi})^{\frac{1}{2}}$, between any two neighboring particles – e.g., A and B above – is constant for all time. Show that a medium undergoes rigid body motion if and only if $\sigma_{\alpha\beta} = 0$ and $\theta = 0$, where $\sigma_{\alpha\beta}$ and θ are the quantities defined in Problem 5.18.

(b) How many independent equations do these conditions constitute? How many degrees of freedom are there in relativistic rigid-body motion?

Problem 14.10. If θ is the expansion of a fluid, derive the Raychaudhuri equation

$$\frac{d\theta}{d\tau} = a^{\alpha}{}_{;\alpha} + 2\omega^2 - 2\sigma^2 - \frac{1}{3}\theta^2 - R_{\alpha\beta}u^{\alpha}u^{\beta}$$

where $\omega^2 = \frac{1}{2}\omega_{\alpha\beta}\omega^{\alpha\beta}$, $\sigma^2 = \frac{1}{2}\sigma_{\alpha\beta}\sigma^{\alpha\beta}$, and the notation is the same as in Problem 5.18.

Problem 14.11. In a certain spacetime fluid flows along geodesics with zero shear and expansion. (See Problem 5.18 for the definitions of shear $\sigma_{\alpha\beta}$ and expansion θ.) Show that the spacetime has a timelike Killing vector.

Problem 14.12. An observer in a closed box (not necessarily in free fall) measures positions and times inside his box with rulers and a clock. Show that the equation of motion for a particle, correct to first order in its measured velocity $v << 1$ and position x^j is

$$\begin{aligned} dv^j/dt &= 2(\underset{\sim}{v}\times\underset{\sim}{\omega})^j - [(\underset{\sim}{x}\times\underset{\sim}{\omega})\times\underset{\sim}{\omega}]^j + (\underset{\sim}{x}\times d\underset{\sim}{\omega}/dt)^j \\ &\quad - a^j(1+\underset{\sim}{a}\cdot\underset{\sim}{x}) - R^j{}_{0k0}x^k \ . \end{aligned}$$

Here $\underset{\sim}{\omega}$ is the angular velocity of the box, $\underset{\sim}{a}$ its acceleration, and $R^j{}_{0k0}$ the Riemann tensor evaluated at the origin.

Problem 14.13. The stress-energy tensor of a massless scalar field is taken to be

$$T_{\mu\nu} = (4\pi)^{-1}(\Phi_{,\mu}\Phi_{,\nu} - \frac{1}{2}g_{\mu\nu}\Phi_{,\alpha}\Phi^{,\alpha}) \ .$$

Derive the equation of motion of this scalar field from $T^{\mu\nu}{}_{;\nu} = 0$.

Problem 14.14. The equation for a scalar field, in flat spacetime, is $\Phi_{,\nu}{}^{,\nu} = \rho_s$ where ρ_s is the density of "scalar charge." We are tempted to conclude that in curved spacetime the equation should be

$$\Phi_{;\nu}{}^{;\nu} = \rho_s \qquad (1)$$

but another possible generalization is

$$\Phi_{;\nu}{}^{;\nu} - \frac{1}{6} R\Phi = \rho_s \tag{2}$$

where R is the Ricci scalar. (Equation (2) is "conformally invariant" while (1) is not.) Does Equation (2), in principle, violate the strong equivalence principle? Find the influence of the R term on the force ($\sim \nabla\Phi$) between two "scalar charged" particles, assuming that $R = 1/a^2$ varies very slowly on a laboratory scale. In a practical laboratory experiment what would be the magnitude of such anomalous R-term forces compared to the ordinary scalar forces?

Problem 14.15. Show that Maxwell's equations $F^{\mu\nu}{}_{;\nu} = 4\pi J^{\mu}$, imply $J^{\mu}{}_{;\mu} = 0$.

Problem 14.16. The generalization of Maxwell's equations to curved spacetime by the "comma-goes-to-semicolon" rule (or principle of equivalence) is not completely unambiguous. Show that the use of this rule with the vector potential A^{μ} can lead to two different results for a relativistic equation.

Problem 14.17. Estimate the fractional error introduced into Maxwell's equations (applied to some earthbound process of characteristic frequency ν and size ℓ) by our ignorance of what curvature coupling terms there may be.

Problem 14.18. Show that except in the case $\underset{\sim}{E}\cdot\underset{\sim}{B} = 0$, the sourceless Maxwell equations $F^{\beta\nu}{}_{;\nu} = 0$ follow from the requirements $\nabla\cdot T = 0$, where T is the electromagnetic stress-energy tensor, and from the fact that $F^{\mu\nu}$ is derived from a vector potential.

Problem 14.19. Show that $H = \frac{1}{2} g^{\mu\nu} (\pi_\mu - eA_\mu)(\pi_\nu - eA_\nu)$ is a Hamiltonian giving the equations of motion of a test particle of charge e. Here π_μ is the canonical momentum. (The canonical momentum π^{μ} is not equal to p^{μ}, the particles 4-momentum unless the 4-potential A^{μ} is zero.)

Problem 14.20. Suppose ξ is a Killing vector for a solution of the Einstein-Maxwell equations. Write down an integral of the motion for charged test particles. (Assume $\mathcal{L}_{\xi} A = 0$ where A is the 4-potential.)

Problem 14.21. Show that Maxwell's equations are invariant under the "conformal transformation"

$$g_{\alpha\beta} \to \tilde{g}_{\alpha\beta} = f g_{\alpha\beta}$$

$$F_{\alpha\beta} \to \tilde{F}_{\alpha\beta} = F_{\alpha\beta}$$

$$J_{\mu} \to \tilde{J}_{\mu} = f^{-2} J_{\mu} ,$$

where f is an arbitrary function of position.

CHAPTER 15
THE SCHWARZSCHILD GEOMETRY

The vacuum ($T^{\mu\nu} = 0$) solution to the Einstein field equations which is spherically symmetric and static is called the Schwarzschild geometry. In "curvature coordinates" (where $2\pi r$ measures the proper circumference of 2-spheres) the Schwarzschild metric has the form

$$ds^2 = -\left(1-\frac{2M}{r}\right) dt^2 + \left(1-\frac{2M}{r}\right)^{-1} dr^2 + r^2(d\theta^2 + \sin^2\theta\, d\phi^2) .$$

(One sometimes abbreviates $d\Omega^2 \equiv d\theta^2 + \sin^2\theta\, d\phi^2$.) The constant M is the mass of the source of the field. If the metric is generated by a spherical star, the Schwarzschild metric holds outside the star and matches smoothly to the star's interior metric at its surface.

——— o ——— o ——— o ——— o ——— o ——— o ——— o ———

Problem 15.1. Prove that the total angular momentum squared

$$L^2 = p_\theta^2 + \sin^{-2}\theta\, p_\phi^2$$

is a constant of motion along any Schwarzschild geodesic.

Problem 15.2.

(a) Prove that all orbits in the Schwarzschild geometry are planar.

(b) Prove that all orbits are *stably* planar.

Problem 15.3. A particle falls radially into a Schwarzschild metric. As measured by proper time at infinity, what is its inward coordinate velocity (dr/dt) at a (curvature-) radius r? What is the *locally-measured* velocity relative to a stationary observer at the same radius?

Problem 15.4. Derive the equations of motion (equations relating t, r and τ) for a particle falling radially in the Schwarzschild geometry. Consider the three cases (i) particle released from rest at $r = R$ (ii) particle released from rest at infinity (iii) particle projected inward from infinity with velocity v_∞.

Problem 15.5. Derive a first-order differential equation for the trajectory (r as a function of ϕ) for equatorial orbits in the Schwarzschild geometry.

Problem 15.6. Show that the trajectory of light rays in the Schwarzschild metric obeys

$$\frac{d^2u}{d\phi^2} + u = 3u^2$$

where $u \equiv M/r$, and r is the Schwarzschild radial coordinate. Denote the minimum value of r along the trajectory by b, the "impact parameter." In the case $M/b << 1$ what is the deflection of a photon as it passes a spherical gravitating body? Give a formula for the deflection angle to lowest nonvanishing order in (M/b).

Problem 15.7.

(a) For a nearly Newtonian planetary orbit (i.e. $M/r << 1$) calculate to lowest order in M/r the advance of the periastron, per orbit, predicted by general relativity.

(b) Suppose that the central star is somewhat oblate or prolate, so that the form of the classical Newtonian potential is $\Phi(r) = -M/r - AM/r^3$, where A is related to the magnitude of the oblateness or prolateness. Calculate the advance (if oblate) of the periastron, to lowest order in A/r^2, per orbit. (This is a purely Newtonian calculation.)

(c) Assume that the oblateness of the sun is so large that the rate of advance of the perihelion due to oblateness and the rate of advance due to general relativity are equal for the orbit of Mercury. Compute the rate of advance of the perihelion (in seconds of arc per century) for the four planets closest to the sun, due to each of the effects. Note: to simplify the calculations assume, throughout the problem, that the orbits are nearly circular – i.e. they all have negligible eccentricity.

Problem 15.8. A rocket ship in a circular orbit of circumference $2\pi r$ around a star of mass M fires a laser gun (rest frequency ν_0). The gun is aimed in the orbital plane, and at an angle α (in the ship's frame) outward from the tangential direction of motion. What is the frequency of the laser as seen by a stationary observer at infinity?

Problem 15.9. A test particle of relativistic velocity v flies past a mass M at an impact parameter b so great that the deflection θ_{grav} is small. Calculate θ_{grav}. In flat space, a test charge e flies with velocity v past a nucleus of charge Ze at an impact parameter b so great that the deflection θ_{EM} is small. Calculate θ_{EM}. Why is the formula for θ_{grav} different from that for θ_{EM}?

Problem 15.10. A radio commentator is describing his radial fall into a Schwarzschild black hole. Just before he crosses the Schwarzschild radius his broadcast frequency starts becoming redshifted enormously with a time dependence exp (–t/constant), where t measures proper time at infinity. From the constant deduce the mass of the hole.

Problem 15.11. Calculate the cross section for capture of particles by a Schwarzschild black hole of mass M, in the limits of very high velocity particles ($v \to c$) and very low velocity particles ($v \ll c$).

Problem 15.12. Suppose that Paul is orbiting a neutron star in a circular orbit at radial coordinate $r = 4M$. Peter, his colleague, has been fired radially from a cannon on the neutron star with less than escape velocity; he flies outward just meeting Paul as he crosses his orbit, reaches a maximum radius and falls back down just happening to meet Paul again. Between their two meetings Paul has completed 10 orbits of the neutron star. Peter and Paul have an obsession about comparing their clocks whenever they meet. They set their clocks to agree on their first meeting as Peter flies outward. When they again compare their clocks, by how much do they disagree?

Problem 15.13. Give the coordinate transformation from Schwarzschild coordinates, in which $ds^2 = -e^{2\phi}dt^2 + e^{2\Lambda}dr^2 + r^2d\Omega^2$ to "isotropic coordinates", in which $ds^2 = -e^{2\phi}dt^2 + e^{2\mu}(d\bar{r}^2 + \bar{r}^2 d\Omega^2)$. Specialize to the vacuum Schwarzschild solution and construct coordinate diagrams showing the relation between (t, r) and $(t, \bar{r})$ coordinates. Is the area A of the surface $\bar{r} = \text{constant}$, $t = \text{constant}$ given by $A = 4\pi\bar{r}^2$? Construct an embedding diagram (see MTW p. 613) for the spacelike hypersurface $t = 0$, for $0 < \bar{r} < \infty$.

Problem 15.14.

(a) Show that in general a boost in the spatial direction $e_{\hat{j}}$ leaves invariant the physical components of the Riemann tensor $R_{\hat{t}\hat{j}\hat{t}\hat{j}}$ which are "parallel" to the boost. This is analogous to the invariance of $E_{\hat{j}}$ and $B_{\hat{j}}$ for a boost in the $e_{\hat{j}}$ direction.

(b) In the Schwarzschild geometry show that *all* the physical components of the Riemann tensor are invariant for a boost in the r-direction, but that all physical components are not invariant for a boost in the θ or ϕ direction.

Problem 15.15. Show that the spacelike slice $v = \text{constant}$ $(|v| > 1)$ of the Schwarzschild geometry in Kruskal coordinates u, v cannot be embedded in a Euclidean 3-space. What is the general condition on the slope dv/du of a spacelike slice of the Schwarzschild geometry that allows it to be embedded in a Euclidean 3-space?

Problem 15.16. Prove that the metric,

$$ds^2 = -dt^2 + \frac{4}{9}\left[\frac{9M}{2(r-t)}\right]^{2/3} dr^2 + \left[\frac{9M}{2}(r-t)^2\right]^{2/3} d\Omega^2$$

(which looks dynamical because the metric coefficients depend on t) is actually static. Show that it is in fact the Schwarzschild geometry.

Problem 15.17. In the preceding problem (15.16) show that the set of coordinate-stationary observers are all in free fall and have zero energy (i.e. they fell in from infinity with zero initial velocity).

Problem 15.18. A perfectly adiabatic gas with equation of state $p = Kn^{\gamma}$, $4/3 \leq \gamma \leq 5/3$, γ constant, accretes spherically onto a Schwarzschild black hole of mass M. The speed of sound in the gas at radial infinity is a_{∞}. At what radius does the inward flow become supersonic? (Give answer only to leading term in a_{∞}/c.)

Problem 15.19. A scalar field satisfies $\Box\Phi = 0$. Show that in the Schwarzschild geometry Φ can be decomposed into spherical harmonic components ($Y_{\ell m}$ = spherical harmonic) as

$$\Phi = r^{-1}\psi(r,t)Y_{\ell m}(\theta,\phi)$$

where ψ satisfies

$$\psi_{,tt} - (1-2M/r)[(1-2M/r)\psi_{,r}]_{,r} + V_{\ell}(r)\psi = 0$$

$$V_{\ell}(r) \equiv (1-2M/r)\left[\frac{2M}{r^3} + \frac{\ell(\ell+1)}{r^2}\right] .$$

Problem 15.20. Show that the Schwarzschild metric is also a solution of the Brans-Dicke theory of gravity. (For the Brans-Dicke field equations, see e.g. MTW p. 1070.)

CHAPTER 16

SPHERICAL SYMMETRY AND RELATIVISTIC STELLAR STRUCTURE

The geometry generated by a nonrotating, perfect fluid star is spherically symmetric. Exterior to the star it is the Schwarzschild geometry, even if the star is nonstatic (radially pulsating or collapsing). If the star is static, the interior metric can be written

$$ds^2 = -e^{2\Phi}\, dt^2 + (1-2m/r)^{-1}\, dr^2 + r^2\, d\Omega^2$$

where $m = \int_0^r 4\pi r^2 \rho\, dr$. The pressure gradient inside the star is given by the OV (Oppenheimer-Volkoff) equation of hydrostatic equilibrium

$$\frac{dp}{dr} = -\frac{(\rho + p)(m + 4\pi r^3 p)}{r(r-2m)} \; .$$

Here p and ρ are the pressure and mass-energy density of the fluid, satisfying the equations of state

$$p = p\,(n, T)$$

$$\rho = \rho\,(n, T) \; .$$

If the entropy per baryon s is constant in the star, then p depends only on ρ: $p = p(\rho)$. The metric function Φ is determined by the Einstein field equations:

$$\frac{d\Phi}{dr} = \frac{m + 4\pi r^3 p}{r(r-2m)} \quad \text{and} \quad e^{2\Phi} = 1 - 2M/R, \text{ at } r = R$$

where R is the radius of the star and $M = m(R)$ is its total mass (mass of the exterior Schwarzschild metric).

Equilibrium stellar models may be unstable against gravitational collapse. In this situation the relevant dynamical equations derive from the Einstein field equations and/or from $T^{\mu\nu}{}_{;\nu} = 0$.

A stationary rotating star is not spherically symmetric, but rather is only axisymmetric. The structural equations for axisymmetric stars are quite complicated. However, certain general properties can be deduced (i) from the symmetries or (ii) when the star is assumed to be *rigidly* rotating.

Problem 16.1. Find basis vectors (and a dual basis of 1-forms) for orthonormal tetrads in a spherical geometry. Take the legs of the tetrad to be in the t, r, θ, and ϕ directions of the isotropic coordinate system whose metric is

$$ds^2 = -e^{2\Phi}dt^2 + e^{2\mu}(dr^2 + r^2 d\Omega^2) .$$

Problem 16.2. Suppose that an observer, at rest at some point inside a spherical relativistic star, measures the radial pressure-buoyant force, F_{buoy}, on a small element of volume V, using the usual laboratory techniques. What value will he find for F_{buoy}, in terms of ρ, p, m, V, and dp/dr? If he equates this buoyant force to an equal and opposite gravitational force, F_{grav}, what will F_{grav} be in terms of ρ, p, m, V, and r? How do these results differ from the corresponding Newtonian results?

Problem 16.3. Prove Birkhoff's Theorem: a spherically symmetric vacuum gravitational field is always static, and is always the Schwarzschild solution.

Problem 16.4. Show that test particles experience no gravitational forces inside a self-gravitating hollow sphere.

Problem 16.5. In the Brans-Dicke theory of gravity (see MTW p. 1070 or Weinberg p. 160 for field equations), show that the only static spherically vacuum solution which is regular at the origin is the flat-space metric η and constant scalar field Φ.

Problem 16.6. How many algebraically independent components of $T^{\mu\nu}$ are there in a spherically symmetric configuration?

Problem 16.7. Evaluate the 4 components of the equation $T^{\alpha\beta}{}_{;\beta} = 0$ for the stress energy tensor describing a static, spherically-symmetric perfect fluid star.

Problem 16.8. Polytropic stars (stars with fluids described by $p = K\rho^{\gamma}$) are unstable in Newtonian theory if $\gamma < 4/3$. Consider the influence of small relativistic effects on this stability criterion. Show that the effect is to increase the unstable range of γ to $\gamma < 4/3 + \varepsilon$ where ε may depend on the mass, radius, and structure of the star.

Problem 16.9. Express the Chandrasekhar and Oppenheimer-Volkoff upper mass limits (for white dwarfs and neutron stars, respectively) as dimensional combinations of fundamental constants and the mass of the nucleon and electron. Similarly express the limiting radii corresponding to these mass limits.

Problem 16.10. The mass $m(r)$ inside radius r for a spherical star appears in the g_{rr} term of the line element

$$ds^2 = -e^{2\Phi}dt^2 + (1-2m(r)/r)^{-1}dr^2 + r^2 d\Omega^2 \ .$$

Express $m(r)$ in a coordinate-independent manner in terms of the surface area and radial separation of spherical surfaces.

Problem 16.11.

(a) What is the form of the Schwarzschild metric in "outgoing Eddington-Finkelstein coordinates", obtained from curvature coordinates by the transformation

$$dt = du + (1-2M/r)^{-1}\, dr \ .$$

(b) Now let M be a function of the null coordinate u in part (a). Show that the space-time is not vacuum, and find the corresponding $T^{\alpha\beta}$. Give a physical interpretation. (This is the "Vaidya" metric.)

Problem 16.12. Solve the relativistic equations of stellar structure for a static, spherically symmetric star of uniform density. Show that the mass and radius of the star satisfy $R/2M > 9/8$. What is the smallest $R/2M$ can be if the dominant energy condition (see Problem 13.7) holds?

Problem 16.13. A static, spherically symmetric star is made out of a zero-temperature Fermi gas with Fermi energies much larger than the particle rest masses. Show that the equations of stellar structure have a solution $m(r) = 3r/14$. Find $\rho(r)$, $p(r)$, and $n(r)$. Although n is infinite at $r = 0$, show that the number of particles out to any radius is finite. Make an embedding diagram of the 3-surface, $t = \text{constant}$. What kind of singularity is at $r = 0$?

Problem 16.14. Calculate the surface stresses in a static self-gravitating shell of mass M and circumference $2\pi R$. What is the proper surface mass density? Compare the stresses to the Newtonian limit when $R \gg M$.

Problem 16.15. What is the smallest possible proper circumference of a self-supporting spherical shell of mass M, if its matter satisfies the dominant energy condition (as do all known forms of matter)?

Problem 16.16. What is the redshift to radial infinity from a thin spherical shell in static equilibrium, in terms of its proper surface density $\Lambda^{\hat{0}}{}_{\hat{0}}$ and proper surface stresses $\Lambda^{\hat{\theta}}{}_{\hat{\theta}}$? What is the largest possible redshift if it satisfies the dominant energy condition?

Problem 16.17. Show that for a rigidly rotating, self-gravitating, perfect fluid star

$$\nabla p = (\rho + p)\nabla \log u^t \ .$$

Here u^t is a component of the 4-velocity of the fluid in the canonical coordinate system adapted to the Killing vectors (i.e. $\xi_{(t)} = \partial/\partial t, \xi_{(\phi)} = \partial/\partial\phi$).

Problem 16.18. Show that in a rigidly rotating, self-gravitating, perfect fluid star, the surfaces of constant p and ρ coincide.

Problem 16.19. Show that the surface of a rigidly rotating star, with angular velocity of rotation Ω as seen at ∞, is given by

$$g_{tt} + 2g_{t\phi}\Omega + g_{\phi\phi}\Omega^2 = \text{constant} .$$

Problem 16.20. Find the Doppler broadening for a spectral line from a rigidly rotating star observed by an astronomer who is infinitely far away along the axis of rotation. (The Doppler broadening is the variation across the star in the Doppler shift:

$$z = \frac{\nu_{\text{emitted}}}{\nu_{\text{observed}}} - 1.)$$

Problem 16.21. Derive the general relativistic criterion for convective stability in a static equilibrium configuration of perfect fluid.

Problem 16.22. Prove the equivalence of isentropy and constant injection energy $[(\rho+p)/(nu^0)]$ for a rigidly rotating configuration.

Problem 16.23. Consider a stationary, axisymmetric star. There are two Killing vectors, $\xi_{(t)}$ and $\xi_{(\phi)}$. Show that

$$M = -\int (2T^\mu{}_\nu - \delta^\mu{}_\nu T)\xi^\nu{}_{(t)} d^3\Sigma_\mu$$

is the mass of the star as measured from infinity. Here $d^3\Sigma_\mu$ is the volume element of the star at some instant of time t (the time coordinate t is chosen such that $\xi_{(t)} = \partial/\partial t$). Similarly, show that

$$J = \int T^\mu{}_\nu \xi^\nu{}_{(\phi)} d^3\Sigma_\mu$$

is the angular momentum as measured from infinity.

Problem 16.24. Show that the integral for M given in Problem 16.23, in the case of a static spherical star made of perfect fluid, is

$$M = \int_0^R (\rho + 3p)\, e^{\Phi+\lambda}\, 4\pi r^2 dr$$

in curvature coordinates ($g_{00} = e^{2\Phi}$, $g_{rr} = e^{2\lambda}$). Show that this is the same as the expression derived from the equation of stellar structure,

$$M = \int_0^R \rho\, 4\pi r^2 dr \ .$$

Problem 16.25. For collapsing spherical stars we can't simultaneously have the three nice properties: (i) radial coordinate is comoving with a fluid shell; (ii) time coordinate is proper time for the fluid; (iii) the metric is diagonal. Prove that we *can* have all three properties if and only if the pressure vanishes.

Problem 16.26. If R is a comoving coordinate, the metric for a spherically symmetric collapsing star (see Problem 16.25) can be written as

$$ds^2 = -e^{2\Phi} dt^2 + e^{2\Lambda} dR^2 + r^2(t,R)\, d\Omega^2$$

where Φ and Λ are functions of R, t. If the star is made of a perfect fluid it is often useful to define the following functions:

$$m \equiv \int_0^R 4\pi r^2 \rho r' dR$$

$$U \equiv e^{-\Phi}\dot{r}$$

$$\Gamma^2 \equiv e^{-2\Lambda}(r')^2 \ .$$

Here primes denote partial differentiation with respect to R, and dots with respect to t. The function m is interpreted as the mass interior to the shell at R, and U is the rate at which a shell is moving with respect to the proper time of a comoving observer.

Prove the following relations:

(a) $$\dot{m} = -4\pi p r^2 \dot{r} \ .$$

[Hint: Use the first law of thermodynamics (Problem 5.19), baryon conservation, the equations of motion, and $G^t{}_R = 0$ (Problem 9.20).]

(b) $$\Gamma^2 = 1 + U^2 - 2m/r \ .$$

[Hint: Use $G^t{}_t = -8\pi\rho$ and $G^t{}_R = 0$ (Problem 9.20).]

Problem 16.27. In a collapsing star made of perfect fluid show that once a mass shell at comoving radius R has collapsed far enough so that $2m(R, t)/r(R, t) > 1$, the mass shell will collapse to $r = 0$ in a finite proper time.

Problem 16.28. For spherically symmetric pressure free collapse show that the fall of a shell is governed by the mass interior to that shell in the same manner as the radial fall of a particle in the Schwarzschild geometry is governed by the Schwarzschild mass:

$$d^2r/d\tau^2 = -M/r^2 \ .$$

Problem 16.29. For pressure free spherically symmetric collapse (see Problem 16.26), show that both m and Γ are independent of time. Solve the resulting dynamical equation

$$\left(\frac{dr}{d\tau}\right)^2 - \frac{2m(R)}{r} = \Gamma^2(R) - 1$$

in the three physically distinct cases $\Gamma^2 - 1$ greater than, less than, and equal to, zero.

Problem 16.30. Consider the gravitational collapse of a spherically symmetric, perfect fluid star of zero pressure and uniform density (i.e. uniform throughout the star as seen by observers comoving with the fluid).

(i) Show that the metric inside the star is locally the Friedmann solution with $k = +1$ if the star collapses from rest at some finite radius, $k = 0$ if the star is at rest at infinity, or $k = -1$ if the star is projected with finite velocity from infinity.

(ii) By Birkhoff's theorem (Problem 16.3), the exterior metric is the Schwarzschild metric. Show that each point on the surface of the star moves along a radial geodesic of the Schwarzschild metric.

(iii) Show that the Friedmann and Schwarzschild metrics match together smoothly at the surface of the star. (It is necessary and sufficient to show that the intrinsic 3-geometry of the surface and the extrinsic curvature of the surface are the same whether measured in the exterior or the interior.)

CHAPTER 17
BLACK HOLES

The Kerr-Newman black hole is an exact solution of the Einstein field equations possessing mass, angular momentum, and (in principle but not in astrophysical cases) charge.

The metric describing this solution is (in "Boyer-Lindquist coordinates"):

$$ds^2 = -\left(1-\frac{2Mr-Q^2}{\Sigma}\right)dt^2 - \frac{(2Mr-Q^2)\,2a\sin^2\theta}{\Sigma}\,dt\,d\phi$$

$$+\frac{\Sigma}{\Delta}\,dr^2 + \Sigma\,d\theta^2 + \left(r^2+a^2+\frac{(2Mr-Q^2)\,a^2\sin^2\theta}{\Sigma}\right)\sin^2\theta\,d\phi^2$$

where

$$a^2 + Q^2 \le M^2 ,$$

$M \equiv$ mass, $Q \equiv$ charge

$a =$ angular momentum per unit mass

$\Delta \equiv r^2 - 2Mr + a^2 + Q^2$

$\Sigma \equiv r^2 + a^2\cos^2\theta.$

The metric coefficients are independent of t and ϕ, so $\xi_{(t)} = \partial/\partial t$ and $\xi_{(\phi)} = \partial/\partial\phi$ are Killing vectors. Among the properties of this solution which follow from the metric are the orbital equations for test particles:

$$\Sigma\dot{r} = \pm(V_r)^{\frac{1}{2}}$$

$$\Sigma\dot{\theta} = \pm(V_\theta)^{\frac{1}{2}}$$

$$\Sigma\dot{\phi} = -(aE - L_z/\sin^2\theta) + \frac{a}{\Delta}P$$

$$\Sigma\dot{t} = -a(aE\sin^2\theta - L_z) + \frac{r^2+a^2}{\Delta}P .$$

Here "dot" indicates derivative with respect to proper time or affine parameter, and

$$P \equiv E(r^2+a^2) - L_z a - eQr$$

$$V_r \equiv P^2 - \Delta[\mu^2 r^2 + (L_z - aE)^2 + \mathcal{Q}]$$

$$V_\theta \equiv \mathcal{Q} - \cos^2\theta\,[a^2(\mu^2 - E^2) + L_z^2/\sin^2\theta],$$

$E \equiv$ conserved total energy

$L_z \equiv$ conserved z component of angular momentum

$\mathcal{Q} \equiv$ conserved quantity related to total angular momentum

$\mu \equiv$ rest mass of particle

$e \equiv$ charge of particle.

A Schwarzschild black hole is the special case for which $a = Q = 0$. A Reissner-Nordström black hole is the special case $a = 0$, $Q \neq 0$; it is spherically symmetric. A Kerr black hole is the case $a \neq 0$, $Q = 0$. The defining property of a black hole is that it have a *horizon*, a surface through which matter can fall, but from which no matter or information can escape to infinity. For a Kerr hole, it is located at r_+, the larger root of the equation $\Delta = 0$. The *stationary limit* of a rotating hole is the surface within which all observers are dragged around the hole. For Kerr, the stationary limit is at r_0, the larger root of $g_{tt} = 0$. The region between the horizon and stationary limit is called the ergosphere.

——— o ——— o ——— o ——— o ——— o ——— o ——— o ———

Problem 17.1. Show that the constant M which occurs in the Kerr metric is the mass of the system, and that the constant a is the angular momentum per mass.

Problem 17.2. A suggested use of small ($\ll M_\oplus$) black holes is to crush junked automobiles into neat spherical balls by allowing them to partially collapse around a black hole. Estimate what mass hole, in orbit around the Earth, would be appropriate for this application. How many wrecks per hour could be processed?

Problem 17.3. Show that once a rocket ship crosses the gravitational radius (horizon) of a Schwarzschild black hole, it will reach $r = 0$ in a proper time $\tau \leq \pi M$, no matter how the engines are fired.

Problem 17.4. Show that Kepler's law

$$\Omega^2 = M/r^3$$

holds exactly for circular orbits around a Schwarzschild black hole, if r is the curvature coordinate radius, and Ω is the angular frequency as measured from infinity. Derive an analogous law for equatorial orbits around a Kerr black hole of specific angular momentum a.

Problem 17.5. An observer in a circular orbit of circumference $2\pi r$ around a charged, spherical, black hole (a Reissner-Nördstrom black hole) of mass M and charge Q measures local electric and magnetic fields. What are their strengths and orientations?

Problem 17.6. By considering the mass, charge, and angular momentum of a "classical" electron, show that it cannot be a Kerr-Newman black hole.

Problem 17.7. For circular orbits in the equatorial plane of a Kerr black hole, prove that the marginally stable orbit has minimum energy E and minimum angular momentum L.

Problem 17.8. An observer (not necessarily freely-falling) orbits a Kerr black hole in the equatorial $(\theta = \pi/2)$ plane.

(a) Let his orbit be at constant r. Define $\Omega = d\phi/dt$ to be his "angular velocity relative to a distant stationary observer." In terms of Ω, r, M, and a, find u^0, u^ϕ, u_0, u_ϕ.

(b) Suppose that the circular orbit lies in the ergosphere (the orbital radius is outside the horizon at r_+ but inside the stationary limit at r_0).

Show that the observer cannot remain at rest with respect to a distant observer. That is, show that Ω for the observer must be nonzero.

(c) If the observer is in the region $r_- < r < r_+$, show that he cannot remain at constant radius.

Problem 17.9. Show that there are negative energy particle trajectories inside the ergosphere of a Kerr black hole (and outside the horizon!). Show that it is possible for a rocket ship to increase its total energy by firing a bullet into the hole during an orbital passage through the ergosphere.

Problem 17.10. Show that as a test particle approaches $r = r_+$, the horizon of a Kerr black hole, it has "angular velocity as seen from infinity" equal to

$$\Omega \equiv \frac{d\phi}{dt} = \frac{a}{2Mr_+} \ .$$

Problem 17.11. Prove that there exist "quasi-circular, polar" orbits in the Kerr geometry, i.e. orbits which pass alternately over the north and south poles at a fixed radial coordinate distance. What is the smallest possible polar radius of these orbits?

Problem 17.12. A Killing horizon is a null hypersurface generated by a Killing vector. An ergosurface ("stationary limit") is an infinite red-shift surface for static observers. Show that for a static black hole the ergosurface is a Killing horizon.

Problem 17.13. Show that the surface area of the horizon of a Kerr-Newman black hole (area of surface $r = r_+$, $t = \text{constant}$, in Boyer-Lindquist coordinates) is $4\pi\,[\{M+(M^2-Q^2-a^2)^{\frac{1}{2}}\}^2+a^2]$.

Problem 17.14. According to Hawking's theorem (that in a collision of two black holes the total surface area must not decrease), what is the minimum mass M_2 of a Schwarzschild black hole that results from the collision of two Kerr black holes of equal mass M_1 and opposite angular momentum parameter, a? Suppose $|a| \approx M_1$; what fraction of the original mass can be radiated away? Are there any other uncharged black-hole collisions that can get this much energy out?

Problem 17.15. Use the theorem that the area of a black hole is non-decreasing (cf. Problem 17.14) to prove that a Kerr black hole amplifies (rather than absorbs) certain modes of an incident radiation field.

Problem 17.16.

(a) Write down the scalar wave equation $\Box\Phi = 0$ in the Kerr geometry in Boyer-Lindquist coordinates.

(b) Show that the equation can be reduced to ordinary differential equations by separation of variables.

(c) Find the asymptotic form of Φ for $r \to \infty$.

(d) Find the asymptotic form of Φ for $r \to r_+$.

(e) What boundary condition on Φ corresponds to ingoing waves as seen by a physical observer on the horizon?

(f) Show that, for a wave of the form $\Phi = \exp(-i\omega t + im\phi) f(r, \theta)$, energy flows *out* of the hole if $0 < \omega/m < a/(2Mr_+)$. Compare to Problem 17.15.

Problem 17.17. Charged particles are dropped radially into a Reissner-Nordström black hole with $Q^2 < M^2$. Show that it is never possible to drop in enough charge to make $Q^2 > M^2$ (a solution which would be a naked singularity, not a black hole).

Problem 17.18. The "Zero Angular Momentum Observers" (ZAMO's) in the Kerr geometry have basis 1-forms

$$
\begin{aligned}
\tilde{\boldsymbol{\omega}}^{\hat{t}} &= |g_{tt} - \omega^2 g_{\phi\phi}|^{\frac{1}{2}}\, \widetilde{dt} \\
\tilde{\boldsymbol{\omega}}^{\hat{\phi}} &= (g_{\phi\phi})^{\frac{1}{2}} (\widetilde{d\phi} - \omega \widetilde{dt}) \\
\tilde{\boldsymbol{\omega}}^{\hat{r}} &= (\Sigma/\Delta)^{\frac{1}{2}}\, \widetilde{dr} \\
\tilde{\boldsymbol{\omega}}^{\hat{\theta}} &= \Sigma^{\frac{1}{2}}\, \widetilde{d\theta}
\end{aligned}
$$

where $\omega \equiv -\, g_{t\phi}/g_{\phi\phi}$.

(a) Show that these basis 1-forms are orthonormal.

(b) Find the dual basis vectors.

(c) The 4-velocity of the ZAMO is $\mathbf{u} = \mathbf{e}_{\hat{t}}$; show that $\mathbf{u}$ has zero rotation.

(d) The ZAMO is not an inertial observer; show that the acceleration is $\mathbf{a} = \frac{1}{2} \nabla \log |g_{tt} - \omega^2 g_{\phi\phi}|$.

Problem 17.19. Calculate the Gaussian curvature of the horizon of a Kerr black hole and show that it becomes negative for $a > 3^{\frac{1}{2}} M/2$. (This shows that the horizon cannot be globally embedded in a Euclidean 3-space if $a > 3^{\frac{1}{2}} M/2$.) Use the Gauss-Bonnet theorem to check that the horizon is topologically a 2-sphere.

Problem 17.20. Show that a primordial, rotating, black hole ($\sim 10^{10}$ years old), of mass $\lesssim 10^{15}$ gm. will have already lost most of its angular momentum to spontaneous quantum emission of photons or gravitons. What fraction of the angular momentum of a $1 M_\odot$ ($\sim 10^{33}$ gm.) rotating hole would be lost in the same time?

Problem 17.21. Consider the vacuum metric

$$ds^2 = -\frac{(1 - \frac{1}{2} m/\rho)^2}{(1 + \frac{1}{2} m/\rho)^2} dt^2 + \frac{(1 - \frac{1}{2} m/\rho)^2}{(1 - \frac{3}{2} m/\rho)^2} (d\rho^2 + \rho^2 d\theta^2 + \rho^2 \sin^2\theta \, d\phi^2)$$

(which is a particular solution to the static, spherically symmetric problem in the Lightman-Lee theory of gravity). Does the above metric describe a black hole; and, if so, how do the hole's properties differ from those of the corresponding hole in general relativity?

CHAPTER 18
GRAVITATIONAL RADIATION

Weak gravitational waves are described by linearized theory (see Chapter 13). The basic equations for waves propagating in vacuum are

$$g_{\mu\nu} = \eta_{\mu\nu} + h_{\mu\nu} \quad (|h_{\mu\nu}| << 1)$$

$$\bar{h}_{\mu\nu} \equiv h_{\mu\nu} - \frac{1}{2} h^{\alpha}{}_{\alpha}\, \eta_{\mu\nu}$$

$$\Box\, \bar{h}_{\mu\nu} \equiv \bar{h}_{\mu\nu}{}^{;\alpha}{}_{\alpha} = 0$$

$$\bar{h}^{\mu\alpha}{}_{;\alpha} = 0 \quad (\text{“Lorentz gauge”})$$

$$h_{\mu 0} = 0,\ h^{\alpha}{}_{\alpha} = 0 \quad (\text{“TT” or “Transverse-Traceless” gauge})\,.$$

The effective stress-energy tensor for gravitational waves is

$$T_{\mu\nu}{}^{(GW)} = \frac{1}{32\pi} < h_{jk,\mu}\, h^{jk}{}_{,\nu} >$$

where < > denotes an average over several wavelengths and h_{jk} is in the TT gauge (see e.g. MTW Section 36.7).

The gravitational wave power L_{GW}, emitted by a nearly-Newtonian, slow-motion $(v << c)$ gravitating source is

$$L_{GW} = \frac{1}{5} \frac{G}{c^5} < \dddot{\mathcal{I}}_{jk}\, \dddot{\mathcal{I}}^{jk} > ,$$

where $\mathcal{I}_{jk}$ is the “reduced quadrupole moment tensor” of the source, given by

$$\mathcal{I}_{jk} \equiv \int \rho (x_j x_k - \frac{1}{3} \delta_{jk} r^2) d^3x ,$$

and < > denotes averaging over several characteristics periods of the source.

Problem 18.1. A Massachusetts motorist shakes his fist angrily at another motorist. What fraction of his expended energy goes into gravitational radiation?

Problem 18.2. A gravitationally bound dynamical system (e.g. a binary star) has, in order of magnitude, mass M and size R. Estimate the time for radiation reaction forces to affect the system substantially, and compare this time-scale with the dynamical time-scale for the system.

Problem 18.3. For an electric dipole, and its radiation pattern, there are three independent orientations, corresponding to the three directions in which the dipole may point. How many independent orientations are there for a traceless quadrupole tensor?

Problem 18.4. Calculate the gravitational radiation luminosity of a spinning thin metal rod of mass M and length ℓ, spinning at frequency ω around a symmetrical perpendicular axis. Estimate the electromagnetic luminosity which would arise from the slight excess of electrons pushed toward the ends by centrifugal force. If the rod has a reasonable density ($10\ \mathrm{gm/cm^3}$) and is rotating at a reasonable frequency (1 kHz) will electromagnetic or gravitational radiation be more important in slowing the rotation?

Problem 18.5. The radiation reaction forces on a slow-motion, weak field source can be derived from an addition to the Newtonian potential

$$\Phi^{\mathrm{react.}} = \frac{1}{5} \mathcal{I}^{(5)}_{jk}\, x^j x^k \ .$$

(Cf. W. Burke, J. Math. Phys. 12, 402 (1971); MTW pp. 993.) Here $\mathcal{I}_{jk}$ is the reduced quadrupole moment of the source $\mathcal{I}_{jk} \equiv \int \rho (x_j x_k - \frac{1}{3}\delta_{jk} r^2) d^3x$ at a given time. The superscript (5) indicates the fifth time derivative. From this potential derive expressions for the time-averaged rates at which the source loses energy and angular-momentum, in terms of derivatives of $\mathcal{I}_{jk}$.

Problem 18.6. Two stars of mass M_1 and M_2 separated by a distance R revolve about each other in a nonrelativistic circular orbit. Due to gravitational radiation reaction, R changes with time. Find R(t).

Problem 18.7. Two point masses m_1 and m_2 are in a Newtonian elliptical orbit with semimajor axis a and eccentricity e. Compute da/dt and de/dt due to gravitational radiation reaction. Show that the elliptical orbit tends to be circularized.

Problem 18.8. A plane gravitational wave propagates through nearly flat empty spacetime, in the x^1 direction (i.e. the metric perturbations $h_{\alpha\beta}$ are functions only of $u = t-x$). Give an explicit coordinate transformation which makes all the $h_{\alpha\beta}$ zero except $h_{23} = h_{32}$ and $h_{22} = -h_{33}$. Show that the same resulting components could have been obtained directly by projection into the transverse traceless gauge.

Problem 18.9. Show that gravitational radiation generated by an axisymmetric system carries no net angular momentum. (Do not assume that the sources have weak internal gravitational fields.)

Problem 18.10. Define Stokes parameters for a plane gravitational wave and show how to calculate the fraction of circular polarization, linear polarization and the orientation of maximum linear polarization from the three Stokes parameters.

Problem 18.11. An initially static source undergoes violent motion which generates gravitational radiation and then, after a finite time, becomes again static. A distant observer detects the gravitational waves by watching the motion of two free particles which are initially at rest with respect to each other. Show that after the passage of the waves the observer sees the particles back in their original positions and at rest with respect to each other, to linear order in the wave amplitude.

Problem 18.12. An elastic rod can be used to detect gravitational waves not only at its lowest normal mode frequency ω_0, but also at harmonics

$\omega_n \equiv n\omega_0$. What is the sensitivity of the nth mode relative to the zeroth, i.e. how does the ratio of maximum squared amplitude of the displacement to energy flux of wave vary with n? (Assume the rod has the same mechanical damping time for all modes.)

Problem 18.13. A (weak) plane gravitational wave travelling in the x-direction is normally incident on a slab of cement. The cement absorbs energy E from the plane wave. Show (e.g. as a result of $T^{\mu\nu}{}_{;\nu} = 0$) that the slab must absorb x-momentum E also, and find the relationship between the rate at which energy and x-momentum are absorbed.

Problem 18.14. In the previous problem it was shown that materials must be able to absorb momentum in the direction of wave propagation. This seems incompatible with the description of gravitational waves as transverse! To investigate this point idealize the cement molecules of the previous problem as being harmonically bound to their equilibrium positions, and having a damping force due to internal friction of the cement. Assume the gravitational wave is monochromatic and linearly polarized. Using the equation of geodesic deviation find the time average force, and hence the rate of momentum absorption, in the direction of wave propagation. Show that this rate of momentum absorption is equal to the rate at which the molecule absorbs energy.

Problem 18.15. A weak, plane gravitational wave of frequency ω and dimensionless amplitude h passes through a "hard-sphere" gas of temperature T. The mean free path of atoms in the gas is ℓ, and the gas is dilute enough so that $\ell >> c/\omega$. Show that in a finite amount of time the particles in the gas will be heated to relativistic temperatures. Estimate this time. Estimate the distance over which the wave is damped by a factor of e in amplitude.

Problem 18.16. Estimate the number of gravitons emitted in an asymmetric explosion of energy E.

Problem 18.17. Roughly how many thermal gravitons does a 100 watt lightbulb emit in its rated lifetime of 1000 hours? What is the approximate wavelength and number of gravitons emitted when the lightbulb is dropped and broken on a cement floor?

Problem 18.18. Calculate in detail the lifetime of a hydrogen atom in the 3d state against decaying to the 1s state by gravitational radiation.

Problem 18.19. By symmetry, the thermal graviton flux from a spherically symmetric star is evidently isotropic; reconcile this fact with Birkhoff's theorem. Approximately what is the multipolarity 2^ℓ of the flux ($\ell = 1 \equiv$ dipole, etc.) for a typical star, such as our sun.

Problem 18.20. Consider the following "gravitational wave modes" representing a gravitational wave travelling in the z direction

$$\Psi_2 = -\frac{1}{6} R_{z0z0} \qquad \Psi_4 = R_{y0y0} - R_{x0x0} + 2i R_{x0y0}$$

$$\Psi_3 = \frac{1}{2}(-R_{x0z0} + i R_{y0z0}) \qquad \bar{\Psi}_4 = R_{y0y0} - R_{x0x0} - 2i R_{x0y0}$$

$$\bar{\Psi}_3 = \frac{1}{2}(-R_{x0z0} - i R_{y0z0}) \qquad \Phi_{22} = -(R_{x0x0} + R_{y0y0}) \ .$$

Which of these waves are transverse?

The *spin* of a wave indicates (among other things) the relation of the orientation of polarization states. For a spin 0 (scalar) wave the manifestations of the wave are symmetric about the direction of propagation. For a spin 1 (vector) wave (e.g. an electromagnetic wave) the independent polarization states are oriented at 90° to each other; a rotation of 180° returns to the original polarization state, with only a sign change. In general for a spin s wave a rotation of π/s brings back the original polarization state. Which of the waves above are spin 0? Spin 1? Spin 2? Which are possible in general relativity?

Problem 18.21. Draw the force field of each of the modes in the preceding problem.

Problem 18.22. Consider the metric $ds^2 = dx^2 + dy^2 - dudv + 2H(x,y,u)du^2$. What form must the function H have for this to represent a (strong) plane gravitational wave propagating in vacuum?

CHAPTER 19
COSMOLOGY

If the universe is homogeneous and everywhere isotropic, its geometry is that of the Robertson-Walker metric

$$ds^2 = -dt^2 + R^2(t)\left[\frac{dr^2}{1-kr^2} + r^2\,d\Omega^2\right],$$

where $k = +1,\ 0,\ -1$ for a closed, marginally open, or open universe. When the Einstein equations are used to determine the time development of $R(t)$ and the value of k, the resulting spacetime is called a Friedmann model (sometimes called a Lemaître model, especially for nonzero cosmological constant). The first two derivatives of $R(t)$ at the present epoch (denoted by subscript 0) are parameterized by the Hubble constant

$$H_0 \equiv (dR/dt)/R \quad \text{at} \quad R = R_0$$

and the "deceleration parameter"

$$q_0 \equiv -[(d^2R/dt^2)R]/(dR/dt)^2 \quad \text{at} \quad R = R_0 \ .$$

The matter in a cosmology is generally in a state of expansion or contraction, so that light received by an observer is generally red- or blue-shifted relative to its source by an amount z,

$$1 + z \equiv \frac{\nu_{\text{emitted}}}{\nu_{\text{observed}}} = \frac{\lambda_{\text{observed}}}{\lambda_{\text{emitted}}} \ .$$

Often the magnitude of z varies monotonically with distance from an observer so one speaks of "an object at redshift z."

If ρ and p are the density and pressure of the smoothed-out mass-energy content of the universe, the universe is said to be "matter dominated" when $\rho \gg p$, and "radiation dominated" when $p \approx \frac{1}{3}\rho$.

The $3.^0$K black-body "cosmic microwave background," when extrapolated back in time, implies high temperatures at early times in the Friedmann model, a "hot big bang." However it is also possible that the large "entropy per baryon" implied by this radiation was generated by some dissipative process during the evolution of our universe.

—— o —— o —— o —— o —— o —— o —— o ——

Problem 19.1. Show that the equations of Newtonian gravity and hydrodynamics do not admit a cosmology which is isotropic, homogeneous, and static (i.e. an unchanging universe filled with a uniform perfect fluid).

Problem 19.2. A spacetime contains no matter and is everywhere isotropic. Prove that it is flat Minkowski space.

Problem 19.3. An object emits black body radiation of temperature T in its own rest frame; we see it at a redshift z and subtending a solid angle Ω. What flux do we measure? What if the redshift is due to doppler motion of a local object instead of a cosmological redshift?

Problem 19.4. Homogeneous, isotropic spatial hypersurfaces must (by spherical symmetry) have a line element of the form

$$d\sigma^2 = a^2\,[f^2(r)\,dr^2 + r^2 d\Omega^2], \qquad a = \text{constant}\,.$$

Show that $f^2(r)$ must have the form $(1-kr^2)^{-1}$, where $k = 0, \pm 1$.

Problem 19.5. Show that the Robertson-Walker metric

$$ds^2 = -\,dt^2 + R^2(t)\left[\frac{dr^2}{1-kr^2} + r^2\,(d\theta^2 + \sin^2\theta\, d\phi^2)\right]$$

can also be written as

$$ds^2 = -\,dt^2 + R^2(t)\,[d\chi^2 + \Sigma^2(\chi)\,(d\theta^2 + \sin^2\theta\, d\phi^2)]$$

or as

$$ds^2 = R^2(\eta)[-d\eta^2 + d\chi^2 + \Sigma^2(\chi)(d\theta^2 + \sin^2\theta\, d\phi^2)]$$

where $\Sigma^2(\chi) = \sin^2\chi$ or χ^2 or $\sinh^2\chi$ $(k = 1, 0, -1)$.

Problem 19.6. Show that the spacelike 3-surfaces of a closed, isotropic, homogeneous universe possess a translation symmetry which leaves no points fixed. (Notice that this is not true in 2 dimensions; a 2 sphere cannot be combed smooth!)

Problem 19.7. A bullet is shot out into an expanding Robertson-Walker universe with a velocity V_1 (relative to cosmological observers). Later, when the universe has expanded by a scale factor $(1+z)^{-1}$, it has a different velocity V_2 with respect to cosmological observers. Find V_2 in terms of z and V_1. Show that in the limit $V_1 \to c$, the formula for the redshift of photons is obtained.

Problem 19.8. Show by an explicit coordinate transformation that the Robertson-Walker metric is conformally flat. Write $R_{\mu\nu\alpha\beta}$ in terms of $g_{\mu\nu}$ and ρ, p and the 4-velocity u^μ of the matter.

Problem 19.9. In a Robertson-Walker metric show that angular diameter distance (d_A), luminosity distance (d_L) and proper motion distance (d_M) are related by

$$(1+z)^2 d_A = (1+z) d_M = d_L .$$

Problem 19.10. Suppose astronomers are able to find a family of objects whose absolute luminosities L are known. Suppose their apparent luminosities ℓ (or equivalently their luminosity distance d_L) and their redshift z are measured. Using the Robertson-Walker line element, find an expression for ℓ (or d_L) as a function of L, z, H_0 and q_0 for small z.

Problem 19.11. Let $n(t_0)$ be the number density at the present epoch of a (mythical) family of identical light or radio sources distributed uniformly throughout the universe.

(a) Show that the number of such sources with redshifts less than z as observed from the Earth today is

$$N(z) = \frac{4\pi}{3}\frac{n(t_0)}{H_0^3} z^3\left(1-\frac{3}{2}z(1+q_0)+\cdots\right).$$

Ignore evolutionary effects, i.e. the number of sources in a unit comoving volume remains constant.

(b) If the sources all have intrinsic luminosity L, show that the number with fluxes (ergs $\sec^{-1}\text{cm}^{-2}$) greater than S as observed from the earth today is

$$N(S) = \frac{4\pi}{3} n(t_0)\left(\frac{L}{4\pi S}\right)^{3/2}\left[1-3H_0\left(\frac{L}{4\pi S}\right)^{1/2}+\cdots\right].$$

Problem 19.12. A ray of light travels along a radial line in the Robertson-Walker metric

$$ds^2 = -dt^2 + R^2(t)\left[\frac{dr^2}{1-kr^2}+r^2d\Omega^2\right].$$

How is the coordinate r related to the affine parameter λ along the ray, namely what is $dr/d\lambda$?

Problem 19.13. By requiring that the Robertson-Walker metric satisfy the Einstein field equations, derive the dynamical equations for a perfect fluid Friedmann cosmology:

$$3\ddot{R} + 4\pi G(\rho+3p)R = 0 \tag{1}$$

$$R\ddot{R} + 2\dot{R}^2 + 2k - 4\pi G(\rho-p)R^2 = 0 . \tag{2}$$

Problem 19.14. Show that the two second-order equations of Problem 19.13 are equivalent to the first order equations

$$\dot{R}^2 + k = \frac{8\pi G}{3}\rho R^2 \tag{1}$$

$$\frac{d}{dR}(\rho R^3) = -3pR^2 . \tag{2}$$

Problem 19.15. For a Friedmann cosmology derive the relations

$$\frac{8\pi G\rho}{3} = \left(\frac{k}{R^2} + H^2\right) \tag{1}$$

$$-8\pi Gp = \frac{k}{R^2} + H^2(1-2q) \ . \tag{2}$$

If the cosmology is matter-dominated $(\rho \gg p)$ show that

$$\frac{k}{R^2} = (2q-1)H^2 \tag{3}$$

$$\frac{8\pi G\rho}{3} = 2qH^2 \ . \tag{4}$$

If it is radiation dominated $(p \approx 1/3\ \rho)$ show that

$$\frac{k}{R^2} = (q-1)H^2 \tag{5}$$

$$\frac{8\pi G\rho}{3} = qH^2 \ . \tag{6}$$

Problem 19.16. For a Friedmann cosmology what equations relating p, ρ, and $R(t)$ result from the equation of energy conservation, $T^{\nu}{}_{\mu;\nu} = 0$?

Problem 19.17. For a $k = -1$ Friedmann cosmology with $p = \rho = 0$ show that the line element becomes

$$ds^2 = -dt^2 + t^2[d\chi^2 + \sinh^2\chi(d\theta^2 + \sin^2\theta\, d\phi^2)] \ .$$

Exhibit an explicit coordinate transformation to show that this metric describes Minkowski space.

Problem 19.18. Solve the first-order Friedmann equation

$$\left(\frac{\dot{R}}{R}\right)^2 = \frac{8\pi G}{3}\rho - \frac{k}{R^2}$$

for $R(t)$ when the density ρ is dominated by a) matter and b) radiation. Express any parameters of the present epoch in terms of H_0 and q_0.

Problem 19.19. A bullet is shot out into an expanding Friedmann universe. When $k = -1$, show that it approaches the same velocity as some cosmological observer, but a position which is a constant proper distance away from him. When $k = 0$, show that the bullet again approaches the velocity of some cosmological observer but that the proper distance between the bullet and that observer becomes arbitrarily large as $t \to \infty$.

Problem 19.20. For a closed $(k=1)$ Friedmann universe in which radiation dominates for only a negligibly short fraction of the life of the universe, how many times can a photon encircle the universe from the moment of the creation of the universe to the moment of its death?

Problem 19.21. If an idealized $k=0$ matter dominated Friedmann cosmology with Hubble constant H_0 contains homogeneously distributed sources of constant luminosity L, and if the local number density of such sources in space is now n, what is the brightness B of the night sky (energy per steradian of sky per collection area per time)? (If the universe were static and infinitely old the brightness would be infinite; this is called Olbers' Paradox.)

Problem 19.22. Suppose that at the time of hydrogen recombination (which, say, occurred at a redshift of $z = 1500$) the deceleration parameter was $q = 0.5002$. What would q_0 be today? Repeat for $q = 0.4998$ at $z = 1500$. (Assume a matter-dominated universe.)

Problem 19.23. A closed $(k=1)$ Friedmann universe has Hubble constant H_0 and deceleration parameter q_0. Assume that the universe has always been matter dominated.

(a) What is the total proper volume of the universe at the present epoch?

(b) What is the total proper volume that we see, looking out into the sky?

(c) What is the total proper volume *now* occupied by the matter which we see, looking out into the sky?

Problem 19.24. What is the apparent angular size of an object of proper diameter ℓ seen at redshift z in a matter dominated Friedmann cosmology with present parameters H_0 and q_0? (Analogous results for apparent proper motion and apparent luminosity follow from Problem 19.9.)

Problem 19.25. In a "hot" Friedmann cosmology, two unrelated important epochs are when matter first begins to dominate radiation ($\rho_{\text{matter}} \approx \rho_{\text{radiation}}$), and when protons and electrons recombine to form hydrogen. Given that these epochs happen to be nearly the same in our universe, deduce a numerical value for the conserved entropy per baryon $\sigma \equiv 4aT^3/3n$ (where T = temperature, a = radiation constant, n = number density of baryons).

Problem 19.26. In terms of the conserved entropy per baryon σ of a hot big bang model, find the temperature at which hydrogen characteristically recombined, i.e. had an equilibrium ionization fraction of 0.5. Evaluate your answer for $\sigma = 10^8,\ 10^9$.

Problem 19.27. In a radiation dominated Friedmann cosmology at times near the big bang singularity, calculate the temperature T as a function of (proper cosmological) time t. Assume that only photons, electrons and positrons contribute to the energy density ρ. How is the answer changed if neutrinos and antineutrinos are also allowed?

Problem 19.28. A Friedmann cosmology has temperature T_1 at expansion scale R_1, and is dominated by relativistic electrons, positrons, muons, photons, and neutrinos in thermal equilibrium. Later, at expansion scale R_2, the muon pairs have annihilated, but the other particles are still relativistic and in equilibrium. Find the temperature T_2 in terms of T_1, R_1, and R_2.

Problem 19.29. Under which of the following suppositions would the hot big bang have produced less He^4 than predicted by the "standard" model? Less H^2 (deuterium)?

(i) Suppose the baryon density in the universe today is larger than we now think.

(ii) Suppose the weak interaction constant is smaller than we now think.

(iii) Suppose there are many more neutrinos than antineutrinos or photons in the cosmic background today.

(iv) Suppose there are many more antineutrinos than neutrinos or photons in the cosmic background today.

(v) Suppose that the gravitational constant G varies with cosmological time and was slightly larger in the past.

Problem 19.30. Suppose that a universe is isotropic, homogeneous, and empty except for a "vacuum polarization" stress energy of the form $8\pi T_{\mu\nu} = \Lambda g_{\mu\nu}$. (In old language: there is a non-zero cosmological constant Λ.) Find a $k=0$ cosmological solution. Find a coordinate system in which it is static. (This is the "de Sitter universe".)

Problem 19.31. A universe is isotropic, homogeneous, and contains only pressureless dust and "vacuum polarization" stress energy, so that $T_{\mu\nu} = \rho_0 u_\mu u_\nu - \frac{\Lambda}{8\pi} g_{\mu\nu}$ where u^μ is the 4-velocity field of the matter. Show that there is a static solution for the metric, but that it is unstable. (This cosmology is called the "Einstein universe".)

Problem 19.32. What is the proper volume of the "Einstein universe" of Problem 19.31, in terms of ρ_0, the density of its dust.

Problem 19.33. At one time observations seemed to suggest that there was an unusual clustering of quasar redshifts around $z=2$. One proposal to explain this is that our universe is a $k=+1$ dust cosmology, with nonzero cosmological constant Λ only slightly greater than the value for a static "Einstein Universe" (Problem 19.31). Show that for this model the universe will expand at a decreasing rate to a certain radius R_m, at which radius it will remain for a long time while expanding very slowly, before expanding again at a rate which asymptotically approaches $H = (\Lambda/3)^{\frac{1}{2}}$.

Suppose quasar formation occurred at the time of nearly constant radius. What does this model predict for ρ_{matter} today? [Use $H_0 = 10^{-28}cm^{-1}$].

Problem 19.34. What is the order of magnitude of the influence of the cosmological constant on the celestial mechanics of the solar system if $\Lambda \sim 10^{-57}cm^{-2}$?

Problem 19.35. Prove that for a physically possible perfect fluid no solution of the Einstein equations is homogeneous, everywhere isotropic, and *static*. (Before Hubble's discovery, Einstein considered this a failing of the theory and introduced the "cosmological constant" term as a remedy.)

Problem 19.36. Prove that no solution of the Einstein equations for a *pressureless* fluid is static. Do not assume isotropy or homogeneity. (The difficulty of this problem depends on the definition of "static." Easy case: Take static to mean time invariant and time reversible, the first definition of "static" in Problem 10.8. Harder case: Use the second definition in Problem 10.8.)

Problem 19.37. Prove that no solution of the Einstein equations for a perfect fluid is static and homogeneous. Do not assume isotropy. (Again as in Problem 19.36 there are easy and harder cases depending on the definition of "static" that is used.)

Problem 19.38. In cosmology one usually uses coordinates comoving with the galaxies. Let (τ, x^i) be such a system, and let the metric take the general form

$$ds^2 = -d\tau^2 + 2g_{0i}\,d\tau\,dx^i + g_{ij}\,dx^i dx^j$$

where g_{0i} and g_{ij} can be functions of τ and x^i. Show

(a) that τ is the proper time measured by a galaxy;

(b) that g_{ij} governs proper distances in the hypersurface of constant τ;

(c) that if g_{0i} and g_{ij} are independent of all x^i, then the universe is homogeneous but that the converse is false;

(d) that if g_{0i} and g_{ij} are independent of τ, then $\sigma_{\alpha\beta} = 0$, $\theta = 0$ (but $\omega_{\alpha\beta} \neq 0$ in general);

(e) that if $g_{0i} = 0$ and $g_{ij} = f(\tau)\overline{g}_{ij}(x^k)$, then $\sigma_{ij} = 0$;

(f) that $g_{0i,0} = 0$ if and only if galaxies fall on geodesics;

(g) that if $\omega_{\alpha\beta} \neq 0$, *no* choice of τ and x^i can make $g_{0i} = 0$ everywhere and that this means that $\omega_{\alpha\beta,0} \neq 0$ implies nongeodesic motion of the galaxies.

Problem 19.39. The distance between two neighboring galaxies is $\delta x^\alpha = Rn^\alpha$ where $\mathbf{n}$ is a unit, purely spatial vector in one galaxy's rest frame. Show that

$$\frac{\dot{R}}{R} = \sigma_{\alpha\beta} n^\alpha n^\beta + \frac{1}{3}\theta$$

(where $\boldsymbol{\sigma}$ is the shear tensor and θ is the scalar expansion) and that averaging over all directions n^α gives

$$\left\langle \frac{\dot{R}}{R} \right\rangle = \frac{1}{3}\theta \ .$$

Problem 19.40. For the congruence of galaxy world lines in a Robertson-Walker cosmology find θ, $\omega_{\alpha\beta}$ and $\sigma_{\alpha\beta}$. (See Problem 5.18 for definitions.) Do the same for the anisotropic metric

$$ds^2 = -dt^2 + e^{2a}dx^2 + e^{2b}dy^2 + e^{2c}dz^2$$

where a, b, and c are functions of t and where x, y and z are coordinates comoving with the galaxies.

Problem 19.41. Consider the homogeneous, anisotropic cosmological model with metric

$$ds^2 = -dt^2 + g_{ij}(t)dx^i dx^j \ .$$

where the space slices $t = \text{constant}$ have a flat geometry. Find the evolution of the g_{ij} when the universe is "gravitation dominated," i.e. set $T^{\mu\nu}$ to zero in the field equations. Show that the volume of the universe goes to zero linearly with t as $t \to 0$ (in contrast to the $t^{3/2}$ or t^2 behavior for radiation- or matter-dominated Friedmann models).

CHAPTER 20
EXPERIMENTAL TESTS

This chapter explores concepts relevant to experimental tests of gravitation theory (light deflection, perihelion shift, etc.). A number of problems elsewhere in this book are similarly relevant: 11.9, 11.11, 12.2-12.4, 12.6, 12.7, 13.2-13.4, 14.12, 15.6, 15.7.

Problem 20.1. Gravitational bending of light: An evacuated tube of length ℓ is set up horizontally in a uniform gravitational field – e.g., in the field of the earth at sea level with $\ell \ll r_{earth}$ so that gravitational inhomogeneities (tidal forces) are negligible. A laser beam passing through the tube is deflected from the horizontal by the uniform gravitational field. Calculate the angle of light deflection measured relative to the axis of the tube. Express the answer in terms of the length of the tube, ℓ, and the acceleration of gravity, g. Discuss the feasibility of performing such an experiment in an earth-based laboratory.

Problem 20.2. Calculate the gravitational light deflection of a ray passing near the sun using Newtonian gravity and the fact that light moves along straight lines in the local frame of a freely falling observer. Since the light is always in a weak gravitational field, the Newtonian approximation seems justifiable. Why doesn't this answer agree with the general relativistic answer?

Problem 20.3. Derive the general expression for the angular deflection of light by the sun's gravitational field, if the light comes from a star which is at an angle α from the sun as seen from earth. Take the earth-sun distance to be R. Do not assume that α is small, but show that in the limit of small α the answer reduces to the conventional result (Problem 15.6) $\delta\alpha = 4M/b$.

Problem 20.4. Show that the sun's angular momentum $\underset{\sim}{J}$ modifies the light deflection formula (Problem 20.3) from $\delta\phi = 4M/b$ to

$$\delta\phi = \frac{4M}{b}\left(1 - \frac{\underset{\sim}{J}\cdot\underset{\sim}{n}}{Mb}\right)$$

where $\underset{\sim}{n} \equiv$ unit vector in direction of angular momentum of the photons about the center of sun.

Problem 20.5. In addition to the general relativistic deflection of electromagnetic waves by the sun, there is a frequency-dependent deflection caused by the solar corona, which must be taken into account in the interpretation of measurements. Estimate the impact parameter at which the general relativistic and coronal deflections of an electromagnetic wave of frequency ν make approximately equal contributions. Take an approximate coronal electron density of

$$\log_{10}\left(\frac{n_e}{1\text{cm}^{-3}}\right) = 8.4 - 6.5 \log_{10}\left(\frac{r}{R_\odot}\right)$$

for $r \lesssim 4R_\odot$. Evaluate your answer, in solar radii, for $\nu = 1000$ MHz.

Problem 20.6. The deflection angle of light passing near the sun is given by $\delta = 1.75''/b$, where b is the impact parameter in solar radii. Design a thin lens (i.e. give thickness as a function of radius) which models this focal behavior. Take the solar disc to be a black mask of 8mm diameter in the center of the lens, so that you can simulate the light deflection experiment by holding the lens at arm's length. Assume an index of refraction appropriate to ordinary crown glass, $n = 1.52$.

Problem 20.7. Calculate the expected perihelion shift of the planet Mercury in terms of the semimajor axis of its orbit a, the eccentricity e, and the mass M of the sun.

Problem 20.8. Newtonian gravitation theory can be modified and made covariant if the force equation on a point particle is written as

$$dp^{\mu} = -\eta^{\mu\nu}\Phi_{,\nu}\,p_{\beta}\,dx^{\beta} + p^{\alpha}\Phi_{,\alpha}\,dx^{\mu}$$

where Φ is a scalar potential which is related to stress energy by

$$\Phi^{;\mu}{}_{;\mu} = 4\pi\,T^{\mu}{}_{\mu}\ .$$

Investigate whether this theory agrees with experiment and observation:

(a) Is this theory in agreement with the experiments of Eotvös and Dicke showing the equivalence of inertial and passive gravitational mass?

(b) Is this theory in agreement with the Pound-Rebka experiment on the gravitational redshift of a photon on the earth's surface?

(c) Does this theory predict the bending of starlight near the sun?

Problem 20.9. A physicist wishes to take advantage of the tremendous precision of current atomic clocks by using them to test both special and general relativity. He places various clocks at different locations on the earth (assumed to be rigidly rotating) and measures their ticking rates with respect to some standard clock. Both the doppler shift, due to the earth's rotation, and the redshift effect, due to the earth's gravitational field, make contributions to deviations in ticking rates. Calculate the measured ticking rate of a clock located at (r, θ) relative to a standard clock of your choice. Take into account the rotational deformation of the earth's surface, assuming the earth is a rigidly rotating perfect fluid.

CHAPTER 21
MISCELLANEOUS

Problems in this chapter deal mostly with variational techniques, thin shells of matter, and spinors.

—— o —— o —— o —— o —— o —— o —— o ——

Problem 21.1. Show that

(i) $\delta(-g)^{\frac{1}{2}} = \frac{1}{2}(-g)^{\frac{1}{2}} g^{\mu\nu} \delta g_{\mu\nu}$

(ii) $\delta g^{\mu\nu} = -g^{\rho\mu} g^{\sigma\nu} \delta g_{\rho\sigma}$.

Problem 21.2. Let $L = L(\Phi^A, g_{\mu\nu})$ be the Lagrangian density for some field or matter distribution. The field is described by the variables Φ^A, where A represents any tensor indices. The action is

$$S = \int L(-g)^{\frac{1}{2}} d^4x \ .$$

The functional derivative $\delta L/\delta\Phi^A$ is defined by making a variation $\Phi^A \to \Phi^A + \delta\Phi^A$ and taking the change in S to be

$$\delta S = \int \frac{\delta L}{\delta \Phi^A} \delta\Phi^A (-g)^{\frac{1}{2}} d^4x \ .$$

Show that $\delta L/\delta\Phi^A = 0$ is the usual Euler-Lagrange equation when L depends on Φ^A and its partial derivatives $\Phi^A{}_{,\alpha}$.

Problem 21.3. If L is a Lagrangian density as in Problem 21.2, the stress-energy tensor can be defined by a variation of $g_{\mu\nu}$ in S:

$$\delta S = \int \frac{\delta(L(-g)^{\frac{1}{2}})}{\delta g_{\mu\nu}} \delta g_{\mu\nu} d^4x \equiv \frac{1}{2} \int T^{\mu\nu} \delta g_{\mu\nu} (-g)^{\frac{1}{2}} d^4x \ .$$

Show that $T^{\mu\nu}{}_{;\nu} = 0$ follows from the equation of motion of the field and the fact that S is a scalar.

Problem 21.4. Consider the action

$$S = (16\pi)^{-1} \int (-g)^{\frac{1}{2}} R d^4x + \int L_{\text{matter}} (-g)^{\frac{1}{2}} d^4x \ ,$$

where R is the Ricci scalar and L_{matter} contains g's but no Γ's (so that Γ's are present only in R).

(a) Treat the g's and the Γ's as independent field variables ("Palatini method"), and show that $\delta S = 0$ leads to the Einstein field equations and the usual formula for the Γ's in terms of the g's. (Assume $\Gamma^{\alpha}{}_{\beta\nu} = \Gamma^{\alpha}{}_{\nu\beta}$.)

(b) Now assume Γ's are Christoffel symbols used to define covariant derivatives in the usual way. Show that $\delta S = 0$ (where now $\delta\Gamma^{\alpha}{}_{\beta\nu}$ is not independent of $\delta g^{\alpha\beta}$) leads to the Einstein field equations.

Problem 21.5. The Lagrangian density for a scalar field is $L = -(8\pi)^{-1}(\Phi_{;\alpha}\Phi^{;\alpha} + m^2\Phi^2)$. Find the equations of motion and the stress-energy tensor. Verify explicitly that the stress-energy tensor has vanishing divergence.

Problem 21.6. The electromagnetic Lagrangian density is $L = -(16\pi)^{-1} F^{\mu\nu} F_{\mu\nu}$ where $F_{\mu\nu} = A_{\nu;\mu} - A_{\mu;\nu}$. Show that the Maxwell equations $F^{\alpha\beta}{}_{;\beta} = 0$ follow from setting to zero the variation of $\int L(-g)^{\frac{1}{2}} d^4x$ with respect to A_μ. Find the stress-energy from the prescription

$$T_{\mu\nu} = -2 \frac{\delta L}{\delta g^{\mu\nu}} + g_{\mu\nu} L \ .$$

Show that an equivalent Lagrangian density is

$$L = -\frac{1}{16\pi} F_{\mu\nu} F^{\mu\nu} - \frac{1}{4\pi} F^{\mu\nu} A_{\mu;\nu}$$

where $F^{\mu\nu}$ is antisymmetric and $F^{\mu\nu}$ and A_μ must be varied independently.

Problem 21.7. A Lagrangian for the Brans-Dicke theory is

$$L = (\Phi R - \omega\Phi_{,\alpha}\Phi^{,\alpha}\Phi^{-1} + 16\pi\, L_{\text{matter}})$$

where Φ = scalar field, R = curvature scalar, ω = coupling constant. Derive the field equations from $\delta\int L(-g)^{\frac{1}{2}}d^4x = 0$ by varying $g_{\alpha\beta}$ and Φ.

Problem 21.8. A surface layer is a timelike 3-surface separating two regions of spacetime. In general relativity the intrinsic geometry of such a 3-surface is well defined, but the extrinsic curvature may be discontinuous. That is, we may get a different extrinsic curvature tensor $\mathbf{K}$ if we evaluate it with respect to the 4-geometry on one side or the other. The surface stress-energy $S^\alpha{}_\beta$ contained in such a surface layer is defined as

$$S^\alpha{}_\beta = \lim_{\varepsilon\to 0}\int_{-\varepsilon}^{+\varepsilon} T^\alpha{}_\beta\, dn$$

where n is proper distance perpendicular to the 3-surface. Use the initial value equations to find the discontinuity in $\mathbf{K}$ in terms of $S^\alpha{}_\beta$.

Problem 21.9. For a surface layer described with Gaussian normal coordinates n and x^i $(i = 1, 2, 3)$ (see solution to Problem 21.8) derive the equation of motion of the surface layer

$$S^i{}_{j|i} + [T^n{}_j] = 0$$

where the square brackets denote a discontinuity across the surface and the slash denotes covariant differentiation with respect to the intrinsic geometry of the 3-surface.

Problem 21.10. A thin shell of dust in vacuum has surface density of mass σ as measured by an observer comoving with the dust. If $\mathbf{u}$ is the 4-velocity of the dust, show that

$$[K_{ij}] = 8\pi\sigma\left(u_iu_j + \frac{1}{2}{}^{(3)}g_{ij}\right)$$

$$\frac{d\sigma}{d\tau} = -\sigma u^i{}_{|i}$$

$$\mathbf{a}^+ - \mathbf{a}^- = 4\pi\sigma\,\mathbf{n}$$

$$\mathbf{a}^+ + \mathbf{a}^- = 0\ .$$

Here a^+ and a^- are the 4-accelerations measured on the outside and inside of the shell, respectively.

Problem 21.11. The vacuum geometries exterior to and interior to a collapsing spherical shell of dust are the Schwarzschild geometry

$$ds^2 = -\left(1-\frac{2M}{r}\right)dt^2 + \left(1-\frac{2M}{r}\right)^{-1}dr^2 + r^2 d\Omega^2$$

for the exterior, and the flat geometry

$$ds^2 = -dT^2 + dr^2 + r^2 d\Omega^2$$

for the interior. [The radial coordinates in these metrics both, clearly, have the property that $4\pi r^2$ is the proper surface area of the spherical surfaces $r = \text{constant}$, and t or $T = \text{constant}$.]

Show that for the collapsing spherical shell of dust the "rest mass of the shell" $\mu \equiv 4\pi R^2(\tau)\sigma$ is constant. Here σ is the surface mass density of the shell and the area of the shell as a function of proper shell time is $4\pi R^2(\tau)$. Derive the equation of motion of the shell

$$M = \mu\left[1+\left(\frac{dR}{d\tau}\right)^2\right]^{\frac{1}{2}} - \frac{\mu^2}{2R}$$

and integrate the equation to find (in implicit form) $R(\tau)$ in the case $dR/d\tau = 0$ at $R = \infty$.

Problem 21.12. Find an instantaneous spatial metric which represents N point masses at arbitrary positions at an instant of time symmetry.

Problem 21.13. Suppose one identifies four-vectors U^α with 2-index spinors $U^{AA'}$ by

$$(U^0, U^1, U^2, U^3) \to 2^{-\frac{1}{2}}\begin{bmatrix} U^0 + U^1 & U^2 + iU^3 \\ U^2 - iU^3 & U^0 - U^1 \end{bmatrix}.$$

What is the analog of the Minkowski metric in spinor language? i.e. Find an $L_{AA'BB'}$ such that $U\cdot V = \eta_{\alpha\beta}U^\alpha V^\beta = L_{AA'BB'}U^{AA'}V^{BB'}$. (Hint: use the spinor $(\varepsilon_{AB}) = (\varepsilon^{AB}) = \begin{pmatrix} 0 & 1 \\ -1 & 0 \end{pmatrix}$.) What is the analog of a Lorentz transformation?

[Note: The spinor notation used here and in the following problems is that of e.g. F. A. E. Pirani in A. Trautman, F. A. E. Pirani, and H. Bondi, *Lectures on General Relativity*, Brandeis 1964 Summer Institute on Theoretical Physics (Prentice-Hall, 1965).]

Problem 21.14. Show that

(a) $\varepsilon_{A[B}\varepsilon_{CD]} = 0$

(b) $\xi_{AB} = \xi_{(AB)} + \frac{1}{2}\varepsilon_{AB}\,\xi_C{}^C$

where ξ_{AB} is an arbitrary 2-spinor. [Note: This problem and the two following were suggested by T. Sejnowski.]

Problem 21.15. Let $T_{ab} = T_{AA'BB'}$. Show that if T_{ab} is antisymmetric then its dual in a spinor representation is $*T_{ab} = \frac{1}{2}\, i\,(T_{ABB'A'} - T_{BAA'B'})$.

Problem 21.16. Let $T_{ab} = T_{AA'BB'}$ in a spinor representation. What tensor corresponds to $T_{BA'AB'}$?

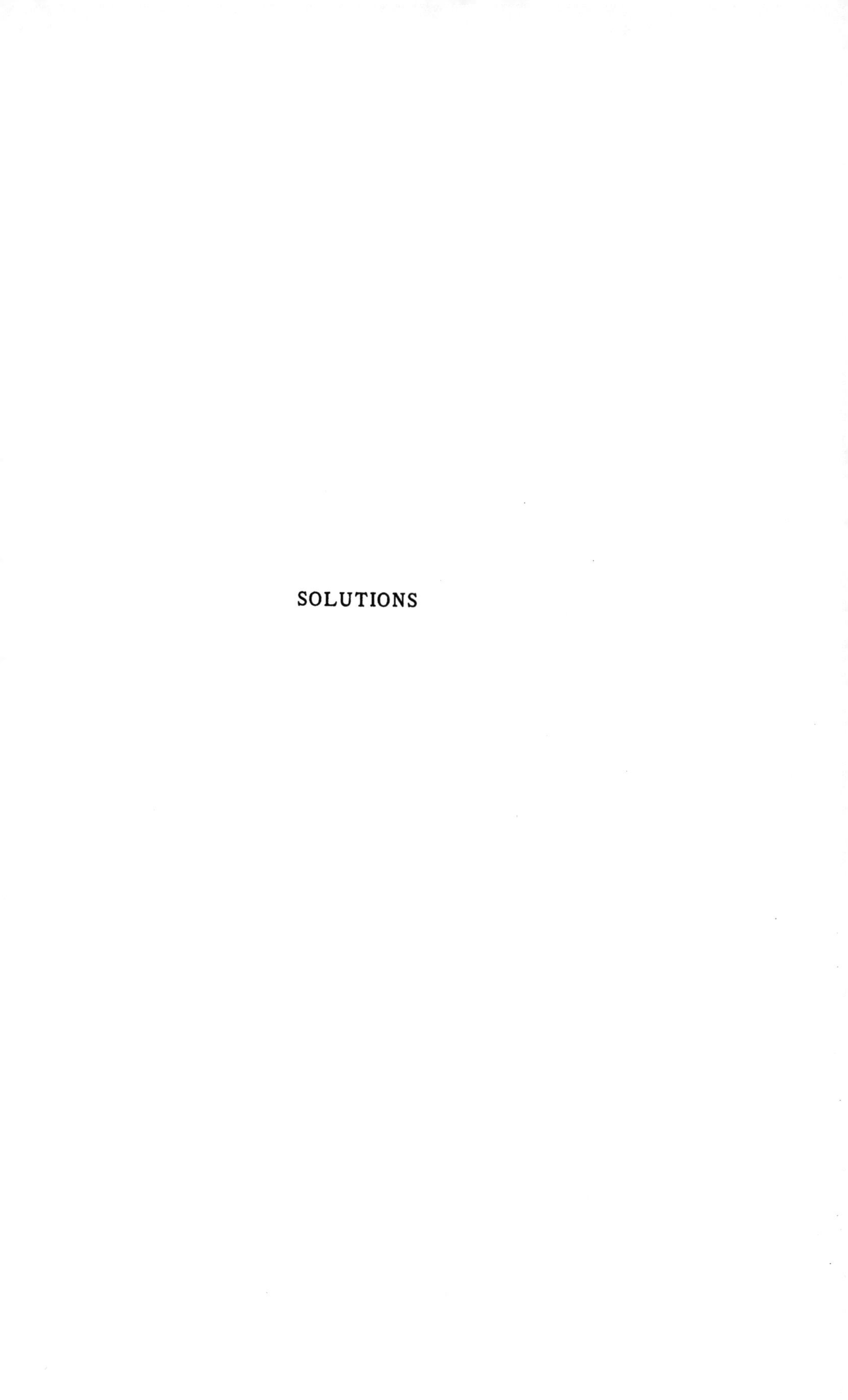

SOLUTIONS

CHAPTER 1: SOLUTIONS

Solution 1.1. $\mathbf{u} = (\gamma, \gamma \underset{\sim}{v})$, where $\gamma \equiv (1-\underset{\sim}{v}^2)^{-\frac{1}{2}} = u^0 = dt/d\tau$. Thus:

(a) $u^0 = (1-\underset{\sim}{v}^2)^{-\frac{1}{2}}$

(b) $u^j = (1-\underset{\sim}{v}^2)^{-\frac{1}{2}} v^j$

(c) $\mathbf{u}\cdot\mathbf{u} = -1$ since it is a 4-velocity, so

$$u^0 = [1+(u^1)^2+(u^2)^2+(u^3)^2]^{\frac{1}{2}} \equiv (1+u^j u_j)^{\frac{1}{2}}$$

(d) $d/d\tau = (dt/d\tau)\, d/dt = (1-\underset{\sim}{v}^2)^{-\frac{1}{2}}\, d/dt$

(e) $v^j = u^j/u^0 = u^j (1+u^i u_i)^{-\frac{1}{2}}$

(f) $|\underset{\sim}{v}| = [1-(u^0)^{-2}]^{\frac{1}{2}}$, from (a).

Solution 1.2. The product of the transformations is

$$\begin{bmatrix} \gamma_x & \gamma_x v_x & 0 & 0 \\ \gamma_x v_x & \gamma_x & 0 & 0 \\ 0 & 0 & 1 & 0 \\ 0 & 0 & 0 & 1 \end{bmatrix} \begin{bmatrix} \gamma_y & 0 & \gamma_y v_y & 0 \\ 0 & 1 & 0 & 0 \\ \gamma_y v_y & 0 & \gamma_y & 0 \\ 0 & 0 & 0 & 1 \end{bmatrix} = \begin{bmatrix} \gamma_x\gamma_y & \gamma_x v_x & \gamma_x\gamma_y v_y & 0 \\ \gamma_x\gamma_y v_x & \gamma_x & \gamma_x\gamma_y v_x v_y & 0 \\ \gamma_y v_y & 0 & \gamma_y & 0 \\ 0 & 0 & 0 & 1 \end{bmatrix},$$

but

$$\begin{bmatrix} \gamma_y & 0 & \gamma_y v_y & 0 \\ 0 & 1 & 0 & 0 \\ \gamma_y v_y & 0 & \gamma_y & 0 \\ 0 & 0 & 0 & 1 \end{bmatrix} \begin{bmatrix} \gamma_x & \gamma_x v_x & 0 & 0 \\ \gamma_x v_x & \gamma_x & 0 & 0 \\ 0 & 0 & 1 & 0 \\ 0 & 0 & 0 & 1 \end{bmatrix} = \begin{bmatrix} \gamma_x\gamma_y & \gamma_y\gamma_x v_x & \gamma_y v_y & 0 \\ \gamma_x v_x & \gamma_x & 0 & 0 \\ \gamma_x\gamma_y v_y & \gamma_x\gamma_y v_x v_y & \gamma_y & 0 \\ 0 & 0 & 0 & 1 \end{bmatrix}$$

which is different.

Solution 1.3. Let $\mathbf{u}_1$ and $\mathbf{u}_2$ be the 4-velocities of the two frames. In frame 1, $\mathbf{u}_1 = (1, \underset{\sim}{0})$, $\mathbf{u}_2 = (\gamma, \gamma \underset{\sim}{v})$ where $\gamma = (1-v^2)^{-\frac{1}{2}}$. Since $\gamma = -\mathbf{u}_1\cdot\mathbf{u}_2$, we just have to evaluate this scalar in a general frame where $\mathbf{u}_1 = (\gamma_1, \gamma_1 \underset{\sim}{v}_1)$, $\mathbf{u}_2 = (\gamma_2, \gamma_2 \underset{\sim}{v}_2)$:

$$\gamma = -\mathbf{u}_1 \cdot \mathbf{u}_2 = \gamma_1\gamma_2 - \gamma_1\gamma_2\, \underset{\sim}{v}_1 \cdot \underset{\sim}{v}_2 = (1-v^2)^{-\frac{1}{2}}$$

$$1 - v^2 = (\gamma_1\gamma_2)^{-2}/(1-\underset{\sim}{v}_1 \cdot \underset{\sim}{v}_2)^2$$

$$v^2 = \frac{(1-\underset{\sim}{v}_1 \cdot \underset{\sim}{v}_2)^2 - (1-v_1{}^2)(1-v_2{}^2)}{(1-\underset{\sim}{v}_1 \cdot \underset{\sim}{v}_2)^2} = \frac{(\underset{\sim}{v}_1 - \underset{\sim}{v}_2)^2 - (\underset{\sim}{v}_1 \times \underset{\sim}{v}_2)^2}{(1-\underset{\sim}{v}_1 \cdot \underset{\sim}{v}_2)^2} .$$

(The solution could also be verified by a painfully tedious application of Lorentz transformations.)

Solution 1.4. We use the fact that in one dimension the "rapidity parameter" $\theta \equiv \tanh^{-1} v$ adds linearly. Thus $\tanh^{-1} v_n = \theta_n = n\theta = n \tanh^{-1} \beta$, or

$$\begin{aligned} v_n &= \tanh\,(n \tanh^{-1} \beta) \\ &= \tanh\ \log \left(\frac{1+\beta}{1-\beta}\right)^{n/2} \qquad \text{(easily verified!)} \\ &= \frac{1 - [(1-\beta)/(1+\beta)]^n}{1 + [(1-\beta)/(1+\beta)]^n} . \end{aligned}$$

As $n \to \infty$, $[(1-\beta)/(1+\beta)]^n \to 0$ and $v_n \to 1$, the speed of light.

Solution 1.5. The usual b/γ answer refers to a measurement made simultaneously in the lab frame, but the photons are not *emitted* simultaneously in the lab frame. They are *received* simultaneously so that photon 2 (see diagram) has to travel for an extra time $b' \cos \alpha$, where b' is the apparent length. From the Lorentz transformation

$$b = \Delta x = \gamma(\Delta x' - \beta \Delta t') = \gamma(b' - \beta b' \cos \alpha) ,$$

so that

$$b' = b/\gamma(1-\beta \cos \alpha) .$$

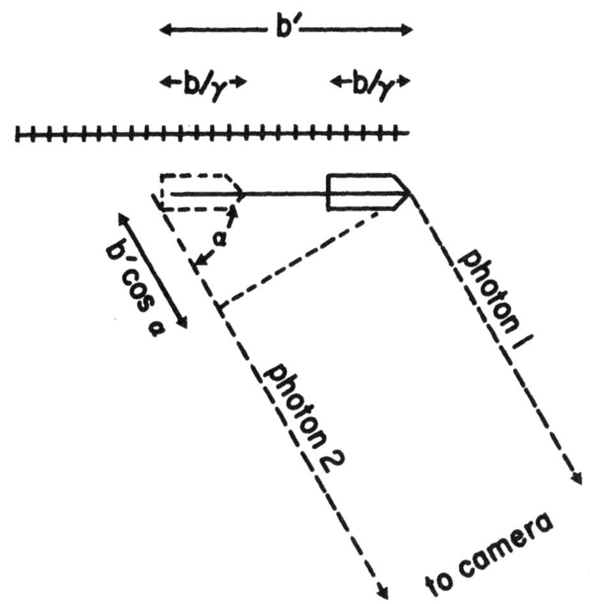

Solution 1.6. When the two observers are at rest in the same Lorentz frame, we have (distance) = (velocity) × (time) and $t_{\text{round trip}} = 2L/u$. Things are more complicated when one is moving. The moving observer's coordinates are

$$t' = \gamma(t - vx)$$
$$x' = \gamma(x - vt) \ ,$$

and the tachyon emitted from the moving frame back towards the stationary observer has velocity

$$dx'/dt' = -u \ .$$

Now $dx' = \gamma(dx - vdt)$, $dt' = \gamma(dt - vdx)$, so substituting and solving for the apparent velocity in the stationary frame gives

$$-\,dx/dt = (u - v)/(1 - uv) \ .$$

Finally we add up times in the stationary frame $t_{\text{total}} = t_{\text{out}} + t_{\text{back}} = L/u + L(1 - uv)/(u - v) = L[1/u + (1 - uv)/(u - v)]$. One can easily calculate that

$$t_{\text{total}} < 0 \quad \text{if} \quad u > [1 + (1 - v^2)^{\frac{1}{2}}]/v \ .$$

A spacetime diagram is helpful in understanding this effect.

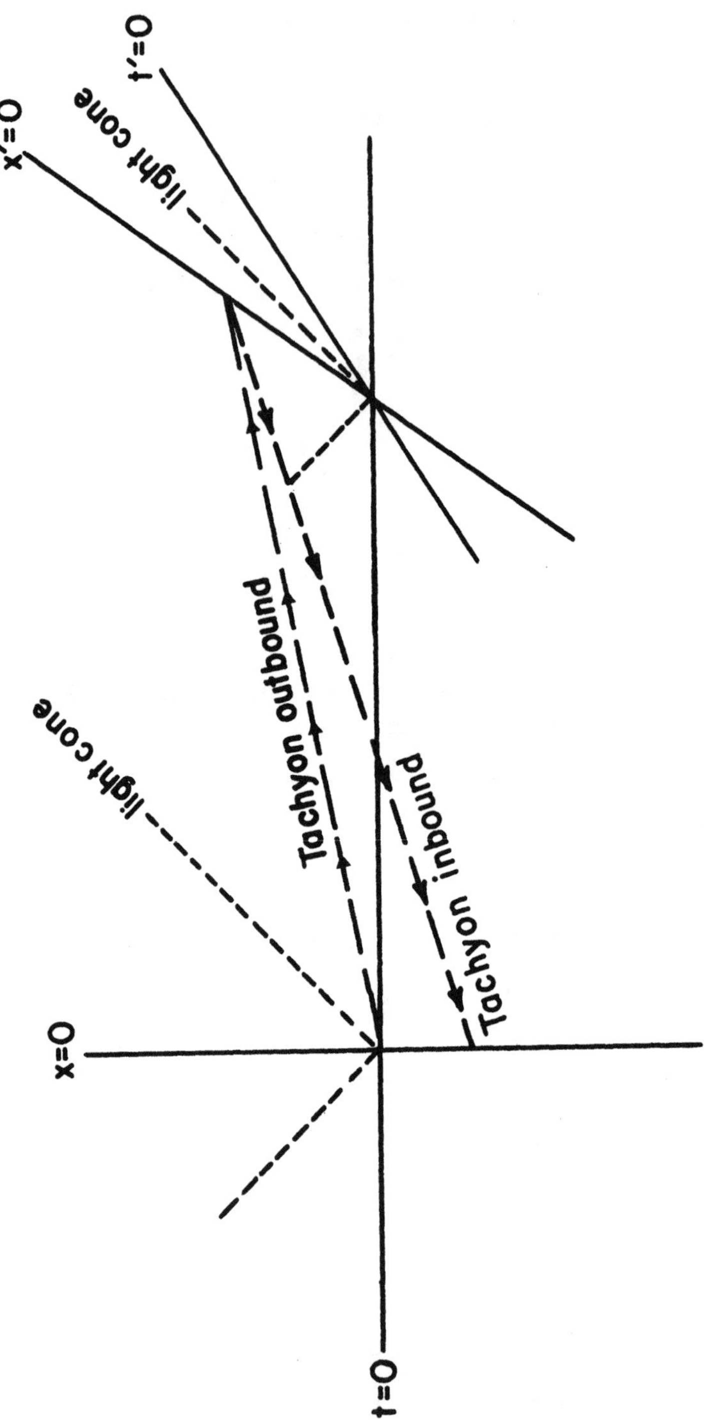
t'=0
x'=0
light cone
light cone
Tachyon outbound
Tachyon inbound
x=0
t=0

Solution 1.7. The rotation comes from the contraction in the x direction but not in the y direction. That is, $\cot\theta = \Delta x/\Delta y$. Because of the motion, $\Delta x = \Delta x'(1-\beta^2)^{\frac{1}{2}}$ while $\Delta y = \Delta y'$. Thus,

$$\cot\theta = (1-\beta^2)^{\frac{1}{2}} \cot\theta' .$$

Solution 1.8. Not only are lengths contracted, but also times are dilated. In S the bullet's velocity is

$$v = (\Delta x/\Delta t, \Delta y/\Delta t) .$$

From the transformation formulas for Δx and Δt

$$v_x = \Delta x/\Delta t = (v_{x'}+\beta)/(1+\beta v_{x'}) .$$

Also, since $\Delta y = \Delta y'$, we have

$$v_y = \Delta y/\Delta t = \Delta y'(1-\beta^2)^{\frac{1}{2}}/(\Delta t'+\beta\Delta x') = v_{y'}(1-\beta^2)^{\frac{1}{2}}/(1+\beta v_{x'}) .$$

The direction of motion then is given by

$$\tan\theta = v_y/v_x = v_{y'}(1-\beta^2)^{\frac{1}{2}}/(v_{x'}+\beta) = \tan\theta'(1-\beta^2)^{\frac{1}{2}}/(1+\beta/v_{x'}) .$$

Except for the factor of $(1-\beta^2)^{\frac{1}{2}}$, this change in direction is identical to the Galilean result. It represents a "funneling" of the motion toward the x axis. For a photon, $v_x^2 + v_y^2 = 1$, $v_x = \cos\theta$, so that

$$\cos\theta = \frac{\cos\theta'+\beta}{1+\beta\cos\theta'} , \qquad \tan\theta = \frac{\tan\theta'(1-\beta^2)^{\frac{1}{2}}}{1+\beta\sec\theta'} .$$

Solution 1.9. If $\bar{\theta}$ is the direction of the motion of a photon from star to observer then the star is observed at angle $\theta = \pi - \bar{\theta}$. The transformation of photon directions was derived in the previous problem. From that result we have here

$$\cos\theta = \frac{\cos\theta'-\beta}{1-\beta\cos\theta'} .$$

The number of stars observed in solid angle $d\Omega$ is then

$$\begin{aligned} dN &= N d\Omega/4\pi = (N/4\pi)\, 2\pi\, d(\cos\theta) \\ &= \frac{1}{2} N \frac{d(\cos\theta)}{d(\cos\theta')} d(\cos\theta') \\ &= \frac{1}{2} N \frac{(1-\beta^2)}{(1-\beta\cos\theta')^2} d(\cos\theta') \,. \end{aligned}$$

Since $dN = 2\pi\, P(\theta',\phi')\, d(\cos\theta')$ the distribution in S' is

$$P(\theta',\phi') = \frac{N}{4\pi} \frac{(1-\beta^2)}{(1-\beta\cos\theta')^2} \,.$$

Checks: $\beta \to 0$ $P(\theta',\phi') \to N/4\pi = P(\theta,\phi)$

$$\int_{\text{sphere}} P(\theta',\phi')\, d\Omega = \frac{N(1-\beta^2)}{4\pi} \int_{-1}^{+1} \frac{2\pi\, d\cos\theta'}{(1-\beta\cos\theta')^2} = N \,.$$

You can easily verify that half the stars are contained between $\theta' = 0$ and $\theta'_{\frac{1}{2}} = \cos^{-1}\beta < \pi/2$. Thus the stars "bunch up" in the forward direction. For $\beta \approx 1$, $\theta'_{\frac{1}{2}} \approx [2(1-\beta)]^{\frac{1}{2}}$ and the bunching is pronounced.

Solution 1.10. The length of A is $|A\cdot A|^{\frac{1}{2}}$ and

$$\begin{aligned} A\cdot A &= (3^{\frac{1}{2}} e_t + 2^{\frac{1}{2}} e_x) \cdot (3^{\frac{1}{2}} e_t + 2^{\frac{1}{2}} e_x) \\ &= 3 e_t\cdot e_t + 2 e_x\cdot e_x + 2\sqrt{6}\, e_t\cdot e_x \\ &= -3 + 2 + 0 = -1 \,. \end{aligned}$$

If θ is the angle between A and e_t then

$$\cos\theta = \frac{A\cdot e_t}{|A\cdot A|^{\frac{1}{2}}\, |e_t\cdot e_t|^{\frac{1}{2}}} = -3^{\frac{1}{2}}$$

which is satisfied by no real θ.

Solution 1.11. By symmetry, the clocks must read the same the next time Adam and Eve meet. An easy way to see this is by considering proper time intervals in a coordinate system fixed to an inertial observer at rest.

In polar coordinates

$$-d\tau^2 = -dt^2 + dr^2 + r^2 d\phi^2 + dz^2 \ . \tag{1}$$

The coordinates of an instantaneously comoving inertial frame attached to Adam are $\phi_A = \omega t$, with r and z constant, and the coordinates of Eve are $\phi_E = -\omega t$, with r and z constant. Thus

$$d\tau_A^2 = d\tau_E^2 = dt^2(1 - r^2\omega^2) \tag{2}$$

and proper time intervals for Adam and Eve are identical.

A second method is to consider a *noninertial* coordinate system attached to Adam. He may define surfaces of simultaneity by extending hypersurfaces orthogonal to his world line, the distance between hypersurfaces being at equal intervals of his proper time. At points where his hypersurfaces intersect Eve's world line, her proper time τ_E is read off and Adam can compute τ_A as a function of τ_E. One proceeds by determining the 4-vector w which connects Adam's and Eve's world lines and which is orthogonal to Adam's 4-velocity:

$$\text{Adam's world line:} \quad \begin{aligned} t &= \gamma\tau_A \\ x &= \sin \omega t &&= \sin \omega \gamma \tau_A \\ y &= \cos \omega t &&= \cos \omega \gamma \tau_A \\ z &= 0 \end{aligned} \tag{3a}$$

$$\text{Eve's world line:} \quad \begin{aligned} t &= \gamma\tau_E \\ x &= -\sin \omega t &&= -\sin \omega \gamma \tau_E \\ y &= \cos \omega t &&= \cos \omega \gamma \tau_E \\ z &= 0 \end{aligned} \tag{3b}$$

where we have taken the ring to have unit radius. From these relations

$$w \equiv x_A - x_E = [\gamma(\tau_A - \tau_E), \sin\omega\gamma\tau_A + \sin\omega\gamma\tau_E, \cos\omega\gamma\tau_A - \cos\omega\gamma\tau_E, 0]$$

$$\mathbf{u}_A = (\gamma, \omega\gamma \cos\omega\gamma\tau_A, -\omega\gamma \sin\omega\gamma\tau_A, 0)$$

so that requiring $w \cdot \mathbf{u}_A = 0$ gives

$$\tau_A - \tau_E = \omega\gamma^{-1} \sin[\omega\gamma(\tau_A + \tau_E)] \ . \tag{4}$$

Equation (4) is transcendental for $\tau_A(\tau_E)$. One may easily see, however, that $\tau_A = \tau_E$ when

$$\sin 2\omega\gamma\tau_A = \sin 2\omega t = 0 \ , \qquad (5)$$

i.e. whenever Adam's and Eve's world lines cross.

Solution 1.12. The Lorentz transformation is

$$t' = \gamma(t - vx)$$
$$x' = \gamma(x - vt) \ ;$$

if $w \equiv it$, $w' \equiv it'$, this becomes

$$w' = \gamma(w - ivx)$$
$$x' = \gamma(x + ivw) \ .$$

To make this look like a rotation

$$w' = \cos\theta w - \sin\theta x$$
$$x' = \cos\theta x + \sin\theta w \ ,$$

we obviously want

$$\sin\theta = iv\gamma$$
$$\cos\theta = \gamma \ .$$

$$\sin^2\theta + \cos^2\theta = \gamma^2(1 - v^2) = 1 \ .$$

So $\theta = \arcsin(iv\gamma) = \arctan(iv) = +\, i \tanh^{-1} v$.

Solution 1.13. Along the curve we have

$$dx^2 + dy^2 + dz^2 - dt^2 = (r^2\cos^2\theta\cos^2\phi + r^2\cos^2\theta\sin^2\phi + r^2\sin^2\theta - r^2)\, d\lambda^2$$
$$= r^2 d\lambda^2(\cos^2\theta + \sin^2\theta - 1) = 0 \ ,$$

so the curve is null.

To be a geodesic (i.e. a straight line) we must have $dz/dt = $ constant, but $dz/dt = \sin\theta$, so θ must be a constant. From the expression for dy/dt or dx/dt we see that ϕ must be constant. These are the requirements for a geodesic; there is no restriction on the function $r(\lambda)$. If, however, λ is to be an affine parameter (e.g. proper time) we must have $dt/d\lambda$ a constant, and hence r a constant.

Solution 1.14. 4-velocities are normalized so that $u^\alpha u_\alpha = -1$, so

$$0 = \frac{d}{d\tau}(u^\alpha u_\alpha) = 2\frac{du^\alpha}{d\tau}u_\alpha = 2a^\alpha u_\alpha .$$

This is one constraint on the 4-components of acceleration a^α. In the momentarily comoving frame $u_\alpha = (-1, \underset{\sim}{0})$, and the constraint requires $a^0 = 0$, with $a^j (j = 1, 2, 3)$ arbitrary. A Newtonian accelerometer can be modeled as follows: Have the observer release a particle in the comoving frame and see how much velocity $d\underset{\sim}{v}$ the observer picks up relative to it in a short time $d\tau$, then calculate $\underset{\sim}{a}_{\text{Newtonian}} = d\underset{\sim}{v}/d\tau$. Of course the particle is really *stationary* in the momentarily comoving inertial frame, and we accelerate relative to it by an amount $du^j = a^j d\tau$. Now $u^j = v^j(1-v^2)^{-\frac{1}{2}}$, so $du^j = (1-v^2)^{-\frac{1}{2}}dv^j + v^j d(1-v^2)^{-\frac{1}{2}}$. But $v = 0$ since the frames are momentarily comoving, and thus $du^j = dv^j$, so finally

$$a^j_{\text{Newtonian}} = dv^j/d\tau = du^j/d\tau = a^j .$$

The 3 independent components of 4-acceleration in a comoving frame are just 3 Newtonian accelerations.

Solution 1.15. Since $\mathbf{a}\cdot\mathbf{u} = 0$, in the observer's local rest frame

$$\mathbf{a} = (0, a^{\hat{j}}) ,$$

where $a^{\hat{j}}$ is the jth component of locally measured acceleration. Then, the squared magnitude of his locally measured acceleration is

$$a^2 \equiv a^{\hat{j}} a_{\hat{j}} = (0, a^{\hat{j}})\cdot(0, a^{\hat{j}}) = \mathbf{a}\cdot\mathbf{a} .$$

Solution 1.16. The transformation of 3-velocity, from Problem 1.8, is

$$u'_\perp = \frac{\gamma^{-1}}{(1-vu_\parallel)}u_\perp , \quad u_\parallel = \frac{u_\parallel - v}{1-vu_\parallel}$$

and the time transformation is

$$t' = \gamma(t - \underset{\sim}{v}\cdot\underset{\sim}{x}) \ .$$

Now take differentials:

$$du'_\perp = \frac{\gamma^{-1}du_\perp}{(1-vu_\|)} + \frac{v\gamma^{-1}u_\perp}{(1-vu_\|)^2}\,du_\| = \frac{\gamma^{-1}[(1-vu_\|)du_\perp + vu_\perp\,du_\|]}{(1-vu_\|)^2}$$

$$du'_\| = \frac{du_\|}{1-vu_\|} - \frac{v(v-u_\|)du_\|}{(1-vu_\|)^2} = \frac{\gamma^{-2}du_\|}{(1-vu_\|)^2}$$

$$dt' = \gamma(dt - \underset{\sim}{v}\cdot d\underset{\sim}{x}) = \gamma dt\,(1-\underset{\sim}{v}\cdot\underset{\sim}{u}) = \gamma dt\,(1-vu_\|)$$

$$\frac{d\underset{\sim}{u}'_\|}{dt'} \equiv \underset{\sim}{a}'_\| = \frac{\gamma^{-3}}{(1-vu_\|)^3}\,\underset{\sim}{a}_\| = \frac{\gamma^{-3}}{(1-\underset{\sim}{v}\cdot\underset{\sim}{u})^3}\,\underset{\sim}{a}_\| \ . \qquad \text{(Answer)}$$

and

$$\frac{d\underset{\sim}{u}'_\perp}{dt'} \equiv \underset{\sim}{a}'_\perp = \frac{\gamma^{-2}}{(1-vu_\|)^3}\,[(1-vu_\|)\underset{\sim}{a}_\perp + v\underset{\sim}{u}_\perp\,a_\|]$$

$$= \frac{\gamma^{-2}}{(1-vu_\|)^3}\,[\underset{\sim}{a}_\perp - v(u_\|\underset{\sim}{a}_\perp - \underset{\sim}{u}_\perp\,a_\|)]$$

$$= \frac{\gamma^{-2}}{(1-\underset{\sim}{v}\cdot\underset{\sim}{u})^3}\,[\underset{\sim}{a}_\perp - \underset{\sim}{v}\times(\underset{\sim}{a}\times\underset{\sim}{u})] \ . \qquad \text{(Answer)}$$

[In the last line we have used the fact that

$$\underset{\sim}{a}\times\underset{\sim}{u} = (\underset{\sim}{a}_\perp + \underset{\sim}{a}_\|)\times(\underset{\sim}{u}_\perp + \underset{\sim}{u}_\|)$$
$$= \underset{\sim}{a}_\perp \times \underset{\sim}{u}_\| + \underset{\sim}{a}_\| \times \underset{\sim}{u}_\perp$$

and the rule for expanding vector triple products.]

Solution 1.17. If the acceleration is in the x direction we can clearly choose $y = \overline{y}$, $z = \overline{z}$. Note that the world lines

$$t = A\sinh g\overline{t} + B$$
$$x = A\cosh g\overline{t} + C$$

represent world lines with constant acceleration. [Easily verified. See also Problem 2.13.] If A, B, C are functions only of $\bar{x}$ then the above equations represent a 4-velocity

$$u = \cosh g\bar{t}\, e_t + \sinh g\bar{t}\, e_x ,$$

independent of $\bar{x}$, and they therefore give us a set of world lines which are parallel on a $\bar{t}$ = constant surface. If, furthermore B and C are constant, **u** is perpendicular to the hyperplane $d\bar{t} = 0$ (i.e. the hyperplanes $\bar{t}$ = constant are hyperplanes of simultaneity for the momentarily comoving inertial frame). It remains only to choose $A(\bar{x})$, B, and C correctly:

$$t = (g^{-1}+\bar{x})\sinh g\bar{t}$$

$$x = (g^{-1}+\bar{x})\cosh g\bar{t} - g^{-1} .$$

Notice that the element of proper time for a coordinate stationary observer is

$$d\tau = (dt^2 - dx^2)^{\frac{1}{2}} = (1+g\bar{x})d\bar{t} .$$

Since it does not depend only on t, clocks will not remain synchronized. i.e. They can only agree on one hyperplane.

Solution 1.18. Without loss of generality take the mirror to be in the xy plane, and take the photon to be in the yz plane. If "before" and "after" mean before and after reflection, we have for a photon in the lab frame:

$$p_{\text{before}} = (E,\ 0,\ E\sin\theta,\ E\cos\theta) ,$$

Lorentz transforming to mirror frame,

$$p_{\text{before}} = [\gamma E(1+v\cos\theta),\ 0,\ E\sin\theta,\ \gamma E(v+\cos\theta)] .$$

Reflecting in mirror:

$$p_{\text{after}} = [\gamma E(1+v\cos\theta),\ 0,\ E\sin\theta,\ -\gamma E(v+\cos\theta)] ,$$

Lorentz transforming back to lab frame:

$$p_{after} = \{\gamma^2 E[(1+v\cos\theta)+v(v+\cos\theta)],\ 0,$$
$$E\sin\theta, \gamma^2 E[-v(1+v\cos\theta)-(v+\cos\theta)]\} \ .$$

So that, finally:

$$\cos\theta|_{after} = \left|\frac{p^3}{p^0}\right| = +\frac{(1+v^2)\cos\theta+2v}{1+2v\cos\theta+v^2}$$

$$E_{after} = p^0_{after} = \left(\frac{1+2v\cos\theta+v^2}{1-v^2}\right)E_{before} \ .$$

Solution 1.19. Let E and $\underset{\sim}{P}$ denote the energy and momentum of the photon. In the frame of the mirror (primed frame) $E'_{in} = E'_{out}$, $P'_{x\ in} = P'_{x\ out}$, $P'_{y\ in} = -P'_{y\ out}$. In the lab frame, we wish to show $\theta_1 = \theta_2$.

$$\frac{\tan\theta_1}{\tan\theta_2} = \frac{(P_x/P_y)_{in}}{(P_x/-P_y)_{out}} = \frac{P_{x\ in}}{P_{x\ out}} \cdot \frac{P'_{y\ in}}{-P'_{y\ out}} = \frac{P_{x\ in}}{P_{x\ out}} \ ,$$

where we have used $P'_y = P_y$. Then, Lorentz transforming P_x,

$$\frac{\tan\theta_1}{\tan\theta_2} = \frac{\gamma(P'_x+\beta E')_{in}}{\gamma(P'_x+\beta E')_{out}} = 1 \ .$$

Hence the angle of incidence equals the angle of reflection in the lab frame as well as the mirror frame.

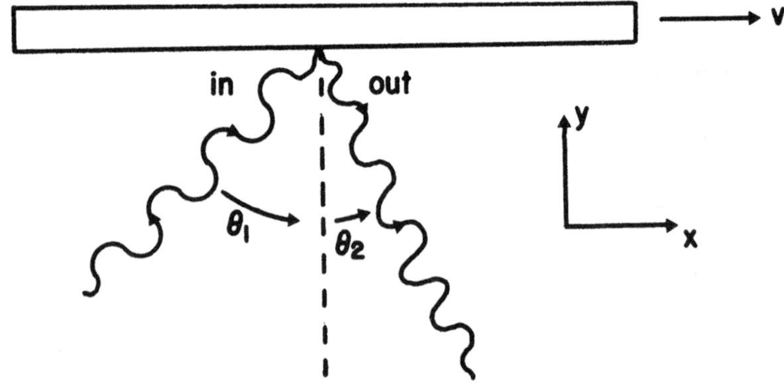

Solution 1.20. In the rest frame of the observer,

$$u^0 = 1, \quad u^j = 0$$
$$p^0 = -p_0 = E$$
$$p^j = p_j = \underset{\sim}{p}$$

so, we verify the invariant formulas in this easiest frame:

(a) $E = -p_0 u^0 = -\mathbf{p}\cdot\mathbf{u}$

(b) $m^2 = E^2 - |\underset{\sim}{p}|^2 = -p_0 p^0 - p_j p^j = -\mathbf{p}\cdot\mathbf{p}$

(c) $|\underset{\sim}{p}| = (E^2 - m^2)^{\frac{1}{2}} = [(\mathbf{p}\cdot\mathbf{u})^2 + \mathbf{p}\cdot\mathbf{p}]^{\frac{1}{2}}$

(d) $|\underset{\sim}{v}| = \dfrac{|\underset{\sim}{p}|}{E} = \left[\dfrac{E^2 - m^2}{E^2}\right]^{\frac{1}{2}} = \left[1 + \dfrac{\mathbf{p}\cdot\mathbf{p}}{(\mathbf{p}\cdot\mathbf{u})^2}\right]^{\frac{1}{2}}$

(e) In the rest frame,

$$v^0 = -u^0 - \frac{p^0}{(-E)} = -1 + 1 = 0$$

$$v^j = -u^j - \frac{p^j}{(-E)} = 0 + \frac{dx^j}{dt} .$$

Solution 1.21. There are three 4-velocities to consider. The laboratory observer has $\mathbf{u}_{lab} = (1, \underset{\sim}{0})$; the nucleus has $\mathbf{u}_{Fe} = (\gamma, \gamma\underset{\sim}{\beta})$, where $\gamma = (1 - \underset{\sim}{\beta}\cdot\underset{\sim}{\beta})^{-\frac{1}{2}}$; the photon has a 4-momentum $\mathbf{p}_\gamma = \text{constant} \times (1, -\underset{\sim}{n})$, since it is traveling toward the observer. Also notice that the photon is null, $\mathbf{p}_\gamma\cdot\mathbf{p}_\gamma \propto -1 + (-\underset{\sim}{n})\cdot(-\underset{\sim}{n}) = 0$. From Problem 1.20, now

$$\frac{\nu_{observed}}{\nu_{emitted}} = \frac{E_{lab}}{E_{Fe}} = \frac{-\mathbf{u}_{lab}\cdot\mathbf{p}_\gamma}{-\mathbf{u}_{Fe}\cdot\mathbf{p}_\gamma} = \frac{1}{\gamma(1+\underset{\sim}{\beta}\cdot\underset{\sim}{n})} . \qquad \text{(Answer).}$$

Solution 1.22. The doppler-shift formula (Problem 1.21) gives

$$1 = \frac{\nu_{obs}}{\nu} = \frac{(1-v^2)^{\frac{1}{2}}}{1+\underset{\sim}{v}\cdot\underset{\sim}{n}}$$

$$\cos\theta = \frac{\underset{\sim}{v}\cdot\underset{\sim}{n}}{v} = \frac{(1-v^2)^{\frac{1}{2}} - 1}{v}$$

$$\theta = \cos^{-1}\left[\frac{(1-v^2)^{\frac{1}{2}} - 1}{v}\right].$$

Solution 1.23. Make an arbitrary boost in a direction which lies in the y-z plane. The photon now may have nonvanishing p^y and p^z as well as p^x. A pure rotation lines up the coordinate frame again so that only p^x is nonvanishing, but p^x does not now have its original magnitude. So, make a final boost along p^x either to redshift it or to blueshift it to the original value. Since $E^2 - p^2 = 0$, E also has its original value. You can easily convince yourself that the product of these transformations is not a pure rotation; there is in general a net boost left over. An example is:

$$\underbrace{\begin{bmatrix} \gamma' & \gamma'v' & 0 & 0 \\ \gamma'v' & \gamma' & 0 & 0 \\ 0 & 0 & 1 & 0 \\ 0 & 0 & 0 & 1 \end{bmatrix}}_{\text{x-boost}} \underbrace{\begin{bmatrix} 1 & 0 & 0 & 0 \\ 0 & (1-v^2)^{\frac{1}{2}} & v & 0 \\ 0 & -v & (1-v^2)^{\frac{1}{2}} & 0 \\ 0 & 0 & 0 & 1 \end{bmatrix}}_{\text{x-y rotation}} \underbrace{\begin{bmatrix} \gamma & 0 & \gamma v & 0 \\ 0 & 1 & 0 & 0 \\ \gamma v & 0 & \gamma & 0 \\ 0 & 0 & 0 & 1 \end{bmatrix}}_{\text{y-boost}} \begin{bmatrix} E \\ E \\ 0 \\ 0 \end{bmatrix} = \begin{bmatrix} E \\ E \\ 0 \\ 0 \end{bmatrix},$$

where v' is chosen to satisfy the equality

$$\gamma'(1+v') = 1/\gamma \ ,$$

which gives

$$v' = -v^2/(2-v^2) \ .$$

Solution 1.24. It is not possible for one inertial frame to jerk back and forth with respect to another. The "paradox" comes from the tacit and erroneous assumption that the cylinder remains rigid when the frogs hit the ends. This cannot be true since the elastic waves that inform one end of the tube that the other has been hit must propagate along the walls of the tube at less than the speed of light. In the freely-falling frame of the cylinder, the two ends are driven outward by the impact of the frogs. Tension waves fly from each end of the cylinder toward the other. Not until the waves meet at the center of the tube do they discover the existence of each other. Then they pass each other and counteract each other's effects, pulling the tube back to its original shape. The tube then pulsates back and forth in its fundamental mode of vibration. During

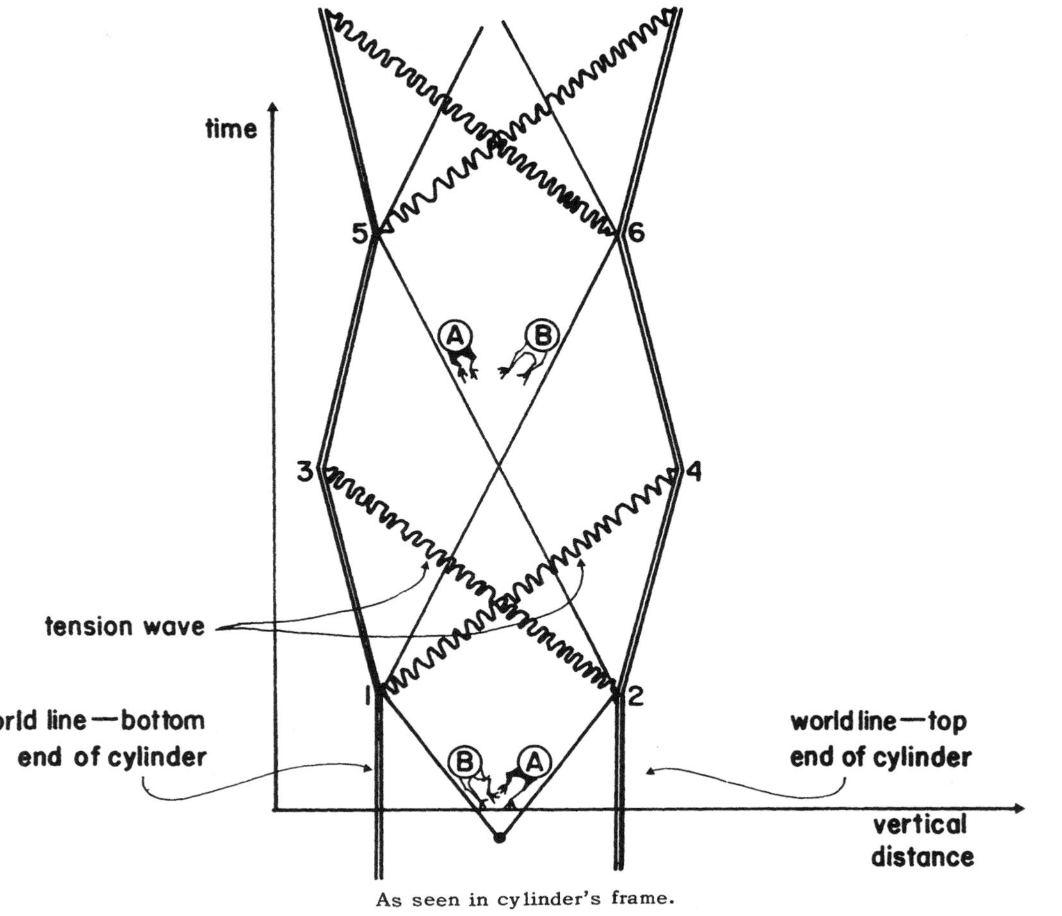

As seen in cylinder's frame.

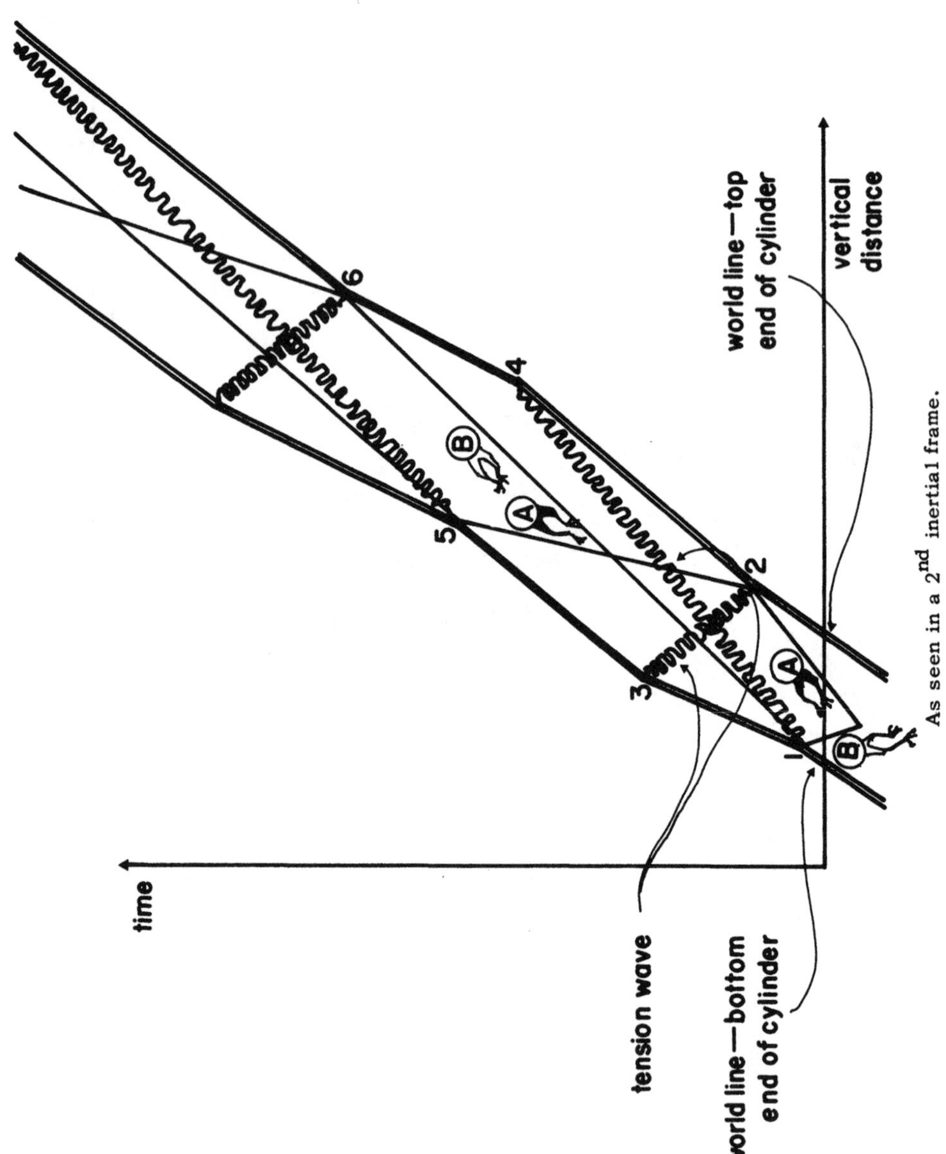

As seen in a 2nd inertial frame.

this pulsation the frogs hit the ends of the tube time and again, each time changing the amplitude and phase of the pulsation.

From another freely-falling frame (perhaps instantaneously comoving with onlooking birds), the cylinder rushes by at speed β, and the frogs do not hit the ends of the tube simultaneously, the pulsations of the two ends will be out of phase with each other, but the overall picture will not be much different. In particular, the center of the tube (which is at rest in the inertial frame of the cylinder) will *not* jerk back and forth in any other inertial frame. The accompanying spacetime diagrams illustrate the phenomena.

Solution 1.25. Let us first derive the forms of the infinitesimal operators by considering their action on a spacetime function $f(x,y,z,t)$:

(i) Rotations about z axis:

$$\begin{aligned} x' &\approx x - y\theta \\ y' &\approx y + x\theta \\ z' &= z \\ t' &= t\ , \end{aligned} \tag{1}$$

$$f(x',y',z') - f(x,y,z) = \theta[-y\partial_x + x\partial_y]f\ , \tag{2}$$

thus

$$J_z = -2i(x\partial_y - y\partial_x)\ , \tag{3}$$

since

$$f(x',y',z') \equiv (1 + iJ_z\theta/2)f(x,y,z)\ .$$

J_y, J_x are obtained by cyclical permutations of Equation (3).

(ii) Boosts in z direction:

$$\begin{aligned} z' &\approx z - vt \\ t' &\approx t - vx \\ y' &= y \\ z' &= z \end{aligned}$$

$$f(x', y', z', t') - f(x, y, z, t) = [-t\partial_z - z\partial_t] v\, f(x, y, z, t)$$

thus

$$K_z = 2i\left(t\frac{\partial}{\partial z} + z\frac{\partial}{\partial t}\right), \quad \text{etc.} \tag{4}$$

(iii) Now, commutation relations may be worked out from Equations (3) and (4) by the usual procedure, for example:

$$[J_x, J_y] = -4[y\partial_z - z\partial_y, z\partial_x - x\partial_z] = 4(x\partial_y - y\partial_x) = 2iJ_z\ , \tag{5}$$

and similarly for all other commutators.

(iv) A representation of the Lorentz group is specified by giving explicit matrices for the generators. We can associate the operators with the Pauli spin matrices

$$\sigma_x = \begin{bmatrix} 0 & 1 \\ 1 & 0 \end{bmatrix} \quad \text{etc.}$$

according to

$$\underset{\sim}{J} \to \underset{\sim}{\sigma}$$

$$\underset{\sim}{K} \to i\underset{\sim}{\sigma}\ .$$

The commutation relations for $\underset{\sim}{J}$ and $\underset{\sim}{K}$ are satisfied since $[\sigma_x, \sigma_y] = 2i\sigma_z$ etc. To find the matrices corresponding to *finite* transformations L, note that for example J_z is defined by

$$\left.\frac{dL}{d\theta}\right|_{\theta=0} = \frac{iJ_z}{2}\ ,$$

so that

$$L(\theta) = \exp(i\theta J_z/2)\ .$$

An arbitrary finite transformation L will have six parameters; in the general case the rotation parameters $\underset{\sim}{\theta}^*$ (here treated as a 3-vector) will be different from the physical rotation angle $\underset{\sim}{\theta}$ and the parameters representing boosts, $\underset{\sim}{v}^*$, will differ from the physical relative velocity $\underset{\sim}{v}$. These are details; the Lorentz group is defined completely by its generators $\underset{\sim}{J}$, $\underset{\sim}{K}$; its parametrization is largely arbitrary. The infinitesimal version of an arbitrary transformation L is

$$\delta L = i(\theta_x^* J_x + \theta_y^* J_y + \theta_z^* J_z + v_x^* K_x + v_y^* K_y + v_z^* K_z)/2 = (i\underset{\sim}{\theta}^* - \underset{\sim}{v}^*)\cdot\underset{\sim}{\sigma}/2 .$$

Let us define

$$\underset{\sim}{q} \equiv (i\underset{\sim}{\theta}^* - \underset{\sim}{v}^*)/2$$

$$q^2 \equiv \underset{\sim}{q}\cdot\underset{\sim}{q} = \frac{1}{4}(-\theta^{*2} - 2i\underset{\sim}{\theta}^*\cdot\underset{\sim}{v}^* + v^2) .$$

(Note that pure boosts and pure rotations correspond respectively to pure real and pure imaginary $\underset{\sim}{q}$.) We have then, for the general finite transformation:

$$L(\underset{\sim}{\theta},\underset{\sim}{v}) = \exp(\underset{\sim}{q}\cdot\underset{\sim}{\sigma}) = \sum_{n=0}^{\infty} (\underset{\sim}{q}\cdot\underset{\sim}{\sigma})^n/n!$$

$$= \sum_{n=0,2,4\cdots}^{\infty} q^n/n! + \sum_{n=1,3,5\cdots}^{\infty} (\underset{\sim}{q}\cdot\underset{\sim}{\sigma})q^{n-1}/n!$$

$$= \cosh q + (\underset{\sim}{q}\cdot\underset{\sim}{\sigma}/q)\sinh q .$$

Here we have used

$$(\underset{\sim}{q}\cdot\underset{\sim}{\sigma})(\underset{\sim}{q}\cdot\underset{\sim}{\sigma}) = q_i q_j (\sigma_i, \sigma_j) = q_i q_j \delta^{ij} = q^2 .$$

Solution 1.26. The hard way to solve the problem is by multiplying 4×4 matrices together and then decomposing into rotations and boosts. An easier way is to use the 2×2 complex unimodular matrix representation of the Lorentz group developed in Problem 1.25. Represent a pure boost of velocity v_1 by

$$L(\underset{\sim}{v}_1) = \exp(-v_1^*\underset{\sim}{n}_1\cdot\underset{\sim}{\sigma}/2) = \cosh(v_1^*/2) + (\underset{\sim}{n}_1\cdot\underset{\sim}{\sigma})\sinh(v_1^*/2) \quad \text{(1a)}$$

where $\underset{\sim}{n}_1$ is a unit vector in the direction of the boost and v_1^* parametrizes the magnitude of the boost. Represent a pure rotation $\underset{\sim}{\theta}$ by

$$L(\underset{\sim}{\theta}_1) = \exp(i\theta^*\underset{\sim}{n}\cdot\underset{\sim}{\sigma}/2) = \cos(\theta^*/2) - i(\underset{\sim}{n}\cdot\underset{\sim}{\sigma})\sin(\theta^*/2) \quad \text{(1b)}$$

where $\underset{\sim}{n}$ is the rotation axis and θ^* parametrizes the magnitude of the rotation.

For pure boosts in the same direction, v^* is in fact the rapidity parameter, $v^* = \tanh^{-1}|v|$, since v^* adds linearly, e.g.

$$L(v_1)L(v_2) = e^{-v_1^*\sigma_x/2}\,e^{-v_2^*\sigma_x/2} = e^{-(v_1^*+v_2^*)\sigma_x/2} .$$

Likewise one can see that for pure rotations in the same direction, θ^* is just the magnitude of the rotation angle, $\theta^* = |\underset{\sim}{\theta}|$.

Using Equations (1a) and (1b) one now solves the relation

$$L(\underset{\sim}{v}_1)L(\underset{\sim}{v}_2) = L(\theta\underset{\sim}{n})L(\underset{\sim}{v}_3) \tag{2}$$

for θ, $\underset{\sim}{n}$, $\underset{\sim}{v}_3$ in terms of $\underset{\sim}{v}_1$ and $\underset{\sim}{v}_2$. A useful identity in multiplying the right hand sides of Equation (2) together is

$$(\underset{\sim}{A}\cdot\underset{\sim}{\sigma})(\underset{\sim}{B}\cdot\underset{\sim}{\sigma}) = A\cdot B + i(\underset{\sim}{A}\times\underset{\sim}{B})\cdot\underset{\sim}{\sigma} .$$

Since the $\underset{\sim}{\sigma}$ and identity 2×2 matrix are all independent, one may equate, in Equation (2), the real and imaginary parts respectively of terms multiplying the identity matrix and of terms multiplying the σ matrices, yielding two scalar and two vector equations:

$$\begin{aligned}\cosh\left(\tfrac{1}{2}v_1^*\right)\cosh\left(\tfrac{1}{2}v_2^*\right) + \sinh\left(\tfrac{1}{2}v_1^*\right)\sinh\left(\tfrac{1}{2}v_2^*\right)(\underset{\sim}{n}_1\cdot\underset{\sim}{n}_2)&\\ = \cosh\left(\tfrac{1}{2}v_3^*\right)\cos\left(\tfrac{1}{2}\theta^*\right)&\end{aligned} \tag{3a}$$

$$\sin\left(\tfrac{1}{2}\theta^*\right)\sinh\left(\tfrac{1}{2}v_3^*\right)(\underset{\sim}{n}\cdot\underset{\sim}{n}_3) = 0 \tag{3b}$$

$$\begin{aligned}\cosh\left(\tfrac{1}{2}v_1^*\right)\sinh\left(\tfrac{1}{2}v_2^*\right)\underset{\sim}{n}_2 + \sinh\left(\tfrac{1}{2}v_1^*\right)\cosh\left(\tfrac{1}{2}v_2^*\right)\underset{\sim}{n}_1 = \cos\left(\tfrac{1}{2}\theta^*\right)\sinh\left(\tfrac{1}{2}v_3^*\right)\underset{\sim}{n}_3&\\ + \sin\left(\tfrac{1}{2}\theta^*\right)\sinh\left(\tfrac{1}{2}v_3^*\right)(\underset{\sim}{n}\times\underset{\sim}{n}_3)&\end{aligned} \tag{3c}$$

$$\sinh\left(\tfrac{1}{2}v_1^*\right)\sinh\left(\tfrac{1}{2}v_2^*\right)(\underset{\sim}{n}_1\times\underset{\sim}{n}_2) = -\sin\left(\tfrac{1}{2}\theta^*\right)\cosh\left(\tfrac{1}{2}v_3^*\right)\underset{\sim}{n}. \tag{3d}$$

In the general case, the solution to Equation (3b) is

$$\underset{\sim}{n}\cdot\underset{\sim}{n}_3 = 0 \tag{4}$$

i.e., the resultant rotation is orthogonal to the resultant boost.

Let γ be the angle between $\underset{\sim}{n}_1$ and $\underset{\sim}{n}_2$. Then, dotting $\underset{\sim}{n}$ into Equation (3d) and combining with Equation (3a), one finds

$$\tan\left(\frac{1}{2}\theta^*\right) = \frac{-\sinh\left(\frac{1}{2}v_2^*\right)\sinh\left(\frac{1}{2}v_2^*\right)\sin\gamma}{\cosh\left(\frac{1}{2}v_1^*\right)\cosh\left(\frac{1}{2}v_2^*\right) + \cos\gamma\,\sinh\left(\frac{1}{2}v_1^*\right)\sinh\left(\frac{1}{2}v_2^*\right)}\,. \tag{5}$$

Note that θ ($=\theta^*$ by above discussion) may take on all values $0 \leqq \theta \leqq 2\pi$ except $\theta = \pi$, and that for $n_1 = n_2$ ($\gamma = 0$), $\theta = 0$ and there is no net rotation as expected.

Solution 1.27. An astronaut looking at the sky (by null photons!) makes an arbitrary Lorentz transformation, and looks at the sky again. There results a *continuous* map of the sky onto itself. Any continuous map of a 2-sphere onto itself has at least one fixed point.

Solution 1.28. From Solution 1.25, any homogeneous proper Lorentz transformation corresponds to a 2×2 complex unimodular matrix L, or a complex 3-vector $\underset{\sim}{P}$, say. We will write the 2×2 matrix as

$$L(\underset{\sim}{P}) = (1+P^2)^{\frac{1}{2}}\, I + \underset{\sim}{P}\cdot\underset{\sim}{\sigma}\;,$$

where $P^2 = \underset{\sim}{P}\cdot\underset{\sim}{P}$, and I is the identity matrix which we will often not write in explicitly. (In Solution 1.25, we used $\underset{\sim}{P} = \underset{\sim}{q}(\sinh q)/q$.) For a pure boost with velocity v and direction $\underset{\sim}{n}$, $\underset{\sim}{P}$ is real: $\underset{\sim}{P} = \underset{\sim}{n}\sinh(\psi/2)$, where $\psi = \tanh v$ is the rapidity. For a pure rotation of magnitude θ about an axis $\underset{\sim}{n}$, $\underset{\sim}{P}$ is imaginary: $\underset{\sim}{P} = i\underset{\sim}{n}\sin(\theta/2)$ and $P^2 \geq -1$.

The composition of two Lorentz transformations represented by $\underset{\sim}{P}$ and $\underset{\sim}{Q}$ is

$$\underset{\sim}{P}\circ\underset{\sim}{Q} = (1+P^2)^{\frac{1}{2}}\,\underset{\sim}{Q} + (1+Q^2)^{\frac{1}{2}}\,\underset{\sim}{P} + i\underset{\sim}{P}\times\underset{\sim}{Q}\;.$$

This follows from multiplying the matrices corresponding to the Lorentz transformations:

$$L(\underset{\sim}{P})L(\underset{\sim}{Q}) = [(1+P^2)^{\frac{1}{2}} + \underset{\sim}{P}\cdot\underset{\sim}{\sigma}][(1+Q^2)^{\frac{1}{2}} + \underset{\sim}{Q}\cdot\underset{\sim}{\sigma}]$$

$$= (1+P^2)^{\frac{1}{2}}(1+Q^2)^{\frac{1}{2}} + [(1+P^2)^{\frac{1}{2}}\underset{\sim}{Q} + (1+Q^2)^{\frac{1}{2}}\underset{\sim}{P}]\cdot\underset{\sim}{\sigma} + \underset{\sim}{P}\cdot\underset{\sim}{Q} + i\underset{\sim}{P}\times\underset{\sim}{Q}\cdot\underset{\sim}{\sigma} .$$

The idea of the solution is first to find a characterization of the product of two pure boosts (Lemma 1), then to find a criterion for the case that an arbitrary Lorentz transformation is the product of three boosts (Lemma 2), and finally to show that this criterion is always satisfied except for the case of a "180° screw", which requires four boosts.

LEMMA 1. $\underset{\sim}{P}$ *is the product of two pure boosts* $\underset{\sim}{C}$ *and* $\underset{\sim}{D}$ *if and only if* P^2 *is real and positive.*

Proof. If $\underset{\sim}{P}$ is the product of $\underset{\sim}{C}$ and $\underset{\sim}{D}$ (real, non-null vectors), then

$$\underset{\sim}{P} = \underset{\sim}{C}\circ\underset{\sim}{D} = (1+C^2)^{\frac{1}{2}}\underset{\sim}{D} + (1+D^2)^{\frac{1}{2}}\underset{\sim}{C} + i\,\underset{\sim}{C}\times\underset{\sim}{D} . \tag{1}$$

$$P^2 = (1+C^2)D^2 + (1+D^2)C^2 + 2(1+C^2)^{\frac{1}{2}}(1+D^2)^{\frac{1}{2}}\underset{\sim}{C}\cdot\underset{\sim}{D} - C^2D^2 + (\underset{\sim}{C}\cdot\underset{\sim}{D})^2$$

$$= [(1+C^2)^{\frac{1}{2}}(1+D^2)^{\frac{1}{2}} + \underset{\sim}{C}\cdot\underset{\sim}{D}]^2 - 1 \tag{2}$$

$$> [(1+2CD+C^2D^2)^{\frac{1}{2}} + \underset{\sim}{C}\cdot\underset{\sim}{D}]^2 - 1 \quad \text{since } C^2 + D^2 > 2CD$$

$$\geq (1+|\underset{\sim}{C}\cdot\underset{\sim}{D}| + \underset{\sim}{C}\cdot\underset{\sim}{D})^2 - 1 \quad \text{since } CD \geq |\underset{\sim}{C}\cdot\underset{\sim}{D}|$$

$$\geq 0 .$$

Conversely, if P^2 is real and positive, then one can write $\underset{\sim}{P} = \underset{\sim}{A} + i\underset{\sim}{B}$, where $\underset{\sim}{A}\cdot\underset{\sim}{B} = 0$ and $A^2 - B^2 > 0$. To construct boosts $\underset{\sim}{C}$ and $\underset{\sim}{D}$, choose a vector $\underset{\sim}{E}$ orthogonal to both $\underset{\sim}{A}$ and $\underset{\sim}{B}$ with E^2 to be determined. As an *ansatz*, take

$$\underset{\sim}{C} = a\underset{\sim}{A} + \underset{\sim}{E}, \quad \underset{\sim}{D} = a\underset{\sim}{A} - \underset{\sim}{E} , \tag{3}$$

where a is a normalization constant to be determined. From Equation (3) $C^2 = D^2 = a^2A^2 + E^2$ and by Equation (1)

$$\underset{\sim}{A} = (1+C^2)^{\frac{1}{2}}\underset{\sim}{D} + (1+D^2)^{\frac{1}{2}}\underset{\sim}{C} = (1+a^2A^2+E^2)^{\frac{1}{2}}2a\underset{\sim}{A}$$

and

$$\underset{\sim}{B} = \underset{\sim}{C}\times\underset{\sim}{D} = 2a\underset{\sim}{E}\times\underset{\sim}{A} \ .$$

Thus the construction will succeed as long as

$$1 = 2a(1+a^2A^2+E^2)^{\frac{1}{2}} \ , \tag{4a}$$

$$B^2 = 4a^2E^2A^2 \ . \tag{4b}$$

Square Equation (4a) and solve for a^2:

$$a^2 = \{[(1+E^2)^2+A^2]^{\frac{1}{2}} - (1+E^2)\}/(2A^2) \ . \tag{5a}$$

Equation (4b) now implies

$$B^2 = 2E^2\{[(1+E^2)^2+A^2]^{\frac{1}{2}} - (1+E^2)\} \ .$$

After squaring and simplifying, this reduces to

$$4(A^2-B^2)E^4 - 4B^2E^2 - B^4 = 0 \ , \tag{5b}$$

which always has a positive solution for E^2 since $A^2 > B^2$. The parameter a is given by (5a) and the proof is now complete.

Now try to express *any* $\underset{\sim}{Q}$ as a product of three boosts. By Lemma 1, this is possible if and only if there exists a boost $\underset{\sim}{C}$ such that P^2 is real and positive, where $\underset{\sim}{P} \equiv \underset{\sim}{Q} \circ (-\underset{\sim}{C})$. By algebra similar to that used to reach Equation (2), we find

$$(1+P^2)^{\frac{1}{2}} = (1+Q^2)^{\frac{1}{2}}(1+C^2)^{\frac{1}{2}} - \underset{\sim}{Q}\cdot\underset{\sim}{C} \ . \tag{6}$$

So we must find $\underset{\sim}{C}$ such that $(1+P^2)^{\frac{1}{2}}$ evaluated from Equation (6) is real ($\underset{\sim}{Q}$ is in general complex) and $(1+P^2)^{\frac{1}{2}} > 1$.

Let $\underset{\sim}{Q} = \underset{\sim}{A} + i\underset{\sim}{B}$ ($\underset{\sim}{A}$ and $\underset{\sim}{B}$ real), and let the boost $\underset{\sim}{D}$ be some linear combination of $\underset{\sim}{A}$ and $\underset{\sim}{B}$.

LEMMA 2. *There exists a boost* $\underset{\sim}{C}$ *satisfying the required properties if and only if there exists a boost* $\underset{\sim}{D}$ *which is a linear combination of* $\underset{\sim}{A}$ *and* $\underset{\sim}{B}$ *such that*

$$d \equiv (1+Q^2)^{\frac{1}{2}}(1+D^2)^{\frac{1}{2}} - \underset{\sim}{Q}\cdot\underset{\sim}{D} \tag{7}$$

with d *real and strictly positive.*

Proof. Let $\underset{\sim}{F}$ be a vector orthogonal to both $\underset{\sim}{A}$ and $\underset{\sim}{B}$ with $F^2 < 1$ to be determined shortly. Define

$$\underset{\sim}{C} = (\underset{\sim}{D}+\underset{\sim}{F})/(1-F^2)^{\frac{1}{2}}$$

so that $1+C^2 = (1+D^2)/(1-F^2)$. From Equation (6),

$$(1+P^2)^{\frac{1}{2}} = [(1+Q^2)^{\frac{1}{2}}(1+D^2)^{\frac{1}{2}} - \underset{\sim}{Q}\cdot\underset{\sim}{D}](1-F^2)^{-\frac{1}{2}} = d(1-F^2)^{-\frac{1}{2}} .$$

Thus if and only if d is real and positive, we can choose F^2 close enough to 1 that $(1+P^2)^{\frac{1}{2}}$ is real and has magnitude greater than 1. Lemma 2 is proven.

Now consider two cases:

Case 1: $\underset{\sim}{A}$ not parallel to $\underset{\sim}{B}$, $A^2 \neq 0$, and $B^2 \neq 0$.

In this case Equation (7) gives

$$(d + \underset{\sim}{Q}\cdot\underset{\sim}{D})^2 = (1+Q^2)(1+D^2) .$$

Substituting $\underset{\sim}{Q} = \underset{\sim}{A} + i\underset{\sim}{B}$ gives two real equations:

$$(d+\underset{\sim}{A}\cdot\underset{\sim}{D})^2 - (\underset{\sim}{B}\cdot\underset{\sim}{D})^2 = (1+A^2-B^2)(1+D^2) , \tag{8a}$$

$$(d+\underset{\sim}{A}\cdot\underset{\sim}{D})\underset{\sim}{B}\cdot\underset{\sim}{D} = \underset{\sim}{A}\cdot\underset{\sim}{B}(1+D^2) . \tag{8b}$$

Solve Equations (8) simultaneously for $d + \underset{\sim}{A}\cdot\underset{\sim}{D}$ and $\underset{\sim}{B}\cdot\underset{\sim}{D}$ to get:

$$(d+\underset{\sim}{A}\cdot\underset{\sim}{D})^2 = \frac{1}{2}(1+D^2)\{[(1+A^2-B^2)^2+4(\underset{\sim}{A}\cdot\underset{\sim}{B})^2]^{\frac{1}{2}}+(1+A^2-B^2)\} , \tag{9a}$$

$$(\underset{\sim}{B}\cdot\underset{\sim}{D})^2 = \frac{1}{2}(1+D^2)\{[(1+A^2-B^2)^2+4(\underset{\sim}{A}\cdot\underset{\sim}{B})^2]^{\frac{1}{2}}-(1+A^2-B^2)\} . \tag{9b}$$

These equations do not determine $\underset{\sim}{D}$ and/or d explicitly, because D^2 occurs on the right-hand side. One possible solution of these equations is to take $\underset{\sim}{D} = b\underset{\sim}{B}$, b to be determined. For this choice Equation (9b) becomes

$$(2b^2B^4)/(1+b^2B^2) = [(1+A^2-B^2)^2+4(\underset{\sim}{A}\cdot\underset{\sim}{B})^2]^{\frac{1}{2}} - (1+A^2-B^2) \ . \tag{10}$$

To show that there is always a real solution b of Equation (10), note that the right-hand side is ≥ 0 (obvious) and also $< 2B^2$; this follows since

$$\begin{aligned} 4(\underset{\sim}{A}\cdot\underset{\sim}{B})^2 &< 4A^2B^2 \\ &= (1+A^2+B^2)^2 - (1+A^2-B^2)^2 - 4B^2 \\ &< (1+A^2+B^2)^2 - (1+A^2-B^2)^2 \ . \end{aligned}$$

Because the left-hand side of Equation (10) varies from 0 to $2B^2$ as b^2 varies from 0 to ∞, there is always a real constant b such that Equation (10) holds. Thus Equation (9b) is satisfied by our *ansatz*, and then Equation (9a) (with $\underset{\sim}{D} = b\underset{\sim}{B}$ and with the sign of b chosen appropriately) defines a positive value of d. Thus $\underset{\sim}{Q}$ is the product of 3 boosts.

Case 2: $\underset{\sim}{A}$ parallel to $\underset{\sim}{B}$ (including $A^2 = 0$ or $B^2 = 0$) .

Now Equation (7) becomes a scalar equation since all the vectors are collinear:

$$d = (1+Q^2)^{\frac{1}{2}}(1+D^2)^{\frac{1}{2}} - QD \ . \tag{11}$$

Let

$$\psi = \sinh^{-1}Q = \alpha + i\beta \ ,$$

and make ψ single-valued by the restriction $-\pi/2 < \beta \leq \pi/2$. Also let

$$\phi = \sinh^{-1}D \qquad \text{(real)} \ .$$

Equation (11) then becomes

$$d = \cosh(\psi-\phi) = \cosh(\alpha-\phi)\cos\beta - i\sinh(\alpha-\phi)\sin\beta \ .$$

If $\beta = 0$, we are done ($\underset{\sim}{Q}$ is a pure boost). If $\beta \neq 0$, for d to be real we must choose $\phi = \alpha$ (this fixes D); i.e. $d = \cos\beta$. If $\beta \neq \pi/2$, d is positive and $\underset{\sim}{Q}$ is the product of three boosts. If $\beta = \pi/2$, $d = 0$ and $\underset{\sim}{Q}$ is not the product of three boosts.

The case $\beta = \pi/2$ corresponds to $Q = \sinh(\alpha + i\pi/2) = i\cosh\alpha$. So Q is pure imaginary and $Q^2 \leq -1$. This corresponds to a rotation through 180° composed with a boost of rapidity 2α, a "180° screw." (In particular, a pure rotation through 180° cannot be attained by three boosts.)

A 180° screw can be achieved by four boosts; for, if $\underset{\sim}{C}$ is a boost and $\underset{\sim}{Q} = i\underset{\sim}{B}$ with $B^2 \geq 1$, then

$$\underset{\sim}{R} \equiv \underset{\sim}{Q} \circ \underset{\sim}{C} = i(B^2 - 1)^{\frac{1}{2}}\underset{\sim}{C} + i(1+C^2)^{\frac{1}{2}}\underset{\sim}{B} - \underset{\sim}{B} \times \underset{\sim}{C}$$

is *not* a 180° screw provided $\underset{\sim}{C}$ is not parallel to $\underset{\sim}{B}$. Therefore $\underset{\sim}{R}$ is the product of three boosts and so $\underset{\sim}{Q} = \underset{\sim}{R} \circ (-\underset{\sim}{C})$ is the product of four boosts.

This solution is due to D. M. Eardley.

CHAPTER 2: SOLUTIONS

Solution 2.1. From 4-momentum conservation we have

$$P_e + P_\gamma = P'_e + P'_\gamma \ .$$

(Here γ indicates the photon; prime denotes values after scattering.) Since we are not interested in the electron's final momentum we use a generally useful trick to eliminate it:

$$(P_e + P_\gamma - P'_\gamma)^2 = P^2_{e'} = -m_e^2 \ ,$$

or, since $P_\gamma^2 = 0$,

$$-m_e^2 + 2P_e \cdot P_\gamma - 2P_e \cdot P'_\gamma - 2P_\gamma \cdot P'_\gamma = -m_e^2 \ .$$

In the lab frame, we have

$$\begin{aligned} P_e &= (m_e, \underset{\sim}{0}) \\ P_\gamma &= (h/\lambda, h/\lambda\, \underset{\sim}{e}_i), \quad \underset{\sim}{e}_i = \text{unit 3-vector in incident direction}, \\ P'_\gamma &= (h/\lambda', h/\lambda'\, \underset{\sim}{e}_0), \quad \underset{\sim}{e}_0 = \text{unit 3-vector in outgoing direction}, \end{aligned}$$

so that

$$-\frac{m_e h}{\lambda} + \frac{m_e h}{\lambda'} + \frac{h^2}{\lambda\lambda'} - \frac{h^2}{\lambda\lambda'} \cos\theta = 0 \ .$$

Multiplying through by $\lambda\lambda'$, we find

$$\lambda' - \lambda = (h/m_e)(1 - \cos\theta) \ .$$

Solution 2.2. Use γ to denote the photon, and prime to denote values after scattering. We proceed by eliminating P' from the equation of 4-momentum conservation:

$$(P'_\gamma + P' - P_\gamma)^2 = P^2 = -m^2$$

$$P_\gamma \cdot P'_\gamma = P \cdot (P_\gamma - P'_\gamma) .$$

Maximum energy transfer obviously occurs for a 180° scattering angle. With this condition and $|\underset{\sim}{P}_\gamma| = E_\gamma$, the above equation becomes

$$-E_\gamma E'_\gamma + \underset{\sim}{P}_\gamma \cdot \underset{\sim}{P}'_\gamma = -E(E_\gamma - E'_\gamma) + \underset{\sim}{P} \cdot (\underset{\sim}{P}_\gamma - \underset{\sim}{P}'_\gamma)$$

$$-2E_\gamma E'_\gamma = -E(E_\gamma - E'_\gamma) + P(-E_\gamma - E'_\gamma)$$

so that

$$E'_\gamma = \frac{E_\gamma(E+P)}{2E_\gamma + E - P} \approx \frac{E}{1 + m^2/4E\,E_\gamma} .$$

The second expression for E'_γ is a result of using the approximation

$$P = (E^2 - m^2)^{\frac{1}{2}} \approx E - \frac{1}{2}\, m^2/E$$

appropriate to our assumed condition $E >> m_e$.

For 3^0K, the photon energy is of order $kT \approx 3 \times 10^{-4}$ eV. Using this, $m_{proton} = .938 \times 10^9$ eV, and $E = 10^{20}$eV in our equation we find $E'_\gamma \approx 10^{19}$ eV.

Solution 2.3. By conservation of 4-momentum, we would have to have

$$p_\gamma + p_e = p'_e , \tag{1}$$

where p_γ is the photon momentum, p_e and p'_e are the electron's momentum when the photon is/is not present.

Squaring both sides of Equation (1)

$$p_\gamma \cdot p_\gamma + 2p_e \cdot p_\gamma + p_e \cdot p_e = p'_e \cdot p'_e$$

$$0 + 2p_e \cdot p_\gamma - m_e^2 = -m_e^2$$

$$p_e \cdot p_\gamma = 0 .$$

But in the frame where $p_e = (m, \underset{\sim}{0})$ and $p_\gamma = (E, \underset{\sim}{p})$, this says that the photon energy must be zero, i.e. there is no photon. Thus the process cannot occur.

Solution 2.4. If p is the 4-momentum of the resultant system, and $\gamma \equiv (1-v_1{}^2)^{-\frac{1}{2}}$, we have

$$p = (m_1\gamma, m_1\gamma \underset{\sim}{v}_1) + (m_2, \underset{\sim}{0})$$

and

$$m = (-p\cdot p)^{\frac{1}{2}} = (m_1{}^2 + m_2{}^2 + 2\gamma\, m_1 m_2)^{\frac{1}{2}}$$

$$\underset{\sim}{v} = \frac{\underset{\sim}{p}}{E} = \frac{m_1\gamma \underset{\sim}{v}_1}{m_1\gamma + m_2} = \frac{\underset{\sim}{v}_1}{1 + m_2/m_1\gamma} .$$

Solution 2.5. Let the neutron travel in the x direction. In its rest frame the electron has momentum 4-vector

$$P' = (E', P'\cos\theta, P'\sin\theta, 0), \qquad E' = m_e(1-v_e^2)^{-\frac{1}{2}}$$

(where we have, without loss of generality, oriented our coordinate system so $\underset{\sim}{P}'$ is in the x-y plane), and P' is isotropic. Using the Lorentz transformation to get the lab momentum $\underset{\sim}{P}$, we have:

$$P_x = \gamma\,(P'\cos\theta + \beta E')$$

$$P_y = P'\sin\theta$$

or

$$\left(\frac{P_x - \gamma\beta E'}{\gamma P'}\right)^2 + \left(\frac{P_y}{P'}\right)^2 = 1 .$$

The last equation says that in the lab, the momentum vector in momentum space lies on an ellipse of semi major axis $\gamma P'$, semiminor axis P', and origin $(\gamma\beta E', 0)$. The plot of such an ellipse falls into three cases:

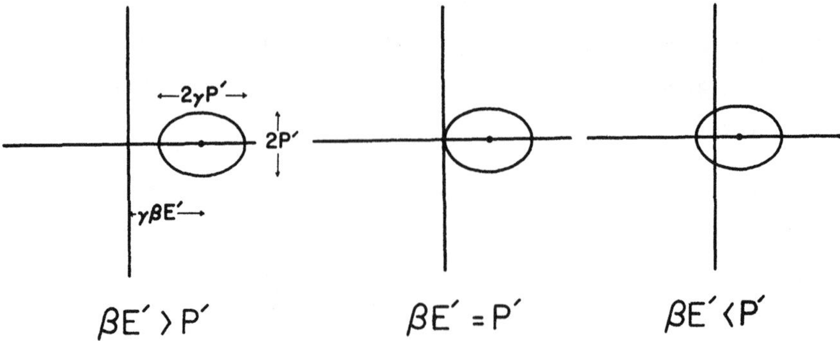

Solution 2.6. Let **P** and **Q** be the 4-momenta of the incident protons, and W the total energy in the C.M. frame. In this frame,

$$W^2 = (P^0 + Q^0)^2 = -(\mathbf{P}+\mathbf{Q})^2 ,$$

which is an invariant expression for W^2 and can be evaluated in any frame. In the conventional experiment, in the lab frame

$$\mathbf{P} = (E, \underset{\sim}{P})$$

$$\mathbf{Q} = (m, \underset{\sim}{0}) ,$$

so

$$\begin{aligned} W^2 &= (E+m)^2 - p^2 \\ &= 2Em + 2m^2 \\ &\approx 2Em \quad \text{for} \quad E \gg m . \end{aligned}$$

For $E = 30$ GeV, $m = .94$ GeV, the available energy is $W \approx 7.5$ GeV.

For the colliding beam experiment, in the lab frame

$$\mathbf{P} = (E, \underset{\sim}{P})$$

$$\mathbf{Q} = (E, -\underset{\sim}{P}) .$$

So $W^2 = 4E^2$, i.e. $W = 2E$. For $E = 15$ GeV, the available energy is $W = 30$ GeV. To achieve $W = 30$ GeV in a conventional experiment would require $E = W^2/2m \approx 480$ GeV.

Solution 2.7. Let subscripts 0 and 1 denote respectively the initially moving, and the initially stationary particle. Balancing 4-momentum before and after the collision, we have

$$P_0 + P_1 = P_{0'} + P_{1'}$$

$$(P_0 + P_1 - P_{0'})^2 = P_{1'}^2$$

$$-3m^2 + 2P_0 \cdot P_1 - 2P_{0'} \cdot (P_0 + P_1) = -m^2 \ .$$

Now we put in

$$P_0 = (E, \underset{\sim}{P}), \quad E = m + T_0$$

$$P_1 = (m, \underset{\sim}{0}), \quad E' = m + T'$$

$$P_{0'} = (E', \underset{\sim}{P}'), \quad \underset{\sim}{P} \cdot \underset{\sim}{P}' = PP' \cos\theta \ ,$$

to find

$$-m^2 - Em + E'(E+m) - PP'\cos\theta = 0$$

i.e.

$$(E^2 - m^2)^{\frac{1}{2}} (E'^2 - m^2)^{\frac{1}{2}} \cos\theta = (E' - m)(E+m) \ .$$

Squaring and factoring the difference of squares, then gives the kinetic energy of the scattered particle:

$$(E-m)(E'+m)\cos^2\theta = (E'-m)(E+m)$$

$$T_0(T'+2m)\cos^2\theta = T'(T_0 + 2m)$$

$$T'(-T_0\cos^2\theta + T_0 + 2m) = 2m\,T_0\cos^2\theta$$

$$T' = \frac{2m\,T_0\cos^2\theta}{2m + T_0\sin^2\theta} \ .$$

Solution 2.8. Conservation of 4-momentum gives us

$$P_\gamma + P_N = P_{N'} + P_\pi \qquad (N' = \text{nucleon coming out of reaction})$$

$$(P_\gamma + P_N)^2 = (P_{N'} + P_\pi)^2 \ .$$

In the lab frame (where $E_\gamma \sim 3^0\mathrm{K} \sim 2.5 \times 10^{-10}\mathrm{MeV}$),

$$P_\gamma = (E_\gamma, \underset{\sim}{P}_\gamma), \quad P_N = (E_N, \underset{\sim}{P}_N) .$$

In the C.M. frame at threshold,

$$P_{N'} + P_\pi = (m_N + m_\pi, \underset{\sim}{0})$$

so that

$$2P_\gamma \cdot P_N - m_N^2 = -(m_N + m_\pi)^2$$

$$-2E_\gamma E_N + 2\underset{\sim}{P}_\gamma \cdot \underset{\sim}{P}_N = -2m_N m_\pi - m_\pi^2 .$$

Now $P_\gamma = E_\gamma$ since $m_\gamma = 0$ and, for a head-on collision $\underset{\sim}{P}_\gamma \cdot \underset{\sim}{P}_N = -P_\gamma P_N$, so that

$$E_N + (E_N^2 - m_N^2)^{\frac{1}{2}} = \frac{2m_N m_\pi + m_\pi^2}{2E_\gamma}$$

$$= \frac{2 \cdot 940 \cdot 140 + 140^2}{2 \cdot (2.5 \times 10^{-10})} \mathrm{MeV} = 6 \times 10^{14} \mathrm{MeV} .$$

Since $E_N >> m_N$, we can replace $(E_N^2 - m_N^2)^{\frac{1}{2}}$ by E_N. Thus $E_N \approx 3 \times 10^{14}$ MeV.

Solution 2.9. We proceed by conserving 4-momentum and removing information about the Λ particle:

$$P_\pi + P_n = P_K + P_\Lambda$$

$$P_\Lambda^2 = -m_\Lambda^2 = (P_\pi + P_n - P_K)^2 = -m_\pi^2 - m_n^2 - m_K^2 + 2P_\pi \cdot P_n - 2P_n \cdot P_K - 2P_\pi \cdot P_K .$$

In the lab

$$P_\pi = (E_\pi, \underset{\sim}{P}_\pi)$$

$$P_n = (m_n, \underset{\sim}{0}), \quad P_K = (E_K, \underset{\sim}{P}_K)$$

so,

$$-m_\pi^2 - m_n^2 - m_K^2 - 2m_n E_\pi + 2m_n E_K + 2E_\pi E_K - 2\underset{\sim}{P}_\pi \cdot \underset{\sim}{P}_K = -m_\Lambda^2 .$$

We now set $\underset{\sim}{P}_\pi \cdot \underset{\sim}{P}_K = 0$ (90° scattering) and find

$$E_\pi = \frac{m_\Lambda^2 - m_\pi^2 - m_n^2 - m_K^2 + 2m_n E_K}{2(m_n - E_K)} .$$

Thus, to make E_π a minimum we must make E_K as small as possible (as you might have guessed!), so we try $E_K = m_K$. Thus

$$E_{\pi(\text{threshold})} = \frac{m_\Lambda^2 - m_\pi^2 - m_n^2 - m_K^2 + 2m_n m_K}{2(m_n - m_K)}$$

$$= \frac{m_\Lambda^2 - m_\pi^2}{2(m_n - m_K)} - \frac{m_n - m_K}{2} = \frac{1115^2 - 140^2}{2(940 - 494)} - \frac{940\text{-}494}{2}$$

$$= 1149 \text{ MeV} ,$$

hence threshold kinetic energy = 1149 – 140 = 1009 MeV.

Solution 2.10.

(a) From conservation of 4-momentum

$$P_C^2 = -m_C^2 = (P_A - P_B)^2 = -m_A^2 - m_B^2 - 2P_A \cdot P_B . \qquad (1)$$

In the lab frame $P_A = (m_A, \underset{\sim}{0})$, $P_B = (E_B, \underset{\sim}{P}_B)$, so

$$-m_C^2 = -m_A^2 - m_B^2 - 2m_A E_B$$

and hence

$$E_B = m_A^2 + m_B^2 - m_C^2/2m_A .$$

(b) Here $m_A = M$, $m_B = 0$, $m_C = M - \delta$, so

$$E_B = h\nu = \frac{M^2 - (M-\delta)^2}{2M} = \delta - \frac{\delta^2}{2M} < \delta .$$

Physically $h\nu \neq \delta$ because some energy goes into the recoil of M which is necessary to conserve momentum. In the Mössbauer effect, the recoil momentum is shared among $\sim 10^{23}$ atoms, so the recoil energy is negligible.

(c) In this case $P_A = (E_A, \underset{\sim}{P}_A)$, $P_B = (E_B, \underset{\sim}{P}_B)$, so (1) becomes

$$-m_A^2 - m_B^2 + 2E_A E_B - 2(E_A^2 - m_A^2)^{\frac{1}{2}}(E_B^2 - m_B^2)^{\frac{1}{2}} \cos\theta = -m_C^2 \ .$$

Solution 2.11.

(a) The total 4-momentum is $P_{total} = P_1 + P_2 = P_3 + P_4$. In the C.M. frame $P_{total} = (E_{total}^{C.M.}, \underset{\sim}{0})$ so that

$$P_{total}^2 = -(E_{total}^{C.M.})^2 = (P_1 + P_2)^2 \ . \tag{1}$$

We can evaluate $(P_1 + P_2)^2$ in the lab frame, where $P_1 = (E_1, \underset{\sim}{P}_1)$, $P_2 = (m_2, \underset{\sim}{0})$:

$$-(E_{total}^{C.M.})^2 = P_1^2 + P_2^2 + 2P_1 \cdot P_2 = -m_1^2 - m_2^2 - 2E_1 m_2 \ .$$

(b) In the equation

$$P_1 \cdot P_{total} = P_1 \cdot (P_1 + P_2)$$

evaluate the left side in the C.M. frame and the right side in the lab frame:

$$\begin{aligned} -E_1^{C.M.} E_{total}^{C.M.} &= -m_1^2 - E_1 m_2 \\ &= -m_1^2 - \frac{1}{2}[(E_{total}^{C.M.})^2 - m_1^2 - m_2^2] \qquad \text{(from (a))} \end{aligned}$$

Hence, the result:

$$E_1^{C.M.} = [(E_{total}^{C.M.})^2 + m_1^2 - m_2^2]/2E_{total}^{C.M.} \ .$$

(c) By using the result of part (b) we have

$$(P_1^{C.M.})^2 = (E_1^{C.M.})^2 - m_1^2 = \left(\frac{m_1^2 + E_1 m_2}{E_{tot}^{C.M.}}\right)^2 - m_1^2$$

$$= \frac{(m_1^2 + E_1 m_2)^2 - m_1^2(m_1^2 + m_2^2 + 2E_1 m_2)}{(E_{tot}^{C.M.})^2} = \frac{m_2^2(E_1^2 - m_1^2)}{(E_{tot}^{C.M.})^2} = \frac{m_2^2 P_1^2}{(E_{tot}^{C.M.})^2} ,$$

and therefore $P_1^{C.M.} = m_2 P_1 / E_{total}^{C.M.}$.

(d) If $\mathbf{u}$ is the 4-velocity of any observer, then it is easy to verify in that observer's frame (hence in all frames) that

$$\mathbf{P}_{(3)} \equiv \mathbf{P} + (\mathbf{P}\cdot\mathbf{u})\mathbf{u}$$

is the 3-momentum measured by that observer. Therefore $\mathbf{u}_{C.M.}$ is defined by

$$\mathbf{P}_{total} + (\mathbf{P}_{total}\cdot\mathbf{u}_{C.M.})\mathbf{u}_{C.M.} = 0 \ .$$

Now if two observers have 4-velocities $\mathbf{u}_1$ and $\mathbf{u}_2$, then $\gamma = -\mathbf{u}_1\cdot\mathbf{u}_2$ is the Lorentz transformation factor between the two frames. (Proof: In rest-frame of 1, $\mathbf{u}_1\cdot\mathbf{u}_2 = (1,0)\cdot(\gamma,\gamma\underset{\sim}{v}) = -\gamma$). So $\gamma_{C.M.} = -\mathbf{u}\cdot\mathbf{u}_{C.M.}$ gives the transformation to the C.M. frame. Thus

$$\mathbf{P}_{tot}\cdot\mathbf{u} + (\mathbf{P}_{total}\cdot\mathbf{u}_{C.M.})\mathbf{u}_{C.M.}\cdot\mathbf{u} = 0$$

$$-E_{total} + E_{total}^{C.M.}\gamma_{C.M.} = 0$$

$$\gamma_{C.M.} = E_{total}/E_{tot}^{C.M.} \ .$$

In our case, $E_{total} = E_1 + m_2$, hence

$$\gamma_{C.M.} = (E_1 + m_2)/E_{total}^{C.M.} \ .$$

(e) From (d), we have

$$1 - v_{C.M.}^2 = [E_{total}^{C.M.}/(E_1 + m_2)]^2 \ ,$$

thus

$$v_{C.M.} = \left[\frac{E_1^2 - m_1^2}{(E_1 + m_2)^2}\right]^{\frac{1}{2}} = \frac{P_1}{E_1 + m_2} \ .$$

Solution 2.12. In the C.M. frame since the collision is elastic, $E_1^{C.M.} = E_{1'}^{C.M.}$, i.e.,

$$\mathbf{P}_1\cdot\mathbf{u}_{C.M.} = \mathbf{P}_{1'}\cdot\mathbf{u}_{C.M.} \ .$$

(The prime denotes values after the collision.) Evaluating this equation in the lab frame yields

$$-E_1 + \underset{\sim}{P}_1 \cdot \underset{\sim}{v}_{C.M.} = -E_{1'} + \underset{\sim}{P}_{1'} \cdot \underset{\sim}{v}_{C.M.} \cdot$$

But $\underset{\sim}{v}_{C.M.} = \underset{\sim}{P}_1/(E_1 + m_2)$, so we have

$$-E_1 + \frac{P_1^2}{E_1 + m_2} = -E_{1'} + \frac{P_{1'} P_1 \cos\theta}{E_1 + m_2} ,$$

or

$$\cos\theta = \frac{E_{1'}(E_1 + m_2) - E_1 m_2 - m_1^2}{P_1 (E_{1'}^2 - m_1^2)^{\frac{1}{2}}} . \tag{1}$$

(This equation could also have been derived without using the center of mass frame.) We can find the minimum value of $\cos\theta$ (maximum value of $\sin\theta$) by setting $d\cos\theta/dE_{1'} = 0$, solving for $E_{1'}$, and then substituting into Equation (1). A quicker way is to consider the intersection of the graph of $\cos\theta\,(E_{1'})$ with the straight line $\cos\theta = K$. The points of intersection are given by

$$K^2 P_1^2 (E_{1'}^2 - m_1^2) = E_{1'}^2 (E_1 + m_2)^2 - 2(E_1 + m_2)(E_1 m_2 + m_1^2) E_{1'} + (E_1 m_2 + m_1^2)^2 . \tag{2}$$

At the minimum value of $\cos\theta$, the straight line is tangent to the curve and the discriminant of Equation (2), a quadratic in $E_{1'}$, is zero. This gives

$$0 = -(E_1 + m_2)^2 K^2 P_1^2 m_1^2 + K^2 P_1^2 (E_1 m_2 + m_1^2)^2 + K^4 P_1^4 m_1^2 ,$$

or

$$K^2 = \frac{(E_1 + m_2)^2 m_1^2 - (E_1 m_2 + m_1^2)^2}{m_1^2 P_1^2} = \frac{m_1^2 - m_2^2}{m_1^2}$$

and, finally

$$\cos\theta_{min} = \frac{(m_1^2 - m_2^2)^{\frac{1}{2}}}{m_1} .$$

$$\sin\theta_{max} = m_2/m_1 .$$

Solution 2.13.

(a) Take the rocket's motion to be along the x-axis. We have the following constraints on its 4-velocity $\mathbf{u}$ and 4-acceleration $\mathbf{a}$:

$$\mathbf{u}\cdot\mathbf{u} = -1 = -(u^t)^2 + (u^x)^2 \quad \text{(normalization of } \mathbf{u}) \tag{1}$$

$$\mathbf{a}\cdot\mathbf{u} = 0 = -a^t u^t + a^x u^x \quad \text{(a orthogonal to } \mathbf{u}) \tag{2}$$

$$\mathbf{a}\cdot\mathbf{a} = g^2 = -(a^t)^2 + (a^x)^2 \quad \text{(Proper acceleration is } g)\ . \tag{3}$$

From these equations, we get

$$a^t = a^x\left(\frac{u^x}{u^t}\right), \qquad (a^x)^2[1-(u^x/u^t)^2] = g^2$$

which imply

$$a_x = gu^t \tag{4}$$

$$a^t = gu^x\ . \tag{5}$$

By differentiating Equation (4) we can get a differential equation for u^x

$$\frac{d^2u^x}{d\tau^2} = \frac{da^x}{d\tau} = g\,\frac{du^t}{d\tau} = ga^t = g^2u^x\ ,$$

with solutions $u^x = A \sinh g\tau + B \cosh g\tau$. Since the initial conditions on the motion are $u^x = 0$, $du^x/d\tau = g$, at $\tau = 0$, $\mathbf{u}$ must be

$$u^x = dx/d\tau = \sinh g\tau\ . \tag{6}$$

$$u^t = dt/d\tau = \cosh g\tau \qquad \text{(from Equation (1))}$$

The integrals of these equations (with $x=t=0$ at $\tau=0$) give us

$$x = g^{-1}(\cosh g\tau - 1) \qquad t = g^{-1}(\sinh g\tau)\ . \tag{7}$$

In units with $c = 1$, it turns out by numerical coincidence that $g(=980\ \text{cm/sec}^2)$ is almost exactly equal to one inverse year (as a time) or one inverse light-year (as a distance). Thus for 40 years as measured on Earth ($t=40$ yr), (7) gives

$$\tau \approx [\sinh^{-1}(40)]\ \text{yr} \approx 4.38\ \text{yr}$$

and

$$x \approx [\cosh[\sinh^{-1}(40)] - 1]\ \text{L.Y.} \approx 39.01\ \text{L.Y.} \tag{8}$$

For 40 years in the rocket $\tau = 40$ yr and from (7)

$$x \approx [\cosh(40) - 1]\ \text{L.Y.} \approx 10^{17}\ \text{L.Y.} \tag{9}$$

(b) To travel halfway requires $x = 15{,}000$ L.Y., or

$$\tau \approx \cosh^{-1}(15{,}000 + 1)\ \text{yr} \approx 10.3\ \text{yr}\ . \tag{10}$$

The deceleration half of the trip is identical, so the total time is 20.6 years.

(c) Denote the rest mass of the rocket (which changes) by M. The change in energy of the rocket equals the energy radiated

$$d(Mu^0) = -dE_{rad.}\ . \tag{11}$$

Since the energy is radiated as photons

$$dE_{rad.} = dP_{rad.} \tag{12}$$

and by conservation of momentum,

$$dP_{rad.} = dP \tag{13}$$

where dP is the change in momentum of the rocket. Combining Equations (11) - (13) gives

$$d(Mu^0) = -dP = -d(Mu^x)$$

$$(dM)\,u^0 + M du^0 = -(dM)\,u^x - M du^x$$

$$dM/M = -d(u^0 + u^x)/(u^0 + u^x)$$

so that

$$M = M_0/(u^0 + u^x) = M_0\, e^{-g\tau} \tag{14}$$

where we have used Equations (6) to give us $u^0(\tau)$ and $u^x(\tau)$.

From part (b), $e^{g\tau} = 30{,}000$ for half the trip. Thus, $M_{\frac{1}{2}} = M_0/30{,}000$ and

$$M_{\text{final}} = M_0/(30{,}000)^2 \approx 10^{-9}\, M_0 \;.$$

Solution 2.14. A cyclotron applies an accelerating potential with a fixed frequency, to the gap between two "dees." The frequency is chosen so that the potential is always in the right direction when the electron crosses the gap, i.e. it is the cyclotron frequency $\omega_0 = eB/mc$. The reason that there is a maximum energy is that the electron actually circles at the "synchrotron frequency" $\omega = eB/\gamma mc = \omega_0 m/E$ (where E = electron energy), and as the electron becomes relativistic this gets out of phase with the fixed cyclotron frequency. Eventually the electron arrives at the gap when the potential is 90° out of phase and the accelerating process breaks down. Quantitatively if the phase of the accelerating potential is α when the electron passes through the gap, the electron receives energy $V\cos\alpha$, where V is the maximum accelerating potential expressed in electron volts. In dN cycles of the electron motion then

$$\frac{d(\text{electron energy})}{dN} = \frac{dE}{\omega dt/2\pi} = 2V\cos(\phi - \omega_0 t) \;.$$

[Note that the electron passes through the gap twice per cycle, hence the factor of 2.] Here ϕ is the angular distance traveled by the electron

$$\phi = \int \omega dt = \omega_0 \int_0^t (m/E)\, dt$$

and

$$\frac{dE}{dt} = \frac{\omega}{2\pi} 2V \cos\left[\omega_0 \int_0^t \frac{m}{E}\, dt - \omega_0 t\right]$$

$$= \frac{V}{\pi E}\,\omega_0 m \cos\left[\omega_0 \int_0^t \left(\frac{m}{E} - 1\right) dt\right] .$$

$$\frac{dE^2}{dt} = a \cos\left[\omega_0 \int_0^t \left(\frac{m}{E} - 1\right) dt\right]$$

where $a \equiv 2V\,\omega_0 m/\pi$. Differentiating we have

$$\frac{d^2(E^2)}{dt^2} = -a \sin\left[\omega_0 \int_0^t \left(\frac{m}{E} - 1\right) dt\right] \omega_0 \left(\frac{m}{E} - 1\right)$$

$$= \left[a^2 - \left(\frac{dE^2}{dt}\right)^2\right]^{\frac{1}{2}} \omega_0 \left(\frac{m}{E} - 1\right) .$$

The negative square root has been used because the sine function above is clearly negative.

We now find a first integral of this equation. Let $q \equiv dE^2/dt$ so that $d^2E^2/dt^2 = q\,dq/dE^2$, then our differential equation becomes

$$-q \frac{dq}{dE^2} = [a^2 - q^2]^{\frac{1}{2}} \omega_0 \left[1 - \frac{m}{(E^2)^{\frac{1}{2}}}\right] .$$

This can be integrated with the initial condition that $q = a$ at $E = m$ (i.e. at $t = 0$):

$$(a^2 - q^2)^{\frac{1}{2}} = \omega_0 (E - m)^2 .$$

The energy keeps increasing until $q = 0$, at which time the energy must be

$$E_{max} = m + a/\omega_0 = m + (2Vm/\pi)^{\frac{1}{2}} .$$

Solution 2.15. If $\mathbf{F}$ is not to be identically zero then for *some* 4-velocity $\mathbf{u}$, the product $\mathbf{a}\cdot\mathbf{u}$ must be nonzero. This contradicts the requirement that $\mathbf{a}\cdot\mathbf{u} = \frac{1}{2}\, d(\mathbf{u}\cdot\mathbf{u})/d\tau = 0$.

CHAPTER 3: SOLUTIONS

Solution 3.1. The neatest way to do this problem is by using spacetime diagrams.

First rotate the coordinate system until both events lie on the x-axis and the first event occurs at $x = t = 0$.

The two events A and B at spacelike separation $(\Delta x, \Delta t)$ are depicted in Figure (i), with the world line of a light ray indicated as the dotted 45° line. A Lorentz transformation, shown as the t', x' axes in Figure (ii), can obviously be made to make A and B simultaneous. (The magnitude of the Lorentz boost is $\beta = \Delta t/\Delta x$ and $\tanh^{-1} \tan\theta$ in Figure (ii) is the rapidity parameter.) Since the t' axis cannot be rotated below the light cone line in a Lorentz transformation, the two events cannot be made to occur at the same location.

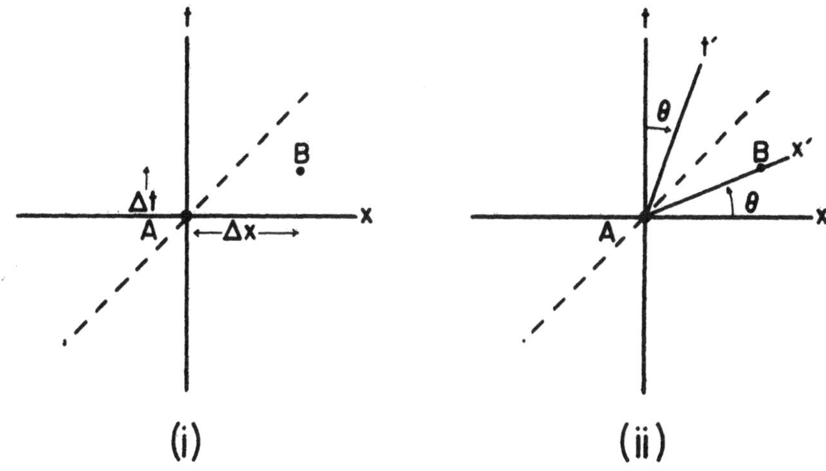

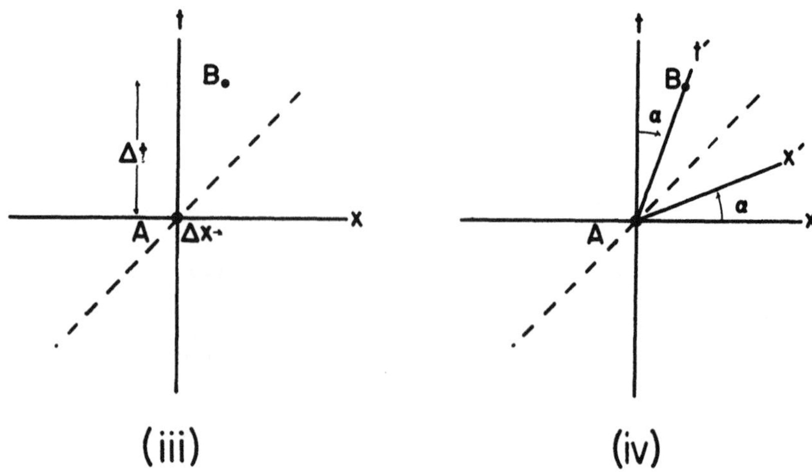

(iii) (iv)

Figure (iii) depicts events A and B at timelike separation. As shown in Figure (iv), a Lorentz transformation can obviously be made to bring A and B to the same location in the primed frame. The magnitude of the required boost is $a = \Delta x/\Delta t$. Since the x' axis cannot be rotated above the light cone line, the two events cannot be made simultaneous.

Solution 3.2. Four linearly independent null vectors are e.g.

$$e_z + e_t, \ e_z - e_t, \ e_x + e_t, \ e_y + e_t \ .$$

Suppose there existed 4 null vectors A, B, C, D which were mutually orthogonal. Linear independence implies that *any* vector can be written

$$V = aA + bB + cC + dD$$

but the length of this vector is clearly zero if A, B, C, D are null and mutually orthogonal. Hence 4 such vectors cannot exist.

Solution 3.3. Without loss of generality, choose coordinates to make the null vector $V = e_x + e_t$. A general vector is $S = Ae_t + Be_x + Ce_y + De_z$. The dot product is $S \cdot V = A(e_t \cdot e_t) + B(e_x \cdot e_x) = B - A$ so A must equal B if S is orthogonal to V. Now the vector S is spacelike unless

$A^2 \geq B^2+C^2+D^2$ which means $C^2+D^2=0$ and hence $C=D=0$, therefore $S = A(e_x+e_t)$, which is a multiple of $\mathbf{V}$.

Solution 3.4. Take flat Minkowski space, so that $\hat{x}$, $\hat{y}$, $\hat{z}$ are spacelike unit vectors; $\hat{t}$ is a timelike unit vector; $\hat{t}\pm\hat{z}$, $\hat{t}\pm\hat{y}$ are null, etc. The following table gives examples in the order: spacelike sum, null sum, timelike sum.

	spacelike	null	timelike
spacelike	$\hat{x}+\hat{y}$ $(\hat{x}+\varepsilon\hat{t})+(-\hat{x}+\varepsilon\hat{z})$ $(\hat{x}+\varepsilon\hat{t})+(-\hat{x})$	$\hat{x}+(\hat{x}-\hat{t})$ $(\hat{x}+\hat{t})+(-2\hat{x})$ $(-\hat{x})+(\hat{x}-\hat{t})$	$\hat{x}+(\varepsilon\hat{t})$ $\hat{x}+\hat{t}$ $(\varepsilon\hat{x})+\hat{t}$
null		$(\hat{x}-\hat{t})+(\hat{x}+\hat{t})$ $(\hat{x}+\hat{t})+(\hat{x}+\hat{t})$ $(\hat{t}-\hat{x})+(\hat{t}+\hat{x})$	$(\hat{x}-\hat{t})+\hat{t}$ $(\hat{x}-\hat{t})+2\hat{t}$ $(\hat{x}+\hat{t})+\hat{t}$
timelike		spacelike sum → null sum → timelike sum →	$(\hat{t}+\varepsilon\hat{x})+(-\hat{t})$ $(\hat{t}+\varepsilon\hat{x})+(-\hat{t}+\varepsilon\hat{t})$ $\hat{t}+\hat{t}$

Here ε represents any small constant, e.g. 0.1. Not all possibilities would be allowed if one required the timelike vectors to be "future directed." i.e. $\mathbf{u}\cdot\hat{t} < 0$ for timelike vectors $\mathbf{u}$. (The reader can work this out.)

Solution 3.5. Let $\mathbf{k}$ be the null vector along which the light is traveling. A given observer takes a little square element of the beam whose sides are the vectors $\mathbf{A}$ and $\mathbf{B}$, which are purely spacelike vectors in his frame, and orthogonal to the light beam, so $\mathbf{A}\cdot\mathbf{k} = \mathbf{B}\cdot\mathbf{k} = 0$. Also $\mathbf{A}\cdot\mathbf{B}=0$ since the element is square. The area of the element is $|\mathbf{A}|\ |\mathbf{B}|$.

A different observer can unambiguously identify the same element of the beam, say by the rays making up its corners, but he slices it differently in time: the old vectors $\mathbf{A}$ and $\mathbf{B}$ are not orthogonal to his 4-velocity $\mathbf{u}$, so are not a purely spacelike cross section in his frame.

Now for any constants α and β, the vectors

$$\mathbf{A}' = \mathbf{A} + \alpha\mathbf{k}$$

$$\mathbf{B}' = \mathbf{B} + \beta\mathbf{k}$$

span the same beam element as $\mathbf{A}$ and $\mathbf{B}$ (the "tip" of each vector is just moved to a different point along the ray by adding a multiple of the wave vector $\mathbf{k}$). The new observer requires $\mathbf{A}'\cdot\mathbf{u} = \mathbf{B}'\cdot\mathbf{u} = 0$, which is always possible, choosing $\alpha = -\mathbf{A}\cdot\mathbf{u}/\mathbf{k}\cdot\mathbf{u}$ and $\beta = -\mathbf{B}\cdot\mathbf{u}/\mathbf{k}\cdot\mathbf{u}$. ($\mathbf{k}\cdot\mathbf{u} \neq 0$, since $\mathbf{k}$ is null and $\mathbf{u}$ is timelike.) Notice that $\mathbf{A}'\cdot\mathbf{k} = \mathbf{B}'\cdot\mathbf{k} = \mathbf{A}'\cdot\mathbf{B}' = 0$ since $\mathbf{k}\cdot\mathbf{k} = 0$, so $\mathbf{A}'$ and $\mathbf{B}'$ are orthogonal vectors which are the sides of the new observer's cross section. The new area is $|\mathbf{A}'|\ |\mathbf{B}'| = (\mathbf{A}'\cdot\mathbf{A}')^{\frac{1}{2}}(\mathbf{B}'\cdot\mathbf{B}')^{\frac{1}{2}} = |\mathbf{A}|\ |\mathbf{B}|$, the same area as seen by the original observer.

Solution 3.6. It is easy to show by example that $\sum_\mu D^{\mu\mu}$ is not invariant. Let $D^{0x} = 1$ and all other components be zero. For a boost in the x-direction $D^{0'0'} = -\beta\gamma$, $D^{x'x'} = -\beta\gamma$ so that $\sum_\mu D^{\mu\mu} = 0$ but $\sum_{\mu'} D^{\mu'\mu'} = -2\beta\gamma$. The proof for $\sum_\mu D_{\mu\mu}$ is similar.

For a Lorentz transformation $\Lambda^{\mu'}_{\ \nu}$, we know $D^{\nu'}_{\ \mu'} = \Lambda^{\alpha}_{\ \mu'}\Lambda^{\nu'}_{\ \beta} D^{\beta}_{\ \alpha}$ so that

$$\sum_{\mu'} D^{\mu'}_{\ \mu'} = \sum_{\mu'} \Lambda^{\alpha}_{\ \mu'}\Lambda^{\mu'}_{\ \beta} D^{\alpha}_{\ \beta} = \delta^{\alpha}_{\beta} D^{\beta}_{\ \alpha} = \sum_{\mu} D^{\mu}_{\ \mu} ,$$

and hence this summation is invariant.

Solution 3.7. Since $\eta^{\alpha\beta}$ is constant we have

$$\begin{aligned} F_{\mu\ \ ,\beta}^{\ \alpha} F^{\beta}_{\ \ \alpha} &= (F_{\mu\gamma}\,\eta^{\gamma\alpha})_{,\beta}(F^{\beta\sigma}\eta_{\sigma\alpha}) \\ &= F_{\mu\gamma,\beta}\, F^{\beta\sigma}(\eta^{\gamma\alpha}\eta_{\sigma\alpha}) = F_{\mu\gamma,\beta}\, F^{\beta\sigma}\ \delta^{\alpha}_{\sigma} = F_{\mu\gamma,\beta}\, F^{\beta\gamma} \\ &= -F_{\mu\alpha,\beta}\, F^{\alpha\beta} , \end{aligned}$$

where the last equality follows from relabelling dummy indices, and by antisymmetry.

Solution 3.8. We express the differentials in terms of the new coordinates:

$$\begin{aligned} ds^2 &= \eta_{\alpha\beta}\, dx^\alpha dx^\beta \\ &= \eta_{\alpha\beta}(\partial x^\alpha/\partial \bar{x}^\mu)\, d\bar{x}^\mu (\partial x^\beta/\partial \bar{x}^\nu)\, d\bar{x}^\nu \\ &= \left(\eta_{\alpha\beta}\frac{\partial x^\alpha}{\partial \bar{x}^\mu}\frac{\partial x^\beta}{\partial \bar{x}^\nu}\right) d\bar{x}^\mu d\bar{x}^\nu \ . \end{aligned}$$

If we write the line element as $ds^2 = g_{\bar{\mu}\bar{\nu}} d\bar{x}^\mu d\bar{x}^\nu$, it follows that

$$g_{\bar{\mu}\bar{\nu}} \equiv \eta_{\alpha\beta}\frac{\partial x^\alpha}{\partial \bar{x}^\mu}\frac{\partial x^\beta}{\partial \bar{x}^\nu} \ .$$

A vector transforms by the following rule

$$U^\alpha = \frac{\partial x^\alpha}{\partial \bar{x}^\beta} U^{\bar{\beta}} \ ,$$

therefore

$$\begin{aligned} U\cdot V &= U^\sigma V^\lambda \eta_{\sigma\lambda} \\ &= \left(U^{\bar{\alpha}}\frac{\partial x^\sigma}{\partial \bar{x}^\alpha}\right)\left(V^{\bar{\beta}}\frac{\partial x^\lambda}{\partial \bar{x}^\beta}\right)\eta_{\sigma\lambda} = U^{\bar{\alpha}} V^{\bar{\beta}}\frac{\partial x^\sigma}{\partial \bar{x}^\alpha}\frac{\partial x^\lambda}{\partial \bar{x}^\beta}\eta_{\sigma\lambda} \\ &= U^{\bar{\alpha}} V^{\bar{\beta}} g_{\bar{\alpha}\bar{\beta}} \ . \end{aligned}$$

Solution 3.9. Under the coordinate transformation $x^\mu \to \bar{x}^\mu(x^\nu)$ the transformation of $g_{\alpha\beta}$ is

$$g_{\bar{\mu}\bar{\nu}} = g_{\alpha\beta}\frac{\partial x^\alpha}{\partial \bar{x}^\mu}\frac{\partial x^\beta}{\partial \bar{x}^\nu} \ ,$$

and therefore the transformation of g is

$$\bar{g} = \det(g_{\bar{\mu}\bar{\nu}}) = \det(g_{\alpha\beta})\det\left(\frac{\partial x^\alpha}{\partial \bar{x}^\mu}\right)\det\left(\frac{\partial x^\beta}{\partial \bar{x}^\nu}\right) = g\left[\det\left(\frac{\partial x^\alpha}{\partial \bar{x}^\mu}\right)\right]^2 \ .$$

Since $\bar{g} \neq g$, $\bar{g}$ is not a scalar.

Solution 3.10. Suppose one coordinate transformation is

$$\bar{x}^\mu = \bar{x}^\mu(x^\nu), \quad \text{so} \quad \Lambda^\alpha_\beta = \partial\bar{x}^\alpha/\partial x^\beta \tag{1}$$

and the other is

$$\tilde{x}^\mu = \tilde{x}^\mu(x^\nu), \quad \text{so} \quad \tilde{\Lambda}^\alpha_\beta = \partial\tilde{x}^\alpha/\partial x^\beta \ . \tag{2}$$

The product matrix is then

$$\Lambda^\alpha_\gamma \tilde{\Lambda}^\gamma_\beta = \frac{\partial\bar{x}^\alpha}{\partial x^\gamma}\frac{\partial\tilde{x}^\gamma}{\partial x^\beta} \ . \tag{3}$$

This looks suspiciously like the chain rule for partial derivatives, so one is led to consider the coordinate transformation

$$\bar{x}^\mu(x^\beta) = \bar{x}^\mu[\tilde{x}^\gamma(x^\beta)] \ , \tag{4}$$

whose transformation matrix is obviously

$$\bar{\Lambda}^\mu_\beta = \frac{\partial\bar{x}^\mu}{\partial x^\beta} = \frac{\partial\bar{x}^\mu}{\partial\tilde{x}^\gamma}\frac{\partial\tilde{x}^\gamma}{\partial x^\beta} \ . \tag{5}$$

This differs from (3) only by a twiddle. A little thought shows that the twiddle is not meaningful. Partial derivatives are taken with respect to argument variables; it makes no difference what symbols are used to represent these variables. Thus (4) is in fact the coordinate transformation of the product matrix.

Solution 3.11. In a basis $\mathbf{K}$ is represented by a matrix $K^{\alpha\beta}$. An obviously necessary and sufficient condition for $K^{\alpha\beta}$ to be expressible as $A^\alpha B^\beta$ is that all columns of the matrix are proportional to each other.

In basis-free language, $\mathbf{K}$ can be written as

$$\mathbf{K} = \mathbf{A}\otimes\mathbf{B}$$

if and only if the vector $\mathbf{W} = \mathbf{K}\cdot\mathbf{v}$ (i.e. $W^\alpha = K^{\alpha\beta}v_\beta$) is in the same direction for any choice of $\mathbf{v}$. Proof: In some frame choose 4 basis vectors $\mathbf{e}^0, \mathbf{e}^1, \cdots$ to satisfy $\mathbf{e}^\mu\cdot\mathbf{e}_\nu = \delta^\mu{}_\nu$. By linearity the direction of $\mathbf{W}$

is independent of *any* v if and only if it is independent of the 4 basis vectors e^{μ}. Our condition then is

$$\mathbf{K}\cdot e^0 = \lambda_0 \mathbf{W}, \quad \mathbf{K}\cdot e^1 = \lambda_1 \mathbf{W}, \cdots$$

or

$$K^{\alpha 0} = \lambda_0 W^{\alpha}, \quad K^{\alpha 1} = \lambda_1 W^{\alpha} \cdots .$$

Our basis-free criterion then is equivalent to the columns being proportional.

Solution 3.12. A direct product of two vectors (see Problem 3.11) has all of its columns proportional to each other (similarly for all of its rows). This is not true for a general second rank tensor. If e_i $i=1,n$ are the basis vectors $(0,0,\cdots 1,\cdots 0,0)$ where the 1 is in the ith slot, then the n^2 products $e_i \otimes e_j$ obviously span the space of second rank tensors, since there is one with a 1 uniquely in any row and column. Sums over these times constants obviously can give the general tensor. (Any one *specific* tensor can be represented as a sum of n outer products, as the reader is invited to verify.)

Solution 3.13. The object cannot be a tensor. A 2nd rank covariant tensor is a functional of 2 covariant tensors, i.e. $X^{\mu\nu} V_{\mu} W_{\nu}$ would have to be a scalar, but

$$X^{\mu\nu} V_{\mu} W_{\nu} = (\mathbf{A}\cdot\mathbf{V}) W_{\nu} + (\mathbf{B}\cdot\mathbf{W}) V_{\mu}$$

which is coordinate dependent.

If one defines the two index object by the relation $X^{\mu\nu} = A^{\mu} + B^{\nu}$ in any coordinate system, then the coordinate transformation law is $X^{\mu'\nu'} = \Lambda^{\mu'}_{\alpha} A^{\alpha} + \Lambda^{\nu'}_{\beta} B^{\beta}$. It is not possible to express this as

$$X^{\mu'\nu'} = T^{\mu'\nu'}_{\alpha\beta} X^{\alpha\beta}$$

where T is some transformation. This follows from the fact that there are many choices of A^{μ} and B^{ν} which give, in one coordinate system, the same $X^{\mu\nu}$. In particular, in a given coordinate system $A^{\mu}+C$, $B^{\nu}-C$

give the same $X^{\mu\nu}$ as A^μ and B^ν, but the two choices give very different $X^{\mu'\nu'}$.

Solution 3.14. We need only transform and check the results for antisymmetry:

$$F_{\bar{\mu}\bar{\nu}} = \Lambda^{\alpha}_{\bar{\mu}}\Lambda^{\beta}_{\bar{\nu}}F_{\alpha\beta} = -\Lambda^{\alpha}_{\bar{\mu}}\Lambda^{\beta}_{\bar{\nu}}F_{\beta\alpha} = -\Lambda^{\beta}_{\bar{\mu}}\Lambda^{\alpha}_{\bar{\nu}}F_{\alpha\beta} = -F_{\bar{\nu}\bar{\mu}} \,.$$

$$F^{\mu\nu} = g^{\mu\alpha}g^{\nu\beta}F_{\alpha\beta} = -g^{\mu\alpha}g^{\nu\beta}F_{\beta\alpha} = -g^{\mu\beta}g^{\nu\alpha}F_{\alpha\beta} = -F^{\nu\mu} \,.$$

For variety, we treat the symmetric case by a different argument: Consider the tensor $\mathcal{A}^{\mu\nu} = S^{\mu\nu} - S^{\nu\mu}$, where $S^{\mu\nu}$ is a symmetric tensor. Since $\mathcal{A}^{\mu\nu}$ vanishes (by hypothesis) and since $\mathcal{A}^{\mu\nu}$ is a tensor, it vanishes in all coordinate systems, hence $S^{\mu\nu}$ is symmetric in all coordinate systems.

Solution 3.15. By the given symmetries $A_{\mu\nu}S^{\mu\nu} = -A_{\nu\mu}S^{\nu\mu}$, but μ, ν are dummy indices so we can interchange μ and ν to get

$$A_{\mu\nu}S^{\mu\nu} = -A_{\nu\mu}S^{\mu\nu} = -A_{\mu\nu}S^{\mu\nu} \,.$$

Thus $A_{\mu\nu}S^{\mu\nu} = 0$.

Any tensor $V_{\mu\nu}$ can be written as a sum of its symmetric part $\tilde{V}_{\mu\nu} \equiv \frac{1}{2}(V_{\mu\nu}+V_{\nu\mu})$ and its antisymmetric part $\tilde{\tilde{V}}_{\mu\nu} \equiv \frac{1}{2}(V_{\mu\nu}-V_{\nu\mu})$, thus

$$V^{\mu\nu}A_{\mu\nu} = \tilde{V}^{\mu\nu}A_{\mu\nu} + \tilde{\tilde{V}}^{\mu\nu}A_{\mu\nu} = \tilde{\tilde{V}}^{\mu\nu}A_{\mu\nu} = \frac{1}{2}(V^{\mu\nu}-V^{\nu\mu})A_{\mu\nu}$$

$$V^{\mu\nu}S_{\mu\nu} = \tilde{V}^{\mu\nu}S_{\mu\nu} + \tilde{\tilde{V}}^{\mu\nu}S_{\mu\nu} = \tilde{V}^{\mu\nu}S_{\mu\nu} = \frac{1}{2}(V^{\mu\nu}+V^{\nu\mu})S_{\mu\nu} \,.$$

Solution 3.16.

(a) If there are no symmetries there are clearly n^r components.

(b) For s symmetric indices and r-s asymmetric ones, consider: How many inequivalent ways are there of choosing the values of the s indices from the n possibilities? It equals the number of ways of choosing s things from n possibilities, including repetitions:

$$(n+s-1)!/(n-1)!\,s! \,.$$

[See, e.g. J. Mathews and R. Walker, *Mathematical Methods of Physics* (W. A. Benjamin, 1965) Section 14.3.] The other r-s indices can be chosen in n^{r-s} ways so the number of independent components is

$$n^{r-s}(n+s-1)!/(n-1)!\,s!\ .$$

(c) For a antisymmetric indices and r-a other indices, consider first: How many ways are there of choosing the a indices? This is equal to the number of ways of choosing a things from n, with no repetitions:

$$n!/(n-a)!\,a!\ .$$

The number of independent components is then

$$n^{r-a}\,n!/(n-a)!\,a!\ .$$

Note that if $a = n$ there is only one possibility for the a indices. If $a > n$, there are no possibilities; all components must be zero!

Solution 3.17.

(a)
$$\begin{aligned}
V_{[\mu\nu]} &= \tfrac{1}{2}(V_{\mu\nu}-V_{\nu\mu}) \qquad & T_{[\mu\nu]} &= 0\\
F_{[\mu\nu]} &= F_{\mu\nu} & T_{(\mu\nu)} &= T_{\mu\nu}\\
F_{(\mu\nu)} &= 0 & &
\end{aligned}$$

$$\begin{aligned}
V_{[\alpha\beta\gamma]} &= \tfrac{1}{6}\,[V_{\alpha\beta\gamma}-V_{\alpha\gamma\beta}+V_{\beta\gamma\alpha}-V_{\beta\alpha\gamma}+V_{\gamma\alpha\beta}-V_{\gamma\beta\alpha}]\\
T_{(\alpha\beta,\gamma)} &= \tfrac{1}{3}\,[T_{\alpha\beta,\gamma}+T_{\alpha\gamma,\beta}+T_{\beta\gamma,\alpha}]\\
F_{[\alpha\beta,\gamma]} &= \tfrac{1}{3}\,[F_{\alpha\beta,\gamma}+F_{\gamma\alpha,\beta}+F_{\beta\gamma,\alpha}]\ .
\end{aligned}$$

(b) If $A_{\mu\nu\cdots\sigma}$ is a totally antisymmetric tensor, i.e.

$$(-1)^{\pi}A_{\alpha_{\pi_1}\cdots\alpha_{\pi_p}} = A_{\alpha_1\cdots\alpha_p}$$

then

$$A_{[\alpha_1\cdots\alpha_p]} = \frac{1}{p!}\sum(-1)^{\pi}A_{\pi_1\cdots\alpha_{\pi_p}} = A_{\alpha_1\cdots\alpha_p}\ :$$

But $A_{\alpha_1\cdots\alpha_p} \equiv V_{[\alpha_1\cdots\alpha_p]}$ is totally antisymmetric, so

$$A_{[a_1 \cdots a_p]} = V_{[[a_1 \cdots a_p]]} = V_{[a_1 \cdots a_p]} .$$

The theorem for $V_{((a_1 \cdots a_p))}$ follows analogously.

By symmetry $V_{(a_1 \cdots [a_\ell a_m] \cdots a_p)} = V_{(a_1 \cdots [a_m a_\ell] \cdots a_p)}$ but by antisymmetry $V_{(a_1 \cdots [a_\ell a_m] \cdots a_1)} = -V_{(a_1 \cdots [a_m a_\ell] \cdots a_p)}$. Hence V must vanish.

Antisymmetry on two indices is easily expressed as

$$V_{a_1 \cdots [a_\ell a_m] \cdots a_p} = \frac{1}{2} [V_{a_1 \cdots a_\ell a_m \cdots a_p} - V_{a_1 \cdots a_m a_\ell \cdots a_p}] .$$

Thus

$$V_{[a_1 \cdots [a_\ell a_m] \cdots a_p]} = \frac{1}{2p!} \sum (-1)^\pi \left(V_{a_{\pi_1} \cdots a_{\pi_m} a_{\pi_\ell} \cdots a_{\pi_p}} - V_{\cdots a_{\pi_\ell} a_{\pi_m} \cdots} \right) .$$

But $V_{a_{\pi_1} \cdots a_{\pi_\ell} a_{\pi_m} \cdots a_{\pi_p}} = -V_{a_{\pi_1} \cdots a_{\pi_m} a_{\pi_\ell} \cdots a_{\pi_p}}$ so that

$$V_{[a_1 \cdots [a_\ell a_m] \cdots a_p]} = V_{[a_1 \cdots a_\ell a_m \cdots a_p]} .$$

(c) From part (a),

$$F_{\alpha\beta,\nu} + F_{\beta\nu,\alpha} + F_{\nu\alpha,\beta} = 3F_{[\alpha\beta,\nu]} .$$

But $F_{\alpha\beta} = -A_{[\alpha,\beta]}$ so $F_{[\alpha\beta,\nu]} = -A_{[[\alpha,\beta],\nu]}$, and by part (b)

$$A_{[[\alpha,\beta],\nu]} = A_{[\alpha,\beta,\nu]}$$

which must vanish because $A_{\alpha,\beta,\nu} = A_{\alpha,\nu,\beta}$.

Solution 3.18. The first part is straightforward:

$$X_{\alpha\beta} = \frac{1}{2} (X_{\alpha\beta} + X_{\beta\alpha}) + \frac{1}{2} (X_{\alpha\beta} - X_{\beta\alpha}) = X_{(\alpha\beta)} + X_{[\alpha\beta]} .$$

If a similar relation,

$$Y_{\alpha\beta\gamma} = Y_{(\alpha\beta\gamma)} + Y_{[\alpha\beta\gamma]} , \tag{1}$$

were true for a 3rd rank tensor, we would have:

$$Y_{\beta\alpha\gamma} = Y_{(\beta\alpha\gamma)} + Y_{[\beta\alpha\gamma]} = Y_{(\alpha\beta\gamma)} - Y_{[\alpha\beta\gamma]}$$

$$Y_{\beta\gamma\alpha} = Y_{(\gamma\beta\alpha)} - Y_{[\gamma\beta\alpha]} = Y_{(\alpha\beta\gamma)} + Y_{[\alpha\beta\gamma]} = Y_{\alpha\beta\gamma} \ . \qquad (2)$$

Since $Y_{\alpha\beta\gamma}$ does not in general have the symmetry properties implied by Equation (2), the simple decomposition in Equation (1) cannot be correct.

Solution 3.19. One way is to show that δ^{μ}_{ν} can be combined with the components of two vectors to form a scalar. But $A^{\nu}B_{\mu}\delta^{\mu}{}_{\nu} = \mathbf{A}\cdot\mathbf{B}$, hence $\boldsymbol{\delta}$ is a tensor.

Alternately one can check the transformation properties

$$\frac{\partial x^{\mu'}}{\partial x^{\alpha}}\frac{\partial x^{\beta}}{\partial x^{\nu'}}\delta^{\alpha}_{\beta} = \frac{\partial x^{\mu'}}{\partial x^{\alpha}}\frac{\partial x^{\alpha}}{\partial x^{\nu'}} = \delta^{\mu'}_{\nu'} \ ,$$

where the last equality follow from the fact that $\partial x^{\mu'}/\partial x^{\alpha}$ and $\partial x^{\alpha}/\partial x^{\mu'}$, are matrix inverses of each other. Hence δ^{μ}_{ν} transforms as a tensor.

Solution 3.20. If $\varepsilon_{\alpha\beta\gamma\delta}$ is totally antisymmetric, then given any one component e.g. ε_{0123} any other component (with nonrepeated indices) can be found by permuting the indices on that given component. Components with repeated indices, of course, vanish. The tensor is thus fixed once ε_{0123} is given, i.e. the tensor is unique up to scaling by a constant. With the usual, position independent choice of normalization, we have

$$\varepsilon_{0123} = -\varepsilon_{1023} = \varepsilon_{1032} = \cdots = 1$$

in Minkowski coordinates. If we now transform to other coordinates $x^{\mu'}(x^{\alpha})$ we get

$$\varepsilon_{\mu'\nu'\lambda'\sigma'} = \frac{\partial x^{\alpha}}{\partial x^{\mu'}}\frac{\partial x^{\beta}}{\partial x^{\nu'}}\frac{\partial x^{\gamma}}{\partial x^{\lambda'}}\frac{\partial x^{\delta}}{\partial x^{\sigma'}}\varepsilon_{\alpha\beta\gamma\delta} = \det\left[\frac{\partial x^{\alpha}}{\partial x^{\alpha'}}\right]\varepsilon_{\mu\nu\lambda\sigma} \ .$$

Now

$$g_{\mu'\nu'} = \frac{\partial x^\alpha}{\partial x^{\mu'}} \frac{\partial x^\beta}{\partial x^{\nu'}} \eta_{\alpha\beta}$$

$$\det(g_{\mu'\nu'}) = \left|\det\left(\frac{\partial x^\alpha}{\partial x^{\alpha'}}\right)\right|^2 \det(\eta_{\alpha\beta})$$

so that

$$\det(\partial x^\alpha/\partial x^{\alpha'}) = [-\det(g_{\mu'\nu'})]^{\frac{1}{2}}$$

and finally

$$\varepsilon_{\mu'\nu'\lambda'\sigma'} = [-\det(g_{\mu'\nu'})]^{\frac{1}{2}} \varepsilon_{\mu\nu\lambda\sigma} \, .$$

Solution 3.21. In an orthonormal frame

$$\varepsilon^{\alpha\beta\gamma\delta} = \eta^{\alpha\mu}\eta^{\beta\nu}\eta^{\gamma\sigma}\eta^{\delta\lambda} \varepsilon_{\mu\nu\sigma\lambda} \, .$$

For nonzero components of $\varepsilon^{\alpha\beta\gamma\delta}$, precisely one of the indices $\alpha, \beta, \gamma, \delta$ must equal zero. Since $\eta^{00} = -1$ and $\eta^{jj} = 1$ this introduces precisely one minus sign in what is otherwise an identity transformation. Thus

$$\varepsilon_{\alpha\beta\gamma\delta} = -\varepsilon^{\alpha\beta\gamma\delta} \, .$$

In some other frame, this equation becomes

$$[-\det(g_{\mu\nu})]^{-\frac{1}{2}} \varepsilon_{\alpha\beta\gamma\delta} = -[-\det(g_{\mu\nu})]^{\frac{1}{2}} \varepsilon^{\alpha\beta\gamma\delta}$$

according to the transformation derived in Solution 3.20, and its obvious analog for $\varepsilon^{\alpha\beta\gamma\delta}$. Thus in general

$$\varepsilon_{\alpha\beta\gamma\delta} = \det(g_{\mu\nu}) \varepsilon^{\alpha\beta\gamma\delta} \, .$$

Solution 3.22. Evaluate this scalar in a local orthonormal frame. According to Solution 3.21, $\varepsilon_{\mu\nu\rho\sigma} = -\varepsilon^{\mu\nu\rho\sigma}$ so

$$\varepsilon_{\mu\nu\rho\sigma}\varepsilon^{\mu\nu\rho\sigma} = -\sum_{\mu\nu\rho\sigma} |\varepsilon_{\mu\nu\rho\sigma}|^2 = -\sum |\varepsilon_{0123}|^2$$

where the sum is over permutations of 0123. Since there are $4! = 24$ permutations, $\varepsilon_{\mu\nu\rho\sigma}\varepsilon^{\mu\nu\rho\sigma} = -24$.

Solution 3.23. In an orthonormal frame (with $g_{\alpha\beta} = \eta_{\alpha\beta}$) it should be evident that the antisymmetric nature of ε is precisely what is needed to perform all of the operations in finding the determinant. That is,

$$\varepsilon_{\alpha\beta\gamma\delta} A^{\alpha}_{0} A^{\beta}_{1} A^{\gamma}_{2} A^{\delta}_{3} = \det \|A^{\alpha}_{\beta}\| \ .$$

Generalizing from the $(0,1,2,3)$ to arbitrary (μ,ν,λ,σ) requires placing $\varepsilon_{\mu\nu\lambda\sigma}$ on the right-hand side to account for the minus signs introduced by interchange of rows.

To show that the result is true in all (not just orthonormal) frames, we need to show that it is a tensor equation. This is equivalent to showing that $\det \|A^{\alpha}_{\beta}\|$ is a scalar:

$$\det \|A^{\alpha'}_{\beta'}\| = \det \left\| A^{\alpha}_{\beta} \frac{\partial x^{\alpha'}}{\partial x^{\alpha}} \frac{\partial x^{\beta}}{\partial x^{\beta'}} \right\|$$

$$= \det \|A^{\alpha}_{\beta}\| \det \left\| \frac{\partial x^{\alpha'}}{\partial x^{\alpha}} \right\| \det \left\| \frac{\partial x^{\beta}}{\partial x^{\beta'}} \right\| = \det \|A^{\alpha}_{\beta}\| \ .$$

[The last step follows from the fact that $\partial x^{\alpha'}/\partial x^{\alpha}$ and $\partial x^{\alpha}/\partial x^{\alpha'}$ are matrix inverses.]

Solution 3.24. Suppose $\alpha\mathbf{u} + \beta\mathbf{v} + \gamma\mathbf{w} + \delta\mathbf{x} = 0$ for $\alpha, \beta, \gamma, \delta$ not all zero. If $\alpha \neq 0$, take the wedge product of this equation with $\mathbf{v} \wedge \mathbf{w} \wedge \mathbf{x}$ and get $\mathbf{u} \wedge \mathbf{v} \wedge \mathbf{w} \wedge \mathbf{x} = 0$. If $\alpha = 0$, then form analogous product using the β, γ, or δ term. Conversely, if $\mathbf{u}, \mathbf{w}, \mathbf{x}$ and $\mathbf{v}$ are linearly independent, then any vector can be written as a linear combination of them; in particular, any 4 orthonormal basis vectors can be so written. Thus

$$0 \neq \mathbf{e}_0 \wedge \mathbf{e}_1 \wedge \mathbf{e}_2 \wedge \mathbf{e}_3 = (\alpha_1\mathbf{u} + \beta_1\mathbf{v} + \gamma_1\mathbf{w} + \delta_1\mathbf{x}) \wedge (\alpha_2\mathbf{u} + \cdots)$$

$$\wedge\, (\alpha_3\mathbf{u} + \cdots) \wedge (\alpha_4\mathbf{u} + \cdots) \propto (\mathbf{u} \wedge \mathbf{v} \wedge \mathbf{w} \wedge \mathbf{x}) \ ,$$

since cross terms with a duplicated vector vanish. Hence $\mathbf{u}\wedge\mathbf{v}\wedge\mathbf{w}\wedge\mathbf{x} \neq 0$. Since $\mathbf{u}\wedge\mathbf{v}\wedge\mathbf{w}\wedge\mathbf{x}$ is totally antisymmetric on all "slots" and is a multiple of $\mathbf{e}_0\wedge\mathbf{e}_1\wedge\mathbf{e}_2\wedge\mathbf{e}_3$, it must uniquely be proportional to the ε tensor. See Problems 3.20, 3.21.

Solution 3.25.

$$*F^{\mu\nu} = \frac{1}{2}\varepsilon^{\mu\nu\alpha\beta}F_{\alpha\beta} \tag{1}$$

$$\begin{aligned}
*F_{\mu\nu} &= \frac{1}{2}\eta_{\mu\sigma}\eta_{\nu\tau}\varepsilon^{\sigma\tau\alpha\beta}F_{\alpha\beta} \equiv \frac{1}{2}\varepsilon_{\mu\nu}{}^{\alpha\beta}F_{\alpha\beta} \\
&= \frac{1}{2}(\varepsilon_{\mu\nu\lambda\gamma}\eta^{\alpha\lambda}\eta^{\beta\gamma})F_{\alpha\beta} = \frac{1}{2}\varepsilon_{\mu\nu\lambda\gamma}(\eta^{\alpha\lambda}\eta^{\beta\gamma}F_{\alpha\beta}) \\
&= \frac{1}{2}\varepsilon_{\mu\nu\alpha\beta}F^{\alpha\beta}
\end{aligned} \tag{2}$$

$$*(*F^{\mu\nu}) = \frac{1}{2}\varepsilon^{\mu\nu\alpha\beta}(*F_{\alpha\beta}) = \frac{1}{4}\varepsilon^{\mu\nu\alpha\beta}\varepsilon_{\alpha\beta\lambda\sigma}F^{\lambda\sigma}\ . \tag{3}$$

Now, $\varepsilon^{\mu\nu\alpha\beta}\varepsilon_{\alpha\beta\lambda\sigma} \equiv -2\delta^{\mu\nu}_{\lambda\sigma}$ (this defines the symbol $\delta^{\mu\nu}_{\lambda\sigma}$). One should convince oneself that $\delta^{\mu\nu}_{\lambda\sigma}$ has the following properties:

$$\begin{aligned}
\delta^{\mu\nu}_{\lambda\sigma} &= +1 \text{ if } \lambda = \mu,\ \sigma = \nu \\
&= -1 \text{ if } \lambda = \nu,\ \sigma = \mu \\
&= \ \ 0 \text{ otherwise}\ .
\end{aligned}$$

Thus

$$\delta^{\mu\nu}_{\lambda\sigma} = \delta^{\mu}_{\lambda}\delta^{\nu}_{\sigma} - \delta^{\mu}_{\sigma}\delta^{\nu}_{\lambda}$$

and therefore

$$*(*F^{\mu\nu}) = -\frac{1}{2}\delta^{\mu\nu}_{\lambda\sigma}F^{\lambda\sigma} = -\frac{1}{2}(F^{\mu\nu} - F^{\nu\mu}) = -F^{\mu\nu}\ .$$

Solution 3.26. From the definition $*V^{\alpha\beta\gamma} = V_\lambda\varepsilon^{\lambda\alpha\beta\gamma}$, and

$$*V_{\alpha\beta\gamma}\,*V^{\alpha\beta\gamma} = V^\mu V_\lambda\varepsilon_{\mu\alpha\beta\gamma}\varepsilon^{\lambda\alpha\beta\gamma}\ .$$

Since μ must be a different index than α, β, or γ, and λ must also be, it follows that

$$\varepsilon_{\mu\alpha\beta\gamma}\varepsilon^{\lambda\alpha\beta\gamma} = C\delta^{\lambda}_{\mu}$$

for some constant C. If we sum over μ and λ, use $\delta^{\mu}_{\mu} = 4$, and use the result of Problem 3.22 we see that $C = -6$ so

$$*V_{\alpha\beta\gamma}\, *V^{\alpha\beta\gamma} = -6\delta^{\lambda}_{\mu}V^{\mu}V_{\lambda} = -6V^{\sigma}V_{\sigma}\ .$$

Solution 3.27. Notice that $\delta^{\mu\cdots\lambda}_{\rho\cdots\sigma}$ is antisymmetric in both its upper and lower indices. This follows from the fact that the sign of a determinant changes for an odd permutation of its rows or columns. Since, for example, the upper indices are antisymmetric there can be no repeated index. If there are more than four upper indices there must be a repeated index so $\delta^{\mu\cdots\lambda}_{\rho\cdots\sigma}$ must be identically zero.

Solution 3.28. The indices μ, ν, λ, κ must take on different values from ρ, σ. For given ρ, σ, we can have μ equal to κ (with ν then equal to λ) or μ equal to λ (with ν then equal to κ). All other possibilities vanish because of the antisymmetry of $\varepsilon^{\mu\nu\rho\sigma}$. Furthermore, because of that antisymmetry, the $\mu = \kappa$, $\nu = \lambda$ case has a minus sign. Thus

$$\varepsilon^{\mu\nu\rho\sigma}\varepsilon_{\lambda\kappa\rho\sigma} = C(\delta^{\mu}_{\lambda}\delta^{\nu}_{\kappa} - \delta^{\mu}_{\kappa}\delta^{\nu}_{\lambda}) \equiv C\delta^{\mu\nu}_{\lambda\kappa}$$

for some C. Summing on μ, λ and ν, k and using the results of Problem 3.22 gives $C = -2$. Thus

$$\delta^{\mu\nu}_{\lambda\kappa} = -\frac{1}{2}\varepsilon^{\mu\nu\rho\sigma}\varepsilon_{\lambda\kappa\rho\sigma}\ .$$

In general

$$\varepsilon^{\mu\nu\lambda\tau}\varepsilon_{\iota\kappa\rho\sigma} = -\delta^{\mu\nu\lambda\tau}_{\iota\kappa\rho\sigma}$$

$$\varepsilon^{\mu\nu\lambda\tau}\varepsilon_{\iota\kappa\rho\tau} = -\delta^{\mu\nu\lambda}_{\iota\kappa\rho}\ .$$

Solution 3.29. If $P^{\alpha\beta} = A^{[\alpha}B^{\beta]}$, then

$$\det\begin{bmatrix} p^{\alpha\beta} & p^{\alpha\gamma} & p^{\alpha\delta} \\ A^{\beta} & A^{\alpha} & A^{\delta} \\ B^{\beta} & B^{\gamma} & B^{\delta} \end{bmatrix} = 0$$

since the first row is a linear combination of the other two rows (first row $= -\frac{1}{2} A^{\alpha} \times$ third row $+ \frac{1}{2} B^{\alpha} \times$ second row). Multiplying out the determinant gives the desired result.

Solution 3.30. For each value of μ, there are 3! arrangements of α, β, $\gamma \neq \mu$ which make the ε nonzero. The Jacobian determinant has the same property of changing sign that the ε does, so with the 3! in the denominator, we can just take one representative element of each permutation and get the right answer:

$$\begin{aligned} d^3\Sigma_0 &= [\partial(a,b,c)/\partial(a,b,c)]\, da\, db\, dc = da\, db\, dc \\ d^3\Sigma_1 &= -[\partial(\text{const.},b,c)/\partial(a,b,c)]\, da\, db\, dc = 0 \\ d^3\Sigma_2 &= -[\partial(a,\text{const.},c)/\partial(a,b,c)]\, da\, db\, dc = 0 \\ d^3\Sigma_3 &= -[\partial(a,b,\text{const.})/\partial(a,b,c)]\, da\, db\, dc = 0 \ . \end{aligned}$$

Solution 3.31. Let barred coordinates represent an orthonormal frame where the metric is $\eta_{\mu\nu}$. Then $dV \equiv d^4\bar{x} = \det(\partial\bar{x}/\partial x)\, d^4x$ because the Jacobian determinant relates volumes. Now

$$-g = -\det(g_{\alpha\beta}) = -\det\left(\frac{\partial x^{\bar{\mu}}}{\partial x^{\alpha}} \frac{\partial x^{\bar{\nu}}}{\partial x^{\beta}} \eta_{\bar{\mu}\bar{\nu}}\right) = [\det(\partial\bar{x}/\partial x)]^2 (-\det\eta) \ ,$$

so that

$$(-g)^{\frac{1}{2}} = \det(\partial\bar{x}/\partial x)$$

and hence

$$dV = (-g)^{\frac{1}{2}}\, d^4x \ .$$

Solution 3.32. In a comoving local orthonormal frame we have $d^3V = dx\, dy\, dz$; we want a scalar invariant quantity which reduces to this.

Start with $d^4V = (-g)^{\frac{1}{2}}\, dx\, dy\, dx\, dt$, a scalar in any frame. Multiply by $u^0/u^0 = u^0/(dt/ds)$ and get

$$d^4V = [(-g)^{\frac{1}{2}} u^0\, dx\, dy\, dz]\, ds \ .$$

Since d^4V and ds are invariants, the term in parentheses must also be, and it reduces to $dx\, dy\, dz$ in a comoving orthonormal frame, so

$$d^3V = (-g)^{\frac{1}{2}}\, u^0\, dx\, dy\, dz \ .$$

Solution 3.33. Since the contravariant momentum vector P^α transforms just like the contravariant displacement vector x^α, the 4-dimensional momentum element must be of the same form as the invariant 4-volume element (Problem 3.32), so

$$d^4P = (-g)^{\frac{1}{2}}\, dP^x\, dP^y\, dP^z\, dP^t \ .$$

For the 3-volume element we multiply by a δ-function of the constraint and integrate over P^t:

$$\begin{aligned} d^3P &= \int \delta[(-g_{\alpha\beta}\, P^\alpha\, P^\beta)^{\frac{1}{2}} - m]\, (-g)^{\frac{1}{2}}\, dP^x dP^y dP^z dP^t \\ &= (-g)^{\frac{1}{2}} dP^x dP^y dP^z \times \left[-\frac{1}{2}\,(-g_{\alpha\beta}\, P^\alpha P^\beta)^{-\frac{1}{2}} \cdot 2g_{t\alpha}\, P^\alpha\right]^{-1} . \end{aligned}$$

(Here we have used the identity

$$\int \delta(f(x))\, dx = \frac{1}{|f'(x_1)|}$$

where x_1 is a zero of f.) Rewriting the term in brackets, we get

$$d^3P = (-g)^{\frac{1}{2}}\, dP^x\, dP^y\, dP^z \left(\frac{m}{-P_t}\right) .$$

Notice that in a local orthonormal frame comoving with the particles in momentum space, this reduces to

$$d^3P = dP^x\, dP^y\, dP^z$$

as it should.

Sometimes d^3P is renormalized by dividing by m. This gives an invariant volume element which is valid for massless particles as well.

Solution 3.34. Clearly the number of particles N is invariant. We need to show that $dx\,dy\,dz\,dP^x\,dP^y\,dP^z$ is also invariant. If the particles are moving with 4-velocity u relative to the observer's xyz coordinate system, he measures an invariant 3-volume element occupied by them (see Problem 3.32),

$$d^3V = (-g)^{\frac{1}{2}} u^0\, dx\, dy\, dz$$

and an invariant 3-momentum element occupied by them (see Problem 3.33),

$$d^3P = (-g)^{\frac{1}{2}}\, dP^x\, dP^y\, dP^z \left(\frac{-1}{u_0}\right).$$

Since he measures with an orthonormal frame, we have $-g = 1$ and $u^0 = -u_0$, so

$$dx\, dy\, dz\, dP^x\, dP^y\, dP^z = d^3V\, d^3P$$

which is an invariant.

Solution 3.35.

(a) Choose a 4-volume bounded by x^0_A, by $x^0_B > x^0_A$, and by "side" hypersurfaces at infinite distance from the origin. By Gauss' theorem

$$0 = \int J^\alpha{}_{,\alpha}\, d^4\Omega = \int J^\alpha{}_{,\alpha}\, dt\, dx\, dy\, dz = \int J^\alpha\, d^3\Sigma_\alpha\ .$$

The contributions on the sides can be ignored as they are moved to infinity, so

$$0 = \int_{x^0_\beta} J^\alpha\, d^3\Sigma_\alpha + \int_{x^0_A} J^\alpha\, d^3\Sigma_\alpha = \int_{x^0_\beta} J^0\, dx\, dy\, dz - \int_{x^0_A} J^0\, dx\, dy\, dz\ .$$

(b) The planes x^0 = constant, and $x^{0'}$ = constant intersect forming two 4-dimensional regions: I and II. Close off these regions with spatial hypersurfaces at infinity forming two bounded 4-volumes. By Gauss' theorem, for region I

$$0 = \int_{I} J^{\alpha}{}_{,\alpha} d^4\Omega = 0 = \int_{Ix^0} J^0 d^3\Sigma_0 - \int_{Ix^{0'}} J^{0'} d^3\Sigma_0$$

or

$$\int_{Ix^0} J^0 dx\,dy\,dz = \int_{Ix^{0'}} J^{0'} dx' dy' dz' ,$$

and similarly for region II. The proof is completed by adding the equations for the two regions.

CHAPTER 4: SOLUTIONS

Solution 4.1. Suppose the wire lies on the z-axis of cylindrical polar coordinates and has proper charge density ρ_0. In the rest frame of the wire $J^0 = \rho_0$ and $\underset{\sim}{J} = 0$, and from Gauss' law the only nonvanishing component of the field tensor is $E^r = \rho_0 A/2\pi r = F^{0r}$, where A is the cross sectional area of the wire. If the wire is moving in the +z direction with velocity β relative to the lab, the Lorentz transformation $\Lambda^{0'}{}_0 = \gamma$, $\Lambda^{0'}{}_z = \beta\gamma$ gives

$$J^{0'} = \gamma\rho_0 \qquad J^{z'} = \beta\gamma\rho_0$$

$$B^{\hat{\phi}} = F^{z'r'} = \Lambda^{z'}{}_0 F^{0r} = \beta\gamma\rho_0 A/2\pi r \qquad E^{r'} = \gamma\rho_0 A/2\pi r \ .$$

($B^{\hat{\phi}}$ is the "physical component," i.e. the component related to the *unit* vector $e_{\hat{\phi}}$.) If we now superpose the field of a similar wire of charge density $-\gamma\rho_0$, at rest in the lab, the total charge density and $\underset{\sim}{E}$ field both cancel. The current $I = \int J^{z'} dx\, dy = \beta\gamma\rho_0 A$ produces only a B field $B^{\hat{\phi}} = I/2\pi r$.

Solution 4.2. The invariance of these quantities follows from the fact that they are equal to scalars:

$$B^2 - E^2 = \tfrac{1}{2} F^{\alpha\beta} F_{\alpha\beta}$$

$$\underset{\sim}{E}\cdot\underset{\sim}{B} = \tfrac{1}{4} {}^*F_{\alpha\beta} F^{\alpha\beta} = \tfrac{1}{8}\, \varepsilon_{\mu\nu\alpha\beta} F^{\mu\nu} F^{\alpha\beta} = \det(F^{\mu\nu}) \ .$$

Any invariant must be an invariant for rotations in 3-space. Thus the invariants can only be constructed from the scalars $\underset{\sim}{E}\cdot\underset{\sim}{B}$, $\underset{\sim}{E}\cdot\underset{\sim}{E}$ and $\underset{\sim}{B}\cdot\underset{\sim}{B}$. If there were three independent invariants, then it would follow that e.g. $\underset{\sim}{B}\cdot\underset{\sim}{B}$ would be invariant. This is obviously untrue, since B^2 changes in general under a Lorentz transformation.

Solution 4.3. In any frame the angle is given by

$$\cos\theta_0 = \underset{\sim}{E}\cdot\underset{\sim}{B}/|E|\,|B| \ .$$

Now $\underset{\sim}{E}\cdot\underset{\sim}{B}$ is invariant (see Problem 4.2), but $|E|\,|B|$ is not. Thus, θ_0 can be invariant if and only if $\underset{\sim}{E}\cdot\underset{\sim}{B} = 0$, i.e. $\theta_0 = \pi/2$.

Solution 4.4. The expression is shown to be invariant by relating it to the invariants $E^2 - B^2$ and $\underset{\sim}{E}\cdot\underset{\sim}{B}$:

$$\begin{aligned} 64\pi(\mathcal{E}^2 - |\underset{\sim}{S}|^2) &= (E^2+B^2)^2 - 4(\underset{\sim}{E}\times\underset{\sim}{B})^2 \\ &= [E^4 + 2E^2B^2 + B^4] - 4[E^2B^2 - (\underset{\sim}{E}\cdot\underset{\sim}{B})^2] \\ &= E^4 - 2E^2B^2 + B^4 + 4(\underset{\sim}{E}\cdot\underset{\sim}{B})^2 \\ &= (E^2 - B^2)^2 + 4(\underset{\sim}{E}\cdot\underset{\sim}{B})^2 \ . \end{aligned}$$

Solution 4.5. Let $\underset{\sim}{v} = \alpha(\underset{\sim}{E}\times\underset{\sim}{B})$, then

$$\underset{\sim}{E}'\gamma^{-1} = \underset{\sim}{E} + \underset{\sim}{v}\times\underset{\sim}{B} = (1-\alpha B^2)\underset{\sim}{E} + \alpha(\underset{\sim}{E}\cdot\underset{\sim}{B})\underset{\sim}{B}$$

$$\underset{\sim}{B}'\gamma^{-1} = \underset{\sim}{B} - \underset{\sim}{v}\times\underset{\sim}{E} = (1-\alpha E^2)\underset{\sim}{B} + \alpha(\underset{\sim}{E}\cdot\underset{\sim}{B})\underset{\sim}{E} \ .$$

Case i) If $\underset{\sim}{E}\cdot\underset{\sim}{B} = 0$ and $|E| = |B|$, the transformation is equivalent to the redshift of a plane wave, and $\underset{\sim}{E}$ and $\underset{\sim}{B}$ cannot be made parallel.

Case ii) If $\underset{\sim}{E}\cdot\underset{\sim}{B} = 0$ but $E^2 \neq B^2$ choose $\alpha = 1/\max(E^2, B^2)$ to make $\underset{\sim}{E}'$ or $\underset{\sim}{B}'$ vanish. Then $\underset{\sim}{E}'\times\underset{\sim}{B}' = 0$.

Case iii) If $\underset{\sim}{E}\cdot\underset{\sim}{B} \neq 0$, then $\underset{\sim}{E}'$ and $\underset{\sim}{B}'$ can be made parallel by choosing α to satisfy $\alpha^2[(\underset{\sim}{E}\cdot\underset{\sim}{B})^2 - E^2B^2] = 1 - \alpha(E^2+B^2)$, so that

$$\frac{\underset{\sim}{v}}{1+v^2} = \frac{\alpha(\underset{\sim}{E}\times\underset{\sim}{B})}{1+\alpha^2[(\underset{\sim}{E}\times\underset{\sim}{B})]^2} = \frac{\alpha(\underset{\sim}{E}\times\underset{\sim}{B})}{1+\alpha^2[E^2B^2 - (\underset{\sim}{E}\cdot\underset{\sim}{B})^2]} = \frac{\underset{\sim}{E}\times\underset{\sim}{B}}{E^2+B^2} \ .$$

Solution 4.6. If $E^2 - B^2 > 0$ perform a Lorentz transformation with $\underset{\sim}{v} = \underset{\sim}{E}\times\underset{\sim}{B}/E^2$ to get $\underset{\sim}{B}' = \gamma(\underset{\sim}{B} - \underset{\sim}{v}\times\underset{\sim}{E}) = 0$. The transformation is analogous for $E^2 - B^2 < 0$ and $\underset{\sim}{v} = -\underset{\sim}{B}\times\underset{\sim}{E}/B^2$. If $E^2 - B^2 = 0$ then clearly $|E| = |B|$ after any transformation. A transformation of the form (see

Problem 4.5) $\underset{\sim}{v} = \alpha(\underset{\sim}{E}\times\underset{\sim}{B})$ reduces the magnitude of $\underset{\sim}{E}$ or $\underset{\sim}{B}$ by a factor $\gamma(1-\alpha E^2) = \gamma(1-v) = [(1-v)/(1+v)]^{\frac{1}{2}}$. In the limit $v \to 1$ the magnitude of $\underset{\sim}{E}$ and $\underset{\sim}{B}$ can be made arbitrarily small.

Solution 4.7. The integration in the expression for J^μ can be performed by using

$$\int F(\tau)\delta[t-z^0(\tau)]\,d\tau = \int F(\tau)\delta[t-z^0(\tau)]\,(d\tau/dt)\,dt = F(\tau[t])/u^0 .$$

The result is

$$J^\mu = \sum_k \frac{u^\mu{}_k}{u^0{}_k}\, e_k \delta^3[\underset{\sim}{x} - \underset{\sim}{z}_k(t)] .$$

The time and spatial components obviously reduce to the correct expressions.

Solution 4.8. For this problem it is useful to introduce the totally antisymmetric tensor in three dimensions, ε_{ijk} (with normalization $\varepsilon_{123} = 1$). Note that

$$F^{ij} = \varepsilon^{ijk} B_k \qquad B^i = \frac{1}{2}\,\varepsilon^{ijk} F_{jk}$$

$$(\underset{\sim}{U}\times\underset{\sim}{V})^i = \varepsilon^{ijk} U^j V^k .$$

Now, for the "electric" equations $F^{\mu\nu}{}_{,\nu} = 4\pi J^\mu$:

$$F^{0\beta}{}_{,\beta} = F^{0i}{}_{,i} = \underset{\sim}{\nabla}\cdot\underset{\sim}{E} = 4\pi J^0 = 4\pi\rho$$

$$F^{i\beta}{}_{,\beta} = F^{ij}{}_{,j} + F^{i0}{}_{,0} = \varepsilon^{ijk} B_{k,j} - E^i{}_{,0} = 4\pi J^i .$$

This last equation is the ith component of

$$\underset{\sim}{\nabla}\times\underset{\sim}{B} - \dot{\underset{\sim}{E}} = 4\pi \underset{\sim}{J} .$$

In the "magnetic" equations first assume α, β, γ are all spatial. Take $\alpha = 1$, $\beta = 2$, $\gamma = 3$.

$$F_{12,3} + F_{31,2} + F_{23,1} = B^3{}_{,3} + B^2{}_{,2} + B^1{}_{,1} = \underset{\sim}{\nabla}\cdot\underset{\sim}{B} = 0 .$$

Now take $\alpha = 0$, and β, γ spatial:

$$F_{0i,j} + F_{j0,i} + F_{ij,0} = E_{i,j} + E_{j,i} + \varepsilon_{ijk}\dot{B}^k$$
$$= \varepsilon_{ijk}(\underset{\sim}{\nabla} \times \underset{\sim}{E})^k + \varepsilon_{ijk}\dot{B}^k = 0 \ .$$

Multiplying this equation by ε^{ijm} gives us $\underset{\sim}{\nabla} \times \underset{\sim}{E} + \dot{\underset{\sim}{B}} = 0$.

Solution 4.9. It has already been shown (Problem 4.8) that $F^{\mu\nu}{}_{,\nu} = 0$ is equivalent to the "electric" Maxwell equations (in vacuum). As for the magnetic equations:

$$*F^{\mu\nu}{}_{,\nu} = \frac{1}{2}(F_{\alpha\beta}\varepsilon^{\alpha\beta\mu\nu})_{,\nu} = \frac{1}{2}F_{\alpha\beta,\nu}\varepsilon^{\alpha\beta\mu\nu}$$
$$= \frac{1}{2}F_{[\alpha\beta,\nu]}\varepsilon^{\alpha\beta\mu\nu} \ ,$$

so that $*F^{\mu\nu}{}_{,\nu} = 0$ is equivalent to $F_{\alpha\beta,\nu} + F_{\nu\alpha,\beta} + F_{\beta\nu,\alpha} = 0$.

Solution 4.10. With $\mu = 0$ the Lorentz force equation is

$$dP^0/d\tau = eF^{0i}u_i = eE^i\gamma v_i \ .$$

Since $d\tau = dt/\gamma$ this equation can be written

$$dP^0/dt = eE^i v_i \ .$$

Note that this differs from the nonrelativistic equation only in that P^0 is the relativistic mass-energy.

Solution 4.11. The spatial components of the Lorentz equation are

$$\frac{dP^i}{d\tau} = \gamma\frac{dP^i}{dt} = qF^{i\mu}u_\mu = qF^{i0}u_0 + qF^{ij}u_j = q\gamma E^i + q\gamma\varepsilon^{ijk}B_k v_j \ ,$$

so that

$$d\underset{\sim}{P}/dt = q(\underset{\sim}{E} + \underset{\sim}{v} \times \underset{\sim}{B}) \ .$$

Solution 4.12. The only nonzero components of $F^{\mu}{}_{\nu}$ are $F^{t}{}_{y} = F^{y}{}_{t} = E$. The initial 4-velocity u^{α} of the particle has components $(\gamma, \gamma v, 0, 0)$, $\gamma \equiv (1-v^2)^{-\frac{1}{2}}$; the Lorentz force equations

$$dP^{\mu}/d\tau = qF^{\mu}{}_{\nu}u^{\nu}$$

then give us $dP^x/d\tau = 0$ for $\mu = 1$ and hence

$$u^x = \gamma v \qquad \tau = x/\gamma v \ .$$

The equations for $\mu = 0, 2$,

$$du^t/d\tau = (q/m)\,Eu^y \qquad du^y/d\tau = (q/m)\,Eu^t$$

can be combined to give the equation

$$d^2u^y/d\tau^2 = (qE/m)^2u^y$$

with the solution

$$u^y = \gamma \sinh[(qE/m)\tau] \ ,$$

where the initial conditions on u^t and u^y have been used. Integrating this result with respect to τ, substituting $\tau = x/\gamma v$, and using the initial condition $y=0$ when $x=0$, gives

$$y = \left(\frac{m}{qE}\right)\gamma\left[\cosh\left(\frac{qE}{m}\right)\frac{x}{v\gamma} - 1\right] .$$

Solution 4.13.

(a) From the Lorentz force equation (see Problem 4.11)

$$\omega p = |d\underset{\sim}{p}/dt| = qvB \ ,$$

therefore

$$B = \frac{\omega p}{qv} = \frac{m\omega}{q(1-v^2)^{\frac{1}{2}}} = \frac{m\omega}{q(1-\omega^2R^2)^{\frac{1}{2}}} \ .$$

(b) The 4-velocity components seen in the lab are

$$u^0 = (1-\omega^2R^2)^{-\frac{1}{2}} \cdot \qquad u^x = \omega y\,(1-\omega^2R^2)^{-\frac{1}{2}} \qquad u^y = -\omega x\,(1-\omega^2R^2)^{-\frac{1}{2}} .$$

The components measured by the moving observer are found with a Lorentz transformation, e.g.

$$u^{0'} = \gamma(u^0 - \beta u^x) = \gamma(1-\beta\omega y)(1-\omega^2R^2)^{-\frac{1}{2}} ,$$

where $\gamma \equiv (1-\beta^2)^{-\frac{1}{2}}$.

(c) Since $u^{0'}$ is not constant, energy is seen to increase:

$$\frac{dp^{0'}}{d\tau} = m\,\frac{du^{0'}}{d\tau} = -\,\frac{m\beta\omega\gamma u^y}{(1-\omega^2R^2)^{\frac{1}{2}}} .$$

In the moving observer's frame he observes an $\underset{\sim}{E}$ field $E^{y'} = -\gamma\beta B$. (This is easily computed by Lorentz transforming $F^{\mu\nu}$.) The observer should then expect to see work being done on the particle at a rate

$$\frac{dp^{0'}}{d\tau} = qE^{y'}u^{y'} = -\,\frac{m\omega\gamma u^{y'}}{(1-\omega^2R^2)^{\frac{1}{2}}} = -\,\frac{m\omega\gamma u^y}{(1-\omega^2R^2)^{\frac{1}{2}}} .$$

This resolves the "paradox."

Solution 4.14. From the Lorentz force equation (Problem 4.11),

$$\frac{d\underset{\sim}{p}}{dt} = \left(-q\omega RB + \frac{qQ}{R^2}\right)\underset{\sim}{e}_r = m\,\frac{d}{dt}(\gamma\underset{\sim}{v}) .$$

No work is being done on the particle so $\gamma = (1-\omega^2R^2)^{-\frac{1}{2}}$ is a constant. Since $d\underset{\sim}{v}/dt = -\,\omega^2R\underset{\sim}{e}_r$ we have

$$q\omega RB - qQ/R^2 = m\omega^2R/(1-\omega^2R^2)^{\frac{1}{2}}$$

and hence

$$\frac{q}{m} = \frac{\omega^2}{(1-\omega^2R^2)^{\frac{1}{2}}\,(\omega B - Q/R^3)} .$$

Solution 4.15. The form of the electromagnetic stress-energy is given in the introduction. Taking its divergence we have:

$$4\pi T^{\mu\nu}{}_{,\nu} = F^{\mu\alpha}{}_{,\nu} F^{\nu}{}_{\alpha} + F^{\mu\alpha}F^{\nu}{}_{\alpha,\nu} - \frac{1}{2} F_{\alpha\beta}F^{\alpha\beta,\mu}$$

$$= F^{\mu\alpha}F^{\nu}{}_{\alpha,\nu} + F_{\nu\alpha}(F^{\mu\alpha,\nu} - \frac{1}{2} F^{\nu\alpha,\mu})$$

$$= F^{\mu\alpha} F^{\nu}{}_{\alpha,\nu} + \frac{1}{2} F_{\nu\alpha}(F^{\alpha\nu,\mu} + F^{\mu\alpha,\nu} + F^{\nu\mu,\alpha}) \; .$$

According to the free field Maxwell equations, both $F^{\nu}{}_{\alpha,\nu}$ and the terms in parentheses vanish, hence $T^{\mu\nu}{}_{,\nu} = 0$. Note: If a charge current $\underset{\sim}{J}$ is present this result is modified to $T^{\mu\nu}{}_{,\nu} = -F^{\mu\alpha}J_{\alpha}$, as is easily verified.

Solution 4.16. From the stress-energy $T^{\mu\nu} = (1/4\pi)(F^{\mu\alpha}F^{\nu}{}_{\alpha} - \frac{1}{4}\eta^{\mu\nu}F_{\alpha\beta}F^{\alpha\beta})$ the trace is easily computed to be

$$T^{\mu}{}_{\mu} = (1/4\pi)(F^{\mu\alpha} F_{\mu\alpha} - \frac{1}{4} \cdot 4 \cdot F_{\alpha\beta} F^{\alpha\beta}) = 0 \; .$$

Solution 4.17. It is sufficient to show that this relation is true in one frame.

$$\text{Case 1:} \quad E^2 = B^2, \quad \underset{\sim}{E} \cdot \underset{\sim}{B} = 0 \; .$$

Choose $\underset{\sim}{E} = \varepsilon \underset{\sim}{e}_x$, $\underset{\sim}{B} = \varepsilon \underset{\sim}{e}_y$; then the only nonvanishing components of $T^{\mu\nu}$ are $T^{00} = T^{0z} = T^{zz} = \varepsilon^2/4\pi$ from which it follows that $T^{\mu}{}_{\alpha}T^{\alpha}{}_{\nu} = 0$.

$$\text{Case 2:} \quad (E^2 - B^2)^2 + (\underset{\sim}{E} \cdot \underset{\sim}{B})^2 \neq 0 \; .$$

From Problem 4.5 we can make $\underset{\sim}{E}$ and $\underset{\sim}{B}$ parallel. Choose $\underset{\sim}{E} = E\underset{\sim}{e}_x$, $\underset{\sim}{B} = B\underset{\sim}{e}_x$. The nonvanishing components of the stress-energy are then $T^{00} = -T^{xx} = T^{yy} = T^{zz} = (1/8\pi)(E^2 + B^2)$ and

$$T^{\mu}{}_{\alpha} T^{\alpha}{}_{\nu} = \delta^{\mu}{}_{\nu} \, [E^2 + B^2]^2/(8\pi)^2 = \delta^{\mu}{}_{\nu} \, [(E^2 - B^2)^2 + 4E^2B^2]/(8\pi)^2$$

$$= \delta^{\mu}{}_{\nu} \, [(E^2 - B^2)^2 + (2\underset{\sim}{E} \cdot \underset{\sim}{B})^2]/(8\pi)^2 \; .$$

Solution 4.18. If u^{μ} is the 4-velocity of the conducting element, then the $\underset{\sim}{E}$ field seen by charge carriers in the conducting element is $E^{\mu} = F^{\mu}{}_{\nu}u^{\nu}$. A vector which reduces to $\underset{\sim}{J}$ in the rest frame of the charge carriers is $J^{\mu} + u^{\mu} J_{\nu} u^{\nu}$. Ohm's equation can then be written as

$J^\mu + u^\mu J_\nu u^\nu = \sigma F^\mu{}_\nu u^\nu$. Since this is a tensor equation and is correct in the rest frame of the conducting element, it must be correct in all frames.

Solution 4.19. From Problem 4.7 we know that, for a particle of charge q

$$J^\mu(x) = q \int \delta^4(x - z[\tau])\, u^\mu d\tau \ ,$$

so that the action can be written as

$$\int L d\tau = q \int u^\mu A_\mu d\tau - m \int d\tau$$

and the Lagrangian is:

$$L = qu^\mu A_\mu - m(-\eta_{\alpha\beta} u^\alpha u^\beta)^{\frac{1}{2}} \ .$$

The coordinates z^λ specify the position of the particle parametrized by τ and

$$\frac{\partial L}{\partial(dz^\lambda/d\tau)} = \frac{\partial L}{\partial u^\lambda} = qA_\lambda + mu_\lambda \ .$$

The Euler-Lagrange equations then give us

$$\frac{d}{d\tau}\left(\frac{\partial L}{\partial u^\lambda}\right) = qA_{\lambda,\mu}\frac{dz^\mu}{d\tau} + m\frac{du_\lambda}{d\tau} = \frac{\partial}{\partial z^\lambda} L = qu^\mu A_{\mu,\lambda} \ ,$$

so that the equations of motion are

$$m\, du_\lambda/d\tau = q(A_{\mu,\lambda} - A_{\lambda,\mu})\, u^\mu$$

or

$$dp_\lambda/d\tau = qF_{\lambda\mu} u^\mu \ .$$

Solution 4.20.

(a) From the form of $F^{\mu\nu}$ in terms of $\underset{\sim}{E}$ and $\underset{\sim}{B}$ (see the introduction to this chapter) and from the definition for $*F$ (see Problem 3.25) we have

$$*F^{\alpha\beta} = \frac{1}{2}\,\varepsilon^{\alpha\beta\mu\nu}F_{\mu\nu}$$

$$= \begin{bmatrix} 0 & -B^x & -B^y & -B^z \\ B^x & 0 & E^z & -E^y \\ B^y & -E^z & 0 & E^x \\ B^z & E^y & -E^x & 0 \end{bmatrix},$$

therefore $F \to *F$ corresponds to $\underset{\sim}{E} \to -\underset{\sim}{B}$ and $\underset{\sim}{B} \to \underset{\sim}{E}$.

(b) [Note: Since $*(*F) = -F$, if we treat $*$ as an operator it has the same algebraic properties as i, hence we write $e^{*\alpha}F = F\cos\alpha + *F\sin\alpha$.] According to Problem 4.9 Maxwell's equations, in the absence of sources, are

$$*F^{\mu\nu}{}_{,\nu} = 0 \qquad F^{\mu\nu}{}_{,\nu} = 0\ .$$

Clearly the transformation $F \to *F$ leaves these equations invariant i.e. $*F$ is a solution if F is. Since the equations are linear, a linear combination such as $e^{*\alpha}F$ must also satisfy the equations.

Solution 4.21. Since we know that $F \to *F$ under the transformation

$$\underset{\sim}{E} \to -\underset{\sim}{B};\quad \underset{\sim}{B} \to \underset{\sim}{E}\ ,$$

then K would be interpreted as (minus) a charge-current of "magnetic charge." Note that by the $F \leftrightarrow *F$ duality $\underset{\sim}{\nabla}\cdot\underset{\sim}{B} = -4\pi K^0$ implies the existence of magnetic monopoles.

Solution 4.22. In 4-dimensional Euclidean space described with cartesian coordinates $x^i (i = 1,\cdots,4)$, consider the function $G = 1/\sum_i (x^i)^2$. It is straightforward to show that $\nabla^2 G \equiv \sum_i \partial^2 G/\partial x^{i^2} = 0$ except possibly at the point $x^i = 0$.

If we introduce 4-dimensional spherical coordinates

$$x^4 = U\cos V,\ \ r = U\sin V \quad (0 \le U < \infty,\ 0 \le V \le \pi)$$

$$\theta = \theta \qquad \phi = \phi\ .$$

then $G = U^{-2}$. The integral $\int \nabla^2 G\, d^4x$ can easily be written in 4-spherical coordinates. If the 4-volume is taken to be the pseudo-spherical region $U \leq U_0$, the volume integral can be converted into a surface integral on the 3-spherical surface $U = U_0$

$$\int_{U \leq U^0} \nabla^2 G\, d^4x = \int_0^{\pi} dV \int_0^{\pi} d\theta \int_0^{2\pi} d\phi \left(\frac{\partial G}{\partial U}\right) V^3 \sin^2 V \sin\theta = -4\pi^2 \ .$$

It follows therefore that $\nabla^2 G = -4\pi^2 \delta(x^1)\delta(x^2)\delta(x^3)\delta(x^4)$ and that we can therefore use G as a Green's function to solve the 4-dimensional Poisson problem $\nabla^2 \Phi = -4\pi f(x^i)$ as

$$\Phi(x^i) = \frac{1}{\pi} \int \frac{f(\zeta^i)\, d^4\zeta}{\sum_i (\zeta^i - x^i)^2} \ .$$

Now suppose that f is somehow defined for imaginary ζ^4. Introduce new variables $\zeta^4 = i\tilde{x}^4$, and $\zeta^i = \tilde{x}^i$, $i = 1,2,3$. Define S such that

$$f(\zeta^i) = S(x^i) \ .$$

Notice now that $\nabla^2 \Phi = -4\pi f(x^i)$ becomes $\Box\, \Phi = -4\pi S(x^1, x^2, x^3, t)$. The solution to this problem is the above integral. If we transform variables we have

$$\Phi(\underset{\sim}{x}, t) = \frac{-i}{\pi} \int \frac{S(\hat{x}, \hat{t})\, d^3\hat{x}\, d\hat{t}}{\eta_{\mu\nu}(\hat{x}^\mu - x^\mu)(\hat{x}^\nu - x^\nu)}$$

where it is understood that the t integration extends from $i\infty$ to $-i\infty$. Assume that $S(\underset{\sim}{\hat{x}}, \hat{t})$, the source term in the d'Alembertian equation, is defined and has no poles on the real $\hat{t}$ axis for $\hat{t} < t$. Take the definition of S elsewhere in the complex $\hat{t}$ plane to be defined by analytic continuation. The contour of integration is shown as C in Figure 1. The boundary conditions are specified by how the contour C is deformed to run along the real axis where the source $S(\underset{\sim}{\hat{x}}, \hat{t})$ is defined. The poles of

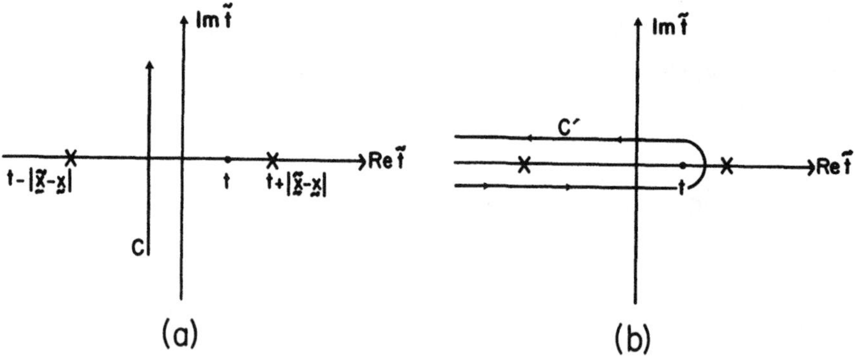

$1/(\hat{x}^\mu - x^\mu)(\hat{x}_\mu - x_\mu)$ are at $\hat{t} = t \pm |\hat{\underset{\sim}{x}} - \underset{\sim}{x}|$. Retarded boundary conditions correspond to deforming C into C′ in Figure 2. This can be verified by using the above result to solve

$$\Box A_\mu = -4\pi J_\mu \,.$$

We get:

$$A_\mu(x^\alpha) = \frac{1}{\pi i} \int_{C'} \frac{J_\mu(\hat{x}^\alpha)\, d^4\hat{x}}{r_\sigma r^\sigma} \,.$$

We now show that this reduces to the familiar retarded integral solution by performing the integral over dt:

$$A_\mu(\underset{\sim}{x},t) = -\frac{1}{\pi i} \int \frac{J_\mu(\hat{\underset{\sim}{x}},\hat{t})\, d^3\hat{x}\, d\hat{t}}{(\hat{t}-t-|\hat{\underset{\sim}{x}}-\underset{\sim}{x}|)(\hat{t}-t+|\hat{\underset{\sim}{x}}-\underset{\sim}{x}|)}$$

$$= -\frac{1}{\pi i} \int 2\pi i \left[\frac{J_\mu(\hat{\underset{\sim}{x}},\hat{t})\, d^3\hat{x}}{\hat{t}-t-|\hat{\underset{\sim}{x}}-\underset{\sim}{x}|}\right]_{\hat{t}=t-|\hat{\underset{\sim}{x}}-\underset{\sim}{x}|} = \int \frac{J_\mu(\hat{x},t-|\hat{\underset{\sim}{x}}-\underset{\sim}{x}|)\, d^3\hat{x}}{|\hat{\underset{\sim}{x}}-\underset{\sim}{x}|} \,.$$

Since $\partial(r_\sigma r^\sigma)^{-1}/\partial x^\nu = -2x^\nu (r_\sigma r^\sigma)^{-2}$, a formal expression for the field tensor is

$$F_{\mu\nu} = 2A_{[\nu,\mu]} = \frac{4}{\pi i} \int \frac{x_{[\nu} J_{\mu]}(\hat{x}^\alpha)\, d^4\hat{x}}{(r_\sigma r^\sigma)^2} \,.$$

Solution 4.23. Since the conductivity is infinite, the Lorentz force on a charge carrier must be zero, so $\underset{\sim}{E} + \underset{\sim}{v} \times \underset{\sim}{B} = 0$. Maxwell's equation $\dot{\underset{\sim}{B}} = -\underset{\sim}{\nabla} \times \underset{\sim}{E}$ gives

$$\dot{\underset{\sim}{B}} = \underset{\sim}{\nabla}\times(\underset{\sim}{v}\times\underset{\sim}{B}) = -\underset{\sim}{B}(\underset{\sim}{\nabla}\cdot\underset{\sim}{v}) + (\underset{\sim}{B}\cdot\underset{\sim}{\nabla})\underset{\sim}{v} - (\underset{\sim}{v}\cdot\underset{\sim}{\nabla})\underset{\sim}{B} . \tag{1}$$

If we now specialize to the instantaneous rest frame of the fluid, $\underset{\sim}{v} = 0$, $t = \tau$ (proper time), then $\gamma = 1$ and has vanishing first derivatives, and $\partial/\partial t$ becomes $\partial/\partial\tau$ (i.e. the convective derivative and the partial derivative are equal), so

$$dB^i/d\tau = -B^i v^j{}_{,j} + B^j v^i{}_{,j} = -B^i u^j{}_{,j} + B^j u^i{}_{,j} .$$

Since $u^i{}_{,j}$ is the projection perpendicular to $\mathbf{u}$, of $u^\mu{}_{,\nu}$ it follows that

$$u_{i,j} = \omega_{ij} + \sigma_{ij} + \frac{1}{3} g_{ij}\theta .$$

The contraction of this 3-tensor is of course $u^i{}_{,i} = \theta$ and the proper time derivative of B is then

$$dB^i/d\tau = -B^i\theta + B_j(\omega^{ij} + \sigma^{ij} + \tfrac{1}{3} g^{ij}\theta) = -\tfrac{2}{3} B^i\theta + (\sigma^{ij} + \omega^{ij})B_j . \tag{2}$$

To Equation (2) may be added any term proportional to the spatial component of the 4-velocity (since that vanishes in the comoving frame). Thus write Equation (2) as

$$dB^i/d\tau = -\tfrac{2}{3} B^i\theta + (\sigma^{ij} + \omega^{ij})B_j + fu^i \tag{3}$$

where f is as yet arbitrary. Now transform Equation (2) into a tensor equation by

(a) Letting indices run from 0-3 instead of just 1-3;

(b) Defining a 4-vector B^α by $B^\alpha u_\alpha = 0$ and the spatial part of $\mathbf{B}$ equal in the comoving frame to $\underset{\sim}{B}$;

(c) Changing ordinary derivatives to covariant derivatives;

(d) Defining f so that $d(\mathbf{B}\cdot\mathbf{u})/d\tau = 0$, i.e. so that the normalization of $\mathbf{B}$ is preserved.

The unique equation satisfying (a), (b), (c), (d) is

$$DB^{\alpha}/d\tau = u^{\alpha}a_{\beta}B^{\beta} + (\omega^{\alpha\beta} + \sigma^{\alpha\beta})B_{\beta} - \frac{2}{3}\theta P^{\alpha}{}_{\beta}B^{\beta} \tag{4}$$

where a is the 4-acceleration, and $P^{\alpha}{}_{\beta}$, is the projection operator defined in Problem 5.18. Equation (4) is obviously a tensor equation and reduces to Equations (3) and (1) in a comoving orthonormal frame, thus it is the correct equation in all frames.

CHAPTER 5: SOLUTIONS

Solution 5.1.

(a) In the rest frame of the particles the only nonzero component is $T^{0'0'} = \rho_0$. By transforming to a frame moving in the x-direction with velocity $-\beta$, we find the nonzero components

$$T^{00} = \rho_0\gamma^2, \quad T^{0x} = T^{x0} = \rho_0\gamma^2\beta, \quad T^{xx} = \rho_0\beta^2\gamma^2 \ ,$$

where $\gamma \equiv (1-\beta^2)^{-\frac{1}{2}}$. In general, if $\mathbf{u}$ is the 4-velocity of the particles $T = \rho_0\, \mathbf{u}\otimes\mathbf{u}$.

(b) Let the circle of motion be in the x-y plane. Consider a continuum of rest mass density ρ_0 at the point $x=0$, $y=a$ moving with velocity $\beta = \omega a$. From part (a), at $x=0$, $y=a$ we have $T^{00} = \rho_0\gamma^2$ etc. Now if we consider a ring of matter, all points on the ring are equivalent so that in polar coordinates the nonzero components of $T^{\mu\nu}$ are

$$T^{00} = \rho_0\gamma^2, \quad T^{0\hat{\phi}} = \rho_0\gamma^2\beta = T^{\hat{\phi}0}, \quad T^{\hat{\phi}\hat{\phi}} = \rho_0\gamma^2\beta^2 \ .$$

We now must relate ρ_0 to the collection of particles. If we assume that N is large enough so that we can view the particles as continuously distributed around the ring, then $\rho_0 \propto \delta(r-a)\delta(z)$. The total energy of the particles must be γNm so

$$T^{00}\, 2\pi\, r\, dr\, dz = \gamma Nm$$

and $\rho_0 = Nm\delta(r-a)\delta(z)/2\pi a\gamma$. The results can now be expressed in cartesian coordinates by making the usual transformation $x = r\cos\phi$, $y = r\sin\phi$:

$$T^{\mu\nu} = \frac{\gamma\, Nm\,\delta([x^2+y^2]^{\frac{1}{2}} - a)\delta(z)}{2\pi a}\begin{bmatrix} 1 & -\beta\sin\phi & \beta\cos\phi & 0 \\ & \beta^2\sin^2\phi & -\beta^2\sin\phi\cos\phi & 0 \\ & & \beta^2\cos^2\phi & 0 \\ (\text{symmetric}) & & & 0 \end{bmatrix}$$

(c) If we have an additional ring rotating in the opposite direction then that ring's stress-energy is the same as that in (b), with the sign of $\beta = \omega a$ reversed. We can then add the two cases. In the sum, terms linear in β disappear; terms quadratic in β are doubled. In polar coordinates, the nonzero components are

$$T^{00} = T^{\hat{\phi}\hat{\phi}}/\beta^2 = \gamma\, Nm\,\delta(r-a)\delta(z)/\pi a\ .$$

Solution 5.2. In Problem 5.1 we saw that for particles with 4-velocity $\mathbf{u}$, the stress-energy is proportional to $\mathbf{u}\otimes\mathbf{u}$. For an isotropic collection with $\mathbf{u} = \gamma(1, v\underset{\sim}{n})$ we have

$$\mathbf{T} = \kappa\langle\mathbf{u}\otimes\mathbf{u}\rangle, \qquad \kappa = \text{some constant},$$

where the average is taken over the orientations of $\underset{\sim}{n}$. By symmetry, off diagonal components average to zero:

$$T^{0i} = \kappa\gamma^2 v\,\langle n^i\rangle = 0$$

$$T^{ij} = \kappa\gamma^2 v^2\,\langle n^i n^j\rangle = 0, \quad i \neq j\ .$$

To evaluate the diagonal, spatial, components note that $\langle n^x n^x\rangle = \langle n^y n^y\rangle = \langle n^z n^z\rangle$ and that

$$\langle n^x n^x\rangle + \langle n^y n^y\rangle + \langle n^z n^z\rangle = \langle n^2\rangle = 1$$

so that

$$\langle n^i n^i\rangle = \frac{1}{3} \qquad T^{ii} = \kappa\gamma^2 v^2/3\ .$$

We know that the moving particles each have energy $m\gamma$ so the density of mass energy must be

$$m\gamma N = T^{00} = \kappa\gamma^2 <1> = \kappa\gamma^2 \ .$$

Thus $\kappa = mN/\gamma$ and the nonvanishing components are

$$T^{00} = m\gamma N \qquad T^{ij} = (m\gamma N v^2/3)\delta^{ij} \ .$$

For photons $v \to c$, $m\gamma \to h\nu$. For cold dust $v=0$ so $T^{00} = mN$ is the only nonvanishing component.

Solution 5.3. In the rest frame of the fluid element $u^0 = 1$, $u^i = 0$ so that

$$T^{\mu\nu} = pg^{\mu\nu} + (\rho + p)\, u^\mu u^\nu \ .$$

Since this is a tensor equation which is true in the rest frame of the fluid element, it must be true in general.

Solution 5.4. We can always choose the spatial coordinates to diagonalize the 3-dimensional stress tensor T^{ij}. It is obvious by symmetry that one of the spatial directions which is singled out is the direction of $\underset{\sim}{B}$, and that the choice of the other two orthogonal directions is arbitrary. If we choose $\underset{\sim}{B} = B\underset{\sim}{e}_z$ it follows that $T^{xx} = T^{yy}$ and that $T^{xy} = T^{xz} = T^{yz} = 0$. Since there is clearly no energy transport in this static problem $T^{0i} = 0$. The energy density of a magnetic field is $T^{00} = B^2/8\pi$.

We must now calculate the "pressures" T^{xx}, T^{yy}, T^{zz}. We do this by considering a box Δx, Δy, Δz of the field and imagining it expanded adiabatically. (An adiabatic expansion keeps constant the magnetic flux $B\Delta x\Delta y$ – i.e. the number of "lines" of B in the volume.) The energy in the box is

$$\mathcal{E} = (B^2/8\pi)\Delta x\,\Delta y\,\Delta z = (B\,\Delta x\,\Delta y)^2\,\Delta z/8\pi\,\Delta x\,\Delta y \ .$$

The pressures can now easily be found:

$$T^{xx} = -\frac{1}{\Delta y\,\Delta z}\frac{d\mathcal{E}}{d(\Delta x)} = B^2/8\pi = T^{yy}$$

$$T^{zz} = -\frac{1}{\Delta x\,\Delta y}\frac{d\mathcal{E}}{d(\Delta z)} = -B^2/8\pi$$

so that

$$T^{\mu\nu} = \frac{B^2}{8\pi}\begin{bmatrix} 1 & & & 0 \\ & 1 & & \\ & & 1 & \\ 0 & & & -1 \end{bmatrix} .$$

To find the stress energy of a chaotic field start by choosing a set of orthogonal axes and average the $\underset{\sim}{B}$ fields in the x, y, and z directions. From our previous result, the averaging gives

$$<T^{\mu\nu}> = \frac{B^2}{8\pi}\begin{bmatrix} 1 & & & \\ & \frac{1}{3} & & \\ & & \frac{1}{3} & \\ & & & \frac{1}{3} \end{bmatrix}$$

where B is the average strength of the field. If we picked any other orthogonal directions we would get the same result because the 3×3 identity matrix block is invariant under rotations; hence the above result also represents the average over *all* directions.

Solution 5.5. We are given the linear mass density as μ, so $T^{00} = \mu/A$. In the rest frame of the rod there is no energy flow so $T^{0i} = 0$. Choose the z direction to be along the rod axis. By symmetry [see the argument in Problem 5.4] the off diagonal stresses T^{xy}, T^{xz}, T^{yz}, must be zero. Since the tension is in the z-direction, $T^{zz} = -F/A$. Since there is no applied stress in the x or y directions (therefore no transport of x or y momentum) $T^{xx} = T^{yy} = 0$.

Solution 5.6. Suppose a force infinitesimally smaller than F is applied. If the cross section is A and the axis of the rope is in the z direction, the components of T in the rest frame of the rod are given in Problem 5.5. An observer moving with velocity $\mathbf{u} = (\gamma, \gamma \underset{\sim}{v})$ sees energy density

$$T^{0'0'} = T^{\alpha\beta}u_\alpha u_\beta = (\gamma^2/A)[\mu - Fv_z^2] .$$

Clearly the observer with the greatest chance of seeing a negative $T^{0'0'}$ is the one moving in the z-direction with $v \to c$. To guarantee that he sees a positive $T^{0'0'}$, the weak energy condition requires $F < \mu$, which is the required upper limit. In physical units, and for good steel rope this would be

$$\frac{F}{A} < \frac{\mu}{A} c^2 = \rho c^2 = 7 \times 10^{13} \text{ Kg} - \text{wt/mm}^2 \ .$$

as compared to its actual breaking strength of ≈ 200 kg − wt/mm². The reason for this difference by a factor of order $10^{11.5}$ is that the strength of steel comes essentially from the energy of the molecular bonds ($\lesssim$ 1eV per atom), not from the total mass density, which is virtually all in the nucleus ($\sim 5 \times 10^{10}$eV).

Solution 5.7. Let S' be the instantaneous rest frame of an element of the rod. In this rest frame the only nonzero components are $T^{0'0'} = \rho$, $T^{x'x'} = p$. If we Lorentz transform to the lab frame we find the nonzero components: $T^{xx} = p$, $T^{yy} = \gamma^2\beta^2\rho$, $T^{00} = \gamma^2\rho$, $T^{0y} = \gamma^2\beta\rho$, where $\beta = \omega r$ and $\gamma \equiv (1-\beta^2)^{-\frac{1}{2}}$. If spherical polar coordinates are used the nonzero components are

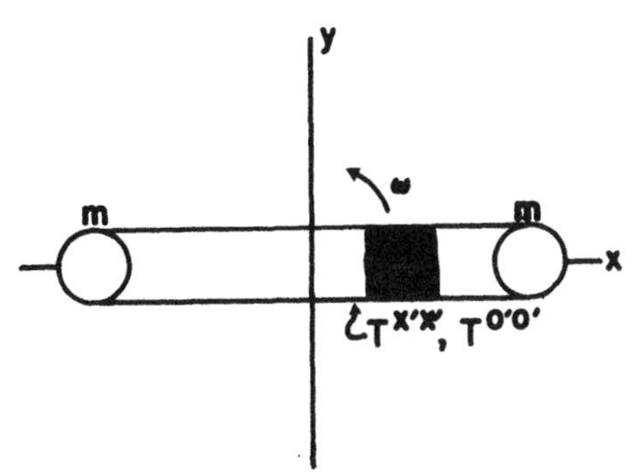

$$T^{rr} = p, \quad T^{\phi\phi} = \gamma^2\beta^2\rho/r^2, \quad T^{0\phi} = \gamma^2\beta\rho/r, \quad T^{00} = \gamma^2\rho \ .$$

To find $p(r)$ we can use the r component of the equation of motion

$$\begin{aligned} 0 = T^{r\nu}{}_{;\nu} &= T^{rr}{}_{,r} + T^{\phi\phi}\Gamma^{r}{}_{\phi\phi} + T^{rr}[\log(-g)^{\frac{1}{2}}]_{,r} \\ &= (T^{rr}r^2)_{,r}/r^2 - r\sin^2\theta \, T^{\phi\phi} \ , \end{aligned}$$

so that

$$(pr^2)_{,r} = r \sin^2\theta\, \gamma^2\beta^2\rho\ .$$

Since the only mass density is in the point masses we have

$$\rho = m\delta(r-a)\delta(\cos\theta)\,[\delta(\phi-\omega t)+\delta(\phi-\omega t-\pi)]/r^2\ .$$

We can now integrate the equation of motion with the boundary condition $p = 0$ for $r > a$, to find p. When this is done the components of the stress-energy tensor (in the lab, described by coordinates t, r, θ, ϕ) are seen to be

$$T^{\mu\nu} = \frac{m}{1-\omega^2 a^2}\,\delta(\cos\theta)\,[\delta(\phi-\omega t)+\delta(\phi-\omega t-\pi)]$$

$$\times \begin{bmatrix} \dfrac{\delta(r-a)}{a^2} & 0 & 0 & \dfrac{\omega\,\delta(r-a)}{a^2} \\ & -\dfrac{\omega^2 a}{r^2} & 0 & 0 \\ & & 0 & 0 \\ \text{symmetric} & & & \dfrac{\omega^2\delta(r-a)}{a^2} \end{bmatrix}\ .$$

Solution 5.8. The electrostatic field for the moving capacitor remains E, since electric fields parallel to the direction of motion do not change. The moving capacitor, due to Lorentz contraction, has thickness d/γ, where $\gamma \equiv (1-v^2)^{-\frac{1}{2}}$. The moving capacitor then has electrostatic energy $E^2 Ad/8\pi\gamma < E^2 Ad/8\pi$.

The pressure of the gas must be equal in magnitude to the (negative) pressure of the E field, so $p = E^2/8\pi$. The nonvanishing components of the stress-energy of the gas in the capacitor rest frame are

$$T^{00} = \rho_0,\quad T^{xx} = T^{yy} = T^{zz} = E^2/8\pi\ .$$

The total rest mass of the capacitor is

$$M = E^2 Ad/8\pi + \rho_0 Ad\ .$$

If the capacitor is moving we can calculate the energy density of gas by using a Lorentz transformation:

$$T^{0'0'} = \gamma^2(T^{00}+\beta^2 T^{xx}) = \gamma^2(\rho_0+\beta^2 E^2/8\pi) \ .$$

If the capacitor is moving, then its total energy is

$$\begin{aligned} \mathcal{E} &= E^2 Ad/8\pi\gamma + \gamma^2(\rho_0+\beta^2E^2/8\pi)\,Ad/\gamma \\ &= (\gamma\rho_0+\gamma E^2/8\pi\,[1/\gamma^2+\beta^2])\,Ad = \gamma(\rho_0+E^2/8\pi)\,Ad = \gamma M \ . \end{aligned}$$

Solution 5.9. For simplicity assume Minkowski coordinates. The stress energy for the particles is

$$T^{\mu\nu}_P(x) = \sum_i m_i \int u^\mu_i(\tau_i)\,u^\nu_i(\tau_i)\,\delta^4[x-x_i(\tau_i)]\,d\tau_i \ .$$

Thus

$$T^{\mu\nu}_{P\ ,\nu} = \sum_i m_i \int u^\mu_i u^\nu_i(\partial/\partial x^\nu)\,\delta^4[x-x_i(\tau_i)]\,d\tau_i \ .$$

The delta function depends only on $x-x_i$ so we can replace $\partial/\partial x^\nu$ by $-\partial/\partial x^\nu_i$. Since $u^\nu_i\,\partial/\partial x^\nu_i = d/d\tau_i$ we have then

$$\begin{aligned} T^{\mu\nu}_{P\ ,\nu} &= -\sum_i m_i \int u^\mu_i(d/d\tau_i)\,\delta^4[x-x_i(\tau_i)]\,d\tau_i \\ &= \sum_i m_i \int (du^\mu_i/d\tau_i)\,\delta^4[x-x_i(\tau_i)]\,d\tau_i \ , \end{aligned}$$

where the second result follows from integration by parts.

Now for the ith particle

$$m_i du^\mu_i/d\tau_i = q_i\,F^\mu{}_\nu(x_i)\,u^\nu_i \ ,$$

so, with the result of Problem 4.7,

$$\begin{aligned} T^{\mu\nu}_{P\ ,\nu} &= \sum_i q_i \int F^\mu{}_\nu\, u^\nu_i\,\delta^4[x-x_i(\tau_i)]\,d\tau_i \\ &= F^\mu{}_\nu J^\mu \ . \end{aligned}$$

But, from Problem 4.15,

$$T^{\mu\nu}_{EM,\nu} = (4\pi)^{-1} F^{\mu\alpha} F^{\nu}{}_{\alpha,\nu} = -F^{\mu\alpha} J_{\alpha}$$

so that $(T^{\mu\nu}_{p} + T^{\mu\nu}_{EM})_{,\nu} = 0.$

Solution 5.10. Consider a small number, dN, of photons with a small spread in frequency and position, moving in a narrow cone, and impinging on a small area in a short time. By definition

$$I_{\nu} = \frac{d\,(\text{energy})}{d\,(\text{frequency})\ d\,(\text{time})\ \ d\begin{bmatrix}\text{solid angle}\\ \text{of cone}\end{bmatrix}\ d\begin{bmatrix}\text{cross-sectional}\\ \text{area}\end{bmatrix}} \ .$$

With no loss of generality take the narrow cone of the motion, whose size is d(solid angle), to be along the z-direction, then

$$\begin{aligned}
&d\,(\text{energy}) = h\nu\, dN\\
&d\,(\text{solid angle}) = dv^x\, dv^y = d(p^x/E)\, d(p^y/E) = dp^x\, dp^y/h^2\nu^2\\
&d\,(\text{time}) = dz/c = dz\\
&d\,(\text{cross sectional area}) = dx\, dy\\
&d\,(\text{frequency}) = d(h\nu)/h = dp^z/h
\end{aligned}$$

so that

$$I_{\nu} = \frac{dN \qquad h^4\nu^3}{dx\, dy\, dz\ dp^x\, dp^y\, dp^z} \ .$$

Number density in phase space, $dN/d^3x\, d^3p$, is a Lorentz invariant (see Problem 3.34) so I_{ν}/ν^3 is invariant.

Solution 5.11. If spherical polar coordinates are used to describe the 3-dimensional space in the frame of the star, then in that frame the radiation has stress energy

$$T^{\mu\nu}(r,\theta,\phi) = (L/4\pi r^2) \begin{bmatrix} 1 & 1 & 0 & 0 \\ 1 & 1 & 0 & 0 \\ 0 & 0 & 0 & 0 \\ 0 & 0 & 0 & 0 \end{bmatrix} .$$

(columns and rows labelled t, r, θ, ϕ)

Let ℓ be the vector (null, of course) from the emission of a photon to the reception of that photon. In the star's frame, if a photon is received at a distance r from the star, $\ell^0 = r$, $\ell^r = r$, $\ell^\theta = \ell^\phi = 0$, and the stress-energy tensor at the event of reception can be written in the frame independent form

$$\mathrm{T} = [L/4\pi(\mathbf{u}_s\cdot\boldsymbol{\ell})^4]\boldsymbol{\ell}\otimes\boldsymbol{\ell}$$

where $\mathbf{u}_s$ is the 4-velocity of the star so that, evaluated in the star's frame, $\mathbf{u}_s\cdot\boldsymbol{\ell} = -r$. It is now simple to use the frame independent expression for T to evaluate the flux the observer sees. Let $\mathbf{n} = (0, \underset{\sim}{n})$ be the spatial vector which, in the observer's frame, points *to* the observed position of the star, so that $\boldsymbol{\ell} = (R, -R\underset{\sim}{n})$. The flux measured by the observer will be, in the observer's coordinates,

$$F_{obs} = -T^{0i}n_i = \mathbf{u}_{obs}\cdot\mathrm{T}\cdot\mathbf{n} .$$

In the observer's coordinates

$$\mathbf{u}_s\cdot\boldsymbol{\ell} = (\gamma, \gamma\underset{\sim}{v})\cdot(R, -R\underset{\sim}{n}) = -\gamma R(1+v\cos\theta) .$$

Furthermore $\boldsymbol{\ell}\cdot\mathbf{u}_{obs} = -R$ and $\boldsymbol{\ell}\cdot\mathbf{n} = -R$ so that

$$F_{obs} = L/[4\pi\gamma^4(1+v\cos\theta)^4R^2] .$$

Another way of arriving at this result is to use the fact that I_ν/ν^3 is a Lorentz invariant. (See Problem 5.10.)

Solution 5.12. Let the 4-velocity of the particle be $\mathbf{u}$. Let the null 4-vector along which the radiation propagates be $\boldsymbol{\ell}$. The particle absorbs a 4-momentum flux $-SA(\mathbf{u}\cdot\boldsymbol{\ell})\boldsymbol{\ell}$, i.e. the stress-energy of the radiation

$T = S\ell\otimes\ell$ multiplied by the effective area of the particle and dotted into its 4-velocity (to get the flux in its frame). The minus sign is due to the $-{+}{+}{+}$ signature. The time component of this in its rest frame is $SA(\mathbf{u}\cdot\ell)(\mathbf{u}\cdot\ell)$; this is the absorbed energy which the particle reradiates in its own rest frame, so the net change in 4-momentum is

$$\frac{d\mathbf{p}}{d\tau} = \frac{m\,d\mathbf{u}}{d\tau} = -SA\,[(\mathbf{u}\cdot\ell)\ell + (\mathbf{u}\cdot\ell)^2\mathbf{u}] \;. \tag{1}$$

This is the equation of motion for $\mathbf{u}$.

To solve, let $W \equiv \mathbf{u}\cdot\ell$, and dot the equation of motion with ℓ to get

$$dW/d\tau = -(SA/m)\,W^3$$

and hence

$$W = -[(2SA\tau/m)+K]^{-\frac{1}{2}} \tag{2}$$

where K is some constant. We now treat W as the independent variable and substitute $d\tau = -dW\,m/SAW^3$ in Equation (1) to get

$$\frac{d\mathbf{u}}{dW} - \frac{\mathbf{u}}{W} = \frac{1}{W^2}\,\ell \;. \tag{3}$$

This has the integrating factor $1/W$ and can easily be integrated:

$$\frac{d}{dW}\left(\frac{1}{W}\,\mathbf{u}\right) = \frac{1}{W^3}\,\ell \tag{4}$$

$$\mathbf{u} = -\frac{1}{2W}\,\ell + W\mathbf{q} \tag{5}$$

where $\mathbf{q}$ is a constant of integration. The conditions $\mathbf{u}\cdot\ell = W$ and $\mathbf{u}\cdot\mathbf{u} = -1$ imply that $\mathbf{q}\cdot\ell = 1$ and $\mathbf{q}\cdot\mathbf{q} = 0$. Substitute for W from Equation (2) and integrate Equation (5) to find

$$\mathbf{x} = +\frac{1}{6\alpha}\,(2\alpha\tau + K)^{\frac{3}{2}}\,\ell - \frac{1}{\alpha}\,(2\alpha\tau + K)^{\frac{1}{2}}\,\mathbf{q} + \mathbf{x}_0$$

where $\alpha = SA/m$. If we choose the constant of integration so that the particle starts from rest at the origin in a radiation field parallel to the x-axis, this is

$$t = \frac{1}{6a}(2a\tau+1)^{\frac{3}{2}} + \frac{1}{2a}(2a\tau+1)^{\frac{1}{2}} - \frac{2}{3a}$$

$$x = \frac{1}{6a}(2a\tau+1)^{\frac{3}{2}} - \frac{1}{2a}(2a\tau+1)^{\frac{1}{2}} + \frac{1}{3a}$$

$$y = 0$$

$$z = 0\ .$$

(The solution here follows that of Robertson and Noonan, pp. 116-118.)

Solution 5.13. An observer on the sphere sees the black body spectrum doppler shifted by different amounts in different directions. Since the black body spectrum has $I_\nu/\nu^3 = \text{constant} \times (e^{h\nu/KT} - 1)^{-1}$, and since I_ν/ν^3 is a Lorentz invariant, the effect of a doppler shift $\nu_0 \to \nu$ is just to change the effective temperature $T_0 \to T = T_0(\nu/\nu_0)$.

What is the doppler shift seen by the comoving observer when he looks at an angle θ to the forward direction of motion? This standard result is easy to rederive with invariants: Let $\mathbf{u}$ be the 4-velocity of the sphere, $\mathbf{u}_0$ be the rest 4-velocity of the radiation, and $\mathbf{p}$ be the 4-momentum of a photon in units of Planck's constant. Then $\nu = -\mathbf{u}\cdot\mathbf{p}$, $\nu_0 = -\mathbf{u}_0\cdot\mathbf{p}$ and $(1-v^2)^{-\frac{1}{2}} \equiv \gamma = -\mathbf{u}\cdot\mathbf{u}_0$. In the sphere's frame,

$$\mathbf{u} = (1, \underset{\sim}{0}), \quad \mathbf{u}_0 = (\gamma, -\gamma\underset{\sim}{v}), \quad \mathbf{p} = (\nu, \nu\underset{\sim}{n})$$

where $\underset{\sim}{n}$ is a unit 3-vector. The unit spatial vector in the direction of $\underset{\sim}{v}$ is

$$\left(0, \frac{\underset{\sim}{v}}{|v|}\right) = -\frac{\mathbf{u}_0 + (\mathbf{u}_0\cdot\mathbf{u})\mathbf{u}}{\gamma v}\ .$$

(This is verified by substituting in components in the sphere's frame.) Similarly, the vector $(0, \underset{\sim}{n})$ is invariantly written as

$$(0, \underset{\sim}{n}) = \frac{\mathbf{p} + (\mathbf{u}\cdot\mathbf{p})\mathbf{u}}{\nu}\ ,$$

so the angle of the photon in the sphere's frame is

$$\cos\theta = (0, \underset{\sim}{n}) \cdot \left(0, \frac{\underset{\sim}{v}}{|v|}\right) = \frac{\nu_0 - \gamma\nu}{\gamma v \nu} .$$

The last line is obtained by dotting together the invariant 4-vector expressions and using $\nu = -u \cdot p$ etc. Solving for ν/ν_0, we have

$$\frac{T}{T_0} = \frac{\nu}{\nu_0} = \frac{1}{\gamma(1 + v\cos\theta)} .$$

Finally, the equilibrium temperature is given by the average over all directions $<T(\theta)^4>^{\frac{1}{4}}$. [This is because the radiation rate, which equals the absorption rate in equilibrium, goes as T^4 (Stefan-Boltzman law).] Thus

$$\left\langle \frac{T}{T_0} \right\rangle^4 = \frac{1}{2} \int_{\cos\theta=-1}^{\cos\theta=1} \frac{d(\cos\theta)}{\gamma^4 (1 + v\cos\theta)^4} = \frac{1}{\gamma^4} \frac{1}{6v} \left[\frac{1}{(1-v)^3} - \frac{1}{(1+v)^3} \right]$$

and

$$T_{\text{equilibrium}} = <T^4>^{\frac{1}{4}} = (\gamma^2(1 + v^2/3))^{\frac{1}{4}}\, T_0 .$$

Solution 5.14. Let the average transfer of energy be $<\Delta E>$. We are given that $E/mc^2 << 1$ and $T/mc^2 << 1$ (units with $k = 1$), so we want only the first nonvanishing terms in the double series expansion

$$\begin{aligned} <\Delta E> = mc^2\ [&\alpha_1 + \alpha_2(E/m) + \alpha_3(T/m) \\ &+ \alpha_4(E^2/m^2) + \alpha_5(ET/m^2) + \alpha_6(T^2/m^2) + \cdots] . \end{aligned} \tag{1}$$

If $T = E = 0$, nothing is happening, so $\alpha_1 = 0$. If $T = 0$, $E \neq 0$, we have standard Compton scattering with a cross section $d\sigma/d\Omega \propto (1 + \cos^2\theta)$ and an energy transfer $\Delta E = (E^2/m)(1 - \cos\theta)$. Since the cross section is forward-backward symmetric, the $\cos\theta$ term cancels out when we average over angles, and

$$<\Delta E> = (E^2/m), \qquad T = 0 .$$

This implies that $\alpha_2 = 0$ and $\alpha_4 = 1$. If $E=0$, $T\neq 0$, the photon has zero energy and is vacuous, so $\alpha_3 = \alpha_6 = 0$.

Finally, we need α_5: Take a thought experiment where there is a dilute black-body flux of photons with the same temperature as the gas so that

$$d(\text{number of photons})/dE = \text{constant} \times E^2 e^{-E/T} .$$

Plugging in Equation (1) with $\alpha_1 = \alpha_2 = \alpha_3 = \alpha_6 = 0$, $\alpha_4 = 1$, and requiring thermal equilibrium, namely $\int_0^\infty \langle\Delta E\rangle\, E^2 e^{-E/T} dE = 0$, gives $0 = (3T/m)(4T + \alpha_5 T)$ or $\alpha_5 = -4$. Expression (1) now takes the final form

$$\langle E\rangle = (E^2 - 4TE)/mc^2 + \cdots .$$

Solution 5.15. The technique here is to use the equations of motion $(T^{\mu\nu}{}_{,\nu} = 0)$ to convert first derivative terms to divergences and then, by Gauss's theorem, to vanishing surface integrals outside the system. By repeated use of the equations of motion we have

$$T^{00}{}_{,00} = -T^{0k}{}_{,k0} = -T^{k0}{}_{,0k} = T^{km}{}_{,mk} .$$

Now multiply by $x^i x^j$:

$$\begin{aligned}(x^i x^j T^{00})_{,00} &= x^i x^j (T^{km})_{,km} \\ &= (T^{km} x^i x^j)_{,km} - 2(x^i x^j)_{,k} T^{km}{}_{,m} - T^{km}(x^i x^j)_{,km} \\ &= (T^{km} x^i x^j)_{,km} - 2(x^j\, T^{im}{}_{,m} + x^i\, T^{jm}{}_{,m}) - 2T^{ij} .\end{aligned}$$

The first derivative terms can be simplified, e.g.

$$x^j\, T^{im}{}_{,m} = (x^j\, T^{im})_{,m} - \delta^j{}_m T^{im} .$$

If we make this simplification, integrate over all space, and ignore all divergences which can be converted to vanishing surface integrals we see

$$\frac{d^2}{dt^2}\int x^i x^j T^{00} d^3x = \int [0 - 2(-T^{ij} - T^{ij}) - 2T^{ij}]\, d^3x$$

$$= 2\int T^{ij} d^3x \ .$$

Solution 5.16. Suppose v is a timelike eigenvector

$$T \cdot v = \alpha v \ .$$

Consider the stress-energy measured by an observer with 4-velocity $u = \pm v/|v|$. (Choose the sign to make u future-pointing.) In that observer's rest frame $u = (1, \underset{\sim}{0})$ and the energy flux vector S is

$$-S = T \cdot u = \frac{1}{|v|} T \cdot v = \frac{\alpha}{|v|} v = \alpha u = (\alpha, \underset{\sim}{0}) \ .$$

Thus the observer sees no energy flux, and α is the negative of the energy density he sees.

Conversely if an observer with 4-velocity u sees no spatial flux, then $-S = T \cdot u = \alpha u$ for some α, and u is thus a timelike eigenvector of T.

Solution 5.17.

(a) If second order terms in velocity are ignored, the general Lorentz transformation is

$$\Lambda^0{}_{0'} = 1, \quad \Lambda^0{}_{i'} = v^i, \quad \Lambda^i{}_{j'} = \delta^{i'}{}_{j'} \ .$$

If $T^{\mu'\nu'}$ are the stress-energy components in the rest frame of the fluid, then $T^{0'i'} = 0$, and

$$g^j \equiv T^{0j} = \Lambda^0{}_{\mu'}\Lambda^j{}_{\nu'}T^{\mu'\nu'} = \Lambda^0{}_{0'}\Lambda^j{}_{0'}T^{0'0'} + \Lambda^0{}_{k'}\Lambda^j{}_{i'}T^{i'k'}$$

$$= v^j T^{0'0'} + v^k T^{j'k'} = v^k(T^{0'0'}\delta^{jk} + T^{j'k'}) \ .$$

For a perfect fluid $T^{0'0'} = \rho$ and $T^{j'k'} = p\,\delta^{jk}$ so that the "inertial mass per unit volume" is

$$m^{jk} = \delta^{jk}(\rho + p) \ .$$

(b) Since

$$M^{ij} \equiv \delta^{ij}\int T^{00}\,dx\,dy\,dz + \int T^{ij}\,dx\,dy\,dz \ ,$$

we need only show that the second term on the right vanishes. Note that the 3-divergence of T^{jk} vanishes

$$T^{jk}{}_{,k} = T^{j\mu}{}_{,\mu} - T^{j0}{}_{,0} = 0 - 0 \ ,$$

and that

$$(x^k T^{ji})_{,i} = \delta^k_i T^{ji} + x^k T^{ji}{}_{,i} = T^{jk} \ .$$

It follows by Gauss' theorem that

$$\int T^{jk}\,dx\,dy\,dz = \int (x^k T^{ji})_{,i}\,dx\,dy\,dz = \int_S x^k T^{ji} n^i\,dS$$

where S is any surface outside the stressed body and $\underset{\sim}{n}$ is the outward normal to S. Since the surface integral obviously vanishes, the proof is complete.

Solution 5.18. When the definitions of $\sigma_{\alpha\beta}$ and $\omega_{\alpha\beta}$ are used, we can see we need only verify

$$u_{\alpha;\beta} = u_{\alpha;\mu}P^{\mu}{}_{\beta} - a_\alpha u_\beta \ .$$

It is straightforward to use the definition of $P^{\mu}{}_{\beta}$ to verify this. Instead we follow a more instructive approach of projecting the equation with $\mathbf{u}$ and with $\mathbf{P}$. Note that

$$\mathbf{u}\cdot\mathbf{P} = 0, \qquad \mathbf{P}\cdot\mathbf{P} = \mathbf{P} \ .$$

The projections with $\mathbf{P}$ and with $\mathbf{u}$ are:

$$u_{\alpha;\beta}P^{\beta}_{\gamma} = u_{\alpha;\mu}P^{\mu}_{\beta}P^{\beta}_{\gamma} - a_\alpha u_\beta P^{\beta}_{\gamma} = u_{\alpha;\mu}P^{\mu}{}_{\gamma} - 0$$

and

$$u_{\alpha;\beta}u^{\beta} = u_{\alpha;\mu}P^{\mu}_{\beta}u^{\beta} - a_\alpha u_\beta u^{\beta} = 0 + a_\alpha \ .$$

Both results check. Since the projection of the equation onto **u** and perpendicular to **u** are true, the equation is verified.

Solution 5.19. If the fluid has mass energy density ρ, pressure P, temperature T and entropy S, then conservation of mass-energy for a fluid element of volume V is

$$d(\rho V) = -PdV + TdS \ .$$

Here S is the entropy in the volume V. This differs from the nonrelativistic law only in that we have replaced energy density by mass-energy density, since mass is not conserved in special relativity. Baryon number *is* conserved so the first law is often written in terms of baryon number density n, and entropy per baryon s, by eliminating $V \equiv$ (number of baryons /n):

$$d(\rho/n) = -Pd(1/n) + Tds$$

or

$$d\rho = (\rho + P)dn/n + nTds \ .$$

Solution 5.20. For a perfect fluid $T^{\mu\nu} = (P+\rho)u^\mu u^\nu + Pg^{\mu\nu}$ so that in the fluid rest frame the 0 component of the equation of motion is

$$u_\mu T^{\mu\nu}{}_{;\nu} = T^{0\nu}{}_{;\nu} = P_{,\nu}g^{0\nu} + Pg^{0\nu}{}_{;\nu} + (P+\rho)_{,\nu}u^\nu u^0$$
$$+ (P+\rho)[u^\nu{}_{;\nu}u^0 + u^0{}_{;\nu}u^\nu] = 0 \ .$$

Now $g^{\mu\nu}{}_{;\nu} = 0$, $u^0 = 1$, and $u^0{}_{;\nu} = u_\alpha u^\alpha{}_{;\nu} = \frac{1}{2}(u_\alpha u^\alpha)_{;\nu} = 0$, so that the equation becomes

$$0 = -\frac{d}{dt}P + \frac{d}{dt}(P+\rho) + (P+\rho)u^\nu{}_{;\nu} \ .$$

If n is the density of baryons in the fluid rest frame then the number-flux vector of baryons n**u** is conserved. In the rest frame then

$$(nu^\nu)_{;\nu} = 0 = n_{,\nu}u^\nu + nu^\nu{}_{;\nu} = \frac{dn}{dt} + nu^\nu{}_{;\nu} \ .$$

When this equation is used to eliminate $u^{\nu}{}_{;\nu}$ the equation of motion becomes

$$\frac{d\rho}{dt} = \frac{(P+\rho)}{n}\frac{dn}{dt} .$$

If we compare this with the first law of thermodynamics (see Problem 5.19) we see that $ds/dt = 0$; the fluid flow is isentropic.

Solution 5.21. From the first law of thermodynamics for a perfect fluid (see Problem 5.20) having equation of state $\rho = \rho(n)$

$$\frac{d\rho}{\rho} = \frac{\rho+P}{\rho}\frac{dn}{n}$$

so that

$$d\log\rho/d\ln n = (\rho+P)/\rho .$$

Thus ρ is greater than $3P$ if and only if $d\log\rho/d\log n < 4/3$. Since

$$T^{\mu}{}_{\mu} = (\rho+P)u^{\mu}u_{\mu} + pg^{\mu}{}_{\mu} = -(\rho+P) + 4P = 3P-\rho$$

for a perfect fluid, the proof is complete.

Solution 5.22. Assume that the acoustical wave is a perturbation (isentropic) in a uniform static fluid with parameters ρ_0, P_0, and n_0. Let the perturbations be ρ_1, P_1, and n_1 and the fluid velocity be $\mathbf{u} \approx (1, \underset{\sim}{v}_1)$ in the rest frame of the unperturbed fluid. The first order perturbation terms in $T^{\mu\nu}{}_{,\nu} = 0$ give us

(i) From $\mu = 0$ $\qquad \nabla\cdot\underset{\sim}{v}_1 = -\frac{\partial\rho_1}{\partial t}(\rho_0+P_0)^{-1}$

(ii) From $\mu = 1,2,3$ $\qquad \frac{\partial\underset{\sim}{v}_1}{\partial t} = -\frac{\underset{\sim}{\nabla}P_1}{(\rho_0+P_0)}$

which can be combined into

$$\partial^2\rho_1/\partial t^2 - \nabla^2 P_1 = 0 .$$

Since P_1 and ρ_1 are related for isentropic flow by

$$P_1 = \partial P/\partial\rho|_s \, \rho_1 \, ,$$

the above equation becomes a wave equation

$$\nabla^2\rho_1 - \frac{1}{v_s^2}\frac{\partial^2\rho_1}{\partial t^2} = 0$$

with characteristic velocity v_s.

For an equation of state $\rho \approx 3P$,

$$v_s^2 = \partial P/\partial\rho = 1/3$$

so that $v_s \approx 3^{-\frac{1}{2}}$ in a highly relativistic gas.

Solution 5.23. From the first law of thermodynamics (see Problem 5.19) with $ds = 0$ we have

$$dn/d\rho = n/(\rho+P) \; .$$

Now we use the definition of Γ_1 and the expression for v_s^2 from Problem 5.21:

$$\frac{v_s^2}{\Gamma_1} = \frac{dP/d\rho}{(ndP)/(Pdn)} = \frac{P}{n}\frac{dn}{d\rho} = \frac{P}{\rho+P} \, .$$

Solution 5.24. At zero temperature all energy states in the Fermi gas are filled up to the Fermi level $E = E_F$. Since there are two spin states for fermions the density of fermions in phase space is $2/h^3$ so that

$$dn/V = (2/h^3)\,d^3p$$

$$\rho = \int_0^{P_F} (p^2+m^2)^{\frac{1}{2}}\,(2/h^3)\,4\pi p^2 dp$$

$$P = \int_0^{P_F} \tfrac{1}{3}\,p^2(p^2+m^2)^{-\frac{1}{2}}\,(2/h^3)\,4\pi p^2 dp$$

where P_F is the momentum corresponding to the Fermi energy. In Problem 5.22 we found $v_s^2 = dP/d\rho$ so

$$v_s^2 = \frac{dP}{d\rho} = \frac{dP/dP_F}{d\rho/dP_F} = \frac{1}{3}\left(\frac{P_F}{E_F}\right)^2 = \frac{1}{3}\left(1 - \frac{m^2}{E_F{}^2}\right).$$

Note that for a highly relativistic Fermi gas $v_s \to 3^{-\frac{1}{2}}$.

Solution 5.25. The relativistic Bernoulli equation, which expresses the conservation of energy along flow lines of a perfect fluid, is (Problem 14.7)

$$(1-v^2)^{-\frac{1}{2}} = \left(\frac{n}{P+\rho}\right) \times \text{constant} .$$

(This equation follows from $T^{0\nu}{}_{,\nu} = 0$, from $(nu^\mu)_{,\mu} = 0$ and from the first law of thermodynamics $d\rho/(P+\rho) = dn/n$.) Since the gas starts in the tank at zero velocity, we have: constant $= (P_0+\rho_0)/n_0$. An upper bound for $n/(P+\rho)$ which will give an upper bound v_{max} for v, occurs when $P \to 0$ implying $\rho = mn$, so

$$(1-v_{max}^2)^{-\frac{1}{2}} = \frac{1}{m}\left(\frac{P_0+\rho_0}{n_0}\right).$$

Now we have to try and write the right hand side only in terms of a, the sound velocity of the gas with P_0, ρ_0, n_0. We have $P = Kn^\gamma$ where K is a constant telling what adiabat the gas is on. The first law reads

$$\frac{d\rho}{dn} = \frac{Kn^\gamma + \rho}{n}$$

which is integrable and (with $\rho \to mn$ when $n \to 0$) gives

$$\rho = mn + \frac{K}{\gamma-1} n^\gamma .$$

The speed of sound a is given by

$$a^2 = \frac{dP}{d\rho} = \frac{dP/dn}{d\rho/dn} = \frac{\gamma K n^{\gamma-1}}{m + \gamma K n^{\gamma-1}/(\gamma-1)} \tag{1}$$

and the first law of thermodynamics gives us

$$\frac{P+\rho}{n} = \frac{d\rho}{dn} = \left(\frac{dP}{dn}\right)\left(\frac{d\rho}{dP}\right) = (\gamma K n^{\gamma-1})(1/a^2) \ . \tag{2}$$

But Equation (1) contains just this same combination $\gamma K n^{\gamma-1}$. So we can solve Equation (1) for

$$\gamma K n^{\gamma-1} = ma^2/[1-a^2/(\gamma-1)]$$

and get

$$\frac{1}{m}\left(\frac{P_0+\rho_0}{n_0}\right) = \frac{1}{1-a^2/(\gamma-1)}$$

and therefore

$$v^2_{max} = 1-[1-a^2/(\gamma-1)]^2 \ .$$

(Notice that $v_{max} \to 1$ in the limit of highly relativistic gas, for which $\gamma \to 4/3$ and $a^2 \to 1/3$.)

Solution 5.26. In the fluid rest frame $T^{0j} = T^{j0} = q^j$ are the only stress-energy components associated with $\mathbf{q}$. Since $\mathbf{u}$ for the fluid is $(1, \underset{\sim}{0})$ in its rest frame, then

$$T^{\alpha\beta} = u^\alpha q^\beta + u^\beta q^\alpha$$

is true in that frame and hence true in general.

Solution 5.27. In the fluid rest frame

$$S^0 = \text{entropy density} = ns$$

$$S^j = \text{entropy flux} = \text{heat flux}/T = q^j/T \ .$$

Since $q^0 = 0$, $u^0 = 1$ in the fluid rest frame $\mathbf{S} = ns\mathbf{u} + \mathbf{q}/T$ is true in the fluid rest frame, and hence in general.

Solution 5.28. For the stress energy of the system we have

$$\mathbf{T} = \mathbf{T}_{fluid} + \mathbf{T}_{heat} = [(P+\rho)\mathbf{u}\otimes\mathbf{u} + P\mathbf{g}] + [\mathbf{q}\otimes\mathbf{u}+\mathbf{u}\otimes\mathbf{q}] \ .$$

We compute the equation expressing energy conservation along the fluid flow:

$$0 = (\nabla\cdot T)\cdot \mathbf{u} = (\nabla\cdot T_{\text{fluid}})\cdot \mathbf{u} + (\nabla\cdot T_{\text{heat}})\cdot \mathbf{u} \ . \tag{1}$$

By a short calculation (see Solution 5.20) the first term is seen to be $-d\rho/d\tau + (dn/d\tau)(P+\rho)/n$ which by the 1st law of thermodynamics equals $-nT\,ds/d\tau$ (see Problem 5.19). The second term is

$$(\nabla\cdot T_{\text{heat}})\cdot \mathbf{u} = (q^{\mu}u^{\nu} + u^{\mu}q^{\nu})_{;\nu}\, u_{\mu} \ .$$

Since $q^{\alpha}u_{\alpha} = 0$ (heat flux is spacelike in comoving frame) and $u^{\alpha}u_{\alpha} = -1$, which imply

$$0 = (q^{\mu}u_{\mu})_{;\nu} = q^{\mu}{}_{;\nu}u_{\mu} + q^{\mu}u_{\mu;\nu}$$

$$0 = (u^{\mu}u_{\mu})_{;\nu} = 2u^{\mu}{}_{;\nu}u_{\mu} \ ,$$

the second term reduces to $-\mathbf{q}\cdot\mathbf{a} - \nabla\cdot\mathbf{q}$. Thus Equation (1) becomes

$$0 = -nT\,ds/d\tau - \mathbf{q}\cdot\mathbf{a} - \nabla\cdot\mathbf{q} \ .$$

Finally using the definition $S = ns\mathbf{u} + \mathbf{q}/T$, we compute

$$\begin{aligned}\nabla\cdot S &= s\boldsymbol{\nabla}\cdot(n\mathbf{u}) + n(\nabla s\cdot\mathbf{u}) + \frac{1}{T}\nabla\cdot\mathbf{q} - \frac{\nabla T}{T^2}\cdot\mathbf{q} \\ &= n\frac{ds}{d\tau} + \frac{1}{T}\nabla\cdot\mathbf{q} - \frac{\nabla T\cdot\mathbf{q}}{T^2} \\ &= -\frac{\mathbf{q}\cdot\mathbf{a}}{T} - \frac{\nabla T}{T^2}\cdot\mathbf{q} \ .\end{aligned}$$

The second term says entropy is generated by heat flow thru a temperature gradient, as in nonrelativistic thermodynamics. The first term says that it is generated by flow along an acceleration, this is a "redshift" effect arising from the fact that in a accelerated system, constant temperature is *not* an equilibrium state: photons from the "forward" side are blueshifted in going backward, thus carrying a net heat flux.

Solution 5.29. An equilibrium state obtains when there is no entropy generation, so according to the results of Problem 5.28

$$-\,\mathbf{q}\cdot\mathbf{a} = \frac{\nabla T}{T}\cdot\mathbf{q}$$

must hold for all heat flows $\mathbf{q}$. Thus we must have $\mathbf{a} = -\nabla \ln T$. The solution to this equation is $\ln T = -\underset{\sim}{a}\cdot\underset{\sim}{x} + \text{constant}$, or

$$T = T_0 \exp(-\underset{\sim}{a}\cdot\underset{\sim}{x}) \; .$$

Solution 5.30. Since $T^{\alpha\beta}u_\alpha = -\rho u^\beta$ there are no energy flow terms, hence $\mathbf{q} = 0$ and $\mathbf{s} = ns\mathbf{u}$. The rate of entropy production is thus

$$S^\alpha{}_{;\alpha} = (nu^\alpha)_{;\alpha}\, s + n\,\frac{ds}{d\tau} = n\,\frac{ds}{d\tau}$$

where the second equality follows from baryon number conservation $\boldsymbol{\nabla}\cdot(n\mathbf{u}) = 0$.

With the first law of thermodynamics

$$\frac{d\rho}{d\tau} = \frac{\rho + p}{n}\frac{dn}{d\tau} + nT\,\frac{ds}{d\tau}$$

and the law of baryon conservation

$$(nu^\alpha)_{;\alpha} = \frac{dn}{d\tau} + n\theta$$

the rate of entropy production can be written as

$$S^\alpha{}_{;\alpha} = \frac{1}{T}\left[\frac{d\rho}{d\tau} + \theta(\rho + p)\right] .$$

To express the right hand side in terms of the viscosity coefficients take the divergence of the vector $T^{\alpha\beta}u_\alpha$:

$$-(\rho u^\beta)_{;\beta} = -\frac{d\rho}{d\tau} - \rho\theta = (T^{\alpha\beta}u_\alpha)_{;\beta} = T^{\alpha\beta}u_{\alpha;\beta} \; .$$

The last term on the right can be calculated using the stress-energy for the viscous fluid and the expansion for $u_{\alpha;\beta}$ given in Problem 5.18. The $T^{\alpha\beta}\omega_{\alpha\beta}$ term vanishes by symmetry; the $T^{\alpha\beta}a_\alpha u_\beta$ term vanishes since $T^{\alpha\beta}u_\beta \propto u^\alpha$ and we are left with

$$T^{\alpha\beta}u_{\alpha;\beta} = [\rho u^\alpha u^\beta + (p-\zeta\theta)P^{\alpha\beta} - 2\eta\sigma^{\alpha\beta}][\sigma_{\alpha\beta} + \tfrac{1}{3}\theta P_{\alpha\beta}]$$
$$= -2\eta\sigma_{\alpha\beta}\sigma^{\alpha\beta} + \theta(p-\zeta\theta)\ .$$

[Here we used the easily verified relations

$$u^\alpha\sigma_{\alpha\beta} = 0, \quad P_{\alpha\beta}\sigma^{\alpha\beta} = 0, \quad P_{\alpha\beta}P^{\alpha\beta} = 3.]$$

The result then is

$$\frac{d\rho}{d\tau} + \theta(\rho+p) = 2\eta\sigma_{\alpha\beta}\sigma^{\alpha\beta} + \zeta\theta^2\ .$$

When this is substituted in our previous formula for $S^\alpha{}_{;\alpha}$ the desired equation results.

Solution 5.31. We know that projecting $T^{\alpha\beta}{}_{,\beta}$ along u_α gives an equation of local energy conservation; to get an equation of motion for the fluid, we project perpendicular to u_α.

$$0 = P^\gamma{}_\alpha T^{\alpha\beta}{}_{,\beta} = P^\gamma{}_\alpha(\rho_{,\beta}u^\alpha u^\beta + \rho u^\alpha{}_{,\beta}u^\beta + \rho u^\alpha u^\beta{}_{,\beta} + p_{,\beta}P^{\alpha\beta}$$
$$+ pP^{\alpha\beta}{}_{,\beta} - 2\eta_{,\beta}\sigma^{\alpha\beta} - 2\eta\sigma^{\alpha\beta}{}_{,\beta} - \zeta_{,\beta}\theta P^{\alpha\beta} - \zeta\theta_{,\beta}P^{\alpha\beta} - \zeta\theta P^{\alpha\beta}{}_{,\beta})\ .$$

We now use the following identities: $P^\gamma{}_\alpha u^\alpha = 0$,

$$P^\gamma{}_\alpha u^\alpha{}_{,\beta} = u^\gamma{}_{,\beta} + u^\gamma u_\alpha u^\alpha{}_{,\beta} = u^\gamma{}_{,\beta}$$

$$P^\gamma{}_\alpha P^{\alpha\beta} = P^{\gamma\beta}$$

$$P^\gamma{}_\alpha P^{\alpha\beta}{}_{,\beta} = P^\gamma{}_\alpha(u^\alpha u^\beta)_{,\beta} = u^\gamma{}_{,\beta}u^\beta$$

$$P^\gamma{}_\alpha\sigma^{\alpha\beta} = \sigma^{\gamma\beta}$$

$$P^\gamma{}_\alpha\sigma^{\alpha\beta}{}_{,\beta} = \sigma^{\gamma\beta}{}_{,\beta} + u^\gamma u_\alpha\sigma^{\alpha\beta}{}_{,\beta} = \sigma^{\gamma\beta}{}_{,\beta} - u^\gamma u_{\alpha,\beta}\sigma^{\alpha\beta} = \sigma^{\gamma\beta}{}_{,\beta} - u^\gamma\sigma_{\alpha\beta}\sigma^{\alpha\beta}.$$

With these identities the projected equations of motion become:

$$0 = (\rho + p)\, u^{\gamma}{}_{,\beta} u^{\beta} + p_{,\beta} P^{\alpha\beta} - 2(\eta\sigma^{\gamma\beta} + \zeta\theta P^{\gamma\beta})_{,\beta}$$
$$+ 2\eta u^{\gamma} \sigma_{\alpha\beta}\sigma^{\alpha\beta} + \zeta\theta^2 u^{\gamma} \ . \tag{1}$$

In the nonrelativistic limit,

$$u^t \approx 1, \quad u^j \approx v^j, \quad p = \mathcal{O}(v^2), \quad \rho = \mathcal{O}(1) \ .$$

Taking the j-component of Equation (1), and working to $\mathcal{O}(v^2)$, we get

$$0 = \rho(v^j{}_{,t} + v^j{}_{,k} v^k) + p_{,j}$$
$$- \left[\eta(v_{j,k} + v_{k,j} - \tfrac{2}{3}\delta_{jk} v^m{}_{,m})\right]_{,k} + (\zeta v^m{}_{,m})_{,j}$$

which is the Navier-Stokes equation.

Solution 5.32. For a perfect Maxwell-Boltzmann gas,

$$p = nkT \tag{1}$$

$$\rho/n = U(T) \tag{2}$$

Equation (2) says that the energy per particle is a function of temperature alone. From the first law of relativistic thermodynamics,

$$\begin{aligned} T ds &= d(\rho/n) + p d(1/n) \\ &= \frac{dU}{dT} dT + p d\left(\frac{1}{n}\right) \end{aligned} \tag{3}$$

we see that

$$c_v = \frac{dU}{dT} \ . \tag{4}$$

Equation (1) implies that

$$p d\left(\frac{1}{n}\right) + \frac{1}{n} dp = k dT \tag{5}$$

so Equation (3) becomes

$$T ds = \left(\frac{dU}{dT} + k\right) dT - \frac{1}{n} dp \tag{6}$$

and thus

$$c_p = \frac{dU}{dT} + k = c_v + k \ . \tag{7}$$

From Equation (1)

$$\Gamma_1 = \left.\frac{\partial \log p}{\partial \log n}\right|_s = 1 + \frac{n}{T}\left.\frac{dT}{dn}\right|_s \tag{8}$$

But Equation (3) implies that, when $s = \text{constant}$,

$$c_v \frac{dT}{dn} = \frac{p}{n^2} = k\frac{T}{n} = (c_p - c_v)\frac{T}{n} \ . \tag{9}$$

Substituting Equation (9) in Equation (8) gives us

$$\Gamma_1 = 1 + \frac{(c_p - c_v)}{c_v} = \gamma \ .$$

Solution 5.33. From Problem 5.32,

$$\gamma = \Gamma_1 = \left.\frac{\partial \log p}{\partial \log n}\right|_s$$

so if $\gamma = \text{constant}$,

$$p = Kn^\gamma \ .$$

For adiabatic changes the first law of thermodynamics is

$$d\rho = \frac{\rho + p}{n}\, dn = \left(\frac{\rho}{n} + Kn^{\gamma-1}\right) dn \ ,$$

or

$$\frac{d}{dn}\left(\frac{\rho}{n}\right) = Kn^{\gamma-2}$$

so that the solution is

$$\frac{\rho}{n} = \frac{Kn^{\gamma-1}}{\gamma-1} + \text{constant} \ .$$

But when $n \to 0$, then $\rho/n \to m$, so

$$\rho = mn + \frac{Kn^\gamma}{\gamma-1} \ .$$

Solution 5.34. We obtain scalar integrals by dotting various combinations of u^{α} into the vector and tensor integrals for J^{μ} and $T^{\mu\nu}$. Thus

$$n = -J^{\mu}u_{\mu} = \int \mathfrak{N} d^3P = \frac{g}{h^3}\int_0^{\infty} \frac{4\pi P^2 dP}{\exp\left[(P^2+m^2)^{\frac{1}{2}}/kT - \theta\right] - \varepsilon},$$

where $g = 2J + 1$. Make the substitution $P = m \sinh \chi$.

Then

$$n = \frac{4\pi g m^3}{h^3}\int_0^{\infty} \frac{\sinh^2\chi \cosh\chi \, d\chi}{\exp(\beta \cosh\chi - \theta) - \varepsilon}, \tag{1}$$

where $\beta = m/kT$. Similarly, we have

$$p = \frac{1}{3}(u_{\mu}u_{\nu} + g_{\mu\nu})T^{\mu\nu} = \frac{1}{3}\int \mathfrak{N} \frac{P^2}{(P^2+m^2)^{\frac{1}{2}}} d^3P$$

$$= \frac{4\pi g m^4}{3h^3}\int_0^{\infty} \frac{\sinh^4\chi \, d\chi}{\exp(\beta \cosh\chi - \theta) - \varepsilon} \tag{2}$$

and finally,

$$\rho - 3p = -g_{\mu\nu}T^{\mu\nu} = m^2 \int \mathfrak{N} \frac{d^3P}{(P^2+m^2)^{\frac{1}{2}}}$$

$$= \frac{4\pi g m^4}{h^3}\int_0^{\infty} \frac{\sinh^2\chi \, d\chi}{\exp(\beta \cosh\chi - \theta) - \varepsilon}. \tag{3}$$

(b) From Equation (2),

$$dp = \frac{4\pi g m^4}{3h^3}\int_0^{\infty} \frac{\sinh^4\chi \, d\chi \, (\beta \cosh\chi \, dT/T + d\theta)\exp(\beta \cosh\chi - \theta)}{[\exp(\beta \cosh\chi - \theta) - \varepsilon]^2}.$$

Now integrate by parts, differentiating the terms $\sinh^3\chi \cosh\chi$ and $\sinh^3\chi$ to get

$$dp = \frac{4\pi g m^4}{3h^3}\frac{1}{\beta}\int_0^{\infty} \frac{d\chi[(3\sinh^2\chi + 4\sinh^4\chi)\beta \, dT/T + 3\sinh^2\chi \cosh\chi \, d\theta]}{\exp(\beta \cosh\chi - \theta) - \varepsilon}.$$

Substitute expressions (1), (2), and (3) for the integrals, and find

$$dp = (\rho+p)\,dT/T + nkT\,d\theta\ . \tag{4}$$

(c) From the definition of μ we have

$$d\mu = d\rho/n + dp/n - (\rho+p)\,dn/n^2 - s\,dT - T\,ds\ ,$$

but

$$d\rho = (\rho+p)\,dn/n + nT\,ds\ ,$$

therefore

$$d\mu = dp/n - s\,dT = dp/n - (\rho+p)\,dT/(nT) + \mu\,dT/T\ .$$

Here we have substituted for s in terms of μ. Comparing this with Equation (4), we identify

$$\mu = kT\theta\ .$$

(d) For $\varepsilon = 0$

$$\frac{p}{n} = \frac{\frac{m}{3}\int_0^\infty \sinh^4\chi\ e^{-\beta\cosh\chi}\,d\chi}{\int_0^\infty \sinh^2\chi\ \cosh\chi\ e^{-\beta\cosh\chi}\,d\chi}\ ,$$

but

$$\int_0^\infty \sinh^4\chi\ \exp(-\beta\cosh\chi)\,d\chi = -\frac{1}{\beta}\sinh^3\chi\ \exp(-\beta\cosh\chi)\Big|_0^\infty$$

$$+\frac{3}{\beta}\int_0^\infty \sinh^3\chi\ \cosh\chi\ \exp(-\beta\cosh\chi)\,d\chi\ .$$

The first term on the right vanishes, so

$$p = nkT\ .$$

(e) For $\varepsilon = 0$, the integrals (1), (2), and (3) can be expressed in terms of modified Hankel functions, since

$$K_n(\beta) = \frac{\beta^n}{(2n-1)!!}\int_0^\infty d\chi\ \sinh^{2n}\chi\ \exp(-\beta\cosh\chi)$$

$$= \frac{\beta^{n-1}}{(2n-3)!!}\int_0^\infty d\chi\ \sinh^{2n-2}\chi\quad \cosh\chi\ e^{-\beta\cosh\chi}\ .$$

Thus for $a = 4\pi g m^3 e^{\theta} h^{-3}$ we have

$$n = aK_2(\beta)/\beta$$

$$p = amK_2(\beta)/\beta^2$$

$$\rho - 3p = amK_1(\beta)/\beta^2 \ .$$

The exact expression for ρ/n is

$$\rho/n = m[K_1(\beta)/K_2(\beta) + 3/\beta] \ .$$

For $kT \ll m$, $\beta \to \infty$. In this limit

$$K_n(\beta) \to \left(\frac{\pi}{2\beta}\right)^{\frac{1}{2}} e^{-\beta}\left[1 + \frac{4n^2 - 1}{8\beta} + \cdots\right] .$$

so

$$\rho/n \to m\left[\frac{1 + 3/(8\beta)}{1 + 15/(8\beta)} + \frac{3}{\beta}\right] = m\left(1 + \frac{3}{2}\frac{kT}{m}\right)$$

For $kT \gg m$, $\beta \to 0$, $K_1(\beta)/K_2(\beta) \to 0$, so $\rho/n \to 3kT$. Whew!

Solution 5.35. From Problem 5.32,

$$\gamma = (c_v + k)/c_v = 1 + \frac{k}{dU/dT}$$

where, from Problem 5.34

$$U(T) = m\{K_1(\beta)/K_2(\beta) + 3/\beta\}, \qquad \beta = m/kT \ .$$

This gives the formal answer. Note that for $kT \ll m$, $U = m + 3kT/2$, $\gamma = 5/3$, and that for $kT \gg m$, $U = 3kT$ and $\gamma = 4/3$.

CHAPTER 6: SOLUTIONS

Solution 6.1.

(a) The analogy with polar coordinates suggests

$$x = v \cosh u \qquad x^2 - t^2 = v^2$$
$$t = v \sinh u \qquad x/t = \coth u \ .$$

Then

$$\begin{aligned} dx^2 &= (dv \cosh u + du\, v \sinh u)^2 \\ dt^2 &= (dv \sinh u + du\, v \cosh u)^2 \\ dx^2 - dt^2 &= dv^2 - v^2\, du^2 \ . \end{aligned} \qquad (1)$$

(b) Solving Equations (1) for du and dv gives

$$dv = dx \cosh u - dt \sinh u$$

$$du = v^{-1}(dt \cosh u - dx \sinh u)$$

and thus for a particle of unit mass,

$$P_u = g_{uu} P^u = -v^2 \frac{du}{d\tau} = -v \cosh u \frac{dt}{d\tau} + v \sinh u \frac{dx}{d\tau} = -x \frac{dt}{d\tau} + t \frac{dx}{d\tau} \ .$$

For an unaccelerated particle $x = \text{const.} + \frac{dx}{dt}\, t$, and $dt/d\tau$, $dx/d\tau$ are also both constant, so

$$P_u = \text{const.} - t \frac{dx}{dt} \frac{dt}{d\tau} + t \frac{dx}{d\tau} = \text{constant.}$$

Now for P_v, use

$$-m^2 = P \cdot P = g^{vv}(P_v)^2 + g^{uu}(P_u)^2 = (P_v)^2 - (P_u)^2/v^2 \ .$$

So, $P_v^2 = P_u^2/v^2 - m^2$, which is not constant since in general v varies along the particle's trajectory. It turns out that whenever the metric coefficient does not depend on some coordinate (in this case u), then the covariant momentum of that coordinate (here P_u) is conserved. (See Problem 7.13 or MTW, p. 651.)

Solution 6.2. The metric in Euclidean 4-space is

$$ds^2 = dx_1^2 + dx_2^2 + dx_3^2 + dx_4^2 \ . \tag{1}$$

The equation of a hypersphere of radius R is

$$x_1^2 + x_2^2 + x_3^2 + x_4^2 = R^2 \ . \tag{2}$$

By analogy with the 3-dimensional case, introduce coordinates on the hypersphere:

$$\begin{aligned} x_4 &= R\cos\alpha \\ x_3 &= R\sin\alpha\cos\theta \\ x_2 &= R\sin\alpha\sin\theta\cos\phi \\ x_1 &= R\sin\alpha\sin\theta\sin\phi. \end{aligned} \tag{3}$$

Then Equation (2) is automatically satisfied; taking the differential of Equation (3) with R constant, and substituting in Equation (1) yields the required metric of the hypersphere.

Solution 6.3.

(a) Introduce x and y coordinates on the cylinder by

$$x = \phi$$

$$y = a\tan\lambda \ .$$

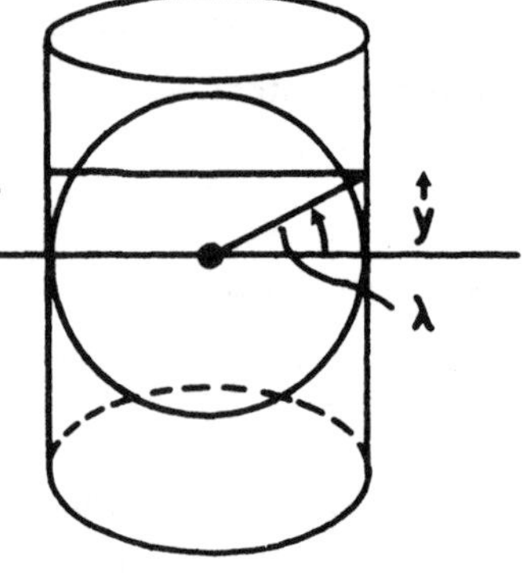

The metric of the sphere in these coordinates is

$$ds^2 = \frac{a^4\,dy^2}{(a^2+y^2)^2} + \frac{a^4\,dx^2}{a^2+y^2} \tag{1}$$

Comparing this with $ds^2 = dx^2 + dy^2$, we see there is the least distortion near $y = 0$, i.e. near the equator.

(b) For the stereographic projection, it is slightly easier to use the usual polar angle $\theta = 90^0 - \lambda$. Let (θ, ϕ) be the coordinates of a point on the sphere, and (θ_0, ϕ_0) the spherical polar coordinates of the projected point. Clearly $\phi_0 = \phi$. The distance of the projected point from the axis is $\rho = 2a \tan(\theta/2)$, so we introduce coordinates

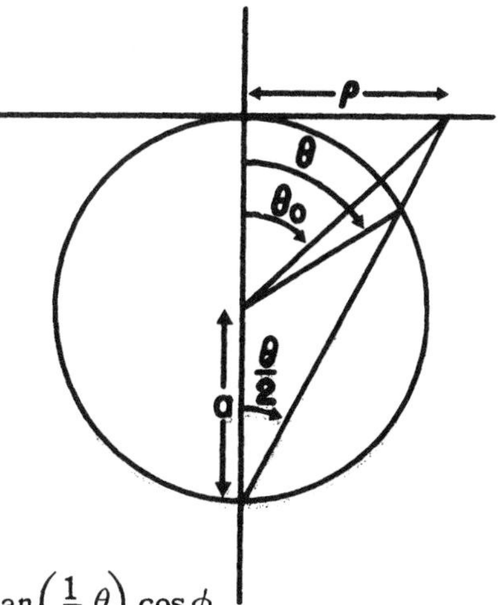

$$x = \rho \cos\phi = 2a \tan\left(\frac{1}{2}\theta\right) \cos\phi$$

$$y = \rho \sin\phi = 2a \tan\left(\frac{1}{2}\theta\right) \sin\phi$$

and it is easily verified that

$$ds^2 = a^2(d\theta^2 + \sin^2\theta\, d\phi^2) = \cos^4\left(\frac{1}{2}\theta\right)(dx^2 + dy^2) .$$

Here the distortion is least near $\theta = 0$, the north pole. This projection is said to be conformal, because $(ds^2)_{sphere} = g(ds^2)_{map}$, where g is some function, in this case equal to $\cos^4(\theta/2)$. A conformal projection preserves angles (see Problem 6.7).

Solution 6.4.

(a) Imagine traveling along some curve $\phi = \phi(\theta)$. The compass bearing ψ is given by

$$\tan\psi = \sin\theta \frac{d\phi}{d\theta} . \tag{1}$$

(Note that ψ is measured clockwise from the y-axis.) On the map we have

$$\frac{dy}{dx} = \tan\left(\psi + \frac{\pi}{2}\right) = -\frac{1}{\tan\psi} . \tag{2}$$

Now we combine this with Equation (1):

$$\sin\theta \frac{d\phi}{d\theta} = -\frac{dx/d\theta}{dy/d\theta} = -\frac{\dfrac{\partial x}{\partial \theta} + \dfrac{\partial x}{\partial \phi}\dfrac{d\phi}{d\theta}}{\dfrac{\partial y}{\partial \theta} + \dfrac{\partial y}{\partial \phi}\dfrac{d\phi}{d\theta}}$$

$$\left(\frac{\partial y}{\partial \theta} + \frac{\partial y}{\partial \phi}\frac{d\phi}{d\theta}\right)\frac{d\phi}{d\theta}\sin\theta = -\frac{\partial x}{\partial \theta} - \frac{\partial x}{\partial \phi}\frac{d\phi}{d\theta} . \qquad (3)$$

Since Equation (3) must hold for arbitrary $d\phi/d\theta$ at the point we are considering, we can equate coefficients of powers of $d\phi/d\theta$ on the two sides of Equation (3). This gives us

$$\partial y/\partial \phi = 0 , \qquad y = y(\theta) \qquad (4)$$

$$\partial x/\partial \theta = 0 , \qquad x = x(\phi) \qquad (5)$$

$$-\sin\theta\, \partial y/\partial \theta = \partial x/\partial \phi . \qquad (6)$$

Equations (4) and (5) imply that the left-hand side of Equation (6) is a function of θ only, while the right-hand side is a function of ϕ only, and so each side must be a constant which we can choose to be 1. Thus the map is given by

$$x = \phi , \qquad y = -\int \frac{d\theta}{\sin\theta} = \log\cot\tfrac{1}{2}\theta . \qquad (7)$$

(b) Take the radius to be 1 for convenience. Then

$$\begin{aligned} ds^2 &= d\theta^2 + \sin^2\theta\, d\phi^2 \\ &= \sin^2\theta\,(dx^2 + dy^2) \\ &= \operatorname{sech}^2 y\,(dx^2 + dy^2) . \end{aligned} \qquad (8)$$

(c) The great circles are the geodesics of the 2-sphere. The geodesic equations are easy to solve, since we have two first integrals. Let a dot denote d/ds; then Equation (8) gives

$$\operatorname{sech}^2 y\,(\dot{x}^2 + \dot{y}^2) = 1 . \qquad (9)$$

Since x is an ignorable coordinate in the metric (8), $g_{xx}\dot{x}$ is constant (see Solution 7.13), i.e.

$$(\operatorname{sech}^2 y)\,\dot{x} = \gamma \,. \tag{10}$$

When neither $\dot{x}$ nor $\dot{y}$ vanishes, we can eliminate s from Equations (9) and (10):

$$\left(\frac{dy}{dx}\right)^2 = \frac{\dot{y}^2}{\dot{x}^2} = \frac{\lambda^2 - \cosh^2 y}{\cosh^2 y} \tag{11}$$

where $\lambda = 1/\gamma$. Equation (11) is easily integrated by putting $z = \sinh y$. The result is

$$\sinh y = \alpha \sin(x+\beta)$$

where $\alpha = (\lambda^2 - 1)^{\frac{1}{2}}$ and β is another constant of integration.

Solution 6.5. We can decide if the space is 3-dimensional or not by evaluating the 3-volume spanned by $dx\,dy\,dz$

$$dV = g^{\frac{1}{2}}\, dx\,dy\,dz$$

$$= \begin{vmatrix} 1-\left(\frac{3}{13}\right)^2 & -\left(\frac{3}{13}\right)\left(\frac{4}{13}\right) & -\left(\frac{12}{13}\right)\left(\frac{3}{13}\right) \\ -\left(\frac{3}{13}\right)\left(\frac{4}{13}\right) & 1-\left(\frac{4}{13}\right)^2 & -\left(\frac{4}{13}\right)\left(\frac{12}{13}\right) \\ -\left(\frac{12}{13}\right)\left(\frac{3}{13}\right) & -\left(\frac{4}{13}\right)\left(\frac{12}{13}\right) & 1-\left(\frac{12}{13}\right)^2 \end{vmatrix}^{\frac{1}{2}} dx\,dy\,dz = 0 \,.$$

Since this is identically zero for all x, y, z, the 3 coordinates are always linearly dependent. The space is thus either two dimensional or one dimensional. We can now throw away one coordinate, say z, by taking z = constant; this is allowed because it is an "ignorable coordinate," i.e. the coefficients of the metric do not depend on z. The metric left is

$$ds^2 = dx^2 + dy^2 - \left(\frac{3}{13}\,dx + \frac{4}{13}\,dy\right)^2 .$$

It is in fact 2-dimensional, not 1-dimensional, because

$$g = \det\begin{pmatrix} 1-\left(\frac{3}{13}\right)^2 & -\left(\frac{3}{13}\right)\left(\frac{4}{13}\right) \\ \left(\frac{3}{13}\right)\left(\frac{4}{13}\right) & 1-\left(\frac{4}{13}\right)^2 \end{pmatrix} \neq 0 \,.$$

It is not difficult to find the coordinate transformation (e.g. by Gram-Schmidt orthogonalization)

$$\xi = \frac{12}{5}\left(\frac{3}{13}x + \frac{4}{13}y\right)$$

$$\eta = \frac{13}{5}\left(-\frac{4}{13}x + \frac{3}{13}y\right)$$

which gives $ds^2 = d\xi^2 + d\eta^2$, obviously the simplest form of the metric. The more sophisticated reader will notice that the original metric was of the form

$$g_{\alpha\beta} = \delta_{\alpha\beta} - V_\alpha V_\beta \ ,$$

where V_α is the Euclidean unit vector $\left(\frac{3}{13}, \frac{4}{13}, \frac{12}{13}\right)$. This metric just projects a 3-dimensional Euclidean space into a 2-dimensional space perpendicular to $\mathbf{V}$, so it is obvious that there *had* to exist ξ and η coordinates which make it a flat-space 2-dimensional metric.

Solution 6.6. Let $\mathbf{A}$ be an arbitrary vector. We want to see if $\mathbf{P}\cdot\mathbf{A}$ is orthogonal to $\mathbf{u}$, i.e. if $\mathbf{u}\cdot\mathbf{P}\cdot\mathbf{A} = 0$:

$$\mathbf{u}\cdot\mathbf{P}\cdot\mathbf{A} = u^\alpha(g_{\alpha\beta} + u_\alpha u_\beta)A^\beta = u^\alpha A_\alpha + (u^\alpha u_\alpha)u_\beta A^\beta = u_\alpha A^\alpha - u_\beta A^\beta = 0 .$$

Also it is easy to see that if $\mathbf{A}\cdot\mathbf{u} = 0$, then $\mathbf{P}\cdot\mathbf{A} = \mathbf{A}$, so $\mathbf{A}$ is unaffected by projection. If $\mathbf{n}$ is a unit spacelike vector, we have

$$\begin{aligned} \mathbf{n}\cdot\mathbf{P}\cdot\mathbf{A} &= n^\alpha(g_{\alpha\beta} - n_\alpha n_\beta)A^\beta \\ &= n^\alpha A_\alpha - (n^\alpha n_\alpha)n_\beta A^\beta \\ &= n_\alpha A^\alpha - n_\beta A^\beta = 0 . \end{aligned}$$

Likewise a vector already orthogonal to $\mathbf{n}$ is unaffected by projection. Suppose now that $\mathbf{P}$ is a projection operator orthogonal to a null vector $\mathbf{k}$, so that $\mathbf{k}\cdot\mathbf{P}\cdot\mathbf{A} = 0$ for all $\mathbf{A}$. It is easy to see that $\mathbf{P}$ + constant $\times$ $\mathbf{k}\otimes\mathbf{k}$ is also a projection operator since $\mathbf{k}\cdot\mathbf{k} = 0$; so $\mathbf{P}$ is not unique. (The set of null projection operators is not empty since it is easy to check that for *any* 4-vector w, one is $\mathbf{P} = g - w \otimes k / k \cdot w$. However, there are no *symmetric* null projection operators at all.)

Solution 6.7. Let $\mathbf{A}$ and $\mathbf{B}$ be two vectors in the metric space. Obviously we want the angle to be related to the dot product $\mathbf{A} \cdot \mathbf{B}$. We also want it to be unaffected by scaling $\mathbf{A}$ and $\mathbf{B}$, so the natural choice is $\mathbf{A} \cdot \mathbf{B} / (|\mathbf{A}|\,|\mathbf{B}|)$, which is $\cos\theta$ in the familiar Euclidean case. Now under $g_{\alpha\beta} \to f(x^\mu)\, g_{\alpha\beta}$ we have

$$\frac{\mathbf{A} \cdot \mathbf{B}}{(|\mathbf{A}|\,|\mathbf{B}|)} = \frac{g_{\alpha\beta} A^\alpha B^\beta}{(g_{\mu\nu} A^\mu A^\nu g_{\rho\sigma} B^\rho B^\sigma)^{\frac{1}{2}}} \to \frac{f(x^\gamma)\, g_{\alpha\beta} A^\alpha B^\beta}{[f(x^\gamma)\, g_{\mu\nu} A^\mu A^\nu\, f(x^\gamma)\, g_{\rho\sigma} B^\rho B^\sigma]^{\frac{1}{2}}}$$

but the f's cancel, giving the same value as before. Null curves remain null curves because the square of their tangent vector remains zero

$$0 = \boldsymbol{\ell} \cdot \boldsymbol{\ell} = g_{\mu\nu} \ell^\mu \ell^\nu \to f(x^\gamma)\, g_{\mu\nu} \ell^\mu \ell^\nu = 0 \ .$$

Solution 6.8. The relative velocity ds between two velocities $\underset{\sim}{v}$ and $\underset{\sim}{v} + d\underset{\sim}{v}$ is (by Problem 1.3) given by

$$ds^2 = \frac{(d\underset{\sim}{v})^2 - (\underset{\sim}{v} \times d\underset{\sim}{v})^2}{(1 - v^2)^2} \ ,$$

and

$$(\underset{\sim}{v} \times d\underset{\sim}{v})^2 = v^2 (d\underset{\sim}{v})^2 - (\underset{\sim}{v} \cdot d\underset{\sim}{v})^2 \ .$$

Let θ and ϕ be polar and azimuthal angles about the direction of $\underset{\sim}{v}$ (i.e. $v_z = v\cos\theta$, $v_x = v\sin\theta\cos\phi$, $v_y = v\sin\theta\sin\phi$). Then

$$(d\underset{\sim}{v})^2 = (dv)^2 + v^2 (d\theta^2 + \sin^2\theta\, d\phi^2)$$

$$\underset{\sim}{v} \cdot d\underset{\sim}{v} = \frac{1}{2}\, d(\underset{\sim}{v} \cdot \underset{\sim}{v}) = v\,dv$$

so

$$ds^2 = \frac{dv^2}{(1 - v^2)^2} + \frac{v^2}{1 - v^2} (d\theta^2 + \sin^2\theta\, d\phi^2) \ .$$

Introduce the "rapidity parameter" $v = \tanh\chi$. Then

$$ds^2 = d\chi^2 + \sinh^2\chi\,(d\theta^2 + \sin^2\theta\, d\phi^2) \ .$$

Note that for small v, $\sinh\chi \sim \chi \sim v$ and velocity space is flat, as we expect it to be in the Newtonian limit.

Solution 6.9. If we had $\sin^2\theta$ instead of $(\theta - \theta^3)$ the metric would just be that of the Euclidean 2-sphere. As given, it is obviously some sort of axially symmetric "warped" 2-sphere with singularities possible at $\theta = 0, \pm 1$. From the given value of $\theta = \frac{1}{2}$, it is obvious that the range of θ can be extended to $0 < \theta < 1$. We must examine the cases $\theta = 0$ and $\theta = 1$ with some care to decide what range χ can take. We know that χ must be periodic because, as the metric shows, the coordinates ($\theta = 0$, all χ) and ($\theta = 1$, all χ) each represent just one point.

For $\theta \approx 0$, we have

$$ds^2 \approx d\theta^2 + \theta^2 d\chi^2 \ .$$

If χ is periodic with period P, then the proper circumference of a small circle of radius $\Delta\theta$ is $\int (\Delta\theta)\, d\chi = (\Delta\theta)\,P$. If we are to avoid a conical singularity this must equal $2\pi\Delta\theta$, so $P = 2\pi$. But now consider $\theta \approx 1$: The metric is

$$ds^2 \approx d\theta^2 + (\theta - 1)^2 (2)^2\, d\chi^2$$

so the condition avoiding a conical singularity, is

$$\int_0^P 2(\Delta\theta)\, d\chi = 2\pi(\Delta\theta) \ ,$$

implies that $P = \pi$. Thus we can *either* take $P = 2\pi$ and have a conical singularity at $\theta = 1$, or $P = \pi$ and have it at $\theta = 0$. These are two different global extensions of the manifold. (This problem is due to Gilbert Miller.)

Solution 6.10. A geometrical object exhibits a particular group symmetry if its *functional* change (i.e. its change in functional form) under the action of the group vanishes. If ξ is an infinitesimal displacement, then under the infinitesimal coordinate transformation $x'^{\mu} = x^{\mu} + \xi^{\mu}$, one can easily show that the functional change of the metric, $\bar{\delta} g_{\mu\nu}$, is

$$\bar{\delta} g_{\mu\nu} \equiv g'_{\mu\nu}(x) - g_{\mu\nu}(x) = - g_{\mu\rho}\xi^{\rho}{}_{,\nu} - g_{\rho\nu}\xi^{\beta}{}_{,\mu} - g_{\mu\nu,\rho}\xi^{\rho}. \quad (1)$$

(See Problem 13.12.) For spherical symmetry, $\bar{\delta} g_{\mu\nu}$ must vanish under the rotation group, a particular realization of which is obtained from the generators

$$\xi^0 = 0, \qquad \xi^i = \varepsilon^{ij} x^j , \quad (2)$$

where $\varepsilon^{ij} = -\varepsilon^{ji}$ are three arbitrary infinitesimal constants. Substituting Equation (2) for the generators ξ^{μ} in Equation (1) and setting the left side to zero, one gets:

a. $\mu = \nu = 0$

$$g_{00,i}\,\varepsilon^{ij} x^j = 0 \quad \text{or} \quad g_{00,i}\, x^j = g_{00,j}\, x^i$$

and hence

$$\begin{aligned} g_{00} &= g_{00}(x^0, r^2) \\ r^2 &\equiv (x^1)^2 + (x^2)^2 + (x^3)^2 . \end{aligned} \quad (3a)$$

b. $\mu = 0, \ \nu \neq 0$

$$g_{0i}\,\varepsilon^{ij} + g_{0j,i}\,\varepsilon^{ik} x^k = 0$$

$$g_{0j} = \Gamma_1(r^2, x^0)\, x^j \quad (3b)$$

where Γ_1 is an arbitrary function.

c. $\mu \neq 0, \ \nu \neq 0$

$$g_{ik}\,\varepsilon^{kj} + g_{\ell j}\,\varepsilon^{\ell i} + g_{ij,\ell}\,\varepsilon^{\ell k} x^k = 0$$

$$g_{ij} = \Gamma_2(r^2, x^0)\,\delta_{ij} + \Gamma_3(r^2, x^0)\, x^i x^j \quad (3c)$$

where Γ_2 and Γ_3 are again arbitrary functions.

From Equation (2) one can see that the above coordinate system (x^0, x^1, x^2, x^3) is "cartesian-like." The group theoretically defined property of spherical symmetry is coordinate independent, however, and to find $g_{\mu\nu}$ in any other coordinate system one merely transforms the metric given in Equations (3) in the usual way.

CHAPTER 7: SOLUTIONS

Solution 7.1. Let the transformation be given by $e_{\mu'} = L^{\sigma}{}_{\mu'} e_{\sigma}$, then

$$\nabla_{e'_\beta} e'_\alpha = \Gamma^{\tau'}{}_{\alpha'\beta'} e_{\tau'} = \nabla_{(L^\lambda{}_{\beta'} e_\lambda)} (L^\mu{}_{\alpha'} e_\mu) = L^\lambda{}_{\beta'} \nabla_{e_\lambda} (L^\mu{}_{\alpha'} e_\mu)$$

$$= L^\lambda{}_{\beta'} (L^\mu{}_{\alpha',\lambda} e_\mu + L^\mu{}_{\alpha'} \Gamma^\tau{}_{\mu\lambda} e_\tau)$$

$$= L^\lambda{}_{\beta'} (L^\mu{}_{\alpha',\lambda} L^{\tau'}{}_\mu e_{\tau'} + L^\mu{}_{\alpha'} L^{\tau'}{}_\gamma \Gamma^\gamma{}_{\mu\lambda} e_{\tau'}) ,$$

and therefore

$$\Gamma^{\tau'}{}_{\alpha'\beta'} = L^\lambda{}_{\beta'} L^\mu{}_{\alpha'} L^{\tau'}{}_\gamma \Gamma^\gamma{}_{\mu\lambda} + L^\lambda{}_{\beta'} L^{\tau'}{}_\mu L^\mu{}_{\alpha',\lambda} .$$

The second term in this equation would not be present in a tensor transformation.

Solution 7.2.

(a) First consider the line $\theta = 0$, $r = s$, where s is the affine parameter (length). The geodesic equation $d^2x^\mu/dr^2 + \Gamma^\mu{}_{rr} = 0$, gives

$$\Gamma^r{}_{rr} = \Gamma^\theta{}_{rr} = 0 .$$

Next, for nonradial lines, reparameterize the geodesic equation using θ as a non-affine parameter:

$$\frac{d^2x^\mu}{d\theta^2} + \frac{d^2\theta/ds^2}{(d\theta/ds)^2}\frac{dx^\mu}{d\theta} + \frac{dx^\alpha}{d\theta}\frac{dx^\beta}{d\theta}\Gamma^\mu{}_{\alpha\beta} = 0 . \tag{1}$$

The general straight line is

$$r\cos(\theta - \alpha) = R_0 \tag{2}$$

where α and R_0 are arbitrary constants. This equation, and $ds^2 = dr^2 + r^2 d\theta^2$, give $ds/d\theta = R_0/\cos^2\Psi$, where $\Psi \equiv \theta - \alpha$, and this in turn yields

$$\frac{d^2\theta}{ds^2} \bigg/ \left(\frac{d\theta}{ds}\right)^2 = -2\tan\Psi \ .$$

The geodesic equation now reads

$$\frac{d^2x^\mu}{d\theta^2} - 2\tan\Psi \frac{dx^\mu}{d\theta} + \frac{dx^\alpha}{d\theta}\frac{dx^\beta}{d\theta}\Gamma^\mu{}_{\alpha\beta} = 0 \ . \tag{3}$$

Consider the point $\theta = \alpha$ (i.e. $\Psi = 0$) and $r = R_0$ on the line. At this point (3) becomes

$$\frac{d^2x^\mu}{d\theta^2} + \Gamma^\mu{}_{\theta\theta} = 0 \ ,$$

which gives $\Gamma^\theta{}_{\theta\theta} = 0$, and $\Gamma^r{}_{\theta\theta} = -r$ (where Equation (2) has been used). Since α and R_0 are arbitrary these expressions are true in general.

Finally, consider an arbitrary point on the line and write out the geodesic equation in full glory, putting in the Γ's known so far. The $\mu = r$ component of the result gives $\Gamma^r{}_{r\theta} = \frac{1}{r}$.

(b) From $r^2 = x^2 + y^2$, and $\cot\theta = x/y$ we get a transformation matrix

$$\begin{array}{cc} \mu = x & \mu = y \end{array}$$

$$L^{\alpha'}{}_\mu = \begin{bmatrix} \cos\theta & \sin\theta \\ -\dfrac{\sin\theta}{r} & \dfrac{\cos\theta}{r} \end{bmatrix}$$

and its inverse

$$\begin{array}{cc} \beta' = r & \beta' = \theta \end{array}$$

$$L^\mu{}_{\beta'} = \begin{bmatrix} \cos\theta & -r\sin\theta \\ \sin\theta & r\cos\theta \end{bmatrix}$$

(Example: $x = r\cos\theta \Longrightarrow dx = dr\cos\theta - r\sin\theta\, d\theta \Longrightarrow$ first row above.)

The transformation law for the Γ's (Problem 7.1) is

$$\Gamma^{\alpha'}{}_{\beta'\gamma'} = L^{\alpha'}{}_{\rho} L^{\mu}{}_{\beta'} L^{\nu}{}_{\gamma'} \Gamma^{\rho}{}_{\mu\nu} + L^{\alpha'}{}_{\mu} L^{\mu}{}_{\beta',\gamma'} .$$

Since $\Gamma^{\rho}{}_{\mu\nu} = 0$ in cartesian coordinates, only the second term contributes, and straightforward differentiation and matrix multiplication gives the desired results.

(c) We now use

$$\Gamma^{\alpha}{}_{\beta\gamma} = \frac{1}{2} g^{\alpha\mu}(g_{\mu\beta,\gamma} + g_{\mu\gamma,\beta} - g_{\beta\gamma,\mu}) .$$

For the metric $ds^2 = dr^2 + r^2\, d\theta^2$ the only nonvanishing derivative is $g_{\theta\theta,r} = 2r$. Thus $\Gamma^{\alpha}{}_{\beta\gamma} = 0$ unless 2 indices are θ's and one is an r:

$$\Gamma^{r}{}_{\theta\theta} = g^{rr}\left(-\frac{1}{2}\, g_{\theta\theta,r}\right) = -r$$

$$\Gamma^{\theta}{}_{r\theta} = g^{\theta\theta}\left(\frac{1}{2}\, g_{\theta\theta,r}\right) = \frac{1}{r} .$$

Solution 7.3.

(a) The geodesic equation is

$$\frac{d^2r}{ds^2} + \Gamma^{r}{}_{\theta\theta}\left(\frac{d\theta}{ds}\right)^2 = 0 \qquad \frac{d^2\theta}{ds^2} + 2\Gamma^{\theta}{}_{r\theta}\left(\frac{dr}{ds}\right)\left(\frac{d\theta}{ds}\right) = 0 .$$

Using the results of Problem 7.2, we have

$$\frac{d^2r}{ds^2} = r\left(\frac{d\theta}{ds}\right)^2$$

$$\frac{d^2\theta}{ds^2} + \frac{2}{r}\left(\frac{dr}{ds}\right)\left(\frac{d\theta}{ds}\right) = 0 .$$

Now $(dr/ds)^2 + r^2(d\theta/ds)^2 = 1$ follows from the metric so

$$\frac{dR_0}{ds} = \frac{d}{ds}\left[r^2 \frac{d\theta}{ds}\right] = r^2 \frac{d^2\theta}{ds^2} + 2r\left(\frac{dr}{ds}\right)\left(\frac{d\theta}{ds}\right)$$

$$= r^2\left[\frac{d^2\theta}{ds^2} + \frac{2}{r}\left(\frac{dr}{ds}\right)\left(\frac{d\theta}{ds}\right)\right] = 0 ,$$

and therefore

$$R_0 \equiv r^2 \frac{d\theta}{ds} = \text{constant} .$$

(b) From one of the first integrals above:

$$\left(\frac{dr}{d\theta}\frac{d\theta}{ds}\right)^2 + r^2\left(\frac{d\theta}{ds}\right)^2 = 1 .$$

Now substitute $d\theta/ds = R_0/r^2$ (R_0 = some constant):

$$\left(\frac{dr}{d\theta}\right)^2 + r^2 = r^4/R_0^4 .$$

(c) The equation for a straight line is

$$r = L/\cos(\theta - a)$$

$$\frac{dr}{d\theta} = \frac{\sin(\theta - a)}{\cos^2(\theta - a)} L$$

$$\left(\frac{dr}{d\theta}\right)^2 + r^2 = L^2\left[\frac{\sin^2(\theta - a)}{\cos^4(\theta - a)} + \frac{1}{\cos^2(\theta - a)}\right] = \frac{L^2}{\cos^4(\theta - a)} = \frac{r^4}{L^2} .$$

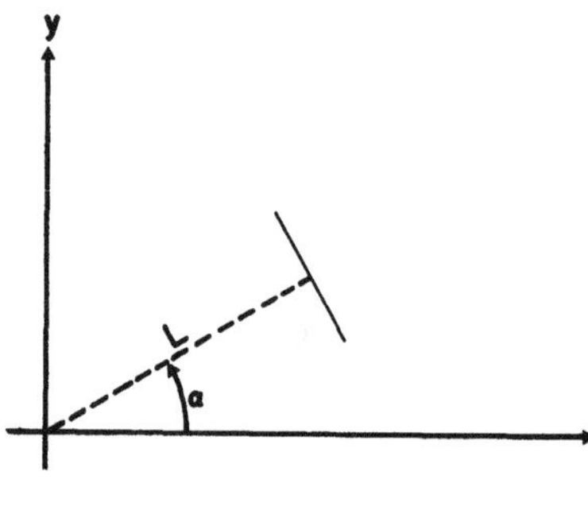

Thus, all straight lines satisfy the geodesic equation.

Solution 7.4. The nonvanishing derivatives of the metric tensor are $g_{xx,t} = -g_{tt,t} = -2/t^3$. Thus the nonvanishing connection coefficients are

$$\Gamma_{ttt} = 1/t^3, \quad \Gamma_{xxt} = \Gamma_{xtx} = -\Gamma_{txx} = -1/t^3 .$$

To find the geodesics it is easiest to proceed directly from their definition as curves of extreme length. Let a geodesic be $x(t)$. Then (dot represents d/dt)

$$0 = \delta \int (ds^2)^{\frac{1}{2}} = \delta \int (1-\dot{x}^2)^{\frac{1}{2}} \frac{dt}{t} .$$

The Euler-Lagrange equation for this extremization is

$$\frac{d}{dt}\left[\frac{\dot{x}}{t(1-\dot{x}^2)^{\frac{1}{2}}}\right] = 0 .$$

This is easily solved by letting $\tanh\theta \equiv \dot{x}$:

$$\frac{\sinh\theta}{t} = \text{constant} ,$$

which integrates to

$$(x-x_0)^2 = t^2 + a^2 .$$

Thus the geodesics are hyperbolas asymptotic to the light cones:

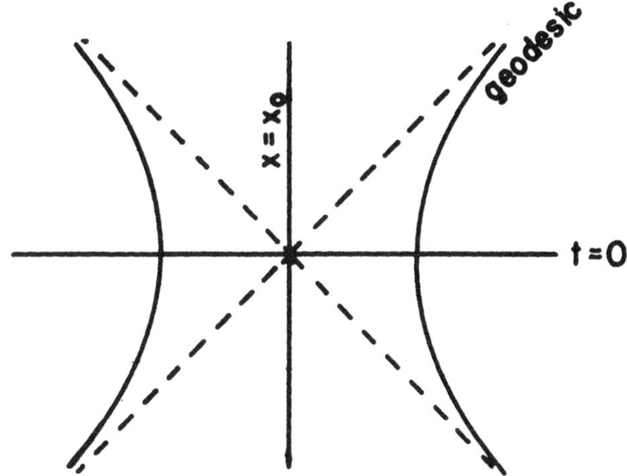

Solution 7.5. The calculation of $g_{\alpha\beta;\gamma}$ is straightforward in a coordinate basis:

$$\begin{aligned} g_{\alpha\beta;\gamma} &= g_{\alpha\beta,\gamma} - g_{\sigma\beta}\Gamma^{\sigma}{}_{\alpha\gamma} - g_{\alpha\sigma}\Gamma^{\sigma}{}_{\beta\gamma} \\ &= g_{\alpha\beta,\gamma} - 2\Gamma_{(\beta\alpha)\gamma} \\ &= g_{\alpha\beta,\gamma} - g_{\alpha\beta,\gamma} = 0 . \end{aligned}$$

Solution 7.6. In a coordinate frame:

$$\Gamma^{\mu}{}_{\nu\lambda} = \frac{1}{2}(g_{\alpha\nu,\lambda} + g_{\alpha\lambda,\nu} - g_{\nu\lambda,\alpha})\,g^{\mu\alpha} \,. \tag{1}$$

(a) If the metric is diagonal, α must be equal to μ. But since $\mu \neq \nu \neq \lambda$, all of the terms in parentheses in Equation (1) obviously vanish.

(b) $\Gamma^{\mu}{}_{\lambda\lambda} = \frac{1}{2}(g_{\alpha\lambda,\lambda} + g_{\alpha\lambda,\lambda} - g_{\lambda\lambda,\alpha})g^{\mu\alpha}$, setting $\nu = \lambda$ in Equation (1).

$$\Gamma^{\mu}{}_{\lambda\lambda} = -\frac{1}{2}g_{\lambda\lambda,\alpha}\,g^{\mu\alpha} = -\frac{1}{2}g_{\lambda\lambda,\alpha}(g_{\mu\alpha})^{-1} = -\frac{1}{2}(g_{\mu\mu})^{-1}\,g_{\lambda\lambda,\mu} \,.$$

Here we have repeatedly used diagonality of the metric [e.g. $g^{\mu\alpha} = (g_{\mu\alpha})^{-1}$].

(c)
$$\begin{aligned}\Gamma^{\mu}{}_{\mu\lambda} &= \frac{1}{2}(g_{\alpha\mu,\lambda} + g_{\alpha\lambda,\mu} - g_{\mu\lambda,\alpha})(g_{\mu\alpha})^{-1} \\ &= \frac{1}{2}(g_{\mu\mu})^{-1}(g_{\mu\mu,\lambda}) = \frac{\partial}{\partial x^{\lambda}}(\log(|g_{\mu\mu}|^{\frac{1}{2}})) \,.\end{aligned}$$

(d) Setting $\lambda = \mu$ in (c) gives

$$\Gamma^{\mu}{}_{\mu\mu} = \frac{\partial}{\partial x^{\mu}}(\log(|g_{\mu\mu}|^{\frac{1}{2}})) \,.$$

Solution 7.7.

(a)
$$\begin{aligned}g_{\alpha\beta,\gamma} &= \nabla_{\gamma}(\mathbf{e}_{\alpha}\cdot\mathbf{e}_{\beta}) = (\nabla_{\gamma}\mathbf{e}_{\alpha})\cdot\mathbf{e}_{\beta} + \mathbf{e}_{\alpha}\cdot(\nabla_{\gamma}\mathbf{e}_{\beta}) \\ &= \Gamma^{\mu}{}_{\alpha\gamma}\mathbf{e}_{\mu}\cdot\mathbf{e}_{\beta} + \Gamma^{\mu}{}_{\beta\gamma}\mathbf{e}_{\mu}\cdot\mathbf{e}_{\alpha} \\ &= \Gamma_{\beta\alpha\gamma} + \Gamma_{\alpha\beta\gamma} \,.\end{aligned}$$

(b)
$$g_{\alpha\mu}\,g^{\mu\beta} = \delta_{\alpha}{}^{\beta}$$

$$g_{\alpha\mu,\gamma}\,g^{\mu\beta} + g_{\alpha\mu}\,g^{\mu\beta}{}_{,\gamma} = 0$$

$$g_{\alpha\mu}\,g^{\mu\beta}{}_{,\gamma} = -g_{\alpha\mu,\gamma}\,g^{\mu\beta} \,.$$

(c)
$$\begin{aligned}g^{\alpha\beta}{}_{,\gamma} &= -g_{\lambda\mu,\gamma}\,g^{\mu\beta}g^{\lambda\alpha} = -(\Gamma_{\lambda\mu\gamma} + \Gamma_{\mu\lambda\gamma})\,g^{\mu\beta}g^{\lambda\alpha} \\ &= -\Gamma^{\alpha}{}_{\mu\gamma}\,g^{\mu\beta} - \Gamma^{\beta}{}_{\lambda\gamma}\,g^{\lambda\alpha} \,.\end{aligned}$$

(Here we have used the result in (a).)

(d) For any matrix $\|g_{\alpha\beta}\|$,

$$(\log \det \|g_{\alpha\beta}\|)_{,\alpha} = \mathrm{Tr}\, \|g_{\alpha\beta}\|^{-1} \|g_{\mu\nu,\alpha}\| \;,$$

therefore

$$(\log g)_{,\alpha} = g^{\mu\nu} g_{\mu\nu,\alpha}$$

and

$$\frac{g_{,\alpha}}{g} = g^{\mu\nu} g_{\mu\nu,\alpha} \;.$$

And so, by (b)

$$g_{,\alpha} = g g^{\mu\nu} g_{\mu\nu,\alpha} = -g g_{\mu\nu} g^{\mu\nu}{}_{,\alpha}$$

(e) In a coordinate frame, $\Gamma^{\mu}{}_{\alpha\beta} = \frac{1}{2} g^{\mu\nu}(g_{\nu\alpha,\beta} + g_{\nu\beta,\alpha} - g_{\alpha\beta,\nu})$, and since the last two terms cancel $\Gamma^{\alpha}{}_{\alpha\beta} = \frac{1}{2} g^{\alpha\nu} g_{\nu\alpha,\beta}$. Thus, by (d)

$$\Gamma^{\alpha}{}_{\alpha\beta} = \frac{1}{2} g_{,\beta}/g = \frac{1}{2} (\log |g|)_{,\beta} = (\log |g|^{\frac{1}{2}})_{,\beta} \,.$$

(f)

$$\begin{aligned} g^{\mu\nu}\Gamma^{\alpha}{}_{\mu\nu} &= -g^{\alpha\beta}{}_{,\beta} - \Gamma^{\beta}{}_{\lambda\beta} g^{\lambda\alpha} && \text{(putting } \beta=\gamma \text{ in (c) and summing)} \\ &= -g^{\alpha\beta}{}_{,\beta} - (\log|g|^{\frac{1}{2}})_{,\lambda} g^{\lambda\alpha} && \text{by (e)} \\ &= -g^{\alpha\nu}{}_{,\nu} - (\log|g|^{\frac{1}{2}})_{,\nu} g^{\alpha\nu} && \text{changing dummy indices} \\ &= -g^{\alpha\nu}{}_{,\nu} - |g|^{\frac{1}{2}}{}_{,\nu}\, g^{\alpha\nu} |g|^{-\frac{1}{2}} \\ &= -\frac{1}{|g|^{\frac{1}{2}}} (g^{\alpha\nu} |g|^{\frac{1}{2}})_{,\nu} \,. \end{aligned}$$

(g)

$$\begin{aligned} A^{\alpha}{}_{;\alpha} &= A^{\alpha}{}_{,\alpha} + \Gamma^{\alpha}{}_{\beta\alpha} A^{\beta} = A^{\alpha}{}_{,\alpha} + \frac{1}{|g|^{\frac{1}{2}}} (|g|^{\frac{1}{2}})_{,\beta} A^{\beta} && \text{by (e)} \\ &= \frac{1}{|g|^{\frac{1}{2}}} (|g|^{\frac{1}{2}} A^{\alpha})_{,\alpha} \end{aligned}$$

(h)
$$A_\alpha{}^\beta{}_{;\beta} = A_\alpha{}^\beta{}_{,\beta} + \Gamma^\beta{}_{\mu\beta} A_\alpha{}^\mu - \Gamma^\lambda{}_{\alpha\beta} A_\lambda{}^\beta$$
$$= A_\alpha{}^\beta{}_{,\beta} + \frac{1}{|g|^{\frac{1}{2}}} (|g|^{\frac{1}{2}})_{,\mu} A_\alpha{}^\mu - \Gamma^\lambda{}_{\alpha\mu} A_\lambda{}^\mu$$
$$= \frac{1}{|g|^{\frac{1}{2}}} (|g|^{\frac{1}{2}} A_\alpha{}^\beta)_{,\beta} - \Gamma^\lambda{}_{\alpha\mu} A_\lambda{}^\mu$$

(i)
$$A^{\alpha\beta}{}_{;\beta} = A^{\alpha\beta}{}_{,\beta} + \Gamma^\alpha{}_{\mu\beta} A^{\mu\beta} + \Gamma^\beta{}_{\mu\beta} A^{\alpha\mu}$$
$$= A^{\alpha\beta}{}_{,\beta} + \Gamma^\alpha{}_{\mu\beta} A^{\mu\beta} + \frac{1}{|g|^{\frac{1}{2}}} (|g|^{\frac{1}{2}})_{,\mu} A^{\alpha\mu}$$
$$= \frac{1}{|g|^{\frac{1}{2}}} (A^{\alpha\beta} |g|^{\frac{1}{2}})_{,\beta} + \Gamma^\alpha{}_{\mu\beta} A^{\mu\beta} \; .$$

But $\Gamma^\alpha{}_{\mu\beta} = \Gamma^\alpha{}_{(\mu\beta)}$ in a coordinate frame, so if $A^{\mu\beta} = A^{[\mu\beta]}$, the last term vanishes, and
$$A^{\alpha\beta}{}_{;\beta} = \frac{1}{|g|^{\frac{1}{2}}} (|g|^{\frac{1}{2}} A^{\alpha\beta})_{,\beta} \; .$$

(j)
$$\Box S = (S_{,\alpha} g^{\alpha\beta})_{;\beta} = \frac{1}{|g|^{\frac{1}{2}}} (|g|^{\frac{1}{2}} S_{,\alpha} g^{\alpha\beta})_{,\beta} \; .$$

(Here (g) has been used.)

Solution 7.8. Under a coordinate transformation,
$$A_{\bar{\mu}\bar{\nu}} = \frac{\partial x^\mu}{\partial x^{\bar{\mu}}} \frac{\partial x^\nu}{\partial x^{\bar{\nu}}} A_{\mu\nu}$$
implies that
$$\bar{A} = \det\left(\frac{\partial x^\mu}{\partial x^{\bar{\mu}}}\right) \det(A_{\mu\nu}) \det\left(\frac{\partial x^\nu}{\partial x^{\bar{\nu}}}\right)^T = J^2 A$$
where J is the Jacobian, $\det(\partial x^\mu/\partial x^{\bar{\mu}})$, and where the superscript T denotes matrix transpose. We want $A_{;\alpha}$ to transform according to the vector analog of the transformation law for A. (A is called a "density of weight 2".) Since $A_{;\alpha}$ must be linear in A, set

$$A_{;\alpha} = A_{,\alpha} + K_\alpha A$$

where K_α must be determined. This can be done by demanding that g, the determinant of the metric tensor, have vanishing covariant derivative:

$$0 = g_{;\alpha} = g_{,\alpha} + K_\alpha g = 2g\Gamma^\beta{}_{\beta\alpha} + K_\alpha g$$

by Problem 7.7. Thus

$$K_\alpha = -2\Gamma^\beta{}_{\beta\alpha} .$$

The generalization to densities of weight W (i.e. they transform with a J^W) can be done by considering powers of g. The result is that

$$K_\alpha = -W\Gamma^\beta{}_{\beta\alpha} .$$

Solution 7.9. A geodesic with tangent vector $\mathbf{u}$ is spacelike, timelike, or null according to whether $\mathbf{u}\cdot\mathbf{u}$ is >0, $=0$, <0. But $\mathbf{u}\cdot\mathbf{u}$ is conserved along the geodesic, since

$$\nabla_{\mathbf{u}}(\mathbf{u}\cdot\mathbf{u}) = 2\mathbf{u}\cdot\nabla_{\mathbf{u}}\mathbf{u} = 0 ,$$

since the geodesic equation is $\nabla_{\mathbf{u}}\mathbf{u} = 0$.

Solution 7.10. The integral for length is

$$\int ds = \int \left(-g_{\alpha\beta}\frac{dx^\alpha}{ds}\frac{dx^\beta}{ds}\right)^{\frac{1}{2}} ds .$$

To extremize length take $\delta \int ds = 0$. According to the Euler-Lagrange equations

$$\frac{d}{ds}(g_{\alpha\beta}u^\alpha) = \frac{1}{2}g_{\alpha\gamma,\beta}u^\alpha u^\gamma \tag{1}$$

where $\left(-g_{\alpha\beta}\frac{dx^\alpha}{ds}\frac{dx^\beta}{ds}\right) = 1$ and $u^\alpha \equiv \frac{dx^\alpha}{ds}$ have been used. Now since

$$\frac{d}{ds}(g_{\alpha\beta}u^\alpha) = g_{\alpha\beta}\frac{du^\alpha}{ds} + \frac{dx^\gamma}{ds}g_{\alpha\beta,\gamma}u^\alpha = g_{\alpha\beta}\frac{du^\alpha}{ds} + u^\gamma u^\alpha g_{\alpha\beta,\gamma}$$

Equation (1) becomes

$$g_{\alpha\beta}\frac{d^2x^\alpha}{ds^2} + u^\alpha u^\gamma\left(g_{\alpha\beta,\gamma} - \frac{1}{2}g_{\alpha\gamma,\beta}\right) = 0\ . \tag{2}$$

Now, use

$$u^\alpha u^\gamma g_{\alpha\beta,\gamma} = u^\alpha u^\gamma \cdot \frac{1}{2}(g_{\alpha\beta,\gamma} + g_{\gamma\beta,\alpha})$$

and multiply Equation (2) by $g^{\beta\tau}$ to obtain

$$\begin{aligned}&\frac{d^2x^\tau}{ds^2} + \frac{1}{2}g^{\beta\tau}(g_{\alpha\beta,\gamma} + g_{\gamma\beta,\alpha} - g_{\alpha\gamma,\beta})u^\alpha u^\gamma \\ &= \frac{d^2x^\tau}{ds^2} + \Gamma^\tau_{\ \alpha\gamma}\frac{dx^\alpha}{ds}\frac{dx^\gamma}{ds} = 0\ .\end{aligned} \tag{3}$$

Solution 7.11. Define a new curve parametrization by the functional relationship $s = f(\lambda)$. Then derivatives are related by

$$\frac{d}{d\lambda} = f'\frac{d}{ds} \qquad \frac{d^2}{d\lambda^2} = f''\frac{d}{ds} + f'^2\frac{d^2}{ds^2}\ , \tag{1}$$

where $f' \equiv \frac{df}{d\lambda}$. With the new parametrization then the geodesic equation becomes

$$\frac{d^2x^\alpha}{ds^2} + \frac{f''}{f'^2}\frac{dx^\alpha}{ds} + \Gamma^\alpha_{\ \beta\gamma}\frac{dx^\beta}{ds}\frac{dx^\gamma}{ds} = 0\ . \tag{2}$$

Equation (2) is in the "standard form" (s an affine parametrization) if the second term vanishes. For s to be an affine parameter, f'' must vanish i.e. s and λ must be linearly related.

Solution 7.12. The components of $\nabla_p p$ are

$$(\nabla_p p)^\beta = mu^\alpha(p^\beta_{\ ,\alpha} + \Gamma^\beta_{\ \sigma\alpha}p^\sigma) = m\left(\frac{dp^\beta}{d\tau} + u^\alpha p^\sigma \Gamma^\beta_{\ \sigma\alpha}\right)\ . \tag{1}$$

In flat space time, one can always find a global coordinate system (Minkowski coordinates) in which all of the Christoffel symbols vanish. In that coordinate system, conservation of 4-momentum, $\frac{dp}{d\tau} = 0$, can be written as

$$\nabla_p p = 0 \tag{2}$$

by Equation (1), but Equation (2) is a tensor equation so it must be the correct expression for momentum conservation in any frame. The momentum vector is proportional to the 4-velocity of particles with mass; for these particles $p \cdot p = -m^2$, so the geodesics are timelike.

Solution 7.13. Let λ be an affine parameter such that $p^\alpha = dx^\alpha/d\lambda$; then for geodesic motion of the particle

$$0 = (\nabla_p p) \cdot e_1 = p_{1;\alpha} p^\alpha = \frac{dp_1}{d\lambda} - p_\sigma \Gamma^\sigma{}_{\alpha 1} p^\alpha ,$$

and therefore

$$\frac{dp_1}{d\lambda} = p^\sigma p^\alpha \Gamma_{(\sigma\alpha)1} = p^\sigma p^\alpha \frac{1}{2} g_{\sigma\alpha,1} = 0 .$$

Solution 7.14. Start with the geodesic equation

$$\frac{d^2 x^\alpha}{d\lambda^2} + \Gamma^\alpha{}_{\beta\gamma} \frac{dx^\beta}{d\lambda} \frac{dx^\gamma}{d\lambda} = 0 .$$

Change variables from affine parameter λ to coordinate time t and use

$$0 = dt^2 + (g_{ij}/g_{00}) dx^i dx^k \tag{1}$$

to get

$$g_{jk} \frac{d^2 x^k}{dt^2} + \Gamma_{jk\ell} \frac{dx^k}{dt} \frac{dx^\ell}{dt} - \Gamma_{j00} \frac{g_{k\ell}}{g_{00}} \frac{dx^k}{dt} \frac{dx^\ell}{dt} + \frac{d^2t/d\lambda^2}{(dt/d\lambda)^2} g_{jk} \frac{dx^k}{dt} = 0 .$$

Combine this with the time part of the geodesic equation

$$\frac{(d^2t/d\lambda^2)}{(dt/d\lambda)^2} = -2\Gamma_{0k0} \frac{dx^k/dt}{g_{00}}$$

and use the expression for the Γ's in terms of the metric to get

$$\gamma_{jk} \frac{d^2 x^k}{dt^2} + \frac{1}{2} (\gamma_{jk,\ell} + \gamma_{j\ell,k} - \gamma_{k\ell,j}) \frac{dx^k}{dt} \frac{dx^\ell}{dt} = 0 , \tag{2}$$

where $\gamma_{jk} \equiv -g_{jk}/g_{00}$. Notice that this is just a geodesic equation with affine parameter t in the 3-dimensional manifold with metric γ_{jk}. From Equation (1) it is clear then that the solution of Equation (2) extremizes $\int dt$; Fermat's principle holds. [See also C. Moller, *The Theory of Relativity* (Oxford University Press, 2nd ed., 1972), p. 308.]

Solution 7.15.

(a) The geodesic is the path between two velocities which minimizes the arc-length between them; but arc-length in the velocity space is just the magnitude of a small change of velocity. Since a rocket expends fuel monotonically for the boost it requires, the geodesics of velocity space are paths of minimum fuel use.

(b) We need to find the velocity geodesic connecting $\underset{\sim}{V}_1$ and $\underset{\sim}{V}_2$. In principle, we could solve the geodesic equation for the metric in Problem 6.8, but this is very tedious. An easier way is to notice that, by symmetry, a geodesic passing through the origin of the velocity space coordinates is $\theta = \phi =$ constant, $\chi = s$ (where s is an affine parameter). To get a more general geodesic let us view this geodesic from a moving frame. Since $\chi = \tanh^{-1}\overline{V}$ is a monotonic function of $\rho = \gamma\overline{V}$ [where $\gamma \equiv (1-\overline{V}^2)^{-\frac{1}{2}}$], we can write the geodesics in terms of the non-affine parameter $\rho(s)$. In terms of this parameter $\gamma = (1+\rho^2)^{\frac{1}{2}}$ and, if we choose the geodesic to be in the x-direction,

$$\mathbf{u} = [(1+\rho^2)^{\frac{1}{2}}, \rho, 0, 0] \qquad -\infty < \rho < +\infty .$$

A boost in the x-direction just takes this trajectory onto itself so we need only find its transformation under a perpendicular boost, say in the y-direction. Under a boost by β in the y-direction $\mathbf{u}$ becomes:

$$\mathbf{u}' = [\gamma'(1+\rho^2)^{\frac{1}{2}}, \rho, (1+\rho^2)^{\frac{1}{2}}\gamma'\beta, 0] \tag{1}$$

where $\gamma' = (1-\beta^2)^{-\frac{1}{2}}$.

The 4-velocity in Equation (1) may be represented now in terms of the corresponding 3-velocity

$$\underset{\sim}{V}/(1-V^2)^{\frac{1}{2}} = \rho \underset{\sim}{n} + (1+\rho^2)^{\frac{1}{2}} \gamma' \beta \underset{\sim}{m} \tag{2}$$

where $\underset{\sim}{n}$ and $\underset{\sim}{m}$ are perpendicular unit vectors, and may be manipulated as if they lived in a cartesian space. We can solve for V^2 and write Equation (2) as

$$\underset{\sim}{V} = \left[\frac{1-\beta^2}{1+\rho^2}\right]^{\frac{1}{2}} \rho \underset{\sim}{n} + \beta \underset{\sim}{m} . \tag{3}$$

This equation gives us the general geodesic of velocity space. One picks a $\beta\,(|\beta|<1)$, two perpendicular vectors m, n and gets a geodesic parameterized by $\rho\,(-\infty<\rho<+\infty)$. From Equation (3) it is clear that the general geodesic is a "straight line." (This does not mean the space is flat!)

For our problem there are two cases. In case (i), in which either angle $O V_1 V_2$ or $O V_2 V_1$ is obtuse, the distance of closest approach to the origin O of the geodesic (straight line) connecting $\underset{\sim}{V}_1$ and $\underset{\sim}{V}_2$ is simply $|\underset{\sim}{V}_1|$ or $|\underset{\sim}{V}_2|$. In case (ii) in which both OV_1V_2 and OV_2V_1 are acute, the distance of closest approach β is (by simple geometric arguments)

$$\beta = \frac{|\underset{\sim}{V}_1 \times \underset{\sim}{V}_2|}{|\underset{\sim}{V}_1 - \underset{\sim}{V}_2|} .$$

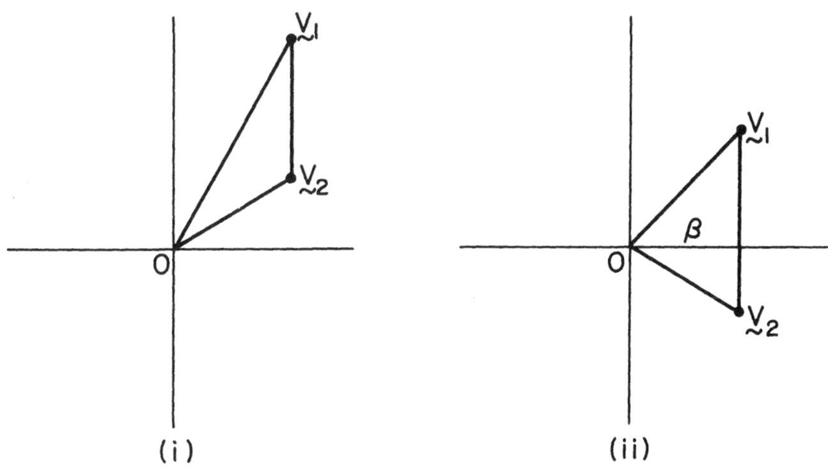

Solution 7.16. A is parallel transported along the ϕ-coordinate line (i.e. θ = constant), so

$$0 = A^{\alpha}{}_{;\phi} = A^{\alpha}{}_{,\phi} + \Gamma^{\alpha}{}_{\beta\phi} A^{\beta} . \qquad (1)$$

The only nonzero Christoffel symbols are $\Gamma^{\theta}{}_{\phi\phi} = -\sin\theta\cos\theta$ and $\Gamma^{\phi}{}_{\theta\phi} = \cot\theta$, so Equation (1) gives

$$A^{\theta}{}_{,\phi} - \sin\theta\cos\theta\, A^{\phi} = 0 \qquad (2)$$

$$A^{\phi}{}_{,\phi} + \cot\theta\, A^{\theta} = 0 . \qquad (3)$$

Throughout this problem, θ is kept constant at θ_0. Equations (2) and (3) are easily solved, e.g., by differentiating Equation (2) with respect to ϕ:

$$A^{\theta}{}_{,\phi\phi} = \sin\theta\cos\theta\, A^{\phi}{}_{,\phi} = -\cos^2\theta\, A^{\theta} .$$

The solution then is

$$A^{\theta} = a\cos(\phi\cos\theta) + \beta\sin(\phi\cos\theta)$$

where a, β are constants. Equation (2) now gives

$$A^{\phi} = -a\sin(\phi\cos\theta)/\sin\theta + \beta\cos(\phi\cos\theta)/\sin\theta .$$

At $\phi = 0$, $A = e_{\theta}$, i.e. $A^{\theta} = 1$, $A^{\phi} = 0$. Thus $a = 1$, $\beta = 0$, and

$$A^{\theta} = \cos(\phi\cos\theta)$$

$$A^{\phi} = -\sin(\phi\cos\theta)/\sin\theta .$$

At $\phi = 2\pi$, after transport around the circle, the vector is

$$A = \cos(2\pi\cos\theta)\, e_{\theta} - \sin(2\pi\cos\theta)/\sin\theta\; e_{\phi} \neq e_{\theta} ,$$

but the magnitude is unchanged

$$(A\cdot A)_{2\pi} = \cos^2(2\pi\cos\theta)\, e_{\theta}\cdot e_{\theta} + \sin^2(2\pi\cos\theta)/\sin^2\theta\; e_{\phi}\cdot e_{\phi} = 1 = (A\cdot A)_0 .$$

Solution 7.17.

(i)
$$\nabla_{\mathbf{u}}(\eta_{\alpha\beta}) = \nabla_{\mathbf{u}}(\mathbf{e}_\alpha \cdot \mathbf{e}_\beta) = (\nabla_{\mathbf{u}}\mathbf{e}_\alpha)\cdot \mathbf{e}_\beta + \mathbf{e}_\alpha \cdot (\nabla_{\mathbf{u}}\mathbf{e}_\beta)$$
$$= A_\alpha{}^\gamma \mathbf{e}_\gamma \cdot \mathbf{e}_\beta + A_\beta{}^\gamma \mathbf{e}_\gamma \cdot \mathbf{e}_\alpha$$
$$= A_{\alpha\beta} + A_{\beta\alpha} \ .$$

The tensor $\mathbf{A}$ must be antisymmetric.

It is conventional to write the transport law in this case as

$$\nabla_{\mathbf{u}}\mathbf{e}_\alpha = A_\alpha{}^\beta \mathbf{e}_\beta = (A^{\gamma\beta}\mathbf{e}_\beta \otimes \mathbf{e}_\gamma)\cdot \mathbf{e}_\alpha \equiv -\boldsymbol{\Omega}\cdot \mathbf{e}_\alpha \qquad (1)$$

where $\Omega^{\beta\gamma} = A^{\beta\gamma}$ is the 4-dimensional version of a 3-dimensional antisymmetric rotation matrix.

(ii) Since $\mathbf{u} = \mathbf{e}_0$ is singled out, decompose $\boldsymbol{\Omega}$ along $\mathbf{u}$ and orthogonal to $\mathbf{u}$:

$$\Omega^{\alpha\beta} = v^\alpha u^\beta - u^\alpha v^\beta + \omega^{\alpha\beta} \qquad (2)$$

where $\omega^{\alpha\beta} = -\omega^{\beta\alpha}$, $\omega^{\alpha\beta}u_\beta = 0$ and v^α is as yet unspecified; without loss of generality we can take $\mathbf{v}\cdot\mathbf{u} = 0$. Now Equation (1) gives

$$\boldsymbol{\Omega}\cdot\mathbf{u} = -\nabla_{\mathbf{u}}\mathbf{u} = -\mathbf{a}$$

where $\mathbf{a}$ is the 4-acceleration of the observer. But Equation (2) gives

$$\boldsymbol{\Omega}\cdot\mathbf{u} = -\mathbf{v}$$

so $\mathbf{v} = \mathbf{a}$ and hence

$$\boldsymbol{\Omega} = \mathbf{a}\otimes\mathbf{u} - \mathbf{u}\otimes\mathbf{a} + \boldsymbol{\omega} \ .$$

(iii) The tensor $\omega^{\alpha\beta}$ has only three independent components and is spatial since $\boldsymbol{\omega}\cdot\mathbf{u} = 0$. It therefore represents a purely spatial rotation of the basis vectors and vanishes if the spatial vectors are non-rotating.

One often uses an angular velocity *vector* $\boldsymbol{\omega}$, with $\boldsymbol{\omega}\cdot\mathbf{u} = 0$, to represent the three degrees of freedom in $\omega^{\alpha\beta}$. The relation between $\boldsymbol{\omega}$ and $\omega^{\alpha\beta}$ is

$$\omega^{\alpha\beta} = \varepsilon^{\alpha\beta\lambda\sigma} u_\lambda \omega_\sigma , \qquad \omega^\alpha = -\frac{1}{2}\, \varepsilon^{\alpha\mu\lambda\sigma} u_\mu \omega_{\lambda\sigma} .$$

Note that a spatially non-rotating frame ($\omega = 0$) satisfies

$$\nabla_{\mathbf{u}} \mathbf{e}_\alpha = (\mathbf{u}\otimes\mathbf{a} - \mathbf{a}\otimes\mathbf{u})\cdot\mathbf{e}_\alpha .$$

Any vector $\mathbf{e}_\alpha$ satisfying an equation of this form is said to be "Fermi-Walker" transported.

Solution 7.18. Let the two vectors be $\mathbf{x}$ and $\mathbf{y}$. Then the Fermi-Walker transport law reads

$$\nabla_{\mathbf{u}} \mathbf{x} = (\mathbf{u}\otimes\mathbf{a} - \mathbf{a}\otimes\mathbf{u})\cdot\mathbf{x}$$

$$\nabla_{\mathbf{u}} \mathbf{y} = (\mathbf{u}\otimes\mathbf{a} - \mathbf{a}\otimes\mathbf{u})\cdot\mathbf{y} ,$$

where $\mathbf{u}$ is the tangent vector to the curve $\mathcal{C}$ and $\mathbf{a} = \nabla_{\mathbf{u}}\mathbf{u}$. Using the product rule we evaluate the change in the dot product along the curve:

$$\begin{aligned}\nabla_{\mathbf{u}}(\mathbf{x}\cdot\mathbf{y}) &= (\nabla_{\mathbf{u}}\mathbf{x})\cdot\mathbf{y} + \mathbf{x}\cdot(\nabla_{\mathbf{u}}\mathbf{y}) \\ &= (\mathbf{a}\cdot\mathbf{x})(\mathbf{u}\cdot\mathbf{y}) - (\mathbf{a}\cdot\mathbf{y})(\mathbf{u}\cdot\mathbf{x}) + (\mathbf{u}\cdot\mathbf{x})(\mathbf{a}\cdot\mathbf{y}) - (\mathbf{a}\cdot\mathbf{x})(\mathbf{u}\cdot\mathbf{y}) = 0 .\end{aligned}$$

The scalar product is unaltered.

Solution 7.19. Fermi-Walker transport has the differential equation

$$\nabla_{\mathbf{u}} \mathbf{x} = (\mathbf{u}\otimes\mathbf{a} - \mathbf{a}\otimes\mathbf{u})\cdot\mathbf{x} ,$$

where $\mathbf{u}$ is the tangent vector to the curve and $\mathbf{a} = \nabla_{\mathbf{u}}\mathbf{u} \equiv D\mathbf{u}/d\tau$. If the curve is a geodesic, it satisfies the geodesic equation $\nabla_{\mathbf{u}}\mathbf{u} = 0$, so the Fermi-Walker transport law becomes

$$\nabla_{\mathbf{u}} \mathbf{x} = 0 ,$$

which is just the parallel transport equation.

Solution 7.20.

(a) $$U_{\alpha;\beta}U^{\beta}U^{\alpha} = U^{\alpha}{}_{;\beta}U^{\beta}U_{\alpha} = (\nabla_{U}U)\cdot U$$

(b) $$V^{\alpha}{}_{;\beta}U^{\beta} - U^{\alpha}{}_{;\beta}V^{\beta} = \nabla_{U}V - \nabla_{V}U \equiv [U,V]$$

(c) $$T_{\alpha\beta;\gamma}V^{\alpha}W^{\beta}U^{\gamma} = V\cdot(\nabla_{U}T)\cdot W$$

(d) $$W^{\alpha;\beta}V_{\beta;\gamma}U^{\gamma} = W^{\alpha}{}_{;\beta}(V^{\beta}{}_{;\gamma}U^{\gamma}) = \nabla_{(\nabla_{U}V)}W$$

(e) Combine the first two terms of the given expression,

$$(W^{\alpha}{}_{;\gamma}U^{\gamma})_{;\beta}U^{\beta} - U^{\alpha}{}_{;\beta}(W^{\beta}{}_{;\gamma}U^{\gamma}) = \nabla_{U}(\nabla_{U}W) - \nabla_{(\nabla_{U}W)}U = [U,\nabla_{U}W]\ .$$

Solution 7.21. The paths of light rays can be obtained from the geometrical optics limit of Maxwell's equations $F^{\alpha\beta}{}_{;\beta} = 0$. With the identity in Problem 7.7(i), these equations can be written as

$$[g^{\alpha\tau}g^{\beta\mu}F_{\tau\mu}(-g)^{\frac{1}{2}}]_{,\beta} = 0\ .$$

Using the spatially isotropic, diagonal form of the metric and noting that $E_i = F_{0i}$, $B_k = \varepsilon^{kj\ell}F_{j\ell}$, one sees that the above equations have the form

$$\nabla\cdot(\epsilon\underset{\sim}{E}) = 0$$

$$\nabla\times(\mu^{-1}\underset{\sim}{B}) = \partial(\epsilon\underset{\sim}{E})/\partial t$$

where $\epsilon = \mu = (f/g_{00})^{\frac{1}{2}}$. Thus light moves as if in a medium with an effective index of refraction

$$n = (\epsilon\mu)^{-\frac{1}{2}} = (g_{00}/f)^{\frac{1}{2}}$$

Solution 7.22. For clarity first consider the Newtonian case. Let $P(\underset{\sim}{v},n)$ be the probability of the velocity being $\underset{\sim}{v}$ after n boosts. We want to find a differential equation for P. To accomplish this note that if the velocity is $\underset{\sim}{v}$ at step n then at step $n-1$ the velocity must have been somewhere in velocity space at a distance Δv from $\underset{\sim}{v}$. By symmetry $P(\underset{\sim}{v},n)$ has equal contributions from points on the sphere of radius

Δv about $\underset{\sim}{v}$ so we can equate $P(\underset{\sim}{v}, n)$ with the isotropic average of P, at $n-1$, over that sphere:

$$\begin{aligned} P(\underset{\sim}{v},n) &= \langle P(\underset{\sim}{v}+\Delta v, n-1)\rangle_{\text{sphere}} \\ &= (1/6)[P(\underset{\sim}{v}+\Delta v\underset{\sim}{e}_x, n-1) + P(\underset{\sim}{v}+\Delta v\underset{\sim}{e}_y, n-1) + P(\underset{\sim}{v}+\Delta v\underset{\sim}{e}_z, n-1) \\ &\quad + P(\underset{\sim}{v}-\Delta v\underset{\sim}{e}_x, n-1) + P(\underset{\sim}{v}-\Delta v\underset{\sim}{e}_y, n-1) + P(\underset{\sim}{v}-\Delta v\underset{\sim}{e}_z, n-1)] \\ &\approx P(\underset{\sim}{v}, n-1) + (\Delta v)^2\nabla^2 P/6 \ . \end{aligned} \tag{1}$$

The differential equation then, is

$$\frac{\partial P}{\partial n} = \frac{(\Delta v)^2}{6}\nabla^2 P \ , \tag{2}$$

the standard diffusion equation. Its solution, with $P(\underset{\sim}{v}, 0) = \delta^3(\underset{\sim}{v})$ is

$$P(v, n) = (4\pi)^{-\frac{3}{2}}(n\Delta v^2/6)^{-\frac{3}{2}}\exp\left(\frac{-3v^2}{2n\Delta v^2}\right) .$$

In the relativistic problem the same arguments hold except that velocities do not add linearly; we must use the relativistic velocity addition formulas. Since $\Delta v \ll c$, however, the proper velocity Δv *does* add linearly in a frame momentarily comoving with the rocket ship. Thus, the diffusion Equation (2) is *locally* valid in velocity space; we only need take into account that the *global* metric of relativistic velocity space is different. This is done by solving Equation (2) with ∇^2 representing the Laplacian in a curved-metric velocity space, with metric (see Problem 6.8)

$$dv^2_{\text{proper}} = d\Psi^2 + \sinh^2\Psi\,(dv_\theta^2 + \sin^2\theta\, dv_\phi^2) \tag{4}$$

where Ψ is the rapidity parameter, that is $\tanh\Psi \equiv |\underset{\sim}{v}|$. Using the relation [Problem 7.7 (j)]

$$P_{;\alpha}{}^{;\alpha} = g^{-\frac{1}{2}}(g^{\frac{1}{2}}g^{\alpha\beta}P_{,\beta})_{,\alpha} \tag{5}$$

and the fact that by spherical symmetry $\dfrac{\partial P}{\partial v_\phi} = \dfrac{\partial P}{\partial v_\theta} = 0$, we get

$$\frac{\partial P}{\partial n} = \frac{\Delta v^2}{6}\left[\frac{1}{\sinh^2\Psi}\frac{\partial}{\partial\Psi}\left(\sinh^2\Psi\,\frac{\partial P}{\partial\Psi}\right)\right]. \tag{6}$$

It is not difficult to check that this has the solution

$$P(\underset{\sim}{v}, n) = (4\pi)^{-\frac{3}{2}}\,e^{-t}\,t^{-\frac{3}{2}}\,\frac{\Psi}{\sinh\Psi}\,e^{-(\Psi^2/4t)}, \tag{7}$$

where $t \equiv n\Delta v^2/6$. This agrees with the Newtonian solution Equation (3) when $t \ll 1$, $\Psi \ll 1$, (no restriction on $\Psi^2/4t$).

The probability of Ψ between Ψ and $\Psi + d\Psi$ is (using the velocity-space metric) $4\pi\sinh^2\Psi\, P d\Psi$ which for $t \gg 1$, $\Psi \gg 1$ has the asymptotic behavior

$$\frac{dP}{d\Psi} \approx (4\pi)^{-\frac{1}{2}}\frac{\Psi}{t^{\frac{3}{2}}}\,e^{-\Psi^2/4t - t + \Psi} = (4\pi)^{-\frac{1}{2}}\frac{\Psi}{t^{\frac{3}{2}}}\,e^{-t\left(1-\frac{\Psi}{2t}\right)^2}.$$

Only the exponential is of significance here, and this tells us that Ψ has a mean value $\langle\Psi\rangle \approx 2t$, and a standard deviation $\Delta\Psi \sim \sqrt{2t}$, so $\Delta\Psi/\langle\Psi\rangle \to 0$ as $t \to \infty$. The mean rapidity $\langle\Psi\rangle$ increases linearly with number of steps, but the average boost per step is $\Delta v^2/3$. A sober pilot would increase his rapidity by Δv each step, $3c/\Delta v$ times as much. Nevertheless it is amazing that the drunk astronaut does as well as he does; in the Newtonian case velocity or rapidity increased only as $n^{\frac{1}{2}}$, not as n. The reason here has to do with the Lorentz transformation: if an observer moving away from you chooses a random direction to fire a bullet, the directions look biased outward from you (headlight effect).

Solution 7.23.

(a) If the hypersurfaces are given by $f = \text{constant}$, then

$$\mathbf{k} \propto \nabla f \qquad k_\mu = h f_{,\mu}$$

and so

$$k_{\mu;\nu} = h_{,\nu} f_{,\mu} + h f_{,\mu;\nu} \tag{1}$$

$$k_{[\mu;\nu}k_{\lambda]} = h_{,[\nu} f_{,\mu} h f_{,\lambda]} + h^2 f_{,[\mu;\nu} f_{,\lambda]}\,. \tag{2}$$

The first term obviously vanishes and the second term vanishes because $f_{,\mu;\nu} = f_{,\nu;\mu}$. The converse of this, that $k_{[\mu;\nu}k_{\lambda]}$ implies that k is hypersurface-orthogonal, is called Frobenius' theorem.

(b) From Equation (1) the extra condition $k_{[\mu;\nu]} = 0$ is equivalent to $\mathbf{k} = \nabla f$ for some f.

Solution 7.24. Let $\mathbf{k}$ be the tangent vector to the congruence of null curves. Since $\mathbf{k}$ is hypersurface orthogonal,

$$k_\alpha = hf_{,\alpha}$$

for some scalar functions f and h (see Problem 7.23). Because $\mathbf{k}$ is null,

$$f_{,\alpha} f^{,\alpha} = 0 \ .$$

From these two equations we have

$$k_{\alpha;\beta}k^\beta = (h_{,\beta}f_{,\alpha} + hf_{,\alpha;\beta})\, hf^{,\beta}$$

$$f_{,\alpha;\beta}f^{,\beta} = f_{,\beta;\alpha}f^{,\beta} = \frac{1}{2}\,(f_{,\beta}f^{,\beta})_{;\alpha} = 0$$

and thus

$$k_{\alpha;\beta}k^\beta = (h_{,\beta}f^{,\beta})\, k_\alpha$$

so the geodesic equation $\nabla_{\mathbf{k}}\mathbf{k} \propto \mathbf{k}$ is satisfied. If $k^\alpha = dx^\alpha/d\lambda$, we can introduce an affine parameter $\lambda' = \lambda'(\lambda)$ to write the equation in the standard form $\nabla_{\bar{\mathbf{k}}}\bar{\mathbf{k}} = 0$ where $\bar{k}^\alpha = dx^\alpha/d\lambda'$.

Solution 7.25. The general case gives an Euler-Lagrange equation

$$0 = \frac{d^2F}{dy^2}\frac{dy}{ds}\frac{\partial y}{\partial \dot{x}} + \frac{dF}{dy}\left[\frac{d}{ds}\left(\frac{\partial y}{\partial \dot{x}} - \frac{\partial y}{\partial x}\right)\right].$$

Since $dy/ds = 0$ (for s an affine parameter) and since $dF/dy \neq 0$, this is the same as the Euler-Lagrange equation obtained from $\delta \int y\, ds = 0$.

CHAPTER 8: SOLUTIONS

Solution 8.1.

(a) The key relations are $<\widetilde{dx}^\alpha, \partial/\partial x^\beta> \equiv \widetilde{dx}^\alpha \cdot \partial/\partial x^\beta = \delta^\alpha_\beta$, $(\partial/\partial x^\alpha)\cdot(\partial/\partial x^\beta) = g_{\alpha\beta}$, and $\widetilde{dx}^\alpha \cdot \widetilde{dx}^\beta = g^{\alpha\beta}$. The answers are:

$$1, \quad 0, \quad g_{01}, \quad g^{01}, \quad g^{00} .$$

(b) The vector $g^{1\alpha}\,\partial/\partial x^\alpha \equiv g^{1\alpha}\,e_\alpha$ corresponds to $\widetilde{dx}^1$ since, for any vector v

$$v \cdot g^{1\alpha} e_\alpha = v^\mu e_\mu \cdot g^{1\alpha} e_\alpha = v^\mu g_{\mu\alpha}\, g^{1\alpha} = v^\mu \delta^1_\mu = v^1 = <\widetilde{dx}^1, v> .$$

Solution 8.2. It is easy to see that f is just the coordinate r, that is $\widetilde{df} = \widetilde{dr}$, since

$$<\widetilde{dr}, e_{\hat{r}}> = <\widetilde{dr}, e_r> = 1$$

$$<\widetilde{dr}, e_{\hat{\theta}}> = r^{-1}<\widetilde{dr}, e_\theta> = 0 .$$

Now suppose there exists a g such that $\widetilde{dg} = \tilde{\omega}^{\hat{\theta}}$ then

$$0 = <\widetilde{dg}, e_{\hat{r}}> = <\widetilde{dg}, e_r> = \partial g/\partial r$$

$$1 = <\widetilde{dg}, e_{\hat{\theta}}> = r^{-1}<\widetilde{dg}, e_\theta> = r^{-1}\,\partial g/\partial\theta .$$

Clearly the conditions $\partial g/\partial r = 0$ and $\partial g/\partial\theta = r$ are incompatible.

Solution 8.3. In a cartesian coordinate basis the condition is that $\sigma_{i,j} = \sigma_{j,i}$. This is equivalent to the basis-free requirement that the curl of $\boldsymbol{\sigma}$ (the vector equivalent of $\tilde{\boldsymbol{\sigma}}$) vanish.

Solution 8.4. Choose a basis $\tilde{\boldsymbol{\omega}}_1 \cdots \tilde{\boldsymbol{\omega}}_N$. Any p-form can be written as a sum $\sum_{i,j} A_{ij} \cdots (\tilde{\boldsymbol{\omega}}_i \wedge \tilde{\boldsymbol{\omega}}_j \cdots)$ in which there are exactly p elements in each term in parentheses. Since the distributive law holds for wedge multiplication and addition, it will be sufficient to prove the identity for monomial p and q forms and to evaluate e.g.

$$(\tilde{\boldsymbol{\omega}}_1 \wedge \tilde{\boldsymbol{\omega}}_2 \wedge \cdots \wedge \tilde{\boldsymbol{\omega}}_p) \wedge (\tilde{\boldsymbol{\omega}}_{1'} \wedge \tilde{\boldsymbol{\omega}}_{2'} \wedge \cdots \wedge \tilde{\boldsymbol{\omega}}_{q'}) \ .$$

Wedge multiplication is associative, so the parentheses are not important. First move $\tilde{\boldsymbol{\omega}}_{1'}$ to the extreme left by interchanging it repeatedly with its left neighbor. Each interchange brings in a minus sign, a total of $(-1)^p$. Now move $\tilde{\boldsymbol{\omega}}_{2'}$ left to the second position. This brings in p more minus signs. And continue... . By the time we are done with $\tilde{\boldsymbol{\omega}}_{q'}$, the sign will have been changed pq times, hence the desired identity.

Solution 8.5. We show that the two definitions are equivalent by showing that $d\boldsymbol{\Omega}$ has the same components in an arbitrary coordinate basis when calculated from either definition. Start with the definition in terms of the covariant derivative. Let the components of $\boldsymbol{\Omega}$ be $\Omega_{\alpha\beta\cdots\gamma} = \Omega_{[\alpha\beta\cdots\gamma]}$ (p indices). Then the components of $d\boldsymbol{\Omega}$ are

$$\begin{aligned}\Omega_{[\alpha\beta\cdots\gamma;\delta]} &= \Omega_{[\alpha\beta\cdots\gamma,\delta]} - \Gamma^{\epsilon}{}_{[\alpha\delta}\Omega_{\beta}{}^{\epsilon}{}_{\cdots\gamma]} \\ &\quad - \Gamma^{\epsilon}{}_{[\beta\delta}\Omega_{\alpha}{}^{\epsilon}{}_{\cdots\gamma} - \cdots - \Gamma^{\epsilon}{}_{[\gamma\delta}\Omega_{\alpha\beta\cdots\epsilon]} = \Omega_{[\alpha\beta\cdots\gamma,\delta]} \ .\end{aligned} \tag{1}$$

(The Γ terms vanish since $\Gamma_{\mu[\nu\sigma]} = 0$ in a coordinate basis.) Now start with

$$\boldsymbol{\Omega} = \Omega_{\alpha\beta\cdots\gamma}\widetilde{dx}^{\alpha} \wedge \widetilde{dx}^{\beta} \wedge \cdots \widetilde{dx}^{\gamma} \ .$$

By property (ii), d does not disturb the summation convention, so

$$\begin{aligned}d\boldsymbol{\Omega} &= (d\Omega_{\alpha\beta\cdots\gamma}) \wedge (\widetilde{dx}^{\alpha} \wedge \widetilde{dx}^{\beta} \wedge \cdots \wedge \widetilde{dx}^{\gamma}) \\ &\quad + \Omega_{\alpha\beta\cdots\gamma} d(\widetilde{dx}^{\alpha} \wedge \widetilde{dx}^{\beta} \wedge \cdots \wedge \widetilde{dx}^{\gamma}) \ ,\end{aligned} \tag{2}$$

where we have used property (iv) with $p = 0$ since $\Omega_{\alpha\beta\cdots\gamma}$ is simply a function (zero form). Now from property (i) we know that

$$\widetilde{df} = f_{,\alpha}\,\widetilde{dx}^{\alpha}$$

for any zero form f. From property (iii), $\langle \widetilde{df}, \partial/\partial x^{\beta} \rangle = \langle f_{,\alpha}\,\widetilde{dx}^{\alpha}, \partial/\partial x^{\beta} \rangle = f_{,\beta}$; thus $\widetilde{d\Omega}_{\alpha\beta\cdots\gamma} = \Omega_{\alpha\beta\cdots\gamma,\delta}\,\widetilde{dx}^{\delta}$. The second term in Equation (2) is zero: According to property (iv) it can be reduced to a sum of terms each of which contains a ddx, which vanishes by property (v). Thus Equation (2) becomes

$$\mathbf{d\Omega} = \Omega_{\alpha\beta\cdots\gamma,\delta}\,\widetilde{dx}^{\delta} \wedge \widetilde{dx}^{\alpha} \wedge \widetilde{dx}^{\beta} \cdots \wedge \widetilde{dx}^{\gamma} \ .$$

Since $\mathbf{d\Omega}$ is a $(p+1)$ form (property (i)), and the wedge product is anti-symmetric, the components of $\mathbf{d\Omega}$ are $\Omega_{[\alpha\beta\cdots\gamma,\delta]}$ as in Equation (1).

Solution 8.6. Define

$$g \equiv \left[x^1 \int_0^1 f(\xi x^1, x^2, \cdots x^n)\, d\xi \right] ,$$

so that $\tilde{\boldsymbol{\beta}} = g\,\widetilde{dx}^2$ and

$$\mathbf{d}\tilde{\boldsymbol{\beta}} = g_{,i}\,\widetilde{dx}^i \wedge \widetilde{dx}^2 = g_{,1}\,\widetilde{dx}^1 \wedge \widetilde{dx}^2 + g_{,3}\,\widetilde{dx}^3 \wedge \widetilde{dx}^2 + \cdots \ .$$

Now

$$g_{,1} = \left[\int_0^1 f(\xi x^1, x^2, \cdots x^n)\, d\xi \right] + \left[x^1 \int_0^1 \xi f'(\xi x^1, x^2, \cdots x^n)\, d\xi \right]$$

where f' indicates a derivative with respect to the first argument. The second term above can be written as

$$\int_0^1 x^1 \xi f'(\xi x^1, x^2 \cdots)\, d\xi = \int_0^1 \xi \frac{\partial}{\partial \xi} f(\xi x^1, x^2, \cdots)\, d\xi$$

$$= \xi f(\xi x^1, x^2, \cdots) \Big|_0^1 - \int_0^1 f(\xi x^1, x^2, \cdots)\, d\xi \ ,$$

where we have integrated by parts; we have then

$$g_{,1} = \xi f(\xi x^1, x^2, \cdots x^n)\Big|_0^1 = f(x^1, x^2, \cdots x^n) .$$

Next, $d\boldsymbol{\alpha} = 0$ tells us that

$$d\boldsymbol{\alpha} = f_{,3}\, dx^3 \wedge dx^1 \wedge dx^2 + f_{,4}\, dx^4 \wedge dx^1 \wedge dx^2 + \cdots = 0 ,$$

which can only be true if $f_{,3} = f_{,4} = f_{,5} = \cdots f_{,n} = 0$, and hence

$$g_{,3} = \left[x^1 \int_0^1 f_{,3}(\xi x^1, x^2, \cdots, x^n)\, d\xi \right] = 0 .$$

Similarly $g_{,4} = g_{,5} = \cdots g_{,n} = 0$, so that finally

$$d\tilde{\boldsymbol{\beta}} = g_{,1}\, \widetilde{dx}^1 \wedge \widetilde{dx}^2 = f(x^1, x^2, \cdots x^n)\, \widetilde{dx}^1 \wedge \widetilde{dx}^2 = \boldsymbol{\alpha} .$$

Solution 8.7. The 2-form is $\mathbf{F} = F_{\mu\nu}\, \widetilde{dx}^\mu \wedge \widetilde{dx}^\nu$ so that

$$0 = d\mathbf{F} = F_{\mu\nu,\lambda}\, \widetilde{dx}^\lambda \wedge \widetilde{dx}^\mu \wedge \widetilde{dx}^\nu$$

gives us the Maxwell's equations $F_{[\mu\nu;\lambda]} = 0$. Similarly $d{*}\mathbf{F} = 0$ gives ${*}F_{[\lambda\sigma;\nu]} = 0$, but this is equivalent to (see Problem 3.25)

$$0 = \varepsilon^{\mu\nu\lambda\sigma}\, {*}F_{\lambda\sigma;\nu} = 2{*}{*}F^{\mu\nu}{}_{;\nu} = -2F^{\mu\nu}{}_{;\nu} ,$$

and so we have the remaining Maxwell's equations $F^{\mu\nu}{}_{;\nu} = 0$.

Solution 8.8. At any point on the surface choose a local orthonormal tetrad $e_{\hat{t}}, e_{\hat{x}}, e_{\hat{y}}, e_{\hat{z}}$. If the surface is spacelike, we can choose the tetrad so that $e_{\hat{t}}$ is the unit normal. The 3 orthogonal, linearly independent vectors lying in the surface are then spacelike $(e_{\hat{x}}, e_{\hat{y}}, e_{\hat{z}})$. Now any other orthonormal tetrad we might have chosen is related to the first one by a Lorentz transformation, which does not affect whether a vector is spacelike or not. Hence the three vectors are always spacelike when the surface is spacelike.

In the null case, choose $e_{\hat{t}} + e_{\hat{x}}$ to be normal to the surface. A null vector is orthogonal to itself, so the 3 orthogonal vectors in the surface are $e_{\hat{y}}$ and $e_{\hat{z}}$ (spacelike) and $e_{\hat{t}} + e_{\hat{x}}$ (null). In the timelike case, choose $e_{\hat{x}}$ to be normal. Then in the surface, $e_{\hat{t}}$ is timelike; $e_{\hat{y}}$ and $e_{\hat{z}}$ are spacelike. Again these results are unchanged by Lorentz transformations.

Solution 8.9. The volume element is

$$d^3\Sigma_\mu(a, b, c) = \frac{1}{3!}\,\varepsilon_{\mu\nu\beta\sigma}\,\frac{\partial(x^\nu, x^\beta, x^\sigma)}{\partial(a, b, c)}\,da\,db\,dc\ .$$

If we change parameterization from a, b, c to α, β, γ then

$$da\,db\,dc \rightarrow \left|\frac{\partial(a, b, c)}{\partial(\alpha, \beta, \gamma)}\right|\,d\alpha\,d\beta\,d\gamma\ .$$

Provided (α, β, γ) has the same orientation as (a, b, c), the Jacobian is positive (definition of same orientation!), so we can drop the absolute value signs, and

$$\begin{aligned} d^3\Sigma_\mu(a, b, c) &\rightarrow \frac{1}{3!}\,\varepsilon_{\mu\nu\rho\sigma}\,\frac{\partial(x^\nu, x^\rho, x^\sigma)}{\partial(a, b, c)}\,\frac{\partial(a, b, c)}{\partial(\alpha,\beta,\gamma)}\,d\alpha\,d\beta\,d\gamma \\ &= \frac{1}{3!}\,\varepsilon_{\mu\nu\rho\sigma}\,\frac{\partial(x^\nu, x^\rho, x^\sigma)}{\partial(\alpha, \beta, \gamma)}\,d\alpha\,d\beta\,d\gamma \\ &= d^3\Sigma_\mu(\alpha, \beta, \gamma) \end{aligned}$$

which completes the proof. With differential forms, the invariance is obvious since $d^3\Sigma_\mu$ can be viewed as a 3-form

$$d^3\Sigma_\mu = \frac{1}{3!}\,\varepsilon_{\mu\nu\rho\sigma}\,\widetilde{dx}^\nu \wedge \widetilde{dx}^\rho \wedge \widetilde{dx}^\sigma\ .$$

Solution 8.10.

(a) The 2-form $\boldsymbol{\theta}$ and its exterior derivative are

$$\boldsymbol{\theta} = f^k d^2 S_k = f^k\,\frac{1}{2}\,\varepsilon_{klm}\,\widetilde{dx}^l \wedge \widetilde{dx}^m$$

$$d\boldsymbol{\theta} = f^k{}_{,n}\,\frac{1}{2}\,\varepsilon_{klm}\,\widetilde{dx}^n \wedge \widetilde{dx}^l \wedge \widetilde{dx}^m + f^k\,\frac{1}{2}\,(d\,\varepsilon_{klm}) \wedge \widetilde{dx}^l \wedge \widetilde{dx}^m\ .$$

Define the symbol $[klm]$ to be $+1(-1)$ if k, l, m is an even (odd) permutation of $1, 2, 3$, and zero otherwise. We then have

$$d\varepsilon_{klm} = d(|g|^{\frac{1}{2}}[klm]) = [klm]\,\frac{1}{2}\,g_{,n}\,\widetilde{dx}^n/|g|^{\frac{1}{2}} = \varepsilon_{klm}\,\frac{1}{2}\,g_{,n}\,\widetilde{dx}^n/g\ .$$

Thus we have

$$\begin{aligned} d\boldsymbol{\theta} &= \frac{1}{2}\left(f^k{}_{,n} + \frac{1}{2}\,g_{,n}/g\right)\varepsilon_{klm}\,\widetilde{dx}^n \wedge \widetilde{dx}^l \wedge \widetilde{dx}^m \\ &= \frac{1}{2}|g|^{-\frac{1}{2}}(|g|^{\frac{1}{2}} f^k)_{,n}\,\varepsilon_{klm}\,\widetilde{dx}^n \wedge \widetilde{dx}^l \wedge \widetilde{dx}^m\ . \end{aligned}$$

But in three dimensions, $\widetilde{dx}^n \wedge \widetilde{dx}^l \wedge \widetilde{dx}^m = |g|^{\frac{1}{2}}\,\varepsilon^{nlm}\,\widetilde{dx}^1 \wedge \widetilde{dx}^2 \wedge \widetilde{dx}^3$ and $\varepsilon_{klm}\,\varepsilon^{nlm} = 2\delta^n{}_k$, so

$$d\boldsymbol{\theta} = f^k{}_{;k}\,|g|^{\frac{1}{2}}\,\widetilde{dx}^1 \wedge \widetilde{dx}^2 \wedge \widetilde{dx}^3\ .$$

Thus, in ordinary vector notation, the theorem becomes

$$\int_V \nabla\cdot\underset{\sim}{f}\,dV = \int_{\partial V} \underset{\sim}{f}\cdot d\underset{\sim}{S}\ .$$

(b) $\boldsymbol{\theta} = f^\mu d^3\boldsymbol{\Sigma}_\mu = f^\mu\,\frac{1}{3!}\,\varepsilon_{\mu\alpha\beta\gamma}\,\widetilde{dx}^\alpha \wedge \widetilde{dx}^\beta \wedge \widetilde{dx}^\gamma$

$$\begin{aligned} d\boldsymbol{\theta} &= \frac{1}{3!}(f^\mu{}_{,\lambda}\,\varepsilon_{\mu\alpha\beta\gamma}\,\widetilde{dx}^\lambda + f^\mu\,d\varepsilon_{\mu\alpha\beta\gamma}) \wedge \widetilde{dx}^\alpha \wedge \widetilde{dx}^\beta \wedge \widetilde{dx}^\gamma \\ &= \frac{1}{3!}\left(f^\mu{}_{,\lambda}\,\varepsilon_{\mu\alpha\beta\gamma}\,\widetilde{dx}^\lambda + f^\mu\,\frac{1}{2}\,\frac{g_{,\lambda}}{g}\,\varepsilon_{\mu\alpha\beta\gamma}\,\widetilde{dx}^\lambda\right) \wedge \widetilde{dx}^\alpha \wedge \widetilde{dx}^\beta \wedge \widetilde{dx}^\gamma \\ &= \frac{1}{3!}|g|^{-\frac{1}{2}}(|g|^{\frac{1}{2}} f^\mu)_{,\lambda}\,\varepsilon_{\mu\alpha\beta\gamma}\,\widetilde{dx}^\lambda \wedge \widetilde{dx}^\alpha \wedge \widetilde{dx}^\beta \wedge \widetilde{dx}^\gamma \\ &= \frac{-1}{3!\,|g|^{\frac{1}{2}}}(|g|^{\frac{1}{2}} f^\mu)_{,\lambda}\,\varepsilon_{\mu\alpha\beta\gamma}\,\varepsilon^{\lambda\alpha\beta\gamma}|g|^{\frac{1}{2}}\,\widetilde{dx}^0 \wedge \widetilde{dx}^1 \wedge \widetilde{dx}^2 \wedge \widetilde{dx}^3 \\ &= \frac{1}{3!\,|g|^{\frac{1}{2}}}(|g|^{\frac{1}{2}} f^\mu)_{,\lambda}\,\delta^\lambda{}_\mu\,3!\,|g|^{\frac{1}{2}}\,\widetilde{dx}^0 \wedge \widetilde{dx}^1 \wedge \widetilde{dx}^2 \wedge \widetilde{dx}^3 \\ &= f^\mu{}_{;\mu}\,|g|^{\frac{1}{2}}\,\widetilde{dx}^0 \wedge \widetilde{dx}^1 \wedge \widetilde{dx}^2 \wedge \widetilde{dx}^3\ . \end{aligned}$$

The theorem then reduces to

$$\int_\Omega \nabla\cdot f\,|g|^{\frac{1}{2}}\,d^4x = \int_{\partial\Omega} f^\mu d^3\Sigma_\mu\ .$$

(c) $$\boldsymbol{\theta} = F^{\mu\nu} d^2\boldsymbol{\Sigma}_{\mu\nu} = F^{\mu\nu} \frac{1}{2} \varepsilon_{\mu\nu\alpha\beta} \widetilde{dx}^{\alpha} \wedge \widetilde{dx}^{\beta}$$

$$d\boldsymbol{\theta} = \frac{1}{2} d(F^{\mu\nu} \varepsilon_{\mu\nu\alpha\beta}) \wedge \widetilde{dx}^{\alpha} \wedge \widetilde{dx}^{\beta}$$
$$= \frac{1}{2} |g|^{-\frac{1}{2}} (|g|^{\frac{1}{2}} F^{\mu\nu})_{,\lambda} \varepsilon_{\mu\nu\alpha\beta} \widetilde{dx}^{\lambda} \wedge \widetilde{dx}^{\alpha} \wedge \widetilde{dx}^{\beta}$$

where we have taken $d\varepsilon_{\mu\nu\alpha\beta}$ as in part b. Now

$$\varepsilon_{\mu\nu\alpha\beta} \widetilde{dx}^{\lambda} \wedge \widetilde{dx}^{\alpha} \wedge \widetilde{dx}^{\beta} = \varepsilon_{\mu\nu\alpha\beta} (1/6)\, \delta^{\lambda\alpha\beta}_{\kappa\sigma\tau} \widetilde{dx}^{\kappa} \wedge \widetilde{dx}^{\sigma} \wedge \widetilde{dx}^{\tau}$$
$$= -(1/6)\, \varepsilon_{\mu\nu\alpha\beta} \varepsilon^{\gamma\lambda\alpha\beta} \varepsilon_{\gamma\kappa\sigma\tau} \widetilde{dx}^{\kappa} \wedge \widetilde{dx}^{\sigma} \wedge \widetilde{dx}^{\tau}$$
$$= -\varepsilon_{\mu\nu\alpha\beta} \varepsilon^{\gamma\lambda\alpha\beta} d^3\boldsymbol{\Sigma}_{\gamma}$$
$$= 2\delta^{\gamma\lambda}_{\mu\nu} d^3\boldsymbol{\Sigma}_{\gamma} ,$$

and by Problem 7.7(i) we have

$$d\boldsymbol{\theta} = |g|^{-\frac{1}{2}} (|g|^{\frac{1}{2}} F^{\mu\nu})_{,\lambda} \delta^{\gamma\lambda}_{\mu\nu} d^3\boldsymbol{\Sigma}_{\gamma}$$
$$= 2F^{\mu\nu}{}_{;\nu} d^3\boldsymbol{\Sigma}_{\mu} ,$$

and finally

$$\int_{\partial\Omega} F^{\mu\nu} d^2\boldsymbol{\Sigma}_{\mu\nu} = 2 \int_{\Omega} F^{\mu\nu}{}_{;\nu} d^3\boldsymbol{\Sigma}_{\mu} .$$

(d) Let $\boldsymbol{\theta} = A_k dx^k = \underset{\sim}{A} \cdot \underset{\sim}{d\ell}$, and let Ω be 2-dimensional; then the exterior derivative is

$$d\boldsymbol{\theta} = A_{k,j} \widetilde{dx}^{j} \wedge \widetilde{dx}^{k}$$
$$= A_{k,j} \delta^{[k}_{m} \delta^{j]}_{n} \widetilde{dx}^{m} \wedge \widetilde{dx}^{n}$$
$$= \frac{1}{2} A_{k,j} \varepsilon^{rkj} \varepsilon_{rmn} \widetilde{dx}^{m} \wedge \widetilde{dx}^{n}$$
$$= (\nabla \times \underset{\sim}{A})^r d^2S_r ,$$

which completes the derivation.

Solution 8.11. Suppose there were such a tensor. What would be its components at some arbitrary point in a freely falling local orthonormal frame? In this frame all the $g_{\alpha\beta,\mu}$ vanish (because all the $\Gamma_{\alpha\beta\mu}$ vanish), and $g_{\alpha\beta} = \eta_{\alpha\beta}$. Thus the tensor must be some algebraic combination of $\eta_{\alpha\beta}$, "zero," and possibly $\delta^{\alpha}{}_{\beta}$ and $\varepsilon^{\alpha\beta\gamma\delta}$. You can very quickly convince yourself that the only nontrivial one is $\boldsymbol{\eta}\otimes\boldsymbol{\eta}\otimes\cdots\otimes\boldsymbol{\eta}$ (the δ only renames indices, e.g. $\delta^{\alpha}{}_{\beta}\,\eta_{\alpha\gamma} = \eta_{\beta\gamma}$ and the $\varepsilon^{\alpha\beta\gamma\delta}$ gives zero by its antisymmetries, or else just gets some of its indices lowered). Since we chose an arbitrary point, the conclusion holds at all points.

Solution 8.12.

(a) The general definition of the Γ's is

$$\Gamma_{\mu\alpha\beta} = \mathbf{e}_{\mu}\cdot\nabla_{\beta}\,\mathbf{e}_{\alpha} \,.$$

In a coordinate frame

$$\begin{aligned}\Gamma_{\mu\alpha\beta} - \Gamma_{\mu\beta\alpha} &= \mathbf{e}_{\mu}\cdot(\nabla_{\beta}\,\mathbf{e}_{\alpha} - \nabla_{\alpha}\,\mathbf{e}_{\beta}) \\ &= \mathbf{e}_{\mu}\cdot[\mathbf{e}_{\beta},\mathbf{e}_{\alpha}] \\ &= 0 \,,\end{aligned}$$

since in a coordinate frame basis vectors commute.

(b) In an orthonormal frame

$$\begin{aligned}\Gamma_{\mu\alpha\beta} + \Gamma_{\alpha\mu\beta} &= \mathbf{e}_{\mu}\cdot\nabla_{\beta}\mathbf{e}_{\alpha} + \mathbf{e}_{\alpha}\cdot\nabla_{\beta}\mathbf{e}_{\mu} \\ &= \nabla_{\beta}(\mathbf{e}_{\mu}\cdot\mathbf{e}_{\alpha}) = \nabla_{\beta}(\eta_{\mu\alpha}) = 0 \,.\end{aligned}$$

Solution 8.13.

(a) We use the fact ("usual rule for a derivative operator") that for any vector field $\mathbf{v}$

$$\mathcal{L}_{\mathbf{x}}\langle\tilde{\mathbf{A}},\mathbf{v}\rangle = \langle\mathcal{L}_{\mathbf{x}}\tilde{\mathbf{A}},\mathbf{v}\rangle + \langle\tilde{\mathbf{A}},\mathcal{L}_{\mathbf{x}}\mathbf{v}\rangle \,. \tag{1}$$

Thus for the components of the 1-form $\mathcal{L}_{\mathbf{x}}\tilde{\mathbf{A}}$ we have

$$
\begin{aligned}
(\pounds_x\tilde{A})_\alpha &\equiv \pounds_x A_\alpha = <\pounds_x\tilde{A}, e_\alpha> \\
&= \pounds_x<\tilde{A}, e_\alpha> - <\tilde{A}, \pounds_x e_\alpha> \qquad (1)\\
&= A_{\alpha,\beta}x^\beta - <\tilde{A}, [x, e_\alpha]> \\
&= A_{\alpha,\beta}x^\beta - A_\gamma<\tilde{\omega}^\gamma, [x^\beta e_\beta, e_\alpha]> \\
&= A_{\alpha,\beta}x^\beta - A_\gamma<\tilde{\omega}^\gamma, (x^\beta c_{\beta\alpha}{}^\delta e_\delta - x^\beta{}_{,\alpha}e_\beta)> \\
&= A_{\alpha,\beta}x^\beta + x^\beta{}_{,\alpha}A_\beta - A_\gamma x^\beta c_{\beta\alpha}{}^\gamma . \qquad (2)
\end{aligned}
$$

The last term vanishes in a coordinate frame. An alternative form is, from Equation (1),

$$
\begin{aligned}
(\pounds_x\tilde{A})_\alpha &= \nabla_x<\tilde{A}, e_\alpha> - <\tilde{A}, (\nabla_x e_\alpha - \nabla_{e_\alpha}x)> \\
&= <\nabla_x\tilde{A}, e_\alpha> + <\tilde{A}, \nabla_{e_\alpha}x> \\
&= A_{\alpha;\beta}x^\beta + x^\beta{}_{;\alpha}A_\beta . \qquad (3)
\end{aligned}
$$

The connection coefficients do not really enter into the definition of the Lie derivative; they cancel out of Equation (3) to leave Equation (2).

(b) For simplicity, work in a coordinate basis:

$$
\begin{aligned}
\pounds_x T &= \pounds_x(T^\alpha{}_\beta e_\alpha \otimes \tilde{\omega}^\beta) \\
&= (\pounds_x T^\alpha{}_\beta) e_\alpha \otimes \tilde{\omega}^\beta + T^\alpha{}_\beta(\pounds_x e_\alpha)\otimes\tilde{\omega}^\beta + T^\alpha{}_\beta e_\alpha \otimes (\pounds_x\tilde{\omega}^\beta) \\
&= T^\alpha{}_{\beta,\gamma}x^\gamma e_\alpha \otimes \tilde{\omega}^\beta - T^\alpha{}_\beta x^\gamma{}_{,\alpha}e_\gamma \otimes \tilde{\omega}^\beta + T^\alpha{}_\beta x^\beta{}_{,\gamma}e_\alpha \otimes \tilde{\omega}^\gamma \\
&= (T^\alpha{}_{\beta,\gamma}x^\gamma - T^\mu{}_\beta x^\alpha{}_{,\mu} + T^\alpha{}_\mu x^\mu{}_{,\beta}) e_\alpha \otimes \tilde{\omega}^\beta .
\end{aligned}
$$

This is often written

$$
\pounds_x T^\alpha{}_\beta = T^\alpha{}_{\beta,\gamma}x^\gamma - T^\mu{}_\beta x^\alpha{}_{,\mu} + T^\alpha{}_\mu x^\mu{}_{,\beta}
$$

and the commas can be replaced by semicolons as in part (a).

Solution 8.14.

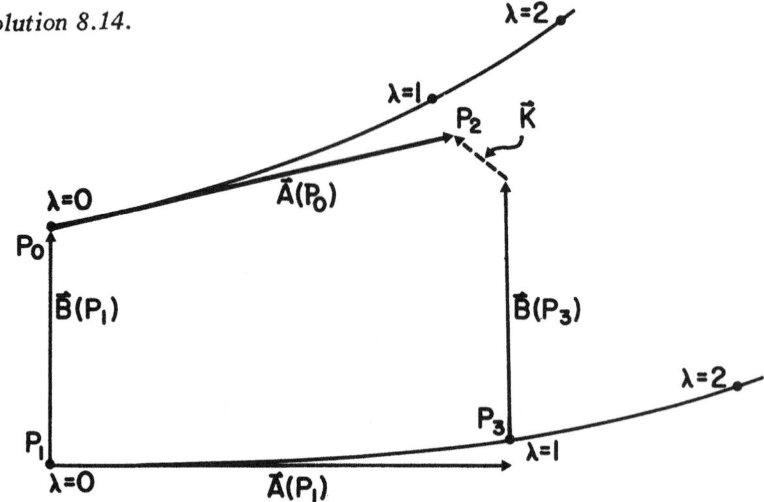

At the points P_0 and P_1 consider two tangent vectors $A(P_0)$ and $A(P_1)$ each extending from $\lambda = 0$ to $\lambda = \Delta\lambda = \mathcal{O}(\epsilon)$. (Note: In the figure $\Delta\lambda$ is taken to be unity so that the **A** vectors are large and the "difference vector" **K** is exaggerated for clarity.) Suppose that at point P_1 the vector $B(P_1)$ connects P_1 and P_0. Let us compute the amount by which the vector field **B** fails to connect points of equal λ along the congruence **A**, i.e. let us compute the vector **K**:

$$\begin{aligned} K &= [B(P_1)+A(P_0)] - [A(P_1)+B(P_3)] \\ &= [B(P_1)-B(P_3)] - [A(P_1)-A(P_0)] \\ &= A^\tau B^\gamma{}_{,\tau}\, e_\gamma + \mathcal{O}(A^2B) - B^\tau A^\gamma{}_{,\tau}\, e_\gamma + \mathcal{O}(B^2A) \\ &= [A, B] + \mathcal{O}(\epsilon^3) \quad \text{if} \quad |A| \sim \mathcal{O}(\epsilon) \sim |B| \ . \end{aligned}$$

Thus in the limit of infinitesimal vectors **A** and **B**,

$$K = [A, B] = \mathcal{L}_A B \ .$$

Thus **K** vanishes if $\mathcal{L}_A B = 0$, i.e. if **B** is "Lie-dragged" along **A**.

Solution 8.15. Contraction may be thought of as multiplication by the *constant* Kronecker-delta tensor $\delta^\mu{}_\nu$. But the Lie derivative of the tensor $\delta^\mu{}_\nu$ is zero, as we can easily prove in a coordinate frame,

$$\mathcal{L}_{\mathbf{x}}(\delta^\mu{}_\nu) = \delta^\mu{}_{\nu,\lambda}x^\lambda + \delta^\mu{}_\lambda x^\lambda{}_{,\nu} - \delta^\lambda{}_\nu x^\mu{}_{,\lambda} = 0 + x^\mu{}_{,\nu} - x^\mu{}_{,\nu} = 0 ,$$

and therefore the contraction operator $\delta^\mu{}_\nu$ can go on either side of $\mathcal{L}_{\mathbf{x}}$.

Solution 8.16. If the operator identity holds for scalar functions and for vector fields, it will hold for arbitrary tensors. This is because the action of the Lie derivative on arbitrary tensors is determined by its action on scalars and vectors (see Problem 8.13). Let f be a scalar function, then

$$\mathcal{L}_{\mathbf{u}}\mathcal{L}_{\mathbf{v}} f - \mathcal{L}_{\mathbf{v}}\mathcal{L}_{\mathbf{u}} f = \mathcal{L}_{\mathbf{u}}\nabla_{\mathbf{v}} f - \mathcal{L}_{\mathbf{v}}\nabla_{\mathbf{u}} f = [\mathbf{u}, \mathbf{v}]f = \mathcal{L}_{[\mathbf{u},\mathbf{v}]}f .$$

Now let w be a vector field, then

$$\mathcal{L}_{\mathbf{u}}\mathcal{L}_{\mathbf{v}}\mathbf{w} - \mathcal{L}_{\mathbf{v}}\mathcal{L}_{\mathbf{u}}\mathbf{w} - \mathcal{L}_{[\mathbf{u},\mathbf{v}]}\mathbf{w} = [\mathbf{u}, [\mathbf{v},\mathbf{w}]] - [\mathbf{v}, [\mathbf{u},\mathbf{w}]] - [[\mathbf{u},\mathbf{v}], \mathbf{w}] = 0 .$$

The vanishing of this last expression (the Jacobi identity for commutators) can easily be checked e.g. by writing out all terms. This completes the proof for scalars and for vectors, and hence in general.

Solution 8.17.

(i) First the scalar case:

$$\begin{aligned}\Phi(P_N) &= \Phi(P_0) + [x^\mu(P_N) - x^\mu(P_0)]\Phi_{,\mu}(P_0) + \cdots \\ &= \Phi(P_0) - \xi^\mu\Phi_{,\mu} ,\end{aligned}$$

and $\overline{\Phi}(P_N) = \Phi(P_N)$ since Φ is a scalar, hence

$$\mathcal{L}_\xi\Phi = \Phi_{,\mu}\xi^\mu .$$

(ii) Since $\overline{x}^\mu(P_N) = x^\mu(P_0) = x^\mu(P_N) + \xi^\mu$, we have for vector fields

$$\overline{A}_\mu(P_N) = (\partial x^\alpha/\partial\overline{x}^\mu)\, A_\mu(P_N) = (\delta^\alpha{}_\mu - \xi^\alpha{}_{,\mu})\, A_\mu(P_N) .$$

Now we combine this with

$$A_\mu(P_N) = A_\mu(P_0) - A_{\mu,\beta}\xi^\beta$$

to get

$$\begin{aligned}\mathcal{L}_{\boldsymbol{\xi}} A_\mu &= A_\mu(P_0) - \bar{A}_\mu(P_N)\\ &= A_\mu(P_0) - (\delta^\alpha{}_\mu - \xi^\alpha{}_{,\mu})(A_\alpha(P_0) - A_{\alpha,\beta}\xi^\beta)\\ &= A_{\mu,\beta}\xi^\beta + A_\alpha\xi^\alpha{}_{,\mu}\ .\end{aligned}$$

(iii) The tensor case is handled similarly:

$$\begin{aligned}\bar{T}^\mu{}_\nu(P_N) &= (\partial\bar{x}^\mu/\partial x^\alpha)(\partial x^\beta/\partial\bar{x}^\nu)\,T^\alpha{}_\beta(P_N)\\ &= (\delta^\mu{}_\alpha + \xi^\mu{}_{,\alpha})(\delta^\beta{}_\nu - \xi^\beta{}_{,\nu})\,[T^\alpha{}_\beta(P_0) - T^\alpha{}_{\beta,\gamma}\xi^\gamma]\\ &= T^\mu{}_\nu(P_0) - T^\mu{}_{\nu,\gamma}\xi^\gamma - T^\mu{}_\beta\xi^\beta{}_{,\nu} + T^\alpha{}_\nu\xi^\mu{}_{,\alpha}\end{aligned}$$

$$\mathcal{L}_{\boldsymbol{\xi}} T^\mu{}_\nu = T^\mu{}_{\nu,\gamma}\xi^\gamma + T^\mu{}_\beta\xi^\beta{}_{,\nu} - T^\alpha{}_\nu\xi^\mu{}_{,\alpha}\ .$$

Solution 8.18. The parallel transport law $u^\alpha v^\beta{}_{;\alpha} = 0$ requires only the affine connection $\Gamma^\alpha{}_{\beta\gamma}$. The Fermi-Walker transport law

$$\nabla_{\mathbf{u}}\mathbf{v} = \mathbf{u}(\mathbf{a}\cdot\mathbf{v}) - \mathbf{a}(\mathbf{u}\cdot\mathbf{v}) \qquad \mathbf{a} \equiv \nabla_{\mathbf{u}}\mathbf{u}$$

requires, in addition, a metric for the dot products. The Lie transport equation $\nabla_{\mathbf{u}}\mathbf{v} - \nabla_{\mathbf{v}}\mathbf{u} = 0$ requires $\mathbf{u}$ to be a vector field defined off the curve, *not* just a tangent vector along the curve, because the gradient of $\mathbf{u}$ in the $\mathbf{v}$-direction must be defined, and $\mathbf{v}$ is not in general along the curve. Although the formula above includes ∇'s, a connection is actually *not* needed, because the Γ's cancel out due to the antisymmetry of the equation. Obviously no metric is required for Lie differentiation.

Solution 8.19. If $\underset{\sim}{u}$ can be returned to A unchanged, then in principle it can be extended into a vector field which is both parallel transported and Lie transported from A to B. Let $\underset{\sim}{u} = (u^x, u^y, u^z)$, and consider the identity

$$\nabla_{\mathbf{u}} \mathbf{v} - \nabla_{\mathbf{v}} \mathbf{u} = [\underset{\sim}{\mathbf{u}}, \underset{\sim}{\mathbf{v}}] = \mathcal{L}_{\mathbf{u}} \mathbf{v} = -\mathcal{L}_{\mathbf{v}} \mathbf{u} \ . \tag{1}$$

Setting the appropriate terms to zero in Equation (1), i.e. $\nabla_{\mathbf{v}} \mathbf{u} = \mathcal{L}_{\mathbf{v}} \mathbf{u} = 0$, and noting that the Christoffel symbols vanish, we have

$$\nabla_{\mathbf{u}} \mathbf{v} = 0 = u^i v^k{}_{,i} = 0 \tag{2}$$

$$-u^y = u^x = \alpha u^z z^{\alpha-1} = 0 \ .$$

So $\underset{\sim}{\mathbf{u}}$ can be nonzero (and have $u^z \neq 0$) only if

$$\alpha = 0 \ . \tag{3}$$

Solution 8.20. Let the vector field be $\mathbf{u}(p)$. If it is parallel propagated along itself, we have $\nabla_{\mathbf{u}} \mathbf{u} = 0$. This is just the geodesic equation, so the field must be a field of tangent vectors to a space-filling congruence of geodesics.

If the field is F-W transported along itself, we have

$$\nabla_{\mathbf{u}} \mathbf{u} = (\mathbf{u} \otimes \mathbf{a} - \mathbf{a} \otimes \mathbf{u}) \cdot \mathbf{u} \tag{1}$$

where $\mathbf{a} = \nabla_{\mathbf{u}} \mathbf{u}$. Thus

$$\mathbf{a} = \mathbf{u}(\mathbf{a} \cdot \mathbf{u}) - \mathbf{a}(\mathbf{u} \cdot \mathbf{u}) \ . \tag{2}$$

Now dot $\mathbf{u}$ into this and get $\mathbf{a} \cdot \mathbf{u} = \mathbf{u}^2 (\mathbf{a} \cdot \mathbf{u}) - (\mathbf{a} \cdot \mathbf{u}) \mathbf{u}^2 = 0$. From (2) we have that either $\mathbf{u}^2 = -1$ or $\mathbf{a} = 0$. The first possibility means that any field of properly normalized 4-velocities is F-W transported along itself. The second says that the tangent vectors of any congruence of geodesics are also, even if they are not normalized to -1.

The Lie transport law is

$$0 = \mathcal{L}_{\mathbf{u}} \mathbf{u} \equiv [\mathbf{u}, \mathbf{u}]$$

which is satisfied by *any* vector field.

Solution 8.21. Consider a 1-form $\tilde{U} = U_\alpha \widetilde{dx}^\alpha$. First calculate $d\mathcal{L}_x\tilde{U}$:

$$\mathcal{L}_x\tilde{U} = (x^\alpha U_{\beta,\alpha} + U_\alpha x^\alpha{}_{,\beta})\widetilde{dx}^\beta$$

$$d(\mathcal{L}_x\tilde{U}) = (x^\alpha U_{\beta,\alpha} + U_\alpha x^\alpha{}_{,\beta})_{,\gamma}\widetilde{dx}^\gamma \wedge \widetilde{dx}^\beta \tag{1}$$

Now calculate $\mathcal{L}_x(d\tilde{U})$:

$$d\tilde{U} = U_{\alpha,\gamma}\widetilde{dx}^\gamma \wedge \widetilde{dx}^\alpha$$

$$\begin{aligned}\mathcal{L}_x(d\tilde{U}) &= \mathcal{L}_x(U_{\alpha,\gamma})\widetilde{dx}^\gamma \wedge \widetilde{dx}^\alpha + U_{\alpha,\gamma}\mathcal{L}_x(\widetilde{dx}^\gamma)\wedge\widetilde{dx}^\alpha \\ &\quad + U_{\alpha,\gamma}\widetilde{dx}^\gamma \wedge \mathcal{L}_x(\widetilde{dx}^\alpha) \\ &= x^\tau U_{\alpha,\gamma\tau}\widetilde{dx}^\gamma \wedge \widetilde{dx}^\alpha + U_{\alpha,\gamma}x^\gamma{}_{,\tau}\widetilde{dx}^\tau \wedge \widetilde{dx}^\alpha + U_{\alpha,\gamma}x^\alpha{}_{,\tau}\widetilde{dx}^\gamma \wedge \widetilde{dx}^\tau \\ &= [(x^\tau U_{\beta,\tau} + U_\alpha x^\alpha{}_{,\beta})_{,\gamma} - U_\alpha x^\alpha{}_{,\beta\gamma}]\widetilde{dx}^\gamma \wedge \widetilde{dx}^\beta .\end{aligned} \tag{2}$$

Equation (2) is obtained merely by relabeling indices. By symmetry of partial derivatives and antisymmetry of wedge product, the last term in brackets on the right hand side of Equation (2) vanishes. Comparing Equations (1) and (2), we have for 1-forms:

$$d(\mathcal{L}_x\tilde{U}) = \mathcal{L}_x(d\tilde{U}) .$$

The generalization of this method from 1-form to p-form is straightforward.

Solution 8.22.

(i) $$(\nabla S)^{\hat{i}} = \tilde{\boldsymbol{\omega}}^{\hat{i}} \cdot \nabla S = h_i \widetilde{dx}^i \cdot \nabla S = h_i\, \partial S/\partial x^i .$$

(ii) The general expression for the curl must be

$$(\nabla \times V)^i = |g|^{-\frac{1}{2}} \varepsilon^{ijk} V_{k;j} = |g|^{-\frac{1}{2}} \varepsilon^{ijk} V_{k,j}$$

since it is a covariant relation and it is true in a locally flat (cartesian) coordinate basis. (See Problems 3.20 and 3.21 for the relevant properties of the antisymmetric tensor ε^{ijk}.) In the orthonormal basis

$$(\nabla\times\mathbf{V})^{\hat{i}} = \tilde{\omega}^{\hat{i}}\cdot(\nabla\times\mathbf{V}) = h_i(\nabla\times\mathbf{V})^i = h_i(h_1h_2h_3)^{-1}\varepsilon^{ijk}\,\partial(h_kV_{\hat{k}})/\partial x^j\ .$$

(iii) Using the result in Problem 7.7(g), we have

$$\nabla\cdot\mathbf{V} = |g|^{-\frac{1}{2}}[|g|^{\frac{1}{2}}V^i]_{,i} = (h_1h_2h_3)^{-1}[(h_1h_2h_3)\,h_i^{-1}V_{\hat{i}}]_{,i}\ .$$

(iv) With the expression in Problem 7.7(j), $\nabla^2 S$ becomes

$$\nabla^2 S = |g|^{-\frac{1}{2}}(|g|^{\frac{1}{2}}g^{\alpha\beta}S_{,\beta})_{,\alpha} = (h_1h_2h_3)^{-1}[h_1h_2h_3h_i^{-2}S_{,i}]_{,i}\ .$$

Solution 8.23. We use results (ii) and (iii) of Problem 8.22 with $x^1 = r$, $x^2 = \theta$, $x^3 = \phi$ and with $h_1 = 1$, $h_2 = r$, $h_3 = r\ \sin\theta$. (For simplicity we follow the usual notation here and omit the carets indicating physical components.)

$$\begin{aligned}\nabla\cdot\underset{\sim}{\mathbf{A}} &= (r^2\sin\theta)^{-1}[r^2\sin\theta\, h_i^{-1}A_i]_{,i}\\ &= (r^2\sin\theta)^{-1}[(r^2\sin\theta\, A_r)_{,r} + (r\sin\theta\, A_\theta)_{,\theta} + rA_{\phi,\phi}]\end{aligned}$$

$$\begin{aligned}(\nabla\times\mathbf{A})^r &= (r\sin\theta)^{-1}[(\sin\theta\, A_\phi)_{,\theta} - A_{\theta,\phi}]\\ (\nabla\times\mathbf{A})^\theta &= (r\sin\theta)^{-1}[A_{r,\phi} - (r\sin\theta\, A_\phi)_{,r}]\\ (\nabla\times\mathbf{A})^\phi &= r^{-1}[(rA_\theta)_{,r} - A_{r,\theta}]\ .\end{aligned}$$

Solution 8.24. If forms and exterior differentiation are used, this problem is trivial. The first equation gives $\mathbf{F} = d\tilde{\mathbf{A}}$ so, $d\mathbf{F} = dd\tilde{\mathbf{A}} = 0$, since dd is always zero. $F_{[\mu\nu;\lambda]}$ is exactly the components of $d\mathbf{F}$, so we are done. The trick is that the total antisymmetrization implicit in the d's keeps any connection coefficient from entering the problem. If components are used, the proof is straightforward but more tedious.

Solution 8.25. In general, $ds^2 = g_{\alpha\beta}\,dx^\alpha dx^\beta$. Since x^i = constant along any of the geodesics, $ds^2 = g_{00}\,dt^2$ along any of the geodesics. But along the geodesics $ds^2 = -d\tau^2$, so $g_{00} = -1$ everywhere. Let $\mathbf{e}_\alpha$ be the coordinate basis vectors, and let $\mathbf{u} = d/d\tau$ be the tangent vector field to the geodesics (i.e. $\mathbf{u} = \mathbf{e}_0$). Since $\mathbf{u}$ is orthogonal to S_I, then $\mathbf{u}\cdot\mathbf{e}_i = \mathbf{e}_0\cdot\mathbf{e}_i = g_{0i} = 0$ at $\tau = 0$, and

$$d(\mathbf{u}\cdot\mathbf{e}_i)/d\tau = \nabla_{\mathbf{u}}(\mathbf{u}\cdot\mathbf{e}_i) = 0 + \mathbf{u}\cdot\nabla_{\mathbf{u}}\mathbf{e}_i = \mathbf{u}\cdot\nabla_{\mathbf{e}_i}\mathbf{u}$$

since the curves are geodesics ($\nabla_{\mathbf{u}}\mathbf{u} = 0$) and since $\mathbf{e}_i$ and $\mathbf{u}$ form a coordinate basis ($[\mathbf{e}_i, \mathbf{u}] = 0$). This last result vanishes

$$\mathbf{u}\cdot\nabla_{\mathbf{e}_i}\mathbf{u} = \frac{1}{2}\nabla_{\mathbf{e}_i}(\mathbf{u}\cdot\mathbf{u}) = 0$$

and therefore $\mathbf{u}\cdot\mathbf{e}_i = g_{0i} = 0$ everywhere.

Solution 8.26.

(a) From the point of view of modern differential geometry, $\bar{\Gamma}^{\lambda}{}_{\mu\nu}$ and $\Gamma^{\lambda}{}_{\mu\nu}$ are the component description of "machines" which map two vectors and a form into a number; i.e.

$$(\text{number}) = \Gamma^{\lambda}{}_{\mu\nu}\,\sigma_{\lambda}\,u^{\mu}\,v^{\nu} \equiv \langle\tilde{\boldsymbol{\sigma}}, \nabla_{\mathbf{v}}\mathbf{u}\rangle$$

where the covariant derivative is taken using the metric corresponding to whichever Γ is used. A tensor is a "machine" which is linear on all "slots" (i.e. in all its arguments). By itself Γ or $\bar{\Gamma}$ is not a tensor since

$$f\langle\tilde{\boldsymbol{\sigma}}, \nabla_{\mathbf{v}}\mathbf{u}\rangle \neq \langle\tilde{\boldsymbol{\sigma}}, \nabla_{\mathbf{v}}(f\mathbf{u})\rangle = f\langle\tilde{\boldsymbol{\sigma}}, \nabla_{\mathbf{v}}\mathbf{u}\rangle + \langle\tilde{\boldsymbol{\sigma}}, \mathbf{u}\rangle\nabla_{\mathbf{v}} f$$

and likewise with "bars" over the ∇'s for $\bar{g}$. If however we subtract the two equations, the linearity-spoiling $\langle\tilde{\boldsymbol{\sigma}}, \mathbf{u}\rangle(\nabla_{\mathbf{v}} - \bar{\nabla}_{\mathbf{v}})f$ term is identically zero since all covariant derivatives acting on a *scalar* function give the same result, i.e. are just directional derivatives. Thus $S^{\lambda}{}_{\mu\nu}$ *is* a tensor.

(b) We cannot compare the geodesic equations of g and $\bar{g}$ directly since they would involve different affine parameters τ and $\bar{\tau}$. If t is an arbitrary function of τ, a change of variable makes the geodesic equation

$$\frac{d^2x^\lambda}{dt^2} + \Gamma^\lambda{}_{\mu\nu}\frac{dx^\mu}{dt}\frac{dx^\nu}{dt} - \frac{dx^\lambda}{dt}\frac{d^2\tau}{dt^2}\frac{dt}{d\tau} = 0 \ .$$

To eliminate the $d\tau/dt$ term, multiply by dx^α/dt and antisymmetrize on α and λ to get

$$u^\alpha \frac{du^\lambda}{dt} - u^\lambda \frac{du^\alpha}{dt} + (\Gamma^\lambda{}_{\mu\nu}u^\alpha - \Gamma^\alpha{}_{\mu\nu}u^\lambda)u^\mu u^\nu = 0 \ ,$$

where $u^\alpha \equiv dx^\alpha/dt$. If the world lines $u^\alpha(t)$ are geodesics for both g and $\bar{g}$ then the above equation must also hold with Γ replaced by $\bar{\Gamma}$. Subtracting these two equations gives

$$(\delta^\alpha{}_\sigma S^\lambda{}_{\mu\nu} - \delta^\lambda{}_\sigma S^\alpha{}_{\mu\nu})u^\sigma u^\mu u^\nu = 0 \ ,$$

which implies that

$$\delta^\alpha{}_{(\sigma}S^\lambda{}_{\mu)\nu} - \delta^\lambda{}_{(\sigma}S^\alpha{}_{\mu)\nu} = 0 \ .$$

When this equation is contracted on α and ν the result is

$$S^\lambda{}_{\sigma\mu} = \delta^\lambda{}_\sigma\Phi_\mu + \delta^\lambda{}_\mu\Phi_\sigma$$

$$\Phi_\alpha \equiv \frac{1}{5}S^\mu{}_{\alpha\mu} \ .$$

[See also L. P. Eisenhart, *Riemannian Geometry* (Princeton University Press, 1962), Section 40.]

Solution 8.27. The difficult way to do this problem is to compute connection coefficients in a coordinate frame. Here we will use the method of "rotation 1-forms." (See MTW, Section 14.5 for details.) Choose an orthonormal frame as follows:

$$\tilde{\omega}^{\hat{t}} = e^\alpha\widetilde{dt}$$

$$\tilde{\omega}^{\hat{r}} = e^\beta\widetilde{dr}$$

$$\tilde{\omega}^{\hat{\theta}} = e^\gamma\widetilde{d\theta}$$

$$\tilde{\omega}^{\hat{\phi}} = e^\gamma \sin\theta\,\widetilde{d\phi} \ .$$

The key formulas are the definition of $\tilde{\omega}^{\hat{a}}{}_{\hat{\beta}}$

$$d\tilde{\omega}^{\hat{a}} = -\tilde{\omega}^{\hat{a}}{}_{\hat{\beta}} \wedge \tilde{\omega}^{\hat{\beta}} ,$$

and the relation of the forms to the Christoffel symbols

$$\Gamma^{\hat{\nu}}{}_{\hat{\mu}\hat{\lambda}}\,\tilde{\omega}^{\hat{\lambda}} = \tilde{\omega}^{\hat{\nu}}{}_{\hat{\mu}} .$$

We now use these formulas for the given metric; let dot denote $\partial/\partial t$ and prime denote $\partial/\partial r$:

$$d\tilde{\omega}^{\hat{t}} = e^{\alpha}\alpha'\,\widetilde{dr} \wedge \widetilde{dt} = \alpha' e^{-\beta}\tilde{\omega}^{\hat{r}} \wedge \tilde{\omega}^{\hat{t}} \tag{1}$$

$$d\tilde{\omega}^{\hat{r}} = e^{\beta}\dot{\beta}\,\widetilde{dt} \wedge \widetilde{dr} = e^{-\alpha}\dot{\beta}\,\tilde{\omega}^{\hat{t}} \wedge \tilde{\omega}^{\hat{r}} \tag{2}$$

$$\begin{aligned} d\tilde{\omega}^{\hat{\theta}} &= e^{\gamma}\dot{\gamma}\,\widetilde{dt} \wedge \widetilde{d\theta} + e^{\gamma}\gamma'\,\widetilde{dr} \wedge \widetilde{d\theta} \\ &= e^{-\alpha}\dot{\gamma}\,\tilde{\omega}^{\hat{t}} \wedge \tilde{\omega}^{\hat{\theta}} + e^{-\beta}\gamma'\,\tilde{\omega}^{\hat{r}} \wedge \tilde{\omega}^{\hat{\theta}} \end{aligned} \tag{3}$$

$$\begin{aligned} d\tilde{\omega}^{\hat{\phi}} &= e^{\gamma}\dot{\gamma}\sin\theta\,\widetilde{dt}\wedge\widetilde{d\phi} + e^{\gamma}\gamma'\sin\theta\,\widetilde{dr}\wedge\widetilde{d\phi} + e^{\gamma}\cos\theta\,\widetilde{d\theta}\wedge\widetilde{d\phi} \\ &= e^{-\alpha}\dot{\gamma}\,\tilde{\omega}^{\hat{t}}\wedge\tilde{\omega}^{\hat{\phi}} + e^{-\beta}\gamma'\,\tilde{\omega}^{\hat{r}}\wedge\tilde{\omega}^{\hat{\phi}} + \cot\theta\, e^{-\gamma}\tilde{\omega}^{\hat{\theta}}\wedge\tilde{\omega}^{\hat{\phi}} \end{aligned} \tag{4}$$

But we also know that $d\tilde{\omega}^{\hat{a}} = -\tilde{\omega}^{\hat{a}}{}_{\hat{\beta}} \wedge \tilde{\omega}^{\hat{\beta}}$ and $\tilde{\omega}_{\hat{a}\hat{\beta}} = -\tilde{\omega}_{\hat{\beta}\hat{a}}$, where indices are raised and lowered with $\eta_{\hat{a}\hat{\beta}}$. For $\hat{a} = \hat{t}$ we have

$$d\tilde{\omega}^{\hat{t}} = -\tilde{\omega}^{\hat{t}}{}_{\hat{r}} \wedge \tilde{\omega}^{\hat{r}} - \tilde{\omega}^{\hat{t}}{}_{\hat{\theta}} \wedge \tilde{\omega}^{\hat{\theta}} - \tilde{\omega}^{\hat{t}}{}_{\hat{\phi}} \wedge \tilde{\omega}^{\hat{\phi}} .$$

Comparing this with Equation (1), we guess that

$$\tilde{\omega}^{\hat{t}}{}_{\hat{r}} = \tilde{\omega}^{\hat{r}}{}_{\hat{t}} = \alpha' e^{-\beta}\tilde{\omega}^{\hat{t}} + K_1\tilde{\omega}^{\hat{r}} \tag{5}$$

$$\tilde{\omega}^{\hat{t}}{}_{\hat{\theta}} = \tilde{\omega}^{\hat{\theta}}{}_{\hat{t}} = K_2\tilde{\omega}^{\hat{\theta}} \tag{6}$$

$$\tilde{\omega}^{\hat{t}}{}_{\hat{\phi}} = \tilde{\omega}^{\hat{\phi}}{}_{\hat{t}} = K_3\tilde{\omega}^{\hat{\phi}} \tag{7}$$

where K_1, K_2, K_3 are functions we must find. Similarly, Equation (2) implies that

$$\tilde{\boldsymbol{\omega}}^{\hat{r}}{}_{\hat{t}} = e^{-\alpha}\dot{\beta}\tilde{\boldsymbol{\omega}}^{\hat{r}} + K_4\tilde{\boldsymbol{\omega}}^{\hat{t}} \tag{8}$$

$$\tilde{\boldsymbol{\omega}}^{\hat{r}}{}_{\hat{\theta}} = -\tilde{\boldsymbol{\omega}}^{\hat{\theta}}{}_{\hat{r}} = K_5\tilde{\boldsymbol{\omega}}^{\hat{\theta}} \tag{9}$$

$$\tilde{\boldsymbol{\omega}}^{\hat{r}}{}_{\hat{\phi}} = -\tilde{\boldsymbol{\omega}}^{\hat{\phi}}{}_{\hat{r}} = K_6\tilde{\boldsymbol{\omega}}^{\hat{\phi}}\,. \tag{10}$$

Equation (3) implies

$$\tilde{\boldsymbol{\omega}}^{\hat{\theta}}{}_{\hat{t}} = e^{-\alpha}\dot{\gamma}\tilde{\boldsymbol{\omega}}^{\hat{\theta}} \tag{11}$$

$$\tilde{\boldsymbol{\omega}}^{\hat{\theta}}{}_{\hat{r}} = e^{-\beta}\gamma'\tilde{\boldsymbol{\omega}}^{\hat{\theta}} \tag{12}$$

$$\tilde{\boldsymbol{\omega}}^{\hat{\theta}}{}_{\hat{\phi}} = -\tilde{\boldsymbol{\omega}}^{\hat{\phi}}{}_{\hat{\theta}} = K_7\tilde{\boldsymbol{\omega}}^{\hat{\phi}}\,, \tag{13}$$

and Equation (4) gives

$$\tilde{\boldsymbol{\omega}}^{\hat{\phi}}{}_{\hat{t}} = e^{-\alpha}\dot{\gamma}\tilde{\boldsymbol{\omega}}^{\hat{\phi}} \tag{14}$$

$$\tilde{\boldsymbol{\omega}}^{\hat{\phi}}{}_{\hat{r}} = e^{-\beta}\gamma'\tilde{\boldsymbol{\omega}}^{\hat{\phi}} \tag{15}$$

$$\tilde{\boldsymbol{\omega}}^{\hat{\phi}}{}_{\hat{\theta}} = \cot\theta\, e^{-\gamma}\tilde{\boldsymbol{\omega}}^{\hat{\phi}}. \tag{16}$$

Comparing Equations (5) - (16), we find

$$\begin{aligned}
\tilde{\boldsymbol{\omega}}^{\hat{t}}{}_{\hat{r}} &= \alpha' e^{-\beta}\tilde{\boldsymbol{\omega}}^{\hat{t}} + \dot{\beta}e^{-\alpha}\tilde{\boldsymbol{\omega}}^{\hat{r}} \\
\tilde{\boldsymbol{\omega}}^{\hat{t}}{}_{\hat{\theta}} &= e^{-\alpha}\dot{\gamma}\tilde{\boldsymbol{\omega}}^{\hat{\theta}} \\
\tilde{\boldsymbol{\omega}}^{\hat{t}}{}_{\hat{\phi}} &= e^{-\alpha}\dot{\gamma}\tilde{\boldsymbol{\omega}}^{\hat{\phi}} \\
\tilde{\boldsymbol{\omega}}^{\hat{r}}{}_{\hat{\theta}} &= -e^{-\beta}\gamma'\tilde{\boldsymbol{\omega}}^{\hat{\theta}} \\
\tilde{\boldsymbol{\omega}}^{\hat{r}}{}_{\hat{\phi}} &= -e^{-\beta}\gamma'\tilde{\boldsymbol{\omega}}^{\hat{\phi}} \\
\tilde{\boldsymbol{\omega}}^{\hat{\theta}}{}_{\hat{\phi}} &= -\cot\theta\, e^{-\gamma}\tilde{\boldsymbol{\omega}}^{\hat{\phi}}\,.
\end{aligned}$$

But $\Gamma^{\hat{\nu}}{}_{\hat{\mu}\hat{\lambda}}\tilde{\boldsymbol{\omega}}^{\hat{\lambda}} = \tilde{\boldsymbol{\omega}}^{\hat{\nu}}{}_{\hat{\mu}}$, so the nonvanishing connection coefficients are

$$\Gamma^{\hat{t}}{}_{\hat{r}\hat{t}} = \Gamma^{\hat{r}}{}_{\hat{t}\hat{t}} = \alpha' e^{-\beta}$$

$$\Gamma^{\hat{t}}{}_{\hat{r}\hat{r}} = \Gamma^{\hat{r}}{}_{\hat{t}\hat{r}} = \dot{\beta} e^{-\alpha}$$

$$\Gamma^{\hat{t}}{}_{\hat{\theta}\hat{\theta}} = \Gamma^{\hat{\theta}}{}_{\hat{t}\hat{\theta}} = \Gamma^{\hat{t}}{}_{\hat{\phi}\hat{\phi}} = \Gamma^{\hat{\phi}}{}_{\hat{t}\hat{\phi}} = \dot{\gamma} e^{-\alpha}$$

$$\Gamma^{\hat{r}}{}_{\hat{\theta}\hat{\theta}} = -\Gamma^{\hat{\theta}}{}_{\hat{r}\hat{\theta}} = \Gamma^{\hat{r}}{}_{\hat{\phi}\hat{\phi}} = -\Gamma^{\hat{\phi}}{}_{\hat{r}\hat{\phi}} = -\gamma' e^{-\beta}$$

$$\Gamma^{\hat{\theta}}{}_{\hat{\phi}\hat{\phi}} = -\Gamma^{\hat{\phi}}{}_{\hat{\theta}\hat{\phi}} = -\cot\theta\, e^{-\gamma} .$$

Solution 8.28. Start with $1+z = \Delta\tau_A/\Delta\tau_B$. To show that this is equivalent to the other definition, recall that p is proportional to $\mathbf{k}$, the wave vector, which is the gradient of the surfaces of constant phase:

$$\mathbf{k} = \nabla\theta .$$

Since $\mathbf{k}$ is a null vector,

$$0 = \mathbf{k}\cdot\mathbf{k} = \mathbf{k}\cdot\nabla\theta = \nabla_{\mathbf{k}}\theta$$

i.e. $\theta =$ constant along the rays. Consider two rays differing in phase by $\Delta\theta$. Observer A computes $\Delta\theta$ as

$$\Delta\theta = \theta(\tau+\Delta\tau_A) - \theta(\tau) = (\Delta\tau_A)\nabla_{\mathbf{u}_A}\theta = (\Delta\tau_A)\mathbf{u}_A\cdot\mathbf{k} .$$

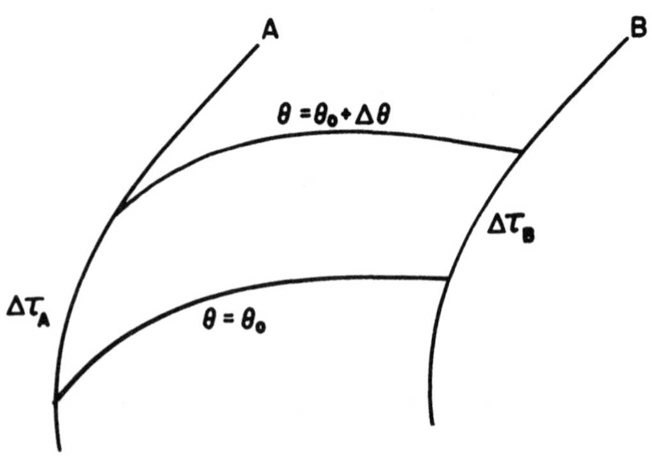

Similarly observer B finds

$$\Delta\theta = (\Delta\tau_B)\,\mathbf{u}_B\cdot\mathbf{k}$$

where $\mathbf{k}$ is now evaluated at B. Since $\Delta\theta$ is the same for A and B, we have

$$\frac{\Delta\tau_A}{\Delta\tau_B} = \frac{(\mathbf{u}\cdot\mathbf{k})_B}{(\mathbf{u}\cdot\mathbf{k})_A} = \frac{(\mathbf{u}\cdot\mathbf{p})_B}{(\mathbf{u}\cdot\mathbf{p})_A}$$

so the two definitions are equivalent.

CHAPTER 9: SOLUTIONS

Solution 9.1. If we construct a coordinate patch from geodesics we can then bisect that coordinate box with a geodesic diagonal, forming two geodesic triangles. The angular excess of a triangle made from great circles is $\pi\,[\text{Area}/a^2]$. Since the sum of the interior angles of the coordinate box will be the sum of the interior angles of the two triangles, the angular excess of the coordinate box is $\pi\,[\text{Area}/a^2]$.

The best coordinate patch constructed of lines of longitude and latitude is that near the equator. Clearly there is no angular excess, and $\overline{AC} = \overline{BD}$, but

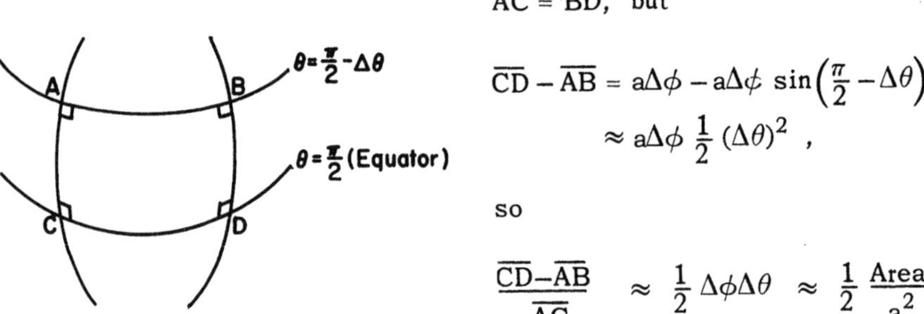

$$\overline{CD} - \overline{AB} = a\Delta\phi - a\Delta\phi \sin\left(\frac{\pi}{2} - \Delta\theta\right)$$
$$\approx a\Delta\phi \, \frac{1}{2}\,(\Delta\theta)^2 \ ,$$

so

$$\frac{\overline{CD}-\overline{AB}}{\overline{AC}} \approx \frac{1}{2}\,\Delta\phi\Delta\theta \approx \frac{1}{2}\,\frac{\text{Area}}{a^2}$$

In both these examples, the departure from flatness is measured by $1/a^2$, the curvature of the sphere.

Solution 9.2. The antisymmetry of $(\alpha\beta)$ and $(\gamma\delta)$ in $R_{\alpha\beta\gamma\delta}$ means that there are $M = \frac{1}{2}\,n(n-1)$ ways of choosing nontrivial pairs $(\alpha\beta)$ and similarly M ways of nontrivially choosing $(\gamma\delta)$. Since the tensor is symmetric with respect to interchange of the pairs $(\alpha\beta)$ and $(\gamma\delta)$ there are $\frac{1}{2}\,M(M+1)$ independent ways of choosing $\alpha\beta\gamma\delta$ when the pair symmetries are considered.

The Riemann tensor also possesses the cyclic symmetry

$$A_{\alpha\beta\gamma\delta} = R_{\alpha\beta\gamma\delta} + R_{\alpha\delta\beta\gamma} + R_{\alpha\gamma\delta\beta} = 0 \ .$$

The pair symmetries guarantee that $A_{\alpha\beta\gamma\delta}$ is totally antisymmetric so that the constraint $A_{\alpha\beta\gamma\delta} = 0$ is trivial unless $\alpha, \beta, \gamma, \delta$ are all distinct. The number of added constraints is then the number of combinations of 4 objects which can be taken from n objects $= n!/(n-4)!\,4!$ (This formula gives zero, and is therefore valid, if $n < 4$ in which case the permutation symmetry gives no additional constraints.) The number of independent components is then

$$\frac{1}{2}\,M(M+1) - n!/(n-4)!\,4! = n^2(n^2-1)/12 \ .$$

Solution 9.3. The problem requires ordering 21 independent components in some canonical way and assigning a number $N \leq 21$ to each nonequivalent quadruple I, J, K, L. The following program accomplishes this and reads the value of R(I, J, K, L) from R(N), $1 \leq N \leq 21$.

```
READ I, J, K, L
SET SIGN = 1
IF I = J OR K = L SET RIEMANN = 0 AND RETURN
IF I > J EXCHANGE I, J AND SET SIGN = -1
IF K > L EXCHANGE K, L AND SET SIGN = - SIGN
SET N1 = (5*I - I*I)/2 + J
SET N2 = (5*K - K*K)/2 + L
IF N1 > N2 EXCHANGE N1, N2
SET N = (13*N1 - N1*N1-12)/2 + N2
SET RIEMANN = R(N)*SIGN
RETURN.
```

Solution 9.4. According to the discussion in Problem 9.2, there is only one independent component. We can choose it to be

$$R_{\theta\phi\theta\phi} = g_{\theta\theta}R^{\theta}{}_{\phi\theta\phi} = g_{\theta\theta}\left\{\Gamma^{\theta}{}_{\phi\phi,\theta} - \Gamma^{\theta}{}_{\theta\phi,\phi} + \Gamma^{\theta}{}_{\theta\alpha}\Gamma^{\alpha}{}_{\phi\phi} - \Gamma^{\theta}{}_{\phi\alpha}\Gamma^{\alpha}{}_{\phi\theta}\right\}.$$

From the metric $g_{\theta\theta} = r^2$, $g_{\phi\phi} = r^2 \sin^2\theta$ we find that the only nonvanishing Christoffel symbols are

$$\Gamma^{\theta}{}_{\phi\phi} = -\sin\theta\cos\theta, \qquad \Gamma^{\phi}{}_{\theta\phi} = \cot\theta ,$$

and, therefore,

$$R_{\theta\phi\theta\phi} = r^2\sin^2\theta = R_{\phi\theta\phi\theta} = -R_{\theta\phi\phi\theta} = -R_{\phi\theta\theta\phi} \cdot$$

Solution 9.5. The only nonvanishing Christoffel symbols are $\Gamma^{v}{}_{uu} = v$ and $\Gamma^{u}{}_{uv} = \Gamma^{u}{}_{vu} = v^{-1}$, so that

$$\begin{aligned} R_{vuvu} = R^{v}{}_{uvu} &= \Gamma^{v}{}_{uu,v} - \Gamma^{v}{}_{vu,u} + \Gamma^{v}{}_{v\alpha}\Gamma^{\alpha}{}_{uu} - \Gamma^{v}{}_{u\alpha}\Gamma^{\alpha}{}_{uv} \\ &= 1 - 0 + 0 - 1 = 0 . \end{aligned}$$

Since there is only one independent Riemann component in two dimensions we conclude that $R_{\alpha\beta\gamma\delta} = 0$, and the spacetime is flat. (This was also shown by a coordinate transformation in Problem 6.1.)

Solution 9.6. From the diagram, two orthogonal infinitesimal displacements have length $a d\phi$ and $(b + a\sin\phi)d\theta$ so $ds^2 = a^2 d\phi^2 + (b + a\sin\phi)^2 d\theta^2$ or $g_{\theta\theta} = (b + a\sin\phi)^2$, $g_{\phi\phi} = a^2$, $g_{\theta\phi} = g_{\phi\theta} = 0$. The formula for the Christoffel symbols now gives the nonvanishing Γ's directly

$$\Gamma^{\phi}{}_{\theta\theta} = -a^{-1}(b + a\sin\phi)\cos\phi$$

$$\Gamma^{\theta}{}_{\theta\phi} = \Gamma^{\theta}{}_{\phi\theta} = a\cos\phi\,(b + a\sin\phi)^{-1} .$$

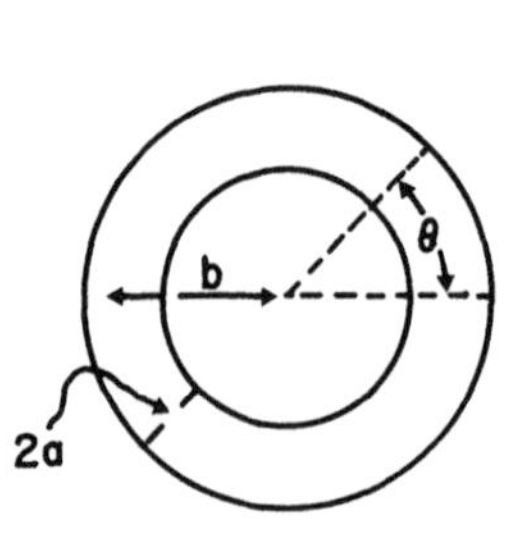

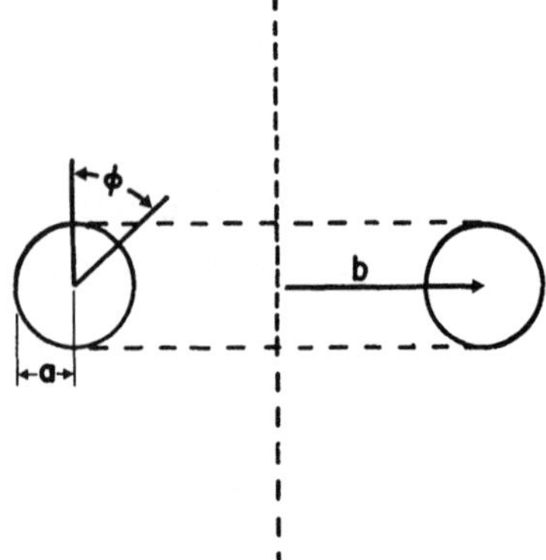

In 2 dimensions the Riemann tensor has only one independent component. Take it to be $R_{\phi\theta\phi\theta}$, then the standard formula gives

$$R_{\phi\theta\phi\theta} = a \sin\phi (b + a \sin\phi) \ .$$

Solution 9.7.

(a) In one dimension there can only be one component R_{1111}, and by the symmetries of the Riemann tensor, this component must vanish.

(b) In two dimensions the relations, due to symmetries, among the components of the Riemann tensor may be summarized in any coordinate system as

$$R_{\alpha\beta\gamma\delta} = (g_{\alpha\gamma} g_{\beta\delta} - g_{\alpha\delta} g_{\beta\gamma}) f$$

but

$$R = R^{\alpha\beta}{}_{\alpha\beta} = (g^{\alpha}{}_{\alpha} g^{\beta}{}_{\beta} - g^{\alpha}{}_{\beta} g^{\beta}{}_{\alpha}) f = (4-2) f = 2f \ .$$

Thus, in any coordinate system

$$R_{\alpha\beta\gamma\delta} = \frac{1}{2} (g_{\alpha\gamma} g_{\beta\delta} - g_{\alpha\delta} g_{\beta\gamma}) R \ .$$

(c) In three dimensions there are six independent components of the Riemann tensor. Since the independent components of the Ricci tensor represent six independent linear combinations of the Riemann components, we should be able to invert such a relation to get the independent Riemann components in terms of the Ricci components.

To have the correct symmetries, we must have

$$\begin{aligned} R_{\mu\nu\lambda\sigma} &= a(g_{\mu\lambda} R_{\nu\sigma} - g_{\nu\lambda} R_{\mu\sigma} - g_{\mu\sigma} R_{\nu\lambda} + g_{\nu\sigma} R_{\mu\lambda}) \\ &\quad + b(g_{\mu\lambda} g_{\nu\sigma} - g_{\mu\sigma} g_{\nu\lambda}) R \ . \end{aligned}$$

Now we find a and b by contracting:

$$R_{\nu\sigma} = a(3R_{\nu\sigma} - R_{\nu\sigma} - R_{\nu\sigma} + g_{\nu\sigma} R) + b(3g_{\nu\sigma} - g_{\nu\sigma}) R \ .$$

So we must have

$$a = 1 \qquad b = -\frac{1}{2} \ .$$

Contract again to check:

$$R = a(R+3R) + b \cdot 6R = (4-3)R \ .$$

Solution 9.8. We give the proof for a tensor of arbitrary rank. Choose locally flat coordinates so that all Christoffel symbols (but not their derivatives!) vanish at a point. At that point

$$R^{\lambda}{}_{\gamma\nu\kappa} = \Gamma^{\lambda}{}_{\gamma\kappa,\nu} - \Gamma^{\lambda}{}_{\gamma\nu,\kappa} = -2\Gamma^{\lambda}{}_{\gamma[\nu,\kappa]}$$

and

$$(T_{\alpha\cdots}{}^{\beta\cdots})_{;\nu\kappa} = (T_{\alpha\cdots}{}^{\beta\cdots}{}_{,\nu} + T_{\alpha\cdots}{}^{\sigma\cdots}\Gamma^{\beta}{}_{\sigma\nu} + \cdots - T_{\sigma\cdots}{}^{\beta\cdots}\Gamma^{\sigma}{}_{\alpha\nu} - \cdots)_{,\kappa}$$
$$= T_{\alpha\cdots}{}^{\beta\cdots}{}_{,\nu\kappa} + T_{\alpha\cdots}{}^{\sigma\cdots}\Gamma^{\beta}{}_{\sigma\nu,\kappa} + \cdots - T_{\sigma\cdots}{}^{\beta\cdots}\Gamma^{\sigma}{}_{\alpha\nu,\kappa} - \cdots \ .$$

It follows that

$$2(T_{\alpha\cdots}{}^{\beta\cdots})_{;[\nu\kappa]} = -T_{\alpha\cdots}{}^{\sigma\cdots}R^{\beta}{}_{\sigma\nu\kappa} - \cdots + T_{\sigma\cdots}{}^{\beta\cdots}R^{\sigma}{}_{\alpha\nu\kappa} + \cdots \ .$$

Since this relation is covariant it is true in any coordinate system.

Solution 9.9. The second derivates of a scalar are

$$S_{;\alpha\beta} = (S_{,\alpha})_{;\beta} = S_{,\alpha\beta} - S_{,\sigma}\Gamma^{\sigma}{}_{\alpha\beta} \ .$$

This expression is obviously symmetric in α and β.

To investigate third derivatives use local flat coordinates as in Problem 9.8. We have then

$$S_{;\alpha\beta\gamma} = (S_{,\alpha\beta} - S_{\sigma}\Gamma^{\sigma}{}_{\alpha\beta})_{,\gamma} = S_{,\alpha\beta\gamma} - S_{\sigma}\Gamma^{\sigma}{}_{\alpha\beta,\gamma} = S_{;(\alpha\beta)\gamma} \ .$$

From this expression and from the result of Problem 9.8 we also have

$$S_{;\alpha[\beta\gamma]} = \frac{1}{2} S_{;\sigma}R^{\sigma}{}_{\alpha\beta\gamma} \ .$$

Solution 9.10. Form the tensor

$$C^{\alpha\beta}{}_{\mu\nu} \equiv A^{\alpha\beta}{}_{;\mu\nu} - A^{\alpha\beta}{}_{;\nu\mu} \ .$$

From Problem 9.8 we have

$$C^{\alpha\beta}{}_{\mu\nu} = -A^{\sigma\beta}R^{\alpha}{}_{\sigma\mu\nu} - A^{\alpha\sigma}R^{\beta}{}_{\sigma\mu\nu}$$

so that

$$C^{\mu\nu}{}_{\mu\nu} = -A^{\sigma\nu}R^{\mu}{}_{\sigma\mu\nu} - A^{\mu\sigma}R^{\nu}{}_{\sigma\mu\nu} = -A^{\sigma\nu}R_{\sigma\nu} + A^{\mu\sigma}R_{\sigma\mu} = 0 .$$

Solution 9.11. Choose locally flat coordinates at the starting point P. Thus $\Gamma^{\alpha}{}_{\beta\gamma} = 0$ at P but $\Gamma^{\alpha}{}_{\beta\gamma} = \Gamma^{\alpha}{}_{\beta\gamma,\delta}u^{\delta}$ after displacement **u**. When we displace **A** from P by **u** we find that its components become

$$\begin{aligned}[\mathbf{A}\ (\text{displaced by }\mathbf{u})]^{\alpha} &= A^{\alpha} + A^{\alpha}{}_{,\beta}u^{\beta} \\ &= A^{\alpha} + A^{\alpha}{}_{;\beta}u^{\beta} - \Gamma^{\alpha}{}_{\gamma\beta}A^{\gamma}u^{\beta} = A^{\alpha} .\end{aligned}$$

When we displace, again, by **v** the new components are

$$\begin{aligned}[\mathbf{A}\ (\text{displaced by }\mathbf{u}\text{ then }\mathbf{v})]^{\alpha} &= A^{\alpha} + A^{\alpha}{}_{;\beta}v^{\beta} - \Gamma^{\alpha}{}_{\gamma\beta}A^{\gamma}v^{\beta} \\ &= A^{\alpha} - \Gamma^{\alpha}{}_{\gamma\beta}A^{\gamma}v^{\beta} = A^{\alpha} - \Gamma^{\alpha}{}_{\gamma\beta,\delta}A^{\gamma}v^{\beta}u^{\delta} .\end{aligned}$$

We could have reached this same point by travelling around the other side of the parallelogram to find, similarly

$$[\mathbf{A}\ (\text{displaced by }-\mathbf{v}\text{ then }-\mathbf{u})]^{\alpha} = A^{\alpha} - \Gamma^{\alpha}{}_{\gamma\beta,\delta}A^{\gamma}u^{\beta}v^{\delta} .$$

The change in **A** in making the complete circuit is our first result minus our second

$$\delta A^{\alpha} = -\Gamma^{\alpha}{}_{\gamma\beta,\delta}A^{\gamma}v^{\beta}u^{\delta} + \Gamma^{\alpha}{}_{\gamma\beta,\delta}A^{\gamma}u^{\beta}v^{\delta} = -R^{\alpha}{}_{\gamma\delta\beta}A^{\gamma}v^{\beta}u^{\delta} .$$

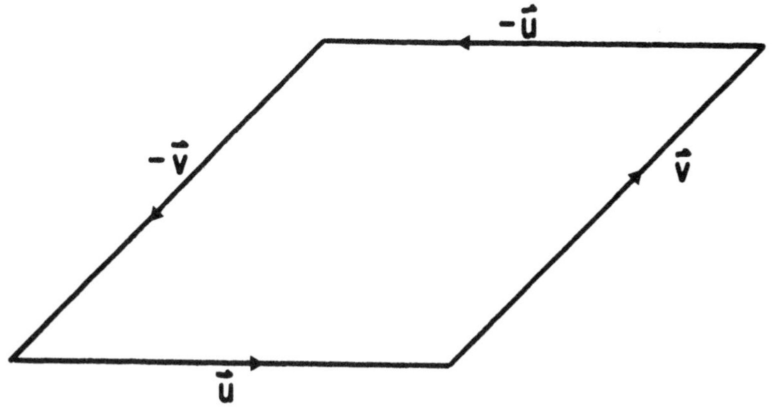

Solution 9.12.

(a) First consider the linearity in A. It is clear that

$$R(A_1 + A_2, B)C = R(A_1, B)C + R(A_2, B)C \ .$$

Furthermore,

$$\nabla_{fA}\nabla_B - \nabla_B\nabla_{fA} - \nabla_{[fA,B]}$$

$$= f\nabla_A\nabla_B - f\nabla_B\nabla_A - (\nabla_B f)\nabla_A - \nabla_{f[A,B]} - A\nabla_B f \ . \qquad (1)$$

The two "extra" terms cancel and hence

$$R(fA, B)C = fR(A, B)C \ . \qquad (2)$$

It is easy now to prove that the value of $R(A, B)C$ at a point P depends only on the value of A at P, *not* on the way in which A varies: Consider $f(P) = 0$; then Equation (2) tells us that $R = 0$ at P whenever the first argument vanishes at P. Now let A_1 and A_2 be two vector fields which are the same at P but different elsewhere. From (1) we know

$$R(A_1, B)C - R(A_2, B)C = R(A_1 - A_2, B)C = 0 \ .$$

This completes the proof for A and, by symmetry, for B. The proof for C follows similarly.

(b) Since only the values of A, B, C *at* point P are important, we can arbitrarily choose all three to be covariantly constant in a neighborhood of P; then we have

$$(R(A, B)C)^\alpha = (\nabla_A\nabla_B C - \nabla_B\nabla_A C)^\alpha$$
$$= 2A^{[\sigma}(B^{\lambda]}C^\alpha{}_{;\lambda})_{;\sigma} = 2A^\sigma B^\lambda C^\alpha{}_{;[\lambda;\sigma]} \ .$$

According to Problem 9.8, this is equal to

$$R^\alpha{}_{\mu\lambda\sigma}C^\mu A^\lambda B^\sigma$$

and the proof is complete.

Solution 9.13. Let $x^\mu = x^\mu(\lambda, n)$ be a family of geodesics with affine parameter λ. The parameter n labels the individual geodesics. The tangent to the geodesics is $u^\alpha \equiv \partial x^\alpha/\partial\lambda$ and the differential vector which connects points of equal affine parameter on adjacent curves is

$$n^\alpha = \partial x^\alpha/\partial n \ .$$

The solution to this problem is most transparent in abstract notation: Write

$$\mathbf{u} = \partial/\partial\lambda, \qquad \mathbf{n} = \partial/\partial n \ .$$

Note that $[\mathbf{u}, \mathbf{n}] = 0$ since the partial derivatives commute. (The vanishing of this commutator in fact is the *definition* of $\mathbf{n}$ as the connecting vector, according to Problem 8.14.) Now use the Riemann operator (see Problem 9.12):

$$R(\mathbf{u},\mathbf{n})\mathbf{u} \equiv \nabla_\mathbf{u}\nabla_\mathbf{n}\mathbf{u} - \nabla_\mathbf{n}\nabla_\mathbf{u}\mathbf{u} - \nabla_{[\mathbf{u},\mathbf{n}]}\mathbf{u} = \nabla_\mathbf{u}\nabla_\mathbf{n}\mathbf{u} = \nabla_\mathbf{u}\nabla_\mathbf{u}\mathbf{n} \ ,$$

where the last equality follows from the vanishing of $[\mathbf{u}, \mathbf{n}]$. In components our result, according to Problem 9.12(b), reads

$$\frac{D^2 n^\alpha}{d\lambda^2} = (\nabla_\mathbf{u}\nabla_\mathbf{u}\mathbf{n})^\alpha = -R^\alpha{}_{\beta\gamma\delta}u^\beta n^\gamma u^\delta \ .$$

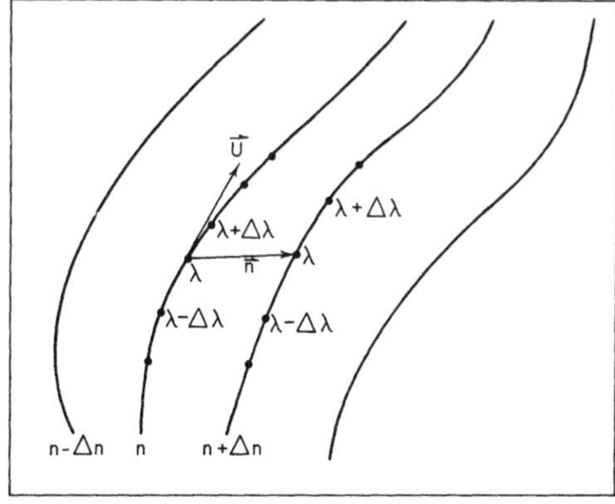

Solution 9.14. It is convenient to transform the spatial coordinates to spherical polars and to write the metric as

$$ds^2 = -(1-2M/r)\,dt^2 + (1+2M/r)(dr^2 + r^2[d\theta^2 + \sin^2\theta\, d\phi^2])\,.$$

For a circular equatorial orbit, $u^r = u^\theta = 0$ so

$$Du^r/d\tau = 0 = u^\alpha \Gamma^r{}_{\alpha\beta} u^\beta = (u^0)^2\, \Gamma^r{}_{00} + (u^\phi)^2\, \Gamma^r{}_{\phi\phi}$$

and therefore

$$(d\phi/dt)^2 = \omega^2 = -\Gamma^r{}_{00}/\Gamma^r{}_{\phi\phi}\,.$$

The Christoffel symbols are easily calculated and we find

$$\omega^2 = (2\pi/\text{Period})^2 = M/r^3\,.$$

This is the same as the Newtonian result. In Problem 9.13 the equation of motion is derived for $\mathbf{n}$, the vector joining points of equal proper time on two geodesics. We are required in this problem to deal with $\underset{\sim}{\xi}$, the separation of points on the two geodesics at a given coordinate time. From the diagram, however, it is clear that the fractional difference between $\underset{\sim}{n}$ and $\underset{\sim}{\xi}$ is proportional to the relative velocity of garbage and the Skylab, and hence can be ignored to lowest order.

Since all the Christoffel symbols in the t, x, y, z system are of order M/r^2, and $d/d\tau$ is of order $\omega \sim (M/r)^{\frac{1}{2}} r^{-1}$, we can approximate

$$\frac{D}{d\tau} = \frac{d}{d\tau} + \Gamma u \approx \frac{d}{d\tau}\,.$$

Furthermore

$$\frac{d}{d\tau} = u^0 \frac{d}{dt} = \left[1 + \mathcal{O}\left(\frac{M}{r}\right)\right] \frac{d}{dt}\,,$$

so we can approximate $D^2/d\tau^2 \approx d^2/dt^2$. Using these approximations and the equation of geodesic deviation we have

$$\frac{d^2\xi^i}{dt^2} + R^i{}_{0j0}\,(u^0)^2\, \xi^j = 0 \qquad i = x, y, z\,.$$

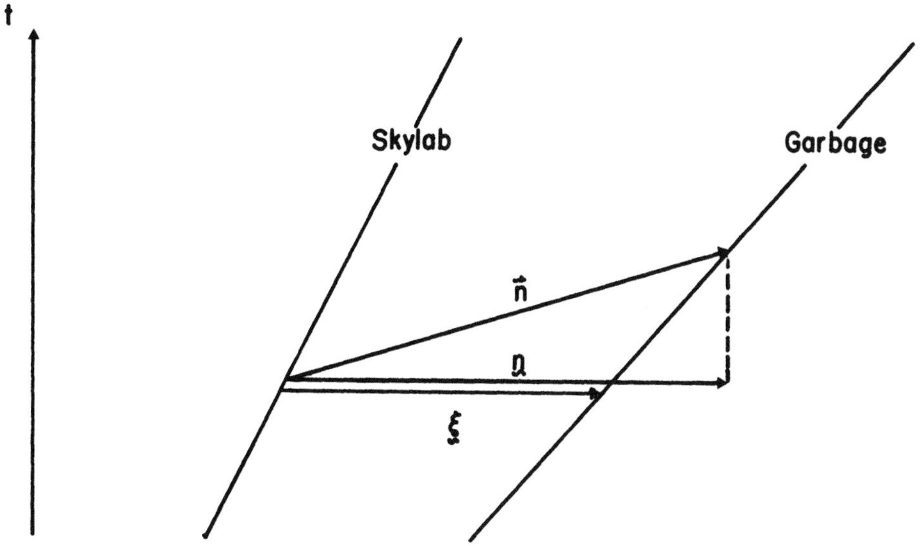

To lowest order the Riemann components are

$$R^i{}_{0j0} \approx \Gamma^i{}_{00,j} - \Gamma^i{}_{0j,0} = -\frac{1}{2}\, g_{00,ij} = \frac{M}{r^3}\left(\delta_{ij} - \frac{3x^i x^j}{r^2}\right).$$

If we choose to describe the Skylab orbit as

$$x = r\cos\omega t, \qquad y = r\sin\omega t\ ,$$

then the equations of motion for $\underset{\sim}{\xi}$ become

$$\ddot{\xi}^z + \omega^2 \xi^z = 0$$

$$\ddot{\xi}^x + \omega^2 \xi^x = 3\omega^2 \cos\omega t\,(\cos\omega t\ \xi^x + \sin\omega t\ \xi^y)$$

$$\ddot{\xi}^y + \omega^2 \xi^y = 3\omega^2 \sin\omega t\,(\cos\omega t\ \xi^x + \sin\omega t\ \xi^y)\ .$$

The relative motion in the z-direction is obviously

$$\xi^z \propto \sin\omega t$$

if the garbage was jettisoned at $t = 0$.

To find the relative motion in the xy plane introduce new variables η^1, η^2 defined by

$$\xi^x = \eta^1 \cos\omega t + \eta^2 \sin\omega t, \qquad \xi^y = \eta^1 \sin\omega t - \eta^2 \cos\omega t$$

and the equations of motion become:

$$\ddot{\eta}^2 - 2\omega\dot{\eta}^1 = 0$$

$$\ddot{\eta}^1 + 2\omega\dot{\eta}^2 - 3\omega^2\eta^1 = 0 \, .$$

These can be solved quite easily to find the four independent solutions:

$$[\eta^1, \eta^2] = [1, \tfrac{3}{2}\,\omega t], \quad [0, 1], \quad [\cos\omega t, 2\sin\omega t], \quad [\sin\omega t, -2\cos\omega t] \, .$$

The linear combination of solutions which correspond to the garbage jettisoned at $t = 0$ (i.e. $\xi^x = \xi^y = 0$ at $t = 0$) is

$$\xi^x = A[\cos 2\omega t - 3 + 2\cos\omega t + 3\omega t \sin\omega t] + B[4\sin\omega t - \sin 2\omega t]$$

$$\xi^y = A[\sin 2\omega t + 2\sin\omega t - 3\omega t\cos\omega t] + B[\cos 2\omega t + 3 - 4\cos\omega t] \, .$$

The constants A and B depend on the x and y components of the velocity with which the garbage was jettisoned. The nonperiodic terms in the solution correspond to the fact that the two orbits are of a slightly different period, so the relative distance will exhibit a secular growth in time.

Solution 9.15. Choose locally flat coordinates so that

$$R^\alpha{}_{\beta\gamma\delta} = 2\Gamma^\alpha{}_{\beta[\delta,\gamma]} \, .$$

In these coordinates then

$$R^\alpha{}_{[\beta\gamma\delta]} = 2\Gamma^\alpha{}_{[\beta[\delta,\gamma]]} = 2\Gamma^\alpha{}_{[\beta\delta,\gamma]} \, .$$

Since the Christoffel symbols are symmetric on their lower indices the right side must vanish, and by lowering the first index we have

$R_{\alpha[\beta\gamma\delta]} = 0$. The Riemann tensor is antisymmetric in its last two indices, so this is equivalent to the cyclic identity.

From Problem 9.8 we have for any U^α that

$$2U^\alpha{}_{;[\beta\gamma]} = -U^\sigma R^\alpha{}_{\sigma\beta\gamma} \; .$$

Take U^α to be fixed at some point and parallel transported along geodesics to all nearby points in some neighborhood, then at the point $U^\alpha{}_{;\beta} = 0$ for all β, and hence

$$2U^\alpha{}_{;[\beta\gamma]\delta} = -U^\sigma R^\alpha{}_{\sigma\beta\gamma;\delta}$$

and

$$2U^\alpha{}_{;[[\beta\gamma]\delta]} = 2U^\alpha{}_{;[\beta\gamma\delta]} = -U^\sigma R^\alpha{}_{\sigma[\beta\gamma;\delta]} \; .$$

From Problem 9.8 we have for the tensor $U^\alpha{}_{;\beta}$ that

$$2U^\alpha{}_{;\beta[\gamma\delta]} = -U^\sigma{}_{;\beta} R^\alpha{}_{\sigma\gamma\delta} + U^\alpha{}_{;\sigma} R^\sigma{}_{\beta\gamma\delta} \; .$$

Since $U^\alpha{}_{;\beta} = 0$ at the point we have $U^\alpha{}_{;\beta[\gamma\delta]} = 0$ and hence $U^\alpha{}_{;[\beta\gamma\delta]} = 0$. We could have chosen the direction of U^α at the point in any way so it follows that

$$R^\alpha{}_{\sigma[\beta\gamma;\delta]} = 0 \; .$$

Since the Riemann tensor is antisymmetric on its last two indices, this is equivalent to the Bianchi identities. (The Bianchi identities can also, of course, be derived by the straightforward but tedious process of differentiating and manipulating the formula for Riemann components in terms of Christoffel symbols. This is best done in locally flat coordinates.)

Solution 9.16. By contracting the Bianchi identities,

$$R^\alpha{}_{\beta\gamma\delta;\epsilon} + R^\alpha{}_{\beta\epsilon\gamma;\delta} + R^\alpha{}_{\beta\delta\epsilon;\gamma} = 0 \; ,$$

first on α and γ then on β and δ we get

$$R_{\beta\delta;\epsilon} + R_{\beta\epsilon;\delta} + R^{\alpha}{}_{\beta\delta\epsilon;\alpha} = 0$$

$$R_{;\epsilon} - R^{\beta}{}_{\epsilon;\beta} - R^{\alpha}{}_{\epsilon;\alpha} = 0 \; .$$

Thus the divergence of the Einstein tensor

$$G^{\nu}{}_{\epsilon;\nu} = \left(R^{\nu}{}_{\epsilon} - \tfrac{1}{2}\,\delta_{\epsilon}{}^{\nu}R\right)_{;\nu} = R^{\nu}{}_{\epsilon;\nu} - \tfrac{1}{2}\,R_{;\epsilon}$$

must vanish.

Solution 9.17. Choose $\mathbf{A}, \mathbf{B}, \mathbf{C}, \mathbf{D}$ to be mutually orthogonal at a point P and to satisfy $\mathbf{A}\cdot\mathbf{A} = -1$, $\mathbf{B}\cdot\mathbf{B} = \mathbf{C}\cdot\mathbf{C} = \mathbf{D}\cdot\mathbf{D} = 1$. Since spacetime is Riemann flat, parallel transport around a closed curve does not change a vector, so we can define $\mathbf{A}, \mathbf{B}, \mathbf{C}, \mathbf{D}$ away from P by parallel transport. If we do this we have e.g.

$$\frac{\partial A_\alpha}{\partial x^\beta} = -\Gamma^{\gamma}{}_{\alpha\beta}\,A_\gamma$$

therefore

$$\frac{\partial A_\alpha}{\partial x^\beta} = \frac{\partial A_\beta}{\partial x^\alpha}$$

so we can choose $A_\alpha = \phi_{,\alpha}$, and similarly for $\mathbf{B}, \mathbf{C}$, and $\mathbf{D}$. We can summarize by defining

$$\mathbf{A} = \mathbf{W}^{(0)}, \quad \mathbf{B} = \mathbf{W}^{(1)}, \quad \mathbf{C} = \mathbf{W}^{(2)}, \quad \mathbf{D} = \mathbf{W}^{(3)}$$

and we can define four functions $\Phi^{(0)}, \Phi^{(1)}, \Phi^{(2)}, \Phi^{(3)}$ such that

$$\mathbf{W}^{(\mu)} = \nabla\Phi^{(\mu)} \; .$$

Since inner products are preserved by parallel transport

$$\mathbf{W}^{(\mu)}\cdot\mathbf{W}^{(\nu)} = [\mathbf{W}^{(\mu)}\cdot\mathbf{W}^{(\nu)}]_P = \eta^{\mu\nu} \; ,$$

and this tells us that

$$g^{\mu\nu}\,\frac{\partial\Phi^{(\alpha)}}{\partial x^\mu}\,\frac{\partial\Phi^{(\beta)}}{\partial x^\nu} = \eta^{\alpha\beta}$$

and hence there exists a coordinate transformation, namely

$$\bar{x}^{\mu} = \Phi^{(\mu)}(x^{\mu})$$

which takes the metric to the Minkowskian form.

Solution 9.18. If the Weyl tensor vanishes, the metric can be written in the conformally flat form

$$ds^2 = e^{2\phi} ds_0^2 = e^{2\phi} \eta_{\mu\nu} dx^{\mu} dx^{\nu}$$

where ϕ is a function of x^{α}. The null geodesics

$$dt = \underset{\sim}{n} \cdot \underset{\sim}{dx}, \qquad \underset{\sim}{n} \cdot \underset{\sim}{n} = 1$$

of ds_0^2 remain null curves for ds^2. If we can show that they remain null geodesics, then clearly a cone of null geodesics, will remain a cone; there will be no preferred direction for squashing of the cross section.

To show the curves $dt = \underset{\sim}{n} \cdot \underset{\sim}{dx}$ are geodesics we need only show that

$$\frac{d^2 x^{\alpha}}{d\lambda^2} = \frac{dp^{\alpha}}{d\lambda} = 0$$

for null geodesics. Note that the Christoffel symbols are

$$\Gamma^{\alpha}{}_{\gamma\beta} = \eta^{\alpha\mu}[-\phi_{,\mu}\eta_{\gamma\beta} + \phi_{,\beta}\eta_{\gamma\mu} + \phi_{,\gamma}\eta_{\mu\beta}]$$

so that

$$p^{\gamma} p^{\beta} \Gamma^{\alpha}{}_{\gamma\beta} = 2p^{\alpha}(\nabla\phi \cdot p)$$

and

$$0 = \frac{dp^{\alpha}}{d\lambda} + \Gamma^{\alpha}{}_{\beta\gamma} p^{\beta} p^{\gamma} = \frac{dp^{\alpha}}{d\lambda} + 2p^{\alpha}(\nabla\phi \cdot p) .$$

By rescaling λ, then we have $dp^{\alpha}/d\lambda = 0$ and the straight null lines of ds_0^2 are null geodesics of ds^2.

Notice that, while beams of light have undistorted cross section the cross sectional area, like all areas, is affected by the scaling factor $e^{2\phi}$.

Since ϕ is in general anistropic, beams propagating in different directions will grow in cross section at different rates.

Section 9.19. In a conformally flat spacetime the Weyl tensor vanishes, so the Riemann tensor can easily be computed from the Ricci tensor

$$R_{\mu\nu} = -(\log |g|^{\frac{1}{2}})_{,\mu\nu} + \Gamma^{a}{}_{\mu\nu,a} + (\log |g|^{\frac{1}{2}})_{,a}\Gamma^{a}{}_{\mu\nu} - \Gamma^{\beta}{}_{\mu a}\Gamma^{a}{}_{\nu\beta} \,.$$

The expressions we need are

$$\Gamma^{a}{}_{\gamma\beta} = -\phi^{,a}\eta_{\gamma\beta} + \phi_{,\beta}\delta^{a}{}_{\gamma} + \phi_{,\gamma}\delta^{a}{}_{\beta}$$

$$\Gamma^{a}{}_{\gamma\beta,a} = -\nabla^{2}\phi\,\eta_{\gamma\beta} + 2\phi_{,\beta\gamma}$$

$$\Gamma^{a}{}_{\gamma\beta}\Gamma^{\gamma}{}_{\mu a} = 6\phi_{,\mu}\phi_{,\beta} - 2(\nabla\phi)^{2}\eta_{\mu\beta}$$

$$-g = e^{8\phi}, \qquad \log |g|^{\frac{1}{2}} = 4\phi$$

$$(\log |g|^{\frac{1}{2}})_{,a}\Gamma^{a}{}_{\gamma\beta} = -4(\nabla\phi)^{2}\eta_{\gamma\beta} + 8\phi_{,\beta}\phi_{,\gamma}$$

$$(\log |g|^{\frac{1}{2}})_{,\gamma\beta} = 4\phi_{,\gamma\beta} \,.$$

In the above expressions raising and lowering have been done with $\eta_{\mu\nu}$, e.g.

$$\phi^{,a} \equiv \eta^{\mu a}\phi_{,\mu}\,; \quad \nabla^{2}\phi \equiv \eta^{\mu a}\phi_{,\mu a}\,; \quad (\nabla\phi)^{2} \equiv \eta^{\mu a}\phi_{,\mu}\phi_{,a} \,.$$

The Ricci tensor and scalar are then

$$R_{\mu\nu} = -2\phi_{,\mu\nu} + 2\phi_{,\mu}\phi_{,\nu} - \eta_{\mu\nu}[\nabla^{2}\phi + 2(\nabla\phi)^{2}]$$

$$R \equiv R^{\mu}{}_{\mu} = e^{-2\phi}\eta^{\mu\nu}R_{\mu\nu} = -6e^{-2\phi}[\nabla^{2}\phi + (\nabla\phi)^{2}]$$

and the Riemann tensor is

$$
\begin{aligned}
R_{\alpha\beta\gamma\delta} &= C_{\alpha\beta\gamma\delta} + \frac{1}{2}(g_{\alpha\gamma}R_{\beta\delta} - g_{\alpha\delta}R_{\beta\gamma} - g_{\beta\gamma}R_{\alpha\delta} + g_{\beta\delta}R_{\alpha\gamma}) \\
&\quad - \frac{1}{6}(g_{\alpha\gamma}g_{\beta\delta} - g_{\alpha\delta}g_{\beta\gamma})R \\
&= \frac{1}{2}e^{2\phi}(\eta_{\alpha\gamma}R_{\beta\delta} - \eta_{\alpha\delta}R_{\beta\gamma} - \eta_{\beta\gamma}R_{\alpha\delta} + \eta_{\beta\delta}R_{\alpha\gamma}) \\
&\quad - \frac{1}{6}e^{4\phi}(\eta_{\alpha\gamma}\eta_{\beta\delta} - \eta_{\alpha\delta}\eta_{\beta\gamma})R \\
&= e^{2\phi}[\eta_{\alpha\gamma}(\phi_{,\beta}\phi_{,\delta} - \phi_{,\beta\delta}) - \eta_{\alpha\delta}(\phi_{,\beta}\phi_{,\gamma} - \phi_{,\beta\gamma}) \\
&\quad - \eta_{\beta\gamma}(\phi_{,\alpha}\phi_{,\delta} - \phi_{,\alpha\delta}) + \eta_{\beta\delta}(\phi_{,\alpha}\phi_{,\gamma} - \phi_{,\alpha\gamma}) \\
&\quad + (\nabla\phi)^2(-\eta_{\alpha\gamma}\eta_{\beta\delta} + \eta_{\alpha\delta}\eta_{\beta\gamma})] \, .
\end{aligned}
$$

Solution 9.20. We use the same orthonormal tetrad as in Problem 8.27:

$$\tilde{\omega}^{\hat{t}} = e^{\alpha}\,\widetilde{dt}\,, \quad \tilde{\omega}^{\hat{r}} = e^{\beta}\,\widetilde{dr}\,, \quad \tilde{\omega}^{\hat{\theta}} = e^{\gamma}\,\widetilde{d\theta}\,, \quad \tilde{\omega}^{\hat{\phi}} = e^{\gamma}\sin\theta\,\widetilde{d\phi}\,. \tag{1}$$

In Problem 8.27, we computed the 6 connection 1-forms $\tilde{\omega}^{\hat{\mu}}{}_{\hat{\nu}}$ from the equations

$$d\tilde{\omega}^{\hat{\mu}} = -\tilde{\omega}^{\hat{\mu}}{}_{\hat{\nu}}\tilde{\omega}^{\hat{\nu}}\,, \qquad \tilde{\omega}_{\hat{\mu}\hat{\nu}} = -\tilde{\omega}_{\hat{\nu}\hat{\mu}}\,. \tag{2}$$

The Riemann tensor is most easily found from the 6 curvature 2-forms $\mathcal{R}^{\hat{\mu}}{}_{\hat{\nu}}$:

$$\mathcal{R}^{\hat{\mu}}{}_{\hat{\nu}} = d\tilde{\omega}^{\hat{\mu}}{}_{\hat{\nu}} + \tilde{\omega}^{\hat{\mu}}{}_{\hat{\alpha}} \wedge \tilde{\omega}^{\hat{\alpha}}{}_{\hat{\nu}} \tag{3}$$

$$\mathcal{R}^{\hat{\mu}\hat{\nu}} = \frac{1}{2}R^{\hat{\mu}\hat{\nu}}{}_{\hat{\alpha}\hat{\beta}}\,\tilde{\omega}^{\hat{\alpha}}\wedge\tilde{\omega}^{\hat{\beta}} \tag{4}$$

The advantage of this method is that only non-zero components of the Riemann tensor are computed. Using the results of Problem 8.27 in Equation (3), we find

$$
\begin{aligned}
\mathcal{R}^{\hat{t}}{}_{\hat{r}} &= d(\alpha' e^{(\alpha-\beta)}\widetilde{dt} + \dot{\beta}e^{\beta-\alpha}\widetilde{dr}) + \tilde{\omega}^{\hat{t}}{}_{\hat{\theta}}\wedge\tilde{\omega}^{\hat{\theta}}{}_{\hat{r}} + \tilde{\omega}^{\hat{t}}{}_{\hat{\phi}}\wedge\tilde{\omega}^{\hat{\phi}}{}_{\hat{r}} \\
&= [\alpha'' + (\alpha')^2 - \alpha'\beta']e^{\alpha-\beta}\,\widetilde{dr}\wedge\widetilde{dt} + (\ddot{\beta} + \dot{\beta}^2 - \dot{\alpha}\dot{\beta})e^{\beta-\alpha}\,\widetilde{dt}\wedge\widetilde{dr} \\
&= [e^{-2\alpha}(\ddot{\beta} + \dot{\beta}^2 - \dot{\alpha}\dot{\beta}) - e^{-2\beta}(\alpha'' + \alpha'^2 - \alpha'\beta')]\,\tilde{\omega}^{\hat{t}}\wedge\tilde{\omega}^{\hat{r}}\,.
\end{aligned} \tag{5}
$$

Here a dot denotes $\partial/\partial t$ and a prime $\partial/\partial r$. Notice how the term $d\tilde{\omega}^{\hat{t}}{}_{\hat{r}}$ is computed by writing $\tilde{\omega}^{\hat{t}}{}_{\hat{r}}$ in the coordinate basis and using $dd = 0$. The remaining curvature 2-forms are

$$\begin{aligned}\mathcal{R}^{\hat{t}}{}_{\hat{\theta}} &= d(e^{\gamma-\alpha}\dot{\gamma}\,\widetilde{d\theta}) + \tilde{\omega}^{\hat{t}}{}_{\hat{r}} \wedge \tilde{\omega}^{\hat{r}}{}_{\hat{\theta}} + \tilde{\omega}^{\hat{t}}{}_{\hat{\phi}} \wedge \tilde{\omega}^{\hat{\phi}}{}_{\hat{\theta}} \\ &= [e^{-2\alpha}(\ddot{\gamma}+\dot{\gamma}^2-\dot{\alpha}\dot{\gamma}) - e^{-2\beta}\alpha'\gamma']\tilde{\omega}^{\hat{t}} \wedge \tilde{\omega}^{\hat{\theta}} \\ &\quad + e^{-(\alpha+\beta)}(\dot{\gamma}\gamma' - \alpha'\dot{\gamma} + \dot{\gamma}' - \gamma'\dot{\beta})\tilde{\omega}^{\hat{r}} \wedge \tilde{\omega}^{\hat{\theta}}\end{aligned} \tag{6}$$

$$\begin{aligned}\mathcal{R}^{\hat{t}}{}_{\hat{\phi}} &= d(e^{\gamma-\alpha}\dot{\gamma}\sin\theta\,\widetilde{d\phi}) + \tilde{\omega}^{\hat{t}}{}_{\hat{r}} \wedge \tilde{\omega}^{\hat{r}}{}_{\hat{\phi}} + \tilde{\omega}^{\hat{t}}{}_{\hat{\theta}} \wedge \tilde{\omega}^{\hat{\theta}}{}_{\hat{\phi}} \\ &= [e^{-2\alpha}(\ddot{\gamma}+\dot{\gamma}^2-\dot{\alpha}\dot{\gamma}) - e^{-2\beta}\alpha'\gamma']\tilde{\omega}^{\hat{t}} \wedge \tilde{\omega}^{\hat{\phi}} \\ &\quad + e^{-(\alpha+\beta)}(\dot{\gamma}\gamma' - \alpha'\dot{\gamma} + \dot{\gamma}' - \gamma'\dot{\beta})\tilde{\omega}^{\hat{r}} \wedge \tilde{\omega}^{\hat{\phi}}\end{aligned} \tag{7}$$

$$\begin{aligned}\mathcal{R}^{\hat{r}}{}_{\hat{\theta}} &= -d(e^{\gamma-\beta}\gamma'\widetilde{d\theta}) + \tilde{\omega}^{\hat{r}}{}_{\hat{t}} \wedge \tilde{\omega}^{\hat{t}}{}_{\hat{\theta}} + \tilde{\omega}^{\hat{r}}{}_{\hat{\phi}} \wedge \tilde{\omega}^{\hat{\phi}}{}_{\hat{\theta}} \\ &= [e^{-2\beta}(\gamma'' + \gamma'^2 - \beta'\gamma') - e^{-2\alpha}\dot{\beta}\dot{\gamma}]\tilde{\omega}^{\hat{\theta}} \wedge \tilde{\omega}^{\hat{r}} \\ &\quad + e^{-(\alpha+\beta)}(\dot{\gamma}\gamma' - \alpha'\dot{\gamma} + \dot{\gamma}' - \gamma'\dot{\beta})\tilde{\omega}^{\hat{\theta}} \wedge \tilde{\omega}^{\hat{t}}\end{aligned} \tag{8}$$

$$\begin{aligned}\mathcal{R}^{\hat{r}}{}_{\hat{\phi}} &= -d(e^{\gamma-\beta}\gamma'\sin\theta\,\widetilde{d\phi}) + \tilde{\omega}^{\hat{r}}{}_{\hat{t}} \wedge \tilde{\omega}^{\hat{t}}{}_{\hat{\phi}} + \tilde{\omega}^{\hat{r}}{}_{\hat{\theta}} \wedge \tilde{\omega}^{\hat{\theta}}{}_{\hat{\phi}} \\ &= [e^{-2\beta}(\gamma'' + \gamma'^2 - \beta'\gamma') - e^{-2\alpha}\dot{\beta}\dot{\gamma}]\tilde{\omega}^{\hat{\phi}} \wedge \tilde{\omega}^{\hat{r}} \\ &\quad + e^{-(\alpha+\beta)}(\dot{\gamma}\gamma' - \alpha'\dot{\gamma} + \dot{\gamma}' - \gamma'\dot{\beta})\tilde{\omega}^{\hat{\phi}} \wedge \tilde{\omega}^{\hat{t}}\end{aligned} \tag{9}$$

$$\begin{aligned}\mathcal{R}^{\hat{\theta}}{}_{\hat{\phi}} &= -d(\cos\theta\,\widetilde{d\phi}) + \tilde{\omega}^{\hat{\theta}}{}_{\hat{t}} \wedge \tilde{\omega}^{\hat{t}}{}_{\hat{\phi}} + \tilde{\omega}^{\hat{\theta}}{}_{\hat{r}} \wedge \tilde{\omega}^{\hat{r}}{}_{\hat{\phi}} \\ &= (e^{-2\gamma} + e^{-2\alpha}\dot{\gamma}^2 - e^{-2\beta}\gamma'^2)\tilde{\omega}^{\hat{\theta}} \wedge \tilde{\omega}^{\hat{\phi}}\,.\end{aligned} \tag{10}$$

We can now read off the non-zero components of the Riemann tensor from Equation (4):

$$
\begin{aligned}
R^{\hat{t}\hat{r}}{}_{\hat{t}\hat{r}} &= A \\
R^{\hat{t}\hat{\theta}}{}_{\hat{t}\hat{\theta}} &= R^{\hat{t}\hat{\phi}}{}_{\hat{t}\hat{\phi}} = B \\
R^{\hat{t}\hat{\theta}}{}_{\hat{r}\hat{\theta}} &= R^{\hat{t}\hat{\phi}}{}_{\hat{r}\hat{\phi}} = -R^{\hat{r}\hat{\theta}}{}_{\hat{t}\hat{\theta}} = -R^{\hat{r}\hat{\phi}}{}_{\hat{t}\hat{\phi}} = C \\
R^{\hat{\theta}\hat{\phi}}{}_{\hat{\theta}\hat{\phi}} &= D \\
R^{\hat{r}\hat{\phi}}{}_{\hat{r}\hat{\phi}} &= R^{\hat{r}\hat{\theta}}{}_{\hat{r}\hat{\theta}} = E
\end{aligned}
\tag{11}
$$

where

$$
\begin{aligned}
A &= e^{-2\alpha}(\ddot{\beta}+\dot{\beta}^2-\dot{\alpha}\dot{\beta}) - e^{-2\beta}(\alpha'' + \alpha'^2 - \alpha'\beta') \\
B &= e^{-2\alpha}(\ddot{\gamma}+\dot{\gamma}^2-\dot{\alpha}\dot{\gamma}) - e^{-2\beta}\alpha'\gamma' \\
C &= e^{-(\alpha+\beta)}(\dot{\gamma}'+\dot{\gamma}\gamma'-\alpha'\dot{\gamma}-\dot{\beta}\gamma') \\
D &= e^{-2\gamma} + e^{-2\alpha}\dot{\gamma}^2 - e^{-2\beta}\gamma'^2 \\
E &= e^{-2\alpha}\dot{\beta}\dot{\gamma} - e^{-2\beta}(\gamma'' + \gamma'^2 - \beta'\gamma') \, .
\end{aligned}
\tag{12}
$$

The Ricci tensor is obtained by contracting the Riemann tensor:

$$
R^{\hat{\alpha}}{}_{\hat{\beta}} = R^{\hat{\gamma}\hat{\alpha}}{}_{\hat{\gamma}\hat{\beta}} \, .
$$

This gives

$$
\begin{aligned}
R^{\hat{t}}{}_{\hat{t}} &= A + 2B \\
R^{\hat{t}}{}_{\hat{r}} &= 2C \\
R^{\hat{r}}{}_{\hat{r}} &= A + 2E \\
R^{\hat{\theta}}{}_{\hat{\theta}} &= R^{\hat{\phi}}{}_{\hat{\phi}} = B + D + E \\
R^{\hat{t}}{}_{\hat{\theta}} &= R^{\hat{t}}{}_{\hat{\phi}} = R^{\hat{r}}{}_{\hat{\theta}} = R^{\hat{r}}{}_{\hat{\phi}} = R^{\hat{\theta}}{}_{\hat{\phi}} = 0 \, .
\end{aligned}
\tag{13}
$$

The scalar curvature is

$$
R = R^{\hat{\alpha}}{}_{\hat{\alpha}} = 2A + 4B + 2D + 4E \, . \tag{14}
$$

The Einstein tensor is found from

$$G^{\hat{\alpha}}{}_{\hat{\beta}} = R^{\hat{\alpha}}{}_{\hat{\beta}} - \tfrac{1}{2}\,\delta^{\hat{\alpha}}{}_{\hat{\beta}} R$$

i.e.

$$\begin{aligned}
G^{\hat{t}}{}_{\hat{t}} &= -(D+2E)\\
G^{\hat{t}}{}_{\hat{r}} &= 2C\\
G^{\hat{r}}{}_{\hat{r}} &= -(D+2B)\\
G^{\hat{\theta}}{}_{\hat{\theta}} &= G^{\hat{\phi}}{}_{\hat{\phi}} = -(A+B+E)\\
G^{\hat{t}}{}_{\hat{\theta}} &= G^{\hat{t}}{}_{\hat{\phi}} = G^{\hat{r}}{}_{\hat{\theta}} = G^{\hat{r}}{}_{\hat{\phi}} = G^{\hat{\theta}}{}_{\hat{\phi}} = 0
\end{aligned} \qquad (15)$$

Solution 9.21. Choose an orthonormal basis $e_{\hat{\alpha}}$ and construct a complex, null tetrad basis oriented so that the wave vector $\nabla u \equiv k$ is one of the tetrad legs:

$$\begin{aligned}
k &= (2)^{-\frac{1}{2}}(e_{\hat{t}} + e_{\hat{z}})\\
l &= (2)^{-\frac{1}{2}}(e_{\hat{t}} - e_{\hat{z}})\\
m &= (2)^{-\frac{1}{2}}(e_{\hat{x}} + ie_{\hat{y}})\\
\bar{m} &= (2)^{-\frac{1}{2}}(e_{\hat{x}} - ie_{\hat{y}})
\end{aligned} \qquad (1)$$

and consider the components of the Riemann tensor in this basis: e.g. $R_{lm\bar{m}l} = R_{\alpha\beta\gamma\delta}\, l^{\alpha} m^{\beta} \bar{m}^{\gamma} l^{\delta}$ etc. Since the Riemann tensor is a function only of retarded time and since the only nonvanishing dot products of the basis vectors in Equation (1) are

$$-k\cdot l = m\cdot\bar{m} = 1\ , \qquad (2)$$

one finds

$$R_{abcd,p} = 0 \qquad (3)$$

where (a, b, c, d) range over $(k, l, m, \bar{m})$ and $(p, q, r\cdots)$ range over $(k, m, \bar{m})$. Now, consider the following subset of Bianchi identities for the Riemann tensor:

$$R_{ab[pq,l]} = 0 \tag{4}$$

where l is the l index of Equation (1). Using Equation (3), we get

$$R_{abpq,l} = 0 \tag{5}$$

from Equation (4), and Equations (5) and (3) imply

$$R_{abpq} = R_{pqab} = 0 \ , \tag{6}$$

aside from a trivial, nonwavelike constant. Consequently, all nonvanishing components of the Riemann tensor must have the form R_{plql} and, taking into account the symmetries of the Riemann tensor, we see that there are only six independent components. These six components correspond to the number of degrees of freedom in the most general gravitational wave in the generic metric theory of gravity. Einstein's theory has only two degrees of freedom in a gravitational wave.

Solution 9.22. In a local Lorentz frame, the acceleration of any particle is related to $F^{\mu\nu}$ by the Lorentz force law:

$$\frac{d^2x^\alpha}{d\tau^2} = \frac{e}{m} F^\alpha{}_\beta u^\beta \ .$$

It is convenient to think of the 6 components of $F^\alpha{}_\beta$ as the vectors $\underset{\sim}{E}$ and $\underset{\sim}{B}$. In the comoving frame of one particle, it has zero velocity and so feels no magnetic forces. Measuring its three components of acceleration gives $\underset{\sim}{E}$. Measuring the acceleration of a second particle as seen in that frame gives only two components of $\underset{\sim}{B}$, since the components of $\underset{\sim}{B}$ along the direction of motion of the second particle produces no acceleration. Thus at least three particles are required to measure all components of $F^\alpha{}_\beta$.

The reader is invited to figure out the answer for the Riemann tensor from the geodesic deviation equation

$$\frac{D^2 \xi^\alpha}{d\tau^2} = - R^\alpha{}_{\beta\gamma\delta} u^\beta \xi^\gamma u^\delta \ .$$

Solution 9.23. Choose a coordinate system geared to the surface. Let x^1, x^2 vary, and x^3, x^4 be constant on the surface. The only contravariant components of **A** and **B** are then the 1 and 2 components. We are interested then only in the Riemann components with indices 1 and 2, but there is only one such independent component, R_{1212}. The non-vanishing Riemann components of this type are related by symmetries which can be expressed as

$$R_{ijkl} \propto (g_{ik} g_{jl} - g_{il} g_{jk}) \qquad i, j, k, l = 1, 2 \ .$$

It follows immediately that in this coordinate system K is independent of **A** and **B**. Since K is clearly coordinate independent the desired result is proved.

Solution 9.24. Let the interior of the circuit be subdivided into infinitesimal parallelograms, and consider the change in angle for one small rectangle. If **u** and **v** represent the sides of one rectangle then from Problems 9.23 and 9.11 we have that the change in **A** after transport is

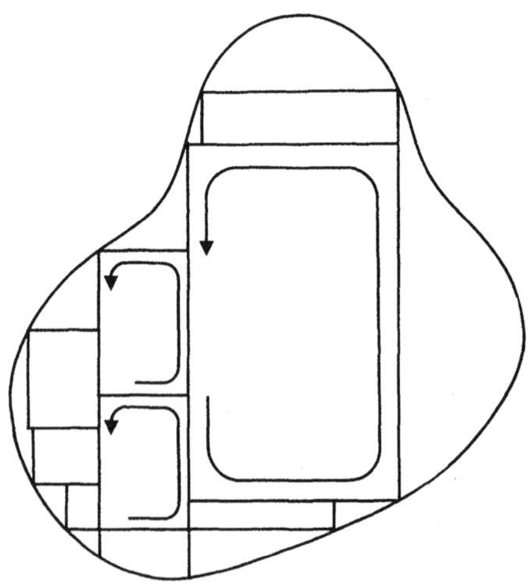

Circuit Σ — approximated with rectangles

$$\delta A_\alpha = -R_{\alpha\beta\mu\nu} A^\beta u^\mu v^\nu = K(g_{\alpha\mu} g_{\beta\nu} - g_{\alpha\nu} g_{\beta\mu}) A^\beta u^\mu v^\nu .$$

Note that $A^\alpha \delta A_\alpha = \frac{1}{2}\delta(A \cdot A) = 0$ so the length of A is unchanged. The change in angle θ between A and **B** can be calculated from

$$B^\alpha \delta A_\alpha = \delta(A^\alpha B_\alpha) = |A|\,|B|\,\delta(\cos\theta) = |A|\,|B| \sin\theta\, \delta\theta$$

and

$$B^\alpha \delta A_\alpha = K[(\mathbf{B}\cdot\mathbf{u})(A\cdot v) - (\mathbf{B}\cdot v)(A\cdot\mathbf{u})] .$$

Now choose a locally flat coordinate system on the surface with $\mathbf{u} \sim e_1$ and $v \sim e_2$ then

$$\begin{aligned} ||A|\,|B| \sin\theta\, \delta\theta| &= K(B^1A^2 - B^2A^1)uv \\ &= K\,|A \times \mathbf{B}|\,uv = K \sin\theta\, |A|\,|B| \cdot (\text{area enclosed}) . \end{aligned}$$

Hence, for transport around a small rectangle $|\delta\theta| = K\delta\Sigma$. If we now consider a more general circuit, the change in A due to transport around that circuit will be equal to the change in A due to transport around all the rectangles into which the circuit is divided, and the area of the circuit is the sum of the areas of the rectangles, thus $|\Delta\theta| = K\Delta\Sigma$.

Solution 9.25. Define

$$W_{\alpha\gamma\beta\delta} \equiv K(g_{\alpha\beta}\, g_{\gamma\delta} - g_{\alpha\delta}\, g_{\beta\gamma}) - R_{\alpha\gamma\beta\delta} \tag{1}$$

and note that $W_{\alpha\gamma\beta\delta}$ has the same symmetries as $R_{\alpha\gamma\beta\delta}$. If K is independent of the vectors A and **B** at the given point, then by the definition of K (Problem 9.23)

$$W_{\alpha\gamma\beta\delta}\, A^\alpha A^\beta B^\gamma B^\delta = 0 \tag{2}$$

for *all* A, **B**, thus

$$W_{\alpha\gamma\beta\delta} + W_{\beta\delta\alpha\gamma} + W_{\alpha\delta\beta\gamma} + W_{\beta\gamma\alpha\delta} = 0 . \tag{3}$$

The symmetries of $W_{\alpha\gamma\beta\delta}$ allow us to rewrite Equation (3) as

$$W_{\alpha\gamma\beta\delta} = W_{\alpha\delta\gamma\beta} \cdot \tag{4}$$

Permute $(\gamma\beta\delta)$ cyclically:

$$W_{\alpha\delta\gamma\beta} = W_{\alpha\beta\delta\gamma} \cdot \tag{5}$$

Substitute Equations (4) and (5) in the cyclic identity

$$W_{\alpha\gamma\beta\delta} + W_{\alpha\delta\gamma\beta} + W_{\alpha\beta\delta\gamma} = 0 \tag{6}$$

and find

$$W_{\alpha\gamma\beta\delta} = 0$$

which is the required result.

Solution 9.26. Since the metric is always covariantly constant

$$R_{\alpha\beta\gamma\delta;\lambda} = K_{,\lambda}(g_{\alpha\gamma}\, g_{\beta\delta} - g_{\alpha\delta}\, g_{\beta\gamma}) \cdot$$

Now substitute in the Bianchi identities

$$0 = R_{\alpha\beta\gamma\delta;\lambda} + R_{\alpha\beta\lambda\gamma;\delta} + R_{\alpha\beta\delta\lambda;\gamma}$$

and contract on α,γ and β,δ to find $K_{,\lambda} = 0$ i.e. K is constant.

Solution 9.27. By contracting we see

$$R_{\mu\kappa} = 3Kg_{\mu\kappa}$$

$$R = 12K \, .$$

The definition of the Weyl tensor is

$$C_{\lambda\mu\nu\kappa} = R_{\lambda\mu\nu\kappa} - \frac{1}{2}\,(g_{\lambda\nu}\, R_{\mu\kappa} - g_{\lambda\kappa}\, R_{\mu\nu} - g_{\mu\nu}\, R_{\lambda\kappa} + g_{\mu\kappa}\, R_{\lambda\nu})$$
$$+ \frac{1}{6}\,(g_{\lambda\nu}\, g_{\mu\kappa} - g_{\lambda\kappa}\, g_{\mu\nu})R \, ,$$

so that

$$C_{\lambda\mu\nu\kappa} = K(g_{\lambda\nu}g_{\mu\kappa} - g_{\lambda\mu}g_{\nu\kappa}) - \frac{1}{2}\cdot 3K(2g_{\lambda\nu}g_{\mu\kappa} - 2g_{\lambda\mu}g_{\nu\kappa})$$
$$+ \frac{1}{6}\cdot 12K(g_{\lambda\nu}g_{\mu\kappa} - g_{\lambda\kappa}g_{\mu\nu}) = 0 .$$

Solution 9.28. Since $\mathbf{n}\cdot\mathbf{u} = 0$ we have immediately

$$u^\alpha(n^\beta u_\beta)_{;\alpha} = 0 = n^\beta{}_{;\alpha}u^\alpha u_\beta + n^\beta u_{\beta;\alpha}u^\alpha$$
$$= -K_{\alpha\beta}u^\alpha u^\beta + 2\mathbf{n}\cdot\boldsymbol{\xi} .$$

Solution 9.29. The metric has the form

$$ds^2 = -d\tau^2 + g_{ij}\,dx^i\,dx^j$$

where

$$g_{ij} = a^2(\tau)\gamma_{ij}(x^k) .$$

The normal vector to the τ = constant surfaces is $\mathbf{n} = \partial/\partial\tau$. Thus

$$K_{ij} = -\mathbf{e}_j\cdot\nabla_i\mathbf{n} = \mathbf{n}\cdot\nabla_i\mathbf{e}_j = \Gamma_{nji} = -\frac{1}{2}\,g_{ij,n} = \frac{-a_{,\tau}}{a}\,g_{ij} .$$

Solution 9.30. From the definition of the Lie derivative (Problem 8.13) we have

$$\mathcal{L}_{\mathbf{n}}P_{\alpha\beta} = P_{\alpha\beta;\gamma}n^\gamma + P_{\gamma\beta}n^\gamma{}_{;\alpha} + P_{\alpha\gamma}n^\gamma{}_{;\beta}$$
$$= (g_{\alpha\beta} - n_\alpha n_\beta)_{;\gamma}\,n^\gamma + (g_{\gamma\beta} - n_\gamma n_\beta)\,n^\gamma{}_{;\alpha} + (g_{\alpha\gamma} - n_\alpha n_\gamma)\,n^\gamma{}_{;\beta}$$
$$= -(n_\alpha n_\beta)_{;\gamma}\,n^\gamma + n_{\beta;\alpha} + n_{\alpha;\beta} \tag{1}$$

where we have used $n_\gamma n^\gamma{}_{;\alpha} = \frac{1}{2}(n_\gamma n^\gamma)_{;\alpha} = 0$. If $\mathbf{e}_i$ are three basis vectors in the hypersurface, then

$$K_{ij} = -\mathbf{e}_j\cdot\nabla_i\mathbf{n} = -n_{i;j} = -\frac{1}{2}(n_{i;j} + n_{j;i}) \tag{2}$$

where we have used the fact that K_{ij} is symmetric. If we are using a general coordinate system (i.e. not one in which i, j lie in the hypersurface), then Equation (2) can be rewritten

$$K_{\alpha\beta} = -\frac{1}{2}(n_{\gamma;\delta} + n_{\delta;\gamma}) P^{\gamma}{}_{\alpha} P^{\delta}{}_{\beta} \tag{3}$$

i.e. as the symmetrized covariant derivative of **n** projected into the hypersurface. Using the explicit form of $P^{\gamma}{}_{\alpha}$ in Equation (3), and comparing with Equation (1), we find

$$K_{\alpha\beta} = -\frac{1}{2}\mathcal{L}_{\mathbf{n}} P_{\alpha\beta} \ .$$

Solution 9.31. Let the surface be $x^3 = 0$, where (x^1, x^2, x^3) are Gaussian normal coordinates (see Problem 8.25) based on this surface. Then the area is

$$A = \int g^{\frac{1}{2}}\, dx^1\, dx^2 \tag{1}$$

where g is the determinant of the two-dimensional metric g_{ij} in the surface. Vary the area by displacing surface elements a distance δx^3 along the normal $\mathbf{n} = \partial/\partial x^3$:

$$\delta A = \int \delta g^{\frac{1}{2}}\, dx^1\, dx^2 \ .$$

In Problem 21.1 it is shown that

$$\delta g^{\frac{1}{2}} = \frac{1}{2} g^{\frac{1}{2}} g^{ij}\, \delta g_{ij} = \frac{1}{2} g^{\frac{1}{2}} g^{ij} g_{ij,3}\, \delta x^3 \ .$$

But we have chosen Gaussian normal coordinates, so

$$K_{ij} = -\frac{1}{2} g_{ij,3}$$

$$\delta A = -\int g^{ij} K_{ij}\, g^{\frac{1}{2}}\, dx^1\, dx^2\, \delta x^3 \ .$$

Since A is a minimum, $\delta A = 0$ for all δx^3, and hence we must have

$$K = g^{ij} K_{ij} = 0 \ .$$

Solution 9.32. Let e_i be a coordinate basis in Σ; the equations to be derived are tensor equations and so will be valid independent of the basis used to derive them. The vectors e_i along with n form a basis for the spacetime, so

$$^{(4)}\nabla_i e_j = \alpha_{ij} n + \beta^k{}_{ij} e_k \,. \tag{1}$$

The coefficients in Equation (1) can be found by dotting in n and e_k ($n \cdot e_i = 0$):

$$\varepsilon \alpha_{ij} = n \cdot {}^{(4)}\nabla_i e_j = {}^{(4)}\nabla_i (n \cdot e_j) - e_j \cdot {}^{(4)}\nabla_i n = 0 + K_{ij}$$

$$\beta_{kij} = e_k \cdot {}^{(4)}\nabla_i e_j = {}^{(4)}\Gamma_{kji} = {}^{(3)}\Gamma_{kji} \,.$$

The last equality follows from the fact that the Christoffel symbol can be computed directly from $^{(3)}g_{ij}$. With these coefficients Equation (1) becomes the "Gauss-Weingarten equation."

$$^{(4)}\nabla_i e_j = \varepsilon K_{ij} n + {}^{(3)}\Gamma^k{}_{ji} e_k \,. \tag{2}$$

The Riemann tensor can be computed from derivatives of the Γ's; it is somewhat quicker to use the curvature operator (see Problem 9.12):

$$R_{\alpha\beta\gamma\delta} = e_\alpha \cdot R(e_\gamma, e_\delta) e_\beta \tag{3}$$

where

$$R(e_\gamma, e_\delta) = [\nabla_\gamma, \nabla_\delta] - \nabla_{[e_\gamma, e_\delta]} \,. \tag{4}$$

Evaluate Equation (4) for e_j and e_k. The last term vanishes because $[e_j, e_k] = 0$ for a coordinate basis, and

$$\begin{aligned} {}^{(4)}\nabla_j {}^{(4)}\nabla_k e_i &= {}^{(4)}\nabla_j (\varepsilon K_{ik} n + {}^{(3)}\Gamma^m{}_{ik} e_m) \\ &= \varepsilon K_{ik,j} n - \varepsilon K_{ik} K_j{}^m e_m + {}^{(3)}\Gamma^m{}_{ik,j} e_m \\ &\quad + {}^{(3)}\Gamma^m{}_{ik} (\varepsilon K_{jm} n + {}^{(3)}\Gamma^n{}_{mj} e_n) \,. \end{aligned} \tag{5}$$

Evaluate Equation (5) with j and k interchanged and subtract the two equations. The result is

$$R(e_j, e_k)e_i = \varepsilon(K_{ik|j} - K_{ij|k})\mathbf{n} + e_m[\varepsilon(K_{ij}K_k{}^m - K_{ik}K_j{}^m) + {}^{(3)}R^m{}_{ijk}] . (6)$$

Dotting Equation (6) with $\mathbf{n}$ and e_n respectively gives the Gauss-Codazzi equations.

The components of the Riemann tensor with two $\mathbf{n}$'s can be found from $R(e_k, \mathbf{n})\mathbf{n}$. Since $\mathbf{n}$ has been chosen to be a coordinate basis vector, $[\mathbf{n}, e_k] = 0$ so the last term in Equation (4) vanishes and thus

$$R(e_k, \mathbf{n})\mathbf{n} = {}^{(4)}\nabla_k {}^{(4)}\nabla_{\mathbf{n}}\mathbf{n} - {}^{(4)}\nabla_{\mathbf{n}} {}^{(4)}\nabla_k \mathbf{n} .$$

But ${}^{(4)}\nabla_{\mathbf{n}}\mathbf{n} = 0$ because $\mathbf{n}$ is tangent to a geodesic (Gaussian normal coordinates), so that

$$e_i \cdot R(e_k, \mathbf{n})\mathbf{n} = e_i \cdot {}^{(4)}\nabla_{\mathbf{n}}(K_k{}^m e_m) = K_{ki,n} - K_k{}^m e_m \cdot \nabla_{\mathbf{n}} e_i .$$

Now $[e, \mathbf{n}] = 0$ implies that

$${}^{(4)}\nabla_{\mathbf{n}} e_i = {}^{(4)}\nabla_i \mathbf{n} = -K_i^j e_j$$

so finally

$$R_{inkn} = \varepsilon R^n{}_{ink} = K_{ki,n} + K_i^j K_{jk} .$$

Solution 9.33. The Riemann components were found in Problem 9.32. Contraction gives

$$\begin{aligned} {}^{(4)}R^i_j &= g^{ik}({}^{(4)}R^n{}_{knj} + {}^{(4)}R^m{}_{kmj}) \\ &= {}^{(3)}R^i_j + \varepsilon\,(g^{ik}K_{kj,n} + 2K^i_m K^m{}_j - K^i_j K) \qquad (1) \end{aligned}$$

where $K \equiv K^i_i$. Now $g^{ik}{}_{,n} = -g^{im}g^{ks}g_{ms,n}$ (see e.g. Problem 21.1) and $-g_{ms,n} = 2K_{ms}$ in Gaussian normal coordinates, so

$$g^{ik}{}_{,n} = 2g^{im}g^{ks}K_{ms} = 2K^{ik} .$$

Equation (1) can now be rewritten:

$$
\begin{aligned}
{}^{(4)}R^i_j &= {}^{(3)}R^i_j + \varepsilon\,(g^{ik}K_{kj,n} + g^{ik}{}_{,n}K_{kj} - K^i_j K) \\
&= {}^{(3)}R^i_j + \varepsilon\,(K^i_{j,n} - K^i_j K)\,. \qquad (2)
\end{aligned}
$$

For the other components of the Ricci tensor we have:

$$ {}^{(4)}R^n{}_j = \varepsilon\,{}^{(4)}R^\alpha{}_{n\alpha j} = -\,{}^{(4)}R^{n\,i}{}_{i\,j} = \varepsilon\,(K_{|j} - K^i_{j|i}) \qquad (3) $$

$$ {}^{(4)}R^n{}_n = \varepsilon\,{}^{(4)}R^i{}_{nin} = \varepsilon\,g^{ij}(K_{ij,n} + K_{im}K^m{}_j) = \varepsilon\,(K_{,n} - K_{im}K^{mi}). \qquad (4) $$

The Ricci scalar is then

$$ {}^{(4)}R = {}^{(4)}R^n{}_n + {}^{(4)}R^i{}_i = {}^{(3)}R + \varepsilon\,(2K_{,n} - K_{im}K^{im} - K^2) \qquad (5) $$

and the components of the Einstein tensor are:

$$ G^n{}_n = {}^{(4)}R^n{}_n - \frac{1}{2}\,{}^{(4)}R = -\frac{1}{2}\,{}^{(3)}R + \frac{1}{2}\,\varepsilon(K^2 - K_{im}K^{im}) \qquad (6) $$

$$ {}^{(4)}G^n{}_j = {}^{(4)}R^n{}_j = \varepsilon(K_{|j} - K^i{}_{j|i}) \qquad (7) $$

$$
\begin{aligned}
{}^{(4)}G^i{}_j &= {}^{(4)}R^i{}_j - \frac{1}{2}\,\delta^i{}_j\,{}^{(4)}R \\
&= {}^{(3)}G^i{}_j + \varepsilon[K^i{}_{j,n} - K^i{}_j K - \frac{1}{2}\,\delta^i{}_j(2K_{,n} - K_{im}K^{im} - K^2)]\,. \qquad (8)
\end{aligned}
$$

Equating (6) and (7) to the corresponding components of the stress-energy tensor gives "initial value equations" for the gravitational field.

Solution 9.34.

(i) Since there is no favored direction in the spherical surface the extrinsic curvature tensor must be $K_{ij} \propto \delta_{ij}$ in an orthonormal basis; this implies that any vector is an eigenvector. From the definition of **K** as the rate of change of **n**, the eigenvalue must be -1/sphere's radius $= -1/a$.

We can give a mathematical development of these intuitive results by going to the usual spherical coordinates r, θ, ϕ. These are clearly Gaussian normal coordinates and

$$ds^2 = {}^{(2)}g_{ik}\, dx^i\, dx^k = r^2(d\theta^2 + \sin^2\theta\, d\phi^2)\ .$$

Now $K_{ij} = -\frac{1}{2}\, g_{ij,n} = -\frac{1}{2}\, g_{ij,r}$ so

$$K_{\hat\theta\hat\theta} = \frac{1}{g_{\theta\theta}}\, K_{\theta\theta} = \frac{1}{r^2}\left(-\frac{1}{2}\, r^2\right)_{,r} = -1/a$$

$$K_{\hat\phi\hat\phi} = \frac{1}{g_{\phi\phi}}\, K_{\phi\phi} = \frac{1}{r^2\sin^2\theta}\left(-\frac{1}{2}\, r^2\sin^2\theta\right)_{,r} = -1/a\ . \tag{1}$$

(ii) Again we can argue intuitively: The "special" directions must be the axial and circumferential directions. From the definition of **K** the values of the curvature (rate of change of **n**) in these directions should be 0 and $-1/a$.

Mathematically: Again the usual coordinates are Gaussian normal and the 2-geometry is

$$ds^2 = r^2\, d\phi^2 + dz^2\ .$$

The orthonormal components of **K** are then

$$K_{\hat\phi\hat\phi} = \frac{1}{g_{\phi\phi}}\left(-\frac{1}{2}\, g_{\phi\phi,r}\right) = -1/a$$

$$K_{\hat z\hat z} = \frac{1}{g_{zz}}\left(-\frac{1}{2}\, g_{zz,r}\right) = 0 \tag{2}$$

and our intuitive answers are correct.

(iii) In cartesian coordinates

$$ds^2 = dx^2 + dy^2 + dz^2 \tag{3}$$

the surface is

$$f(x, y, z) = 0 \tag{4}$$

where

$$f = -\frac{1}{2}(ax^2 + 2bxy + cy^2) + z . \tag{5}$$

The unit normal is

$$\mathbf{n} = \nabla f/|\nabla f| = \frac{1}{N}[-(ax+by)\mathbf{e}_x - (bx+cy)\mathbf{e}_y + (\mathbf{e}_z)] \tag{6}$$

where

$$N \equiv [(ax+by)^2 + (bx+cy)^2 + 1]^{\frac{1}{2}} . \tag{7}$$

Note that since $\mathbf{n} = \mathbf{e}_z$ at the origin, $\mathbf{e}_x$ and $\mathbf{e}_y$ can be used as basis vectors in the surface there. Thus

$$K_{ij} = -\mathbf{e}_j \cdot \nabla_i \mathbf{n} = -n_{j;i} = -n_{j,i}$$

since the connection coefficients vanish for the metric (3). Because $N_{,x} = N_{,y} = 0$ at the origin, we have

$$K_{ij} = \begin{bmatrix} a & b \\ b & c \end{bmatrix} .$$

The secular equation is

$$0 = \det(K_{ij} - Kg_{ij}) = \begin{vmatrix} a-K & b \\ b & c-K \end{vmatrix} \tag{8}$$

where g_{ij} is found by restricting (3) to the surface (5). Since $dz = 0$ at the origin, $g_{ij} = \delta_{ij}$ at the origin. The principal curvatures are found from Equation (8) to be

$$K_{\pm} = \frac{c+a}{2} \pm \frac{c-a}{2}\left[1 + \frac{4b^2}{(c-a)^2}\right]^{\frac{1}{2}} . \tag{9}$$

With these eigenvalues, the principal vectors are found from

$$\begin{bmatrix} a - K_{\pm} & b \\ b & c - K_{\pm} \end{bmatrix}\begin{bmatrix} V^x_{\pm} \\ V^y_{\pm} \end{bmatrix} = 0 . \tag{10}$$

The solutions are:

$$V_{\pm} = (e_x + \alpha_{\pm} e_y)/\beta_{\pm}$$

$$\alpha_{\pm} \equiv \gamma[1 \pm (1+\gamma^{-2})^{\frac{1}{2}}]$$

$$\beta_{\pm} \equiv (1+\alpha_{\pm}^2)^{\frac{1}{2}}$$

$$\gamma \equiv (a-c)/2b .$$

Solution 9.35. From the Gauss-Codazzi equation (Problem 9.32), we have

$${}^{(3)}R^m{}_{ijk} = {}^{(2)}R^m{}_{ijk} + \varepsilon(K_{ij}K_k{}^m - K_{ik}K_j{}^m) .$$

Contract on m and j and on i and k, and set ${}^{(3)}R^m{}_{ijk} = 0$ because the 3-space is flat. Thus find

$${}^{(2)}R = -\varepsilon(K_i{}^j K_j{}^i - K_i{}^i K_j{}^j) . \tag{1}$$

At any point P on Σ, choose the principal directions as axes and let the coordinates x^i measure lengths away from P. Then ${}^{(2)}g_{ij} = \delta_{ij}$ at P and K_{ij} is diagonal, with the principal curvatures as the diagonal elements. Hence

$${}^{(2)}R = -\varepsilon\left[\frac{1}{\rho_1^2} + \frac{1}{\rho_2^2} - \left(\frac{1}{\rho_1} + \frac{1}{\rho_2}\right)^2\right] = \frac{2}{\rho_1\rho_2} ,$$

where we have set $\varepsilon = \mathbf{n}\cdot\mathbf{n} = +1$.

For Σ 3-dimensional in a flat 4-space, Equation (1) gives

$${}^{(3)}R = -\varepsilon\left[\frac{1}{\rho_1^2} + \frac{1}{\rho_2^2} + \frac{1}{\rho_3^2} - \left(\frac{1}{\rho_1} + \frac{1}{\rho_2} + \frac{1}{\rho_3}\right)^2\right]$$

$$= \varepsilon\left(\frac{2}{\rho_1\rho_2} + \frac{2}{\rho_2\rho_3} + \frac{2}{\rho_1\rho_3}\right) .$$

CHAPTER 10: SOLUTIONS

Solution 10.1. The metric is $ds^2 = d\theta^2 + \sin^2\theta\, d\phi^2$; the connection coefficients are $\Gamma^{\phi}{}_{\theta\phi} = \Gamma^{\phi}{}_{\phi\theta} = \cot\theta$; $\Gamma^{\theta}{}_{\phi\phi} = -\sin\theta\cos\theta$; all others are zero. Now:

$$\xi_{\alpha;\beta} + \xi_{\beta;\alpha} = \xi_{\alpha,\beta} + \xi_{\beta,\alpha} - 2\Gamma^{\mu}{}_{\alpha\beta}\xi_{\mu} = 0\ .$$

$$\alpha = \beta = \phi:\ \xi_{\phi,\phi} = -\xi_{\theta}\sin\theta\cos\theta \tag{1}$$

$$\alpha = \beta = \theta : \xi_{\theta,\theta} = 0; \qquad \therefore \xi_{\theta} = f(\phi) \tag{2}$$

$$\alpha = \theta, \beta = \phi:\ \xi_{\theta,\phi} + \xi_{\phi,\theta} = 2\cot\theta\,\xi_{\phi}\ . \tag{3}$$

Substitution of Equation (2) in Equation (1) gives $\xi_{\phi,\phi} = -f(\phi)\sin\theta\cos\theta$ which implies

$$\xi_{\phi} = -F(\phi)\sin\theta\cos\theta + g(\theta) \tag{4}$$

where $F(\phi) \equiv \int f\, d\phi$. Next substitute Equations (4) and (2) in Equation (3):

$$\frac{df}{d\phi} + \frac{dg}{d\theta} - F(\cos^2\theta - \sin^2\theta) = 2\cot\theta\, g - 2\cos^2\theta\, F$$

or

$$\frac{df}{d\phi} + F(\phi) = -\left(\frac{dg}{d\theta} - 2\cot\theta\, g(\theta)\right)\ .$$

Since the left side is a function only of ϕ and the right side only of θ, each side must be a constant:

$$df/d\phi + \int f\, d\phi = b\ , \tag{5}$$

$$dg/d\theta - 2\cot\theta\, g = -b\ . \tag{6}$$

Equation (6) has an integrating factor $\exp[-2\int \cot\theta\, d\theta] = \sin^{-2}\theta$, giving

$$\frac{d}{d\theta}\left(\frac{g}{\sin^2\theta}\right) = \frac{-b}{\sin^2\theta}$$

$$g(\theta) = (b\cot\theta + c)\sin^2\theta \ . \tag{7}$$

Equation (5) is solved by differentiation:

$$\frac{d^2 f}{d\phi^2} + f = 0 \ ,$$

$$f = d\cos\phi + e\sin\phi$$

$$F = d\sin\phi - e\cos\phi \ .$$

(The integration constant is included in $g(\theta)$.) Substituting this result in Equation (5)

$$-d\sin\phi + e\cos\phi + d\sin\phi - e\cos\phi = b$$

shows that $b = 0$, and hence from Equations (2) and (7) that

$$\xi_\theta = d\cos\phi + e\sin\phi = \xi^\theta$$

$$\xi_\phi = c\sin^2\theta - \sin\theta\cos\theta\,(d\sin\phi - e\cos\phi) = \sin^2\theta\,\xi^\phi \ .$$

Thus

$$\xi = (d\cos\phi + e\sin\phi)\frac{\partial}{\partial\theta} + [c - \cot\theta\,(d\sin\phi - e\cos\phi)]\frac{\partial}{\partial\phi}$$

is the most general Killing vector. Note that it is a linear combination of the three Killing vectors

$$\frac{\partial}{\partial\phi}$$

$$-\left(\cos\phi\frac{\partial}{\partial\theta} - \cot\theta\sin\phi\frac{\partial}{\partial\phi}\right)$$

$$\sin\phi\frac{\partial}{\partial\theta} + \cot\theta\cos\phi\frac{\partial}{\partial\phi}$$

which are just the usual angular momentum operators L_z, L_x and L_y, the generators of the rotation group.

Solution 10.2. It is straightforward to prove this statement in component notation. Alternately, "dot" $\mathcal{L}_{\xi} g$ with two arbitrary vector fields A and B:

$$0 = A \cdot \mathcal{L}_{\xi} g \cdot B = \mathcal{L}_{\xi}(A \cdot B) - B \cdot \mathcal{L}_{\xi} A - A \cdot \mathcal{L}_{\xi} B$$
$$= \nabla_{\xi}(A \cdot B) - B \cdot (\nabla_{\xi} A - \nabla_{A} \xi) - A \cdot (\nabla_{\xi} B - \nabla_{B} \xi)$$
$$= B \cdot \nabla_{A} \xi + A \cdot \nabla_{B} \xi = 2 A^{\alpha} B^{\beta} \xi_{(\alpha;\beta)} \, .$$

Since A and B are arbitrary, $\xi_{(\alpha;\beta)}$ must vanish. To interpret this result geometrically recall that the Lie derivative of any geometrical quantity, ϕ_A, is the *functional* change of that quantity, $\delta\phi_A \equiv \phi_A^{\text{new}}(x^{\alpha}) - \phi_A^{\text{old}}(x^{\alpha})$ under a displacement of coordinates by ξ. This is equivalent to the change in ϕ_A upon moving it in the manifold along ξ. For the Lie derivative of g along ξ to vanish, therefore, requires that the geometry to be unchanged as one moves in the ξ direction, i.e. ξ represents a direction of symmetry of spacetime.

Solution 10.3.

(a) We use the equivalence proven in Problem 10.2. If u and v are Killing vectors then $\mathcal{L}_u g = \mathcal{L}_v g = 0$ and, with the result of Problem 8.16,

$$\mathcal{L}_{[u,v]} g = (\mathcal{L}_u \mathcal{L}_v - \mathcal{L}_v \mathcal{L}_u) g = 0 \, .$$

Thus $[u, v]$ is a Killing vector.

(b) If a and b are constants, then

$$(a u_{\alpha} + b v_{\alpha})_{;\beta} = a u_{\alpha;\beta} + b v_{\alpha;\beta} \, .$$

Thus if u and v satisfy Killing's equation, so does $a u + b v$.

Solution 10.4. The three Killing vectors describing rotations are just the angular momentum operators, $J_z = x\partial_y - y\partial_x$, etc. (See Problem 10.1.) At a point (x_0, y_0, z_0)

$$J_z = (-y_0, x_0, 0)$$
$$J_y = (z_0, 0, -x_0)$$
$$J_x = (0, -z_0, y_0) \ ,$$

so $J_y = -(z_0/y_0)\ J_z - (x_0/y_0)\ J_x$ and the three vectors only span a 2-dimensional surface at the point.

Suppose, however, that for some constants a and b

$$a J_x + b J_y + J_z = 0$$

were true at every point, then we would have

$$a[J_x, J_y] + b[J_y, J_y] + [J_z, J_y] = 0 \ .$$

This requires $J_z \propto J_x$, a result which is obviously false. Thus, over the whole sphere, J_x, J_y, J_z span a 3-dimensional space. The explanation is that 2-spheres are 2-dimensional, but *orientations* of 2-spheres are 3-dimensional (e.g. the 3 Euler angles).

Solution 10.5. The commutator must be a Killing field according to Problem 10.3. By hypothesis any Killing vector must be a sum (with constant coefficients) of $\xi_{(t)}$ and $\xi_{(\phi)}$:

$$[\xi_{(\phi)}, \xi_{(t)}] = a\xi_{(\phi)} + b\xi_{(t)} \ .$$

At infinity $\xi_{(\phi)} \to \partial/\partial\phi$ and $\xi_{(t)} \to \partial/\partial t$ so the commutator $\to 0$. This means that a and b must be zero, i.e. that the commutator vanishes everywhere. (B. Carter, [Commun. Math. Phys. 17, 233 (1970)] has proven that $[\xi_{(\phi)}, \xi_{(t)}] = 0$ under very general conditions even when there are other Killing vectors.)

Solution 10.6. For the commutation of second derivatives of any vector

$$\xi_{\mu;\nu\lambda} - \xi_{\mu;\lambda\nu} = R_{\mu\sigma\lambda\nu}\xi^{\sigma} \ .$$

Now we use the Killing equation $\xi_{\mu;\nu} = -\xi_{\nu;\mu}$ and contract on μ and λ

$$\xi^{\nu;\lambda}{}_{;\lambda} + R^{\nu}{}_{\sigma}\xi^{\sigma} = -(\xi^{\mu}{}_{;\mu})^{;\nu} .$$

The term on the right vanishes since $\xi^{\mu}{}_{;\mu} = 0$ by the Killing equation, and we are left with the desired result. The variational principle is easy to write down if we note that $\xi_{\mu;\nu}$ is antisymmetric, just like the electromagnetic field tensor $F_{\mu\nu}$. We know that the Lagrangian $\frac{1}{4} F_{\mu\nu}F^{\mu\nu}$ gives a term $F^{\mu\nu}{}_{;\nu}$ when we write $F_{\mu\nu} = A_{\nu;\mu} - A_{\mu;\nu} = A_{\nu,\mu} - A_{\mu,\nu}$ and vary A_{μ}. Thus the Lagrangian density can be taken to be

$$\mathfrak{L} = \xi_{\mu;\nu}\xi^{\mu;\nu} - \frac{1}{2} R_{\mu\nu}\xi^{\mu}\xi^{\nu}$$

and $\delta \int \mathfrak{L}|g|^{\frac{1}{2}} d^4x = 0$ will give the correct equation. This can be verified by finding the Euler-Lagrange equations for this Lagrangian.

Solution 10.7. For *any* vector ξ

$$\xi_{\sigma;\rho\mu} - \xi_{\sigma;\mu\rho} = R^{\lambda}{}_{\sigma\rho\mu}\xi_{\lambda} \tag{1}$$

by the definition of $R_{\alpha\beta\gamma\delta}$. Add to Equation (1) its two permutations and use the cyclic identity for the Riemann tensor

$$R^{\lambda}{}_{\sigma\rho\mu} + R^{\lambda}{}_{\mu\sigma\rho} + R^{\lambda}{}_{\rho\mu\sigma} = 0$$

to obtain the identity

$$0 = \xi_{\sigma;\rho\mu} - \xi_{\sigma;\mu\rho} + \xi_{\mu;\sigma\rho} - \xi_{\mu;\rho\sigma} + \xi_{\rho;\mu\sigma} - \xi_{\rho;\sigma\mu} . \tag{2}$$

For a Killing vector, Equation (2) becomes

$$0 = \xi_{\sigma;\rho\mu} - \xi_{\sigma;\mu\rho} - \xi_{\mu;\rho\sigma} , \tag{3}$$

and substituting Equation (3) into Equation (1), one obtains

$$\xi_{\mu;\rho\sigma} = R^{\lambda}{}_{\sigma\rho\mu}\xi_{\lambda} . \tag{4}$$

Solution 10.8. We first show that "stationarity" is equivalent to the existence of a time coordinate for which $g_{\alpha\beta,t} = 0$. Choose the time coordinate such that $\boldsymbol{\xi} = \partial/\partial t$, then (see Problems 10.2, 8.13) $\mathcal{L}_{\boldsymbol{\xi}} g = 0$, or

$$g_{\alpha\beta,\gamma}\xi^\gamma + g_{\alpha\gamma}\xi^\gamma{}_{,\beta} + g_{\beta\gamma}\xi^\gamma{}_{,\alpha} = 0 \ .$$

Since $\xi^\gamma = (1,0,0,0)$ this proves that $g_{\alpha\beta,t} = 0$.

Time reversibility (definition (i)) means $g_{\alpha\beta}$ is independent of t and $g_{ti} = 0$. To show equivalence of the definitions it remains to show that $g_{ti} = 0$ is equivalent to the hypersurface orthogonality of $\boldsymbol{\xi} = \partial/\partial t$.

If $g_{ti} = 0$, then $\xi_\alpha = g_{\alpha\beta}\xi^\beta = g_{\alpha t} = 0$ unless $\alpha = t$. Hence ξ_α is proportional to $t_{,\alpha}$ i.e. $\boldsymbol{\xi}$ is orthogonal to the surfaces t = constant. If $\boldsymbol{\xi}$ is hypersurface orthogonal (see Problem 7.23),

$$\xi_{[\alpha;\beta}\xi_{\gamma]} = 0 \ .$$

With Killing's equation $\xi_{(\alpha;\beta)} = 0$, this becomes

$$\xi_{\alpha;\beta}\xi_\gamma + \xi_{\gamma;\alpha}\xi_\beta + \xi_{\beta;\gamma}\xi_\alpha = 0 \ .$$

Dotting with ξ^α and using Killing's equation on the first and third terms gives

$$\tfrac{1}{2}(\xi_{\alpha;\beta} - \xi_{\beta;\alpha})\xi^2 + \tfrac{1}{2}\xi^2{}_{,\alpha}\xi_\beta - \tfrac{1}{2}\xi^2{}_{,\beta}\xi_\alpha = 0$$

$$\xi^2 \equiv \boldsymbol{\xi}\cdot\boldsymbol{\xi}$$

that is

$$(\xi^{-2}\xi_\beta)_{;\alpha} - (\xi^{-2}\xi_\alpha)_{;\beta} = 0 \ .$$

Thus $\xi^{-2}\xi_\alpha$ is a gradient and

$$\xi_\alpha = \xi^2 h_{,\alpha}$$

for some h. Since $\xi^2 = g_{tt}$ and $\xi_\alpha = g_{\alpha t}$, we have $g_{\alpha t} = g_{tt}h_{,\alpha}$. Putting $\alpha = t$ we see that $h_{,t} = 1$ or $h = t + f(x^i)$. By choosing a new time coordinate $t' = t + f(x^i)$, we have

$$g_{it'} = g_{tt} h_{,i} = g_{tt} t'_{,i} = 0 \,,$$

also

$$\xi^{\alpha'} = \frac{\partial x^{\alpha'}}{\partial x^{\beta}} \xi^{\beta} = \frac{\partial x^{\alpha'}}{\partial t} = (1, 0, 0, 0)$$

so $g_{\alpha\beta}$ is still independent of t'.

Solution 10.9. Choose a coordinate system in which the metric is $\eta_{\mu\nu}$ (all of the Christoffel symbols vanish). Then Killing's equation for a Killing vector $\boldsymbol{\xi}$ becomes

$$2\xi_{(\mu;\nu)} = \xi_{\mu,\nu} + \xi_{\nu,\mu} = 0 \,. \tag{1}$$

We now enumerate and classify linearly independent solutions of Equation (1):

(i) Translation Killing vectors:

$$\xi_i^{\ \mu} = a_i^{\ \mu} \qquad i = 1-4 \tag{2}$$

where a_i are constant vectors.

(ii) Rotational Killing vectors:

$$\xi_k^0 = 0, \quad \xi_k^1 = \varepsilon^{1km}\, x_m \qquad k = 1-3 \,. \tag{3}$$

Check:

$$\begin{aligned} \xi_{k(\mu,\nu)} &= \xi_{k(i,j)} = x_{m,(j}\varepsilon_{i)km} \\ &= \delta_{m(j}\varepsilon_{i)km} = -\varepsilon_{k(ij)} = 0 \,. \end{aligned}$$

(iii) Boost Killing vectors:

$$\xi_\mu^{\ k} = \delta_\mu^{[0} x^{k]} \qquad k = 1, 3 \,. \tag{4}$$

Check:

$$\xi^k_{\ (\mu,\nu)} = \frac{1}{2}\left[\delta_\mu^{[0}\delta_\nu^{k]} + \delta_\nu^{[0}\delta_\mu^{k]}\right] = 0 \,.$$

There are ten Killing vectors in Equations (2)-(4), given in the coordinate system in which $g_{\mu\nu} = (-1, 1, 1, 1)$. In any other coordinate system one can find the components of the ξ^μ_i, for $i = 1-10$, by transforming them as vectors.

Solution 10.10. The change in $\mathbf{u}\cdot\boldsymbol{\xi}$ along the geodesic is given by

$$\nabla_{\mathbf{u}}(\mathbf{u}\cdot\boldsymbol{\xi}) = (\nabla_{\mathbf{u}}\mathbf{u})\cdot\boldsymbol{\xi} + \mathbf{u}\cdot(\nabla_{\mathbf{u}}\boldsymbol{\xi}) .$$

Now $\nabla_{\mathbf{u}}\mathbf{u} = 0$ since $\mathbf{u}$ is tangent to a geodesic, and $\mathbf{u}\cdot\nabla_{\mathbf{u}}\boldsymbol{\xi} = 0$ since $\boldsymbol{\xi}$ is a Killing vector. In components,

$$\mathbf{u}\cdot\nabla_{\mathbf{u}}\boldsymbol{\xi} = u^\alpha u^\beta \xi_{\alpha;\beta} = u^\alpha u^\beta \xi_{(\alpha;\beta)} = 0 .$$

Thus $\mathbf{u}\cdot\boldsymbol{\xi}$ = constant along a geodesic.

Solution 10.11. The divergence of $\mathbf{J}$ is

$$\begin{aligned} J^\mu{}_{;\mu} &= (T^{\mu\nu}\xi_\nu)_{;\mu} \\ &= T^{\mu\nu}{}_{;\mu}\xi_\nu + T^{\mu\nu}\xi_{\nu;\mu} \\ &= 0 + T^{\mu\nu}\xi_{(\nu;\mu)} = 0 . \end{aligned}$$

When $\boldsymbol{\xi} = \partial/\partial t$, then $J^\mu = T^\mu{}_0 = -T^{\mu 0}$ at infinity where $g_{00} = -1$. Thus $\mathbf{J}$ is the negative of the energy flux vector for a stationary observer.

Solution 10.12. From Problem 10.11, the integrand $J^\alpha \equiv T^\alpha{}_\beta \xi^\beta$ is a conserved quantity: $J^\alpha{}_{;\alpha} = 0$. The result then follows from the identity $J^\alpha{}_{;\alpha} = (|g|^{\frac{1}{2}} J^\alpha)_{,\alpha} |g|^{-\frac{1}{2}}$ and from Gauss's theorem:

$$\begin{aligned} 0 = \int J^\alpha{}_{;\alpha} |g|^{\frac{1}{2}} d^4x &= \int (|g|^{\frac{1}{2}} J^\alpha)_{,\alpha} d^4x = \oint J^\alpha d^3\Sigma_\alpha \\ &= \int_{F_2} J^\alpha d^3\Sigma_\alpha - \int_{F_1} J^\alpha d^3\Sigma_\alpha . \end{aligned}$$

Here we have assumed that $J \to 0$ sufficiently rapidly at infinity that the contributions to the integral from the edges can be neglected. Thus the integral is independent of the spacelike hypersurface F.

Solution 10.13. From Problem 10.11, we know that for each Killing vector ξ, $J^{\mu} \equiv T^{\mu\nu}\xi_{\nu}$ is a conserved vector, i.e.

$$J^{\mu}{}_{;\mu} = |g|^{-\frac{1}{2}}(|g|^{\frac{1}{2}}J^{\alpha})_{,\alpha} = 0 \,. \tag{1}$$

From Problem 10.9, we know that in flat spacetime there are ten linearly independent Killing vectors $\xi^{\mu}{}_{(i)}$, $i = 1-10$. Thus there are ten conserved vectors

$$J^{\mu}{}_{(i)} \equiv T^{\mu\nu}\xi_{(i)\nu} \,. \tag{2}$$

From Equations (1) and (2), one may construct ten globally conserved quantities

$$Q_{(i)} \equiv \int |g|^{\frac{1}{2}} J^{0}{}_{(i)} d^3x, \qquad i = 1-10 \tag{3}$$

since

$$\frac{dQ_{(i)}}{dt} = \int (|g|^{\frac{1}{2}} J_{(i)}{}^{0})_{,0}\, d^3x = -\int (|g|^{\frac{1}{2}} J_{(i)}{}^{k})_{,k}\, d^3x$$

$$= -\int |g|^{\frac{1}{2}} J_{(i)}{}^{k} d^2\Sigma_k \,.$$

This last expression represents the flux of the $J_{(i)}$ current through the (2-dimensional) surface bounding the integration volume. If T falls off fast enough at large distances, and the integration volume is infinite, this flux term vanishes and $dQ_{(i)}/dt = 0$. [Alternative proof: apply Solution 10.12.]

The four Q's derived from translation Killing vectors correspond to conserved energy and momentum; the three derived from rotation Killing vectors correspond to conserved angular momenta. To see the meaning of the three derived from boost Killing vectors, write (e.g. in flat cartesian coordinates with $\xi_{\nu} = [x, -t, 0, 0]$)

$$Q = \int T^{0\nu}\xi_\nu \, dx\, dy\, dz$$

$$= \int x\, T^{00}\, dx\, dy\, dz - t \int T^{0x}\, dx\, dy\, dz$$

$$= x_{(C.M.)}M_{(C.M.)} - tP_{x_{(C.M.)}} = M_{(C.M.)}(x_{(C.M.)} - v_{(C.M.)}t)$$

which is a component of the conserved "origin of uniform motion" for the center of mass (C.M.) of the system. ($x_{(C.M.)}$ is defined to be $\int x\, T^{00}\, dx\, dy\, dz / \int T^{00}\, dx\, dy\, dz$.)

Solution 10.14. The components of **u** are

$$u_\alpha = \xi_\alpha/(-\xi_\gamma \xi^\gamma)^{\frac{1}{2}}$$

(the minus sign is chosen to make the term in parentheses positive), so

$$a_\alpha = u_{\alpha;\beta}u^\beta = \left[\frac{\xi_{\alpha;\beta}}{(-\xi_\gamma\xi^\gamma)^{\frac{1}{2}}} + \frac{\xi_\alpha \xi_{\gamma;\beta}\xi^\gamma}{(-\xi_\tau\xi^\tau)^{\frac{3}{2}}}\right]\frac{\xi^\beta}{(-\xi_\mu\xi^\mu)^{\frac{1}{2}}} \,.$$

The second term vanishes because $\xi_{\gamma;\beta}\xi^\gamma\xi^\beta = \xi_{(\gamma;\beta)}\xi^\gamma\xi^\beta = 0$. Since $\xi_{\alpha;\beta} = -\xi_{\beta;\alpha}$, the first term gives

$$a_\alpha = \frac{\xi_{\beta;\alpha}\xi^\beta}{+\xi_\gamma\xi^\gamma} = \frac{1}{2}\frac{(\xi_\beta\xi^\beta)_{;\alpha}}{(\xi_\gamma\xi^\gamma)} = \frac{1}{2}\left[\log(-\xi_\beta\xi^\beta)\right]_{,\alpha} \,.$$

Solution 10.15. Consider a particle of rest mass μ and conserved energy $E = -\mathbf{p}\cdot\boldsymbol{\xi}$ where **p** is the 4-momentum and $\boldsymbol{\xi}$ is the time Killing vector. Not all values of E/μ are possible for trajectories through a given point in spacetime. For example, particles at radial infinity must have $E/\mu \geq 1$. To find the bound on E/μ for a general point, pick an orthonormal frame at the point. The 4-velocity of a particle has components $\mathbf{u} = (\gamma, \gamma \underset{\sim}{v})$

with $\underset{\sim}{v}$ a 3-vector and $\gamma = (1-v^2)^{-\frac{1}{2}}$; the time Killing vector has components $\boldsymbol{\xi} = (\xi_0, \underset{\sim}{\xi})$, with $\underset{\sim}{\xi}$ a 3-vector. The particle's ratio of energy to rest mass is given by

$$E/\mu = -\mathbf{u}\cdot\boldsymbol{\xi}_{(t)} = \gamma(\xi_0 - \underset{\sim}{v}\cdot\underset{\sim}{\xi}) \tag{1}$$

where the dot denotes the scalar product in the local Euclidean 3-space. Evidently, a necessary (but not a sufficient) condition for an extremum (hence a bound) on E/μ is

$$\underset{\sim}{v}\cdot\underset{\sim}{\xi} = \pm v\xi . \tag{2}$$

Here $v \equiv |\underset{\sim}{v}|$, $\xi \equiv |\underset{\sim}{\xi}|$. Now we distinguish two cases: If $\boldsymbol{\xi}$ is spacelike (e.g. in the ergosphere of a Kerr blackhole), then we have $\xi_0 < \xi$; and inspection of Equation (1) shows that all values of E/μ are possible, that is $-\infty < E/\mu < +\infty$. This is the answer for regions in which $\boldsymbol{\xi}$ is spacelike. The infinite limits correspond to $v \to 1$ with the two signs of Equation (2). If, instead, $\boldsymbol{\xi}$ is timelike (e.g. at radial infinity), so that $\xi_0 > \xi$, then the right-hand side of Equation (1) is always positive, and there is a nontrivial lower bound on E/μ. Writing Equation (1) and using Equation (2) with the upper sign, we obtain

$$(\xi^2 + E^2/\mu^2)v^2 - 2\xi\xi_0 v + (\xi_0^2 - E^2/\mu^2) = 0 .$$

The extremum in E/μ is obtained by setting the discriminant of this equation, a quadratic in v, equal to zero; this gives

$$0 = (E/\mu)^2\,[(E/\mu)^2 - \xi^2 + \xi_0^2] .$$

The root $E/\mu = 0$ is spurious, and the lower bound on E/μ is

$$(E/\mu)^2 = \xi_0^2 - \xi^2 = -\boldsymbol{\xi}\cdot\boldsymbol{\xi} .$$

We see that the allowed range of E/μ at a point depends only on the norm of the time Killing vector at that point:

$$(-\boldsymbol{\xi}\cdot\boldsymbol{\xi})^{\frac{1}{2}} \leq E/\mu < +\infty$$

for $\boldsymbol{\xi}$ timelike.

Solution 10.16. For a test electromagnetic field, all we have to check is that Maxwell's equations are satisfied in the given background metric. If A^{μ}, the vector potential, is a Killing vector, then

$$A^{\mu}{}_{;\mu} = 0$$

i.e. the Lorentz condition is satisfied, and the wave equation

$$A^{\mu;\nu}{}_{\nu} - R^{\mu}{}_{\nu}A^{\nu} = 0$$

is satisfied by Problem 10.6 and the fact that $R^{\mu}{}_{\nu} = 0$ in vacuum.

If in Minkowski space in spherical polar coordinates $A \propto \partial/\partial\phi$, i.e. only A^{ϕ} = constant is nonzero, then $A^{\hat{\phi}} \propto r\sin\theta$, or in conventional 3-dimensional notation,

$$\underset{\sim}{A} = a\, r\sin\theta\, \underset{\sim}{e}_{\hat{\phi}} \qquad (a = \text{constant})\ .$$

Hence $\underset{\sim}{E} = 0$ and $\underset{\sim}{B} = \nabla\times\underset{\sim}{A} = 2a(\cos\theta\, \underset{\sim}{e}_{\hat{r}} - \sin\theta\, e_{\hat{\theta}}) = 2a\, \underset{\sim}{e}_{\hat{z}}$. Thus there is a uniform magnetic field parallel to the z axis. (This problem is due to Robert M. Wald.)

CHAPTER 11: SOLUTIONS

Solution 11.1.

(i)
$$dJ/d\tau = \frac{d(\mathbf{\Delta}x)}{d\tau}\otimes p + \mathbf{\Delta}x\otimes\frac{dp}{d\tau} - \frac{dp}{d\tau}\otimes\mathbf{\Delta}x - p\otimes\frac{d(\mathbf{\Delta}x)}{d\tau}$$
$$= u\otimes p - p\otimes u$$
$$= \frac{1}{m}\,p\otimes p - p\otimes\frac{1}{m}\,p = 0\,.$$

(ii) Consider times immediately before and immediately after the collision. Since all particles are at the same point at the collision,

$$\mathbf{\Delta}x_{(k)}\big|_{\text{before}} = \mathbf{\Delta}x_{(k)}\big|_{\text{after}} = \mathbf{\Delta}x\ ,$$

(where $\mathbf{\Delta}x$ is the displacement to the point) and thus

$$\sum_{(k)} J_{(k)}\big|_{\text{after}} = \mathbf{\Delta}x\otimes\sum_{(k)} p_{(k)}\big|_{\text{after}} - \Big(\sum_{(k)} p_{(k)}\big|_{\text{after}}\Big)\otimes\mathbf{\Delta}x$$
$$= \mathbf{\Delta}x\otimes\sum_{(k)} p_{(k)}\big|_{\text{before}} - \Big(\sum_{(k)} p_{(k)}\big|_{\text{before}}\Big)\otimes\mathbf{\Delta}x$$
$$= \sum_{(k)} J_{(k)}\big|_{\text{before}}\ ,$$

since total momentum $\sum_{(k)} p_{(k)}$ is conserved.

Solution 11.2.

(a) For the "angular momentum density" $J^{\alpha\beta\gamma}$ defined by

$$J^{\alpha\beta\gamma} \mathrel{\dot=} 2x^{[\alpha}T^{\beta]\gamma} = x^{\alpha}T^{\beta\gamma} - x^{\beta}T^{\alpha\gamma}$$

we have

$$J^{\alpha\beta\gamma}{}_{,\gamma} = \delta^{\alpha}{}_{\gamma} T^{\beta\gamma} + x^{\alpha} T^{\beta\gamma}{}_{,\gamma} - \delta^{\beta}{}_{\gamma} T^{\alpha\gamma} - x^{\beta} T^{\alpha\gamma}{}_{,\gamma} = T^{\beta\alpha} - T^{\alpha\beta} = 0 .$$

Thus $J^{\alpha\beta} \equiv \int J^{\alpha\beta\gamma} d^3\Sigma_{\gamma}$ is a conserved quantity, by Gauss' theorem.

(b) $$J^{\alpha\beta}(x^{\sigma}+a^{\sigma}) = \int (a^{\alpha} T^{\beta\gamma} - a^{\beta} T^{\alpha\gamma}) d^3\Sigma_{\gamma} + J^{\alpha\beta}(x^{\sigma}) .$$

Thus $J^{\alpha\beta}$ is not invariant under coordinate translations (different angular momentum about different points).

(c) Consider the time derivative of the spin:

$$\frac{dS_{\alpha}}{dt} = -\frac{1}{2} \varepsilon_{\alpha\beta\gamma\delta} \left(\frac{dJ^{\beta\gamma}}{dt} u^{\delta} + J^{\beta\gamma} \frac{du^{\delta}}{dt} \right) .$$

The first term on the right vanishes from part (a). The second term vanishes because the system has no forces acting on it ($du^{\delta}/dt = 0$). Thus $dS^{\alpha}/dt = 0$.

(d) From (b),

$$J^{\alpha\beta}(x^{\sigma}+a^{\sigma}) = a^{\alpha} P^{\beta} - a^{\beta} P^{\alpha} + J^{\alpha\beta}(x^{\sigma}) .$$

and thus

$$S_{\alpha}(x^{\sigma}+a^{\sigma}) = -\frac{1}{2} \varepsilon_{\alpha\beta\gamma\delta} (a^{\alpha}P^{\beta} - a^{\beta}P^{\alpha}) P^{\delta}/|P| + S_{\alpha}(x^{\sigma}) .$$

The first term on the right vanishes (ε is totally antisymmetric and P^{μ} occurs quadratically) so $S_{\alpha}(x^{\sigma}+a^{\sigma}) = S_{\alpha}(x^{\sigma})$.

Solution 11.3. From the definition of S_{α},

$$u^{\alpha} S_{\alpha} = -\frac{1}{2} \varepsilon_{\alpha\beta\gamma\delta} J^{\beta\gamma} u^{\delta} u^{\alpha} = 0$$

by antisymmetry of ε and symmetry of $u^{\delta}u^{\alpha}$.

Solution 11.4. Consider an observer in the local inertial frame comoving with the center of mass of the gyroscope. Because there are no torques, he sees no precession of the spin axis, so $d\underset{\sim}{S}/dt = 0$. But in this frame the 4-velocity of the gyroscope is $\mathbf{u} = (1, \underset{\sim}{0})$, so we can write the condition of zero torque as

$$\nabla_{\mathbf{u}} \mathbf{S} = g\mathbf{u}$$

where g is some constant of proportionality. We find g from the fact that $\mathbf{S} \cdot \mathbf{u} = 0$:

$$\begin{aligned} 0 = \nabla_{\mathbf{u}}(\mathbf{S}\cdot\mathbf{u}) &= (\nabla_{\mathbf{u}}\mathbf{S})\cdot\mathbf{u} + \mathbf{S}\cdot(\nabla_{\mathbf{u}}\mathbf{u}) \\ &= g\mathbf{u}\cdot\mathbf{u} + \mathbf{S}\cdot\mathbf{a} \\ &= -g + \mathbf{S}\cdot\mathbf{a} \ . \end{aligned}$$

Thus

$$\nabla_{\mathbf{u}} \mathbf{S} = (\mathbf{S}\cdot\mathbf{a})\,\mathbf{u}$$

which is the equation for Fermi-Walker transport

$$\nabla_{\mathbf{u}} \mathbf{S} = (\mathbf{S}\cdot\mathbf{a})\,\mathbf{u} - (\mathbf{S}\cdot\mathbf{u})\,\mathbf{a}$$

with $\mathbf{S}\cdot\mathbf{u} = 0$.

Solution 11.5.

(a) About the center of mass in the center-of-momentum frame: $\int x^i T^{00} d^3x = 0$ (center of mass) and $\int T^{i0} d^3x = 0$ (center of momentum). Thus we have

$$J^{i0} = \int d^3x\,(x^i T^{00} - t\, T^{i0}) = 0 \ ,$$

or, in frame independent notation $J^{\alpha\beta} u_\beta = 0$.

(b) From the definition of S_δ:

$$-\,\varepsilon^{\alpha\beta\gamma\delta} S_\gamma u_\delta = \frac{1}{2}\, \varepsilon^{\alpha\beta\gamma\delta}\, \varepsilon_{\gamma\mu\nu\sigma} J^{\mu\nu} u^\sigma u_\delta \ .$$

Now perform the summation over γ [see Problems 3.27, 3.28]:

$$-\frac{1}{2}\, \delta^{\alpha\beta\delta}_{\mu\nu\sigma}\, J^{\mu\nu} u^\sigma u_\delta = -\,J^{\alpha\beta} u^\delta u_\delta = J^{\alpha\beta} \equiv S^{\alpha\beta} .$$

[Only the $J^{\alpha\beta}$ terms are kept in the sum; by part (a) we cannot have a δ index on J due to the presence of u_δ.]

Solution 11.6. First we calculate the total angular momentum. We work in the center-of-momentum frame and take as our origin 0 the event at which the center of mass of A and that of B would collide. About this origin both systems A and B have only intrinsic angular momentum (see Problem 11.5), so

$$\underset{A+B}{J}^{\alpha\beta}(0) = \underset{A}{S}^{\alpha\beta} + \underset{B}{S}^{\alpha\beta} = -\varepsilon^{\alpha\beta\mu\nu}(\underset{A}{S}_{\mu}\underset{A}{u}_{\nu} + \underset{B}{S}_{\mu}\underset{B}{u}_{\nu})$$

and by conservation of angular momentum this is equal to $\underset{C}{J}^{\alpha\beta}(0)$. We can now use this result in the definition of the spin vector to find

$$\begin{aligned}\underset{C}{S}_{\sigma} &= -\frac{1}{2}\,\varepsilon_{\sigma\alpha\beta\lambda}\,\underset{C}{J}^{\alpha\beta}\,\underset{C}{u}^{\lambda} \\ &= \frac{1}{2}\,\varepsilon_{\sigma\alpha\beta\lambda}\,\varepsilon^{\alpha\beta\mu\nu}(\underset{A}{S}_{\mu}\underset{A}{u}_{\nu} + \underset{B}{S}_{\mu}\underset{B}{u}_{\nu})\,\underset{C}{u}^{\lambda}\ .\end{aligned}$$

According to Problem 3.28 (see also Problem 3.27)

$$\varepsilon_{\sigma\alpha\beta\lambda}\,\varepsilon^{\alpha\beta\mu\nu} = -2\delta^{\mu\nu}_{\sigma\lambda}$$

so that

$$\begin{aligned}\underset{C}{S}_{\sigma} &= -\delta^{\mu\nu}_{\sigma\lambda}(\underset{A}{S}_{\mu}\underset{A}{u}_{\nu} + \underset{B}{S}_{\mu}\underset{B}{u}_{\nu})\,\underset{C}{u}^{\lambda} \\ &= -\underset{A}{S}_{\sigma}(\underset{A}{u}\cdot\underset{C}{u}) - \underset{B}{S}_{\sigma}(\underset{B}{u}\cdot\underset{C}{u}) + \underset{A}{u}_{\sigma}(\underset{A}{S}\cdot\underset{C}{u}) + \underset{B}{u}_{\sigma}(\underset{B}{S}\cdot\underset{C}{u})\ .\end{aligned}$$

We can now substitute $(\underset{A}{P}+\underset{B}{P})/|\underset{A}{P}+\underset{B}{P}|$ for $\underset{C}{\mathbf{u}}$ to complete the solution.

Solution 11.7. The basic equation (see Problem 11.4) is

$$dS/d\tau = \mathbf{u}(a\cdot S)\ .$$

In the lab frame, the world line of the particle is

$$x = r\cos\omega t$$

$$y = r\sin\omega t$$

(where ω and r are constants) and hence

$$u^0 = \gamma, \quad u^x = \gamma \frac{dx}{dt} = -\omega r \gamma \sin\omega t, \quad u^y = \gamma\omega r \cos\omega t, \quad u^z = 0$$

where $\gamma \equiv (1 - r^2\omega^2)^{-\frac{1}{2}}$. Now $\mathbf{a} = d\mathbf{u}/d\tau$ so

$$a^0 = 0\,(\gamma = \text{const}), \quad a^x = \gamma \frac{du^x}{dt}, \quad a^y = -\gamma^2\omega^2 r \sin\omega t, \quad a^z = 0$$

and hence

$$S\cdot a = S^x a^x + S^y a^y = -\omega^2 r \gamma^2 \cos\omega t\, S^x - \omega^2 r \gamma^2 \sin\omega t\, S^y .$$

We have therefore

$$\frac{dS^0}{d\tau} = \gamma \frac{dS^0}{dt} = u^0(a\cdot S) = \gamma(a\cdot S) \tag{1}$$

$$\frac{dS^x}{d\tau} = \gamma \frac{dS^x}{dt} = u^x(a\cdot S) = -\omega r \gamma \sin\omega t(a\cdot S) \tag{2}$$

$$\frac{dS^y}{d\tau} = \gamma \frac{dS^y}{dt} = u^y(a\cdot S) = \omega r \gamma \cos\omega t(a\cdot S) \tag{3}$$

$$\frac{dS^z}{d\tau} = 0 . \tag{4}$$

Introduce radial and tangential components:

$$S^x = S^r \cos\omega t - S^\theta \sin\omega t$$

$$S^y = S^r \sin\omega t + S^\theta \cos\omega t .$$

Equations (2) and (3) imply:

$$dS^r/dt = \omega S^\theta \tag{5}$$

$$dS^\theta/dt = -\omega\gamma^2 S^r . \tag{6}$$

Equations (5) and (6) imply:

$$d^2S^r/dt^2 = -\omega^2\gamma^2 S^r \implies S^r = A\cos(\omega\gamma t + \alpha), \quad A, \alpha \text{ const.} \tag{7}$$

and from Equation (5)

$$S^{\theta} = -\gamma A \sin(\omega\gamma t+\alpha) . \tag{8}$$

From Equations (4), (7), (8) we obtain

$$S^x = A[\cos\omega t \cos(\omega\gamma t+\alpha) + \gamma \sin\omega t \sin(\omega\gamma t+\alpha)] \tag{9a}$$

$$S^y = A[\sin\omega t \cos(\omega\gamma t+\alpha) - \gamma \cos\omega t \sin(\omega\gamma t+\alpha)] \tag{9b}$$

$$S^z = \text{const.} \tag{9c}$$

Consider initial conditions $S^x = \hbar(2)^{-\frac{1}{2}}$, $S^y = 0$, $S^z = \frac{1}{2}\hbar$. (Although we are not treating the electron spin quantum mechanically, we want $S^2 = 3\hbar^2/4$.) These conditions imply $\alpha = 0$, $A = \hbar(2)^{-\frac{1}{2}}$; and Equations (9) can be written as

$$S^x + iS^y = 2^{-\frac{1}{2}}\hbar\,[e^{-i(\gamma-1)\omega t} + i(1-\gamma)\sin(\omega\gamma t)e^{i\omega t}] . \tag{10}$$

The first term on the right hand side of Equation (10) and Equation (9c) indicate a precession around the z-axis with angular velocity

$$\omega_{\text{Thomas}} = (\gamma-1)\omega \approx \frac{1}{2}v^2\omega \tag{11}$$

while the second term in Equation (10) is small $(1-\gamma \approx -\frac{1}{2}v^2)$ for an electron in an atom.

Solution 11.8. Do the computation in the local comoving Lorentz frame of the center of mass of the body. Treating the center of mass as the reference point in the equation of geodesic deviation, the relative acceleration of a mass element at position $x^{\hat{j}}$ is

$$\frac{d^2x^{\hat{j}}}{dt^2} = -R^{\hat{j}}{}_{\hat{0}\hat{k}\hat{0}}x^{\hat{k}} . \tag{1}$$

The ith component of the torque per unit volume is then $-\varepsilon_{\widetilde{\hat{i}\hat{l}\hat{j}}}x^{\hat{l}}\rho R^{\hat{j}}{}_{\hat{0}\hat{k}\hat{0}}x^{\hat{k}}$, where ρ is the mass density at $x^{\hat{j}}$. The total torque, which is just the time derivative of the intrinsic angular momentum, is

$$\frac{dS_{\hat{i}}}{dt} = -\varepsilon_{\hat{i}\hat{l}\hat{j}} R^{\hat{j}}{}_{\hat{0}\hat{k}\hat{0}} \int \rho x^{\hat{l}} x^{\hat{k}} d^3x \tag{2}$$

if we approximate $R^{\hat{j}}_{\hat{0}\hat{k}\hat{0}}$ as constant over the body. Because of the symmetry properties of $\varepsilon_{\hat{i}\hat{l}\hat{j}} R^{\hat{j}}{}_{\hat{0}\hat{k}\hat{0}}$, Equation (2) is equivalent to

$$\frac{dS_{\hat{i}}}{dt} = -\varepsilon_{\hat{i}\hat{l}\hat{j}} t_{\hat{l}\hat{k}} R_{\hat{j}\hat{0}\hat{k}\hat{0}} , \tag{3}$$

where

$$t_{\hat{l}\hat{k}} \equiv \int \rho \, (x^{\hat{l}} x^{\hat{k}} - \tfrac{1}{3} r^2 \delta^{\hat{l}\hat{k}}) \, d^3x .$$

If one now defines the quadrupole tensor, $t^{\alpha\beta}$ to be such that $t^{\alpha\beta} u_\beta = 0$ i.e. to have only spatial components in a comoving local frame, then the invariant tensor expression of Equation (3) is

$$\frac{DS^\kappa}{d\tau} = \varepsilon^{\kappa\beta\alpha\mu} u_\mu u^\sigma u^\lambda t_{\beta\eta} R^\eta{}_{\sigma\alpha\lambda} . \tag{4}$$

Note that either in the case of a spherical body, or a body small enough so that Riemann $\times$ (size of body)^2 ≈ 0, Equation (4) reduces to $DS^\kappa/d\tau = 0$. (Cf. Problem 11.4 with $a = 0$.)

Solution 11.9. We shall use two coordinate systems: the XYZ coordinates are the spatial coordinates of a local Lorentz frame with spatial origin at the center of mass of the Earth. The ecliptic plane is the XY plane, and we shall assume that the Sun and the Moon move in circular orbits in this plane as seen from the Earth. The xyz coordinates are the spatial coordinates of a similar Lorentz frame, but with the z-axis parallel to the Earth's angular momentum $\underset{\sim}{J}$. Choose the x-axis parallel to the X-axis. The relation between basis vectors for the two coordinate systems is

$$\begin{aligned} \underset{\sim}{e}_x &= \underset{\sim}{e}_X \\ \underset{\sim}{e}_y &= \cos\psi \, \underset{\sim}{e}_Y - \sin\psi \, \underset{\sim}{e}_Z \\ \underset{\sim}{e}_z &= \sin\psi \, \underset{\sim}{e}_Y + \cos\psi \, \underset{\sim}{e}_Z . \end{aligned} \tag{1}$$

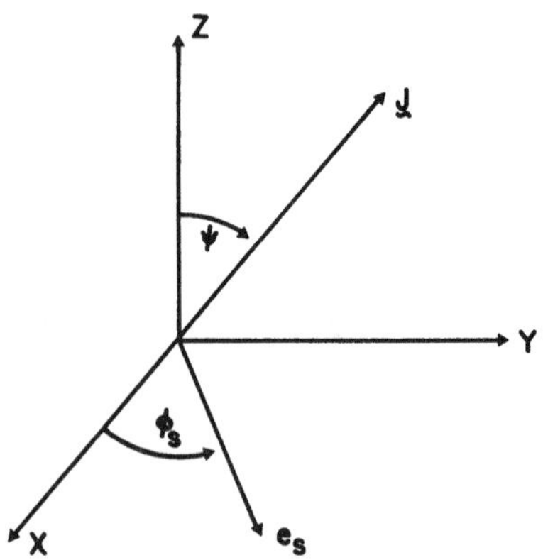

Here $\psi \approx 23\frac{1}{2}^0$ is the angle between $\underset{\sim}{J}$ and the Z-axis, which remains constant if we ignore the small nutation. As time goes on, $\underset{\sim}{J}$ precesses about the Z-axis with a period T which we wish to find.

Let $\underset{\sim}{e}_s$ be a unit vector pointing to the Sun. If the sun has polar coordinates (r_s, θ, ϕ) with respect to the xyz coordinates and $(r_s, \frac{\pi}{2}, \phi_s)$ with respect to the XYZ coordinates, then

$$\begin{aligned}\cos\theta &= \underset{\sim}{e}_z \cdot \underset{\sim}{e}_s \\ &= (\sin\psi\, \underset{\sim}{e}_Y + \cos\psi\, \underset{\sim}{e}_Z)\cdot(\cos\phi_s\, \underset{\sim}{e}_X + \sin\phi_s\, \underset{\sim}{e}_Y) \\ &= \sin\psi\, \sin\phi_s\end{aligned}$$

and similarly,

$$\sin\theta\,\sin\phi = \underset{\sim}{e}_y \cdot \underset{\sim}{e}_s = \cos\psi\,\sin\phi_s$$

$$\sin\theta\,\cos\phi = \underset{\sim}{e}_x \cdot \underset{\sim}{e}_s = \cos\phi_s\ .$$

From Problem 11.8, the equation of motion of the spin vector $\underset{\sim}{J}$ of the Earth in an external gravitational field is

$$\frac{d\underset{\sim}{J}}{dt} = \underset{\sim}{N}$$

where the torque is

$$N_i = -\varepsilon_{ijk} t_{jm} R_{k0m0} .$$

Here the "reduced quadrupole moment tensor" is

$$t_{jm} = \int \rho (x^j x^m - \tfrac{1}{3} r^2 \delta^{jm}) d^3x .$$

Take the Earth to be spheroidal, with moment of inertia C about the rotation axis and A about an equatorial axis. Then in xyz coordinates

$$C = \int \rho (x^2 + y^2) d^3x$$

$$A = \int \rho (x^2 + z^2) d^3x = \int \rho (y^2 + z^2) d^3x .$$

The nonzero components of t_{jm} are

$$t_{zz} = -\tfrac{2}{3} (C - A)$$

$$t_{xx} = t_{yy} = \tfrac{1}{2} (C - A)$$

so that

$$N_x = -(C - A) R_{z0y0}$$

$$N_y = (C - A) R_{z0x0}$$

$$N_z = 0 .$$

In the weak-field, slow-motion limit,

$$R_{z0i0} = \frac{\partial^2 \Phi}{\partial z \, \partial x^i}$$

where Φ is the Newtonian gravitational potential:

$$\Phi = \Phi_{SUN} + \Phi_{MOON} .$$

Since

$$\Phi_{SUN}(x, y, z) = \frac{M_s}{[(x-x_s)^2 + (y-y_s)^2 + (z-z_s)^2]^{\frac{1}{2}}}$$

we find

$$\left.\frac{\partial^2 \Phi}{\partial z\, \partial x}\right|_0 = -\frac{3M_s \sin\theta \cos\theta \cos\phi}{r_s^{\,3}}$$

$$\left.\frac{\partial^2 \Phi}{\partial z\, \partial y}\right|_0 = -\frac{3M_s \sin\theta \cos\theta \sin\phi}{r_s^{\,3}} \; .$$

Thus $\underset{\sim}{N}_{SUN}$ has (x, y, z) components

$$\begin{aligned} \underset{\sim}{N}_{SUN} &= 3a(C-A)\sin\theta \cos\theta (\sin\phi, -\cos\phi, 0) \\ &= 3a(C-A)\sin\psi \sin\phi_s (\cos\psi \sin\phi_s, -\cos\phi_s, 0) \end{aligned}$$

where $a \equiv M_s/r_s^{\,3}$. Since the Sun's angular velocity about the Earth is much greater than the precession angular velocity we wish to find, we can average $\underset{\sim}{N}_{SUN}$ over the Sun's orbit. The term $\sin^2\phi_S$ gives $\frac{1}{2}$, while $\sin\phi_s \cos\phi_s$ averages to zero. Thus the only nonzero component of $\underset{\sim}{N}_{SUN}$ is

$$N_x^{SUN} = N_X^{SUN} = \frac{3}{2}\, a(C-A) \sin\psi \cos\psi \; .$$

A similar expression holds for the average torque exerted by the Moon, with a replaced by $b = M_M/r_M^3$. Thus, at $t = 0$, there is a torque in the X direction of magnitude

$$N = \frac{3}{2}(a+b)(C-A) \sin\psi \cos\psi \; .$$

In a small time dt therefore $\underset{\sim}{J}$ changes by an amount $dJ = Ndt$ perpendicular to itself. Thus $\underset{\sim}{J}$ rotates about the Z axis by an amount

$$d\chi = \frac{dJ}{J \sin\psi} = \frac{Ndt}{J \sin\psi} \; .$$

The precession period is the time for χ to change by 2π:

$$T = \frac{2\pi J \sin\psi}{N} .$$

Put $J = C\omega$, where ω is the angular velocity of rotation of the Earth. Then

$$T = \frac{4\pi}{3} \frac{C}{C-A} \frac{1}{\cos\psi} \frac{\omega}{a+b} .$$

Putting

$$C/(C-A) = 305.3$$

$$\psi = 23.45^0$$

$$\omega = 7.292 \times 10^{-5} \sec^{-1}$$

$$M_s = 1.989 \times 10^{33} g$$

$$r_s = 1.496 \times 10^{13} cm$$

$$M_M = 7.349 \times 10^{25} g$$

$$r_M = 3.844 \times 10^{10} cm$$

we find

$$T = 25{,}600 \text{ years} .$$

This agrees with observation to better than 1%; some error in the calculation arises because we treated the orbits as circular and because the Moon's orbit is not exactly in the ecliptic.

Solution 11.10. The observers have 4-velocity $\mathbf{u} = \boldsymbol{\xi}/|\boldsymbol{\xi}|$, where $|\boldsymbol{\xi}| = (-\boldsymbol{\xi}\cdot\boldsymbol{\xi})^{\frac{1}{2}}$. Thus

$$\mathcal{L}_{\boldsymbol{\xi}} \mathbf{e}_{\hat{0}} = \mathcal{L}_{\boldsymbol{\xi}} \mathbf{u} = [\boldsymbol{\xi}, \mathbf{u}] = [\nabla_{\boldsymbol{\xi}} (-\boldsymbol{\xi}\cdot\boldsymbol{\xi})^{-\frac{1}{2}}]\boldsymbol{\xi}$$

where we have used the fact that $[\boldsymbol{\xi}, \boldsymbol{\xi}] = 0$. But

$$\nabla_{\boldsymbol{\xi}}(\boldsymbol{\xi}\cdot\boldsymbol{\xi}) = (\xi^\alpha \xi_\alpha)_{;\beta}\xi^\beta = 2\xi_{\alpha;\beta}\xi^\alpha\xi^\beta = 0 ,$$

since $\xi_{(\alpha;\beta)} = 0$ by Killing's equation. Thus $\mathcal{L}_{\boldsymbol{\xi}} \mathbf{e}_{\hat{0}} = 0$. Also $\mathcal{L}_{\boldsymbol{\xi}} \mathbf{e}_{\hat{j}} = 0$

since $e_{\hat{j}}$ connects points of equal t along any observer's world line. (See Problem 8.14.)

(2) $$\mathcal{L}_{\xi} Q = \mathcal{L}_{\xi} (Q^{\hat{\alpha}\cdots\hat{\beta}} e_{\hat{\alpha}} \otimes \cdots \otimes e_{\hat{\beta}})$$

$$= (\mathcal{L}_{\xi} Q^{\hat{\alpha}\cdots\hat{\beta}}) e_{\hat{\alpha}} \otimes \cdots \otimes e_{\hat{\beta}} + Q^{\hat{\alpha}\cdots\hat{\beta}} (\mathcal{L}_{\xi} e_{\hat{\alpha}}) \otimes \cdots \otimes e_{\hat{\beta}}$$

$$+ \cdots + Q^{\hat{\alpha}\cdots\hat{\beta}} e_{\hat{\alpha}} \otimes \cdots \otimes (\mathcal{L}_{\xi} e_{\hat{\beta}}) .$$

Only the first term on the right-hand side is nonvanishing by part (1), so

$$(\mathcal{L}_{\xi} Q)^{\hat{\alpha}\cdots\hat{\beta}} = \mathcal{L}_{\xi} (Q^{\hat{\alpha}\cdots\hat{\beta}}) = \nabla_{\xi} Q^{\hat{\alpha}\cdots\hat{\beta}} = \frac{d}{dt} Q^{\hat{\alpha}\cdots\hat{\beta}}$$

(since $Q^{\hat{\alpha}\cdots\hat{\beta}}$ is just a scalar function). Measured in units of proper time, the rate is

$$\frac{d}{d\hat{t}} Q^{\hat{\alpha}\cdots\hat{\beta}} = \nabla_{\mathbf{u}} Q^{\hat{\alpha}\cdots\hat{\beta}} = \frac{1}{|\xi|} \nabla_{\xi} Q^{\hat{\alpha}\cdots\hat{\beta}} = \frac{1}{|\xi|} \frac{d}{dt} Q^{\hat{\alpha}\cdots\hat{\beta}} .$$

(3) In the observer's local rest frame, the precession equation is

$$\frac{dS_{\hat{j}}}{d\hat{t}} = \varepsilon_{\hat{j}\hat{k}\hat{l}} \omega^{\hat{k}} S^{\hat{l}}$$

where S is the gyroscope spin vector. Now $\varepsilon_{\hat{j}\hat{k}\hat{l}} = \varepsilon_{\hat{0}\hat{j}\hat{k}\hat{l}} = u^{\hat{\alpha}} \varepsilon_{\hat{\alpha}\hat{j}\hat{k}\hat{l}}$, so the 4-dimensional version of the precession equation is

$$\frac{dS^{\hat{\beta}}}{d\hat{t}} = u_{\hat{\alpha}} \varepsilon^{\hat{\alpha}\hat{\beta}\hat{\gamma}\hat{\delta}} \omega_{\hat{\gamma}} S_{\hat{\delta}}$$

that is

$$\frac{dS^{\hat{\beta}}}{dt} = \xi_{\hat{\alpha}} \varepsilon^{\hat{\alpha}\hat{\beta}\hat{\gamma}\hat{\delta}} \omega_{\hat{\gamma}} S_{\hat{\delta}} . \tag{1}$$

We wish to derive an equation with the form of Equation (1) and read off $\boldsymbol{\omega}$. We know that S is Fermi-Walker transported (see Problem 11.4), i.e.

$$\nabla_{\mathbf{u}} S = (S \cdot a) \mathbf{u} \tag{2}$$

and

$$\mathbf{S}\cdot\mathbf{u} = 0 \,. \tag{3}$$

Equation (2) becomes

$$\nabla_{\boldsymbol{\xi}}\mathbf{S} = (\mathbf{S}\cdot\nabla_{\boldsymbol{\xi}}\boldsymbol{\xi})\boldsymbol{\xi}/|\boldsymbol{\xi}|^2 \tag{4}$$

and Equation (3) implies

$$\mathbf{S}\cdot\boldsymbol{\xi} = 0 \,. \tag{5}$$

From part (2),

$$\frac{d\mathbf{S}}{dt} = \mathcal{L}_{\boldsymbol{\xi}}\mathbf{S} = \nabla_{\boldsymbol{\xi}}\mathbf{S} - \nabla_{\mathbf{S}}\boldsymbol{\xi} \,. \tag{6}$$

Since $\boldsymbol{\xi}$ plays the part of a 4-velocity (unnormalized), it is convenient to decompose $\nabla\boldsymbol{\xi}$ analogously to the decomposition of $\nabla\mathbf{u}$ (see Problem 5.18). Since $\boldsymbol{\xi}$ is a Killing vector, $\nabla\boldsymbol{\xi}$ is antisymmetric, so we write

$$\nabla\boldsymbol{\xi} = \boldsymbol{\omega} + \mathbf{A}\otimes\boldsymbol{\xi} - \boldsymbol{\xi}\otimes\mathbf{A} \tag{7}$$

where $\boldsymbol{\omega}$ is antisymmetric and $\boldsymbol{\omega}\cdot\boldsymbol{\xi} = 0$. Since

$$(\nabla\boldsymbol{\xi})\cdot\boldsymbol{\xi} = \nabla_{\boldsymbol{\xi}}\boldsymbol{\xi} = -\boldsymbol{\xi}\cdot(\nabla\boldsymbol{\xi})$$

we must have

$$\mathbf{A} = (\nabla_{\boldsymbol{\xi}}\boldsymbol{\xi})\,(\boldsymbol{\xi}\cdot\boldsymbol{\xi})^{-1} \,.$$

Thus Equations (6) and (4) give

$$\frac{d\mathbf{S}}{dt} = -(\mathbf{S}\cdot\nabla_{\boldsymbol{\xi}}\boldsymbol{\xi})\boldsymbol{\xi}(\boldsymbol{\xi}\cdot\boldsymbol{\xi})^{-1} + \mathbf{S}\cdot\nabla\boldsymbol{\xi} = \mathbf{S}\cdot\boldsymbol{\omega} \,. \tag{8}$$

(In the first equality we used $\nabla_{\mathbf{S}}\boldsymbol{\xi} = (\nabla\boldsymbol{\xi})\cdot\mathbf{S} = -\mathbf{S}\cdot\nabla\boldsymbol{\xi}$.) We now want to express $\boldsymbol{\omega}$ in terms of a vector as in Equation (1). It is convenient to work in component notation to keep track of which slots are contracted into which. Equation (7) is

$$\xi_{\alpha;\beta} = \omega_{\alpha\beta} + A_\alpha\xi_\beta - \xi_\alpha A_\beta$$

and from this we can get a vector independent of $\mathbf{A}$:

$$B^{\lambda} \equiv \varepsilon^{\lambda\alpha\beta\gamma}\xi_{\alpha;\beta}\xi_{\gamma} = \varepsilon^{\lambda\alpha\beta\gamma}\omega_{\alpha\beta}\xi_{\gamma}\,. \tag{9}$$

Express $\omega_{\alpha\beta}$ in terms of $\mathbf{B}$:

$$\varepsilon_{\lambda\rho\sigma\kappa}B^{\lambda} = -\omega_{\alpha\beta}\xi_{\gamma}\delta^{\alpha\beta\gamma}_{\rho\sigma\kappa}$$

$$\varepsilon_{\lambda\rho\sigma\kappa}B^{\lambda}\xi^{\kappa} = -\omega_{\alpha\beta}\xi_{\gamma}\delta^{\alpha\beta\gamma}_{\rho\sigma\kappa}\xi^{\kappa}$$

Since $\omega_{\alpha\beta}\xi^{\beta} = 0$, the only non-vanishing terms occur when $\gamma = \kappa$, so

$$\varepsilon_{\lambda\rho\sigma\kappa}B^{\lambda}\xi^{\kappa} = -2\omega_{\rho\sigma}\xi_{\kappa}\xi^{\kappa}\,.$$

Substitute in Equation (8),

$$\frac{dS^{\hat{\beta}}}{dt} = S_{\hat{\alpha}}\omega^{\hat{\alpha}\hat{\beta}} = -\frac{1}{2}S_{\hat{\alpha}}\varepsilon^{\hat{\gamma}\hat{\alpha}\hat{\beta}\hat{\kappa}}B_{\hat{\gamma}}\xi_{\hat{\kappa}}(\xi_{\hat{\delta}}\xi^{\hat{\delta}})^{-1}$$

which is of the form of Equation (1) with

$$\omega^{\gamma} = \frac{1}{2}B^{\gamma}(\xi_{\alpha}\xi^{\alpha})^{-1}$$

and, by Equation (9) this is equal to $\varepsilon^{\gamma\alpha\beta\sigma}\xi_{\alpha;\beta}\xi_{\sigma}/2\xi_{\alpha}\xi^{\alpha}$.

(4) The vanishing of $\boldsymbol{\omega}$ occurs if and only if $\xi_{[\alpha;\beta}\xi_{\sigma]} = 0$. But this is the condition that $\boldsymbol{\xi}$ be hypersurface-orthogonal (Problem 7.23) i.e. that the metric be static. (See Problem 10.8.)

Solution 11.11. The equation of transport for the gyroscope spin is

$$\frac{DS^{\alpha}}{d\tau} = (S\cdot a)u^{\alpha}, \qquad S^{\alpha}u_{\alpha} = 0\,. \tag{1}$$

The locally measured time derivative of the jth component (relative to a local Lorentz frame) of the spin vector is

$$\frac{d}{d\tau}S_{\hat{j}} = \frac{d}{d\tau}(S\cdot e_{\hat{j}}) = \frac{D}{d\tau}(S\cdot e_{\hat{j}}) = S\cdot\frac{D}{d\tau}e_{\hat{j}}$$

$$= S\cdot\Gamma^{\hat{\alpha}}_{\hat{j}\hat{0}}e_{\hat{\alpha}} = \Gamma^{\hat{\alpha}}_{\hat{j}\hat{0}}S_{\hat{\alpha}} = \Gamma_{\hat{1}\hat{j}\hat{0}}S^{\hat{1}} \tag{2}$$

where we have used Equation (1) and the definition of the Christoffel symbols. Now, using the fact that the Christoffel symbols, in an orthonormal frame, are antisymmetric in the first two indicates, Equation (2) may be written as

$$\frac{dS^{\hat{j}}}{dt} = \varepsilon_{\hat{j}\hat{k}\hat{l}}\Omega_{\hat{k}} S_{\hat{l}} \,, \tag{3a}$$

where

$$\Omega_{\hat{k}} \equiv \frac{1}{2}\varepsilon_{\hat{k}\hat{l}\hat{j}}\Gamma_{\hat{l}\hat{j}\hat{0}} \,. \tag{3b}$$

We must now compute the Christoffel symbols, $\Gamma_{\hat{l}\hat{j}\hat{0}}$. The approximate metric of the earth is

$$ds^2 = -(1+2\phi)dt^2 + (1-2\phi)\delta_{jk}dx^i dx^k - 4h_j dx^j dt \,. \tag{4}$$

Here

$$\phi = -M/r = 0(\epsilon^2)$$

$$\underset{\sim}{h} = \frac{\underset{\sim}{J}\times\underset{\sim}{r}}{r^3} = 0(\epsilon^3) \tag{5}$$

where $\underset{\sim}{J}$ is the angular momentum of the earth. We are working in the Newtonian limit, where velocities are $\mathcal{O}(\epsilon)$, and will keep terms of order ϵ^3. A coordinate-stationary observer in the metric (4) has orthonormal basis 1-forms

$$\tilde{\omega}^{\bar{0}} = (1+\phi)\widetilde{dt} + 2h_j\widetilde{dx}^j, \qquad \tilde{\omega}^{\bar{j}} = (1-\phi)\widetilde{dx}^j \,. \tag{6}$$

The dual basis vectors, found from the relation $\langle\tilde{\omega}^{\bar{\alpha}}, e_{\bar{\beta}}\rangle = \delta^{\bar{\alpha}}{}_{\bar{\beta}}$ are

$$e_{\bar{0}} = (1-\phi)\partial/\partial t\,, \quad e_{\bar{j}} = (1+\phi)\frac{\partial}{\partial x^j} - 2h_j\,\partial/\partial t \,. \tag{7}$$

If the gyroscope has coordinate velocity v_j, related to the 4-velocity by

$$u^j = v_j u^0, \quad u^0 = 1-\phi+\frac{1}{2}v^2 \quad (\text{from } \mathbf{u}\cdot\mathbf{u} = -1)\,, \tag{8}$$

then in the stationary observer's frame

$$v_{\bar{j}} = \frac{u^{\bar{j}}}{u^{\bar{0}}} = \langle\tilde{\omega}^{\bar{j}}, u\rangle / \langle\tilde{\omega}^{\bar{0}}, u\rangle = (1-2\phi)v_j \,. \tag{9}$$

The basis vectors in the orthonormal frame comoving with the gyroscope are obtained from those of Equation (7) by a Lorentz transformation, a boost $-v_{\bar{j}}$:

$$e_{\hat{\alpha}} = \Lambda^{\bar{\alpha}}{}_{\hat{\alpha}} e_{\bar{\alpha}}$$

where

$$\Lambda^{\bar{0}}{}_{\hat{0}} = \gamma \equiv 1 + \frac{1}{2}v^2 , \qquad \Lambda^{\bar{0}}{}_{\hat{j}} \equiv \gamma v_{\bar{j}} = \Lambda^{\hat{j}}{}_{\hat{0}} ,$$
$$\Lambda^{\bar{j}}{}_{\hat{k}} = \delta^{jk} + (\gamma-1)v^{\bar{j}}v^{\bar{k}}/v^2 \,. \tag{10}$$

This gives

$$e_{\hat{0}} = \left(1-\phi+\frac{1}{2}v^2\right)\partial/\partial t + \left(1-\phi+\frac{1}{2}v^2\right)v^k\,\partial/\partial x^k$$
$$e_{\hat{j}} = \left[\left(1-3\phi+\frac{1}{2}v^2\right)v_j - 2h_j\right]\partial/\partial t + \left[\delta_{jk}(1+\phi)+\frac{1}{2}v_jv_k\right]\partial/\partial x^k \,. \tag{11}$$

The Christoffel symbols in an orthonormal frame can be found from the formula

$$\Gamma_{\hat{\mu}\hat{\nu}\hat{\alpha}} = \frac{1}{2}\left(c_{\hat{\mu}\hat{\nu}\hat{\alpha}} + c_{\hat{\mu}\hat{\alpha}\hat{\nu}} - c_{\hat{\nu}\hat{\alpha}\hat{\mu}}\right)$$
$$c_{\hat{\mu}\hat{\nu}\hat{\alpha}} = [e_{\hat{\mu}}, e_{\hat{\nu}}]\cdot e_{\hat{\alpha}} \,. \tag{12}$$

Using Equation (11) and working to $\mathcal{O}(\epsilon^3)$, we find

$$[e_{\hat{j}}, e_{\hat{0}}] = \left(-\phi_{,j} + \frac{1}{2}v^2{}_{,j} - v_{j,t} - \frac{1}{2}v^2 v_{j,t} - v_k v_{j,k}\right)\partial/\partial t$$
$$+\left[v_{k,j} - \phi_{,j}v_k + \frac{1}{2}v^2{}_{,j}v_k + \frac{1}{2}v^2 v_{k,j} + v_j v_{k,t}\right.$$
$$\left. + \frac{1}{2}v_j v_m v_{k,m} - \frac{1}{2}(v_j v_k)_{,t} - v_m\phi_{,m}\delta_{jk} - v_m\frac{1}{2}(v_j v_k)_{,m}\right]\partial/\partial x^k.$$

Since $(\partial/\partial x^{\alpha})\cdot(\partial/\partial x^{\beta}) = g_{\alpha\beta}$, we get

$$[e_{\hat{j}},e_{\hat{0}}]\cdot e_{\hat{k}} = \left(1-\phi+\frac{1}{2}v^2\right)v_{k,j} + v_k v_{j,t} + v_j v_{k,t} + \frac{1}{2}v_j v_m v_{k,m} + v_m v_k v_{j,m}$$

$$+\frac{1}{2}v_m v_k v_{m,j} - \frac{1}{2}(v_j v_k)_{,t} - v_m \phi_{,}\ \delta_{jk} - v_m \frac{1}{2}(v_j v_k)_{,m} .$$

Thus

$$\begin{aligned}[e_{\hat{j}},e_{\hat{0}}]\cdot e_{\hat{k}} - [e_{\hat{k}},e_{\hat{0}}]\cdot e_{\hat{j}} &= \left(1-\phi+\frac{1}{2}v^2\right)(v_{k,j}-v_{j,k}) \\ &\quad + \frac{1}{2}v_m(v_k v_{j,m} - v_j v_{k,m} + v_k v_{m,j} - v_j v_{m,k})\end{aligned} \tag{13}$$

and

$$\begin{aligned}[e_{\hat{j}},e_{\hat{k}}] &= \left[\left(1-2\phi+\frac{1}{2}v^2\right)v_{k,j} + v_j v_{k,t} - 3\phi_{,j}v_k + \frac{1}{2}v^2_{,j}v_k - 2h_{k,j}\right. \\ &\quad \left. + \frac{1}{2}v_j v_m v_{k,m}\right]\partial/\partial t \\ &\quad + \left[\phi_{,j}\delta_{km} + \frac{1}{2}(v_k v_m)_{,j} + \frac{1}{2}v_j(v_k v_m)_{,t}\right]\partial/\partial x^m - \{j \leftrightarrow k\}\end{aligned}$$

so

$$\begin{aligned}[e_{\hat{j}},e_{\hat{k}}]\cdot e_{\hat{0}} &= \left(1-\phi+\frac{1}{2}v^2\right)(v_{j,k}-v_{k,j}) + 4(\phi_{,j}v_k - \phi_{,k}v_j) + 2(h_{k,j}-h_{j,k}) \\ &\quad + v_k v_{j,t} - v_j v_{k,t} + \frac{1}{2}v_m(v_j v_{m,k} - v_k v_{m,j} + v_k v_{j,m} - v_j v_{k,m}) .\end{aligned} \tag{14}$$

The Christoffel symbols are then

$$\begin{aligned}\Gamma_{\hat{j}\hat{k}\hat{0}} &= \frac{1}{2}([e_{\hat{j}},e_{\hat{k}}]\cdot e_{\hat{0}} + [e_{\hat{j}},e_{\hat{0}}]\cdot e_{\hat{k}} - [e_{\hat{k}},e_{\hat{0}}]\cdot e_{\hat{j}}) \\ &= 2(\phi_{,j}v_k - \phi_{,k}v_j) + h_{k,j} - h_{j,k} + \frac{1}{2}\left(v_k\frac{dv_j}{d\tau} - v_j\frac{dv_k}{d\tau}\right)\end{aligned} \tag{15}$$

where

$$\frac{dv_j}{d\tau} \equiv \frac{\partial v_j}{\partial t} + v_m\frac{\partial v_j}{\partial x^m} = -\phi_{,j} + a_j . \tag{16}$$

Here $\underset{\sim}{a}$ is the acceleration produced by non-inertial forces; we get $\underset{\sim}{a} = 0$ for geodesic motion. Thus

$$\Gamma_{\hat{j}\hat{k}\hat{0}} = \frac{3}{2}(\phi_{,j}v_k - \phi_{,k}v_j) + h_{k,j} - h_{j,k} \tag{17}$$

and Equation (3b) can be written

$$\underset{\sim}{\Omega} = \underset{\sim}{\nabla} \times \underset{\sim}{h} + \frac{3}{2} \underset{\sim}{\nabla} \phi \times \underset{\sim}{v} \ . \tag{18}$$

Had we not set $\underset{\sim}{a} = 0$, there would have been an extra term $\frac{1}{2} \underset{\sim}{a} \times \underset{\sim}{v}$ which is the Thomas precession.

The term $\frac{3}{2} \underset{\sim}{\nabla} \phi \times \underset{\sim}{v}$ is called the "geodetic precession." For a particle in a circular orbit of radius r,

$$\underset{\sim}{\Omega}_{\text{geodetic}} = \frac{3}{2} \frac{M}{r^2} (v_{\bar{\theta}} e_{\bar{\phi}} - v_{\bar{\phi}} e_{\bar{\theta}})$$

where

$$v_{\bar{\theta}} = \left(\frac{M}{r}\right)^{\frac{1}{2}} \frac{(\sin^2\theta - \sin^2\alpha)^{\frac{1}{2}}}{\sin\theta}$$

$$v_{\bar{\phi}} = \left(\frac{M}{r}\right)^{\frac{1}{2}} \frac{\sin\alpha}{\sin\theta}$$

are the spherical components of the particle's velocity and α is the inclination of the orbital plane to the polar axis. To this order, it does not matter what we use, indices with or without carets. In order of magnitude,

$$\Omega_g \sim \frac{3}{2} \left(\frac{M}{R}\right)^{\frac{1}{2}} \frac{M}{R^2} \sim 8'' \text{ per year}$$

where R is the radius of the earth. The term independent of v is the Lens-Thirring precession:

$$\underset{\sim}{\Omega}_{\text{L.T.}} = \frac{1}{r^3} \left(-\underset{\sim}{J} + \frac{3(\underset{\sim}{J} \cdot \underset{\sim}{r}) \underset{\sim}{r}}{r^2} \right) \sim \frac{J}{R^3} \sim 0.1'' \text{ per year} .$$

This term could also have been obtained from Problem 11.10 since, to order ϵ^3,

$$\Omega^{\hat{j}} \sim \Omega^j \approx \frac{\varepsilon^{j0kl} \xi_0 \xi_{[k,l]}}{2 \xi^\gamma \xi_\gamma} .$$

Here

$$\xi^0 = 1, \quad \xi_0 = -(1+2\phi) \quad \text{and} \quad \xi_j = -2h_j ,$$

so

$$\Omega^{\hat{j}} = \varepsilon_{jkl} h_{l,k} , \qquad \underset{\sim}{\Omega} = \underset{\sim}{\nabla} \times \underset{\sim}{h} .$$

CHAPTER 12: SOLUTIONS

Solution 12.1. Equating gravitational acceleration Gm/r^2 to centripetal acceleration $\omega^2 r$ gives Kepler's law: $\omega = (Gm/r^3)^{\frac{1}{2}}$. The average density inside the sphere of radius r is $\bar{\rho} = 3m/4\pi r^3$, so $\omega = (4\pi G\bar{\rho}/3)^{\frac{1}{2}}$. This result means, for example, that a grain of sand would orbit a steel ball bearing at the ball bearings surface with about the same 90 min. period as a satellite orbiting just above the earth's surface.

Solution 12.2. Spring tides occur when the sun and the moon are in the same line as the earth (new moon or full moon); neap tides occur when the sun and the moon are at right angles. Let the height of the tide be h. For a rough estimate, take the sun and the moon to be in the equatorial plane. An ocean element on the equator at high tide is in equilibrium between the earth's gravitational acceleration $g(r) = -M_\oplus/r^2$ at a height $r = r_\oplus + h$ ($M_\oplus$ = mass of earth, $r_\oplus$ = radius of earth) and the tidal accelerations of the sun and moon relative to the center of the earth:

$$0 = g(r_\oplus + h) + r_\oplus R^{\text{sun}}_{2020} + r_\oplus R^{\text{moon}}_{2020} \; . \tag{1}$$

An element 90° away in longitude experiences low tide:

$$0 = g(r_\oplus - h) + r_\oplus R^{\text{sun}}_{1010} + r_\oplus R^{\text{moon}}_{1010} \; . \tag{2}$$

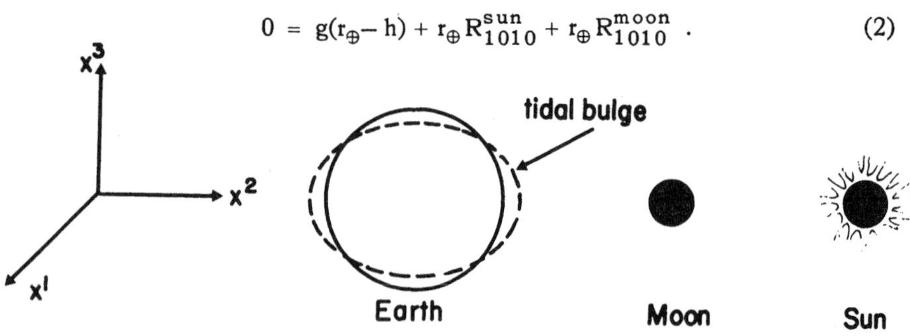

The difference of these two equations gives us

$$0 = 2hg'(r_\oplus) + r_\oplus(R^{sun}_{2020} - R^{sun}_{1010} + R^{moon}_{2020} - R^{moon}_{1010}) . \qquad (3)$$

Note that if we had put in centrifugal forces, they would have cancelled in Equation (3). Compute the Riemann tensor in the Newtonian limit (see Problem 12.12):

$$R^{sun}_{1010} = \left.\frac{\partial^2\Phi^{sun}}{\partial^2 x^2}\right|_{x=y\approx 0} = -\frac{\partial^2}{\partial x^2}\left[\frac{M_\odot}{[(x-x_{sun})^2+(y-y_{sun})^2+(z-z_{sun})^2]^{\frac{1}{2}}}\right]$$

and similarly for the other components. This gives

$$R^{sun}_{1010}(x_{sun}=z_{sun}=0,\ y_{sun}=R) = M_\odot/R^3$$

$$R^{sun}_{1010}(y_{sun}=z_{sun}=0,\ x_{sun}=R) = -2M_\odot/R^3$$

$$R^{sun}_{2020}(x_{sun}=z_{sun}=0,\ y_{sun}=R) = -2M_\odot/R^3$$

$$R^{sun}_{2020}(y_{sun}=z_{sun}=0,\ x_{sun}=R) = M_\odot/R^3$$

and similar expressions for the moon. Spring tides occur when the sun and the moon are in a line, say both on the y-axis. Equation (3) then gives

$$h_{spring} = \frac{3}{4}\left(\frac{M_\odot}{R^3_\odot} + \frac{M_{moon}}{R^3_{moon}}\right)\frac{r^4_\oplus}{M_\oplus} = 39\text{ cm} .$$

Neap tides occur when e.g. the moon is on the y-axis and the sun on the x-axis. This gives

$$h_{neap} = \frac{3}{4}\left(\frac{M_{moon}}{R^3_{moon}} - \frac{M_\odot}{R^3_\odot}\right)\frac{r^4_\oplus}{M_\oplus} = 15\text{ cm} .$$

The actual tides are, of course, considerably larger than this in many places because of hydrodynamical effects (like "sloshing" of water in shallow seas and against irregular coastlines).

Solution 12.3. To good approximation the solid earth responds linearly to the tidal driving force of the Riemann tensor. (This is not true for the hydrodynamic ocean tides.) Thus the fourier spectrum of the earth tides will be the same as that of the Riemann tensor at earth. Since (in Newtonian order at least) the tidal force is linear in its sources we can treat the sun and moon separately. The gravitational potential of a mass M at a position x, y, z in geocentric coordinates is

$$U = M/(x^2+y^2+z^2)^{\frac{1}{2}} .$$

The Newtonian components of the Riemann tensor are thus

$$\begin{aligned} R_{x0x0} &= \partial^2 U/\partial x^2 = \frac{M}{r^3}\left(\frac{3x^2}{r^2} - 1\right) \\ R_{x0y0} &= \partial^2 U/\partial x \partial y = \frac{M}{r^3}\left(\frac{3xy}{r^2}\right) \end{aligned} \tag{1}$$

etc. for yy, zz, yz, xz. In the frame of the rotating earth, the apparent orbit of the sun or moon is given by the standard formula of Keplerian spherical astronomy (see e.g. MTW, p. 648)

$$\begin{aligned} x &= r\cos(\bar{\omega}+v)\cos\phi t - r\sin(\bar{\omega}+v)\cos\epsilon\sin\phi t \\ y &= r\sin(\bar{\omega}+v)\cos\epsilon\cos\phi t + r\cos(\bar{\omega}+v)\sin\phi t \\ z &= r\sin(\bar{\omega}+v)\sin\epsilon . \end{aligned} \tag{2}$$

Here $\bar{\omega}$ and ϵ are constants, the longitude of perigee and inclination to the equatorial plane, respectively. The angle ϕt is the hour angle of the rotating earth,

$$\phi t = \frac{2\pi}{1 \text{ sidereal day}}\, t$$

r is the "radius vector" and v is the "true anomaly," given by

$$r = a(1-e\cos E), \qquad \cos v = (\cos E - e)(1-e\cos E)^{-1}$$

where a is the semimajor axis, e is the eccentricity and E is the eccentric anomaly defined by

$$E - e\sin E = \frac{2\pi}{T}t \equiv \Omega t$$

and where T is the period of the orbit. Working only to first order in the (small) eccentricity e, we compute

$$\begin{aligned} E &= \Omega t + e\sin\Omega t \\ \cos E &= \cos\Omega t - e\sin^2\Omega t \\ r^{-3} &= a^{-3}(1+3e\cos\Omega t) \\ \cos v &= \cos\Omega t + e\cos 2\Omega t - e\ . \end{aligned} \tag{3}$$

Now substituting Equation (3) into Equation (2) and substituting the result into Equation (1), we get

$$R_{j0k0} = \frac{M}{a^3} \times (\text{constant terms} + \text{time varying terms})\ ;$$

the generic time-varying term has the form (here scos means either cos or sin and an exponent mN means any integer between zero and m inclusive)

$$3(\text{scos}^2\Omega t - 2e\,\text{scos}\,\Omega t + 2e\,\text{scos}\,\Omega t\,\text{scos}\,2\Omega t) \times \begin{cases} \sin^2\epsilon & \text{or} \\ \sin\epsilon\cos^N\epsilon\,\text{scos}\,\phi t & \text{or} \\ \cos^{2N}\epsilon\,\text{scos}^2\phi t\ . \end{cases}$$

This expression is analyzed into individual frequencies by repeated use of the trigonometric identities for products of scos's in terms of sum and difference frequencies. The following components result:

Angular Frequency		Amplitudes (leading order in e)
(i)	2ϕ	$\frac{3}{4}\cos^{2N}\epsilon$
(ii)	ϕ	$\frac{3}{2}\sin\epsilon\cos^N\epsilon$
(iii)	$2\Omega \pm 2\phi$	$\frac{3}{8}\cos^{2N}\epsilon$
(iv)	$2\Omega \pm \phi$	$\frac{3}{4}\sin\epsilon\cos^N\epsilon$
(v)	2Ω	$\frac{3}{2}\sin^2\epsilon$ and $\frac{3}{4}\cos^{2N}\epsilon$
(vi)	$\Omega \pm 2\phi$	$3e\cos^{2N}\epsilon$
(vii)	$\Omega \pm \phi$	$6e\sin\epsilon\cos^N\epsilon$
(viii)	Ω	$6e(\sin^2\epsilon$ or $\frac{1}{2}\cos^{2N}\epsilon)$
(ix)	$3\Omega \pm 2\phi$	$\frac{3}{2}e\cos^{2N}\epsilon$
(x)	$3\Omega \pm \phi$	$3e\sin\epsilon\cos^N\epsilon$
(xi)	3Ω	$3e\sin^2\epsilon$

For the Earth-Sun system: $\epsilon = 23\frac{1}{2}^\circ$, $e = .017$, $\Omega = 2\pi/1$ year, and $\phi = 2\pi/1$ sidereal day. For the Earth-Moon system $e = .054$, $\Omega = 2\pi/1$ sidereal month, $\phi = 2\pi/1$ sidereal day, and ϵ varies between $23\frac{1}{2}^\circ \pm 5^\circ$ due to the 18.6 year nutation of the moon's node. (This induces further splitting of the spectrum that we have neglected.) With these values, and using the fact that (M/a^3) is 2.2 times larger for the moon's orbit than for the sun's, we get the following estimate for the ordering of former components by strength (M = moon, S = sun):

$$\text{M(i), M(v), M(ii), M(iii), S(i), S(v), M(iv), S(ii), S(iii), M(vi), M(viii)} \cdots .$$

Solution 12.4. Laplace first realized that there must be a difference in positions. The position of the sun in the sky is displaced by the finite velocity of light; in other words, is aberrated by the velocity v of the earth in its orbit, by an angle $\theta = v/c$. The Coulomb gravitational field is not aberrated: If the force of gravity pointed in this direction, there would be a component of solar acceleration $(GM_\odot/r^2)(v/c)$ in the direction of the earth's motion, so the earth's energy would increase at a rate given by

$$\frac{d(E/M_\oplus)}{dt} = \frac{GM_\odot}{r^2}\frac{v^2}{c}$$

but $E/M_\oplus = -\frac{1}{2}GM_\odot/r$, so the energy loss rate implies $dr/dt = 2v^2/c$. Since $v^2 = GM_\odot/r$ this is easily integrated to give

$$t - t_0 = \frac{c}{4GM}(r^2 - r_0^2) .$$

In particular, the earth's orbit has $r = 1.5 \times 10^{13}$ cm, $v = 30$ km/sec, the radius of the sun is $r_0 = 7 \times 10^{10}$ cm, so $\theta \approx 10^{-4}$, and $t - t_0 \approx 1.3 \times 10^{10}$ sec ≈ 400 years. This is much less than the known geological time during which the radius of the earth's orbit is known to have been constant.

Solution 12.5. Light pressure within the star is given by $P_{rad.} = (1/3)U_{rad.}$, where $U_{rad.}$ is the energy density of radiation. Since the light diffuses outward in many scatterings, $U_{rad.}$ satisfies a diffusion equation and the radiative flux F is proportional to a gradient in $U_{rad.}$

$$F = \frac{-c}{3\kappa\rho} \nabla U_{rad.}$$

where κ is the opacity (given here by the Thompson value $\kappa_T = \sigma_T/m_p = .4\ \text{cm}^2/\text{gm}$). The relation $L = 4\pi r^2 F$ between flux and total luminosity L of the star gives

$$dP_{rad.} = -\frac{\kappa\rho}{4\pi c}\frac{L dr}{r^2}$$

which is the net light pressure on a unit area slab of matter of thickness dr. We equate this to the downward gravitational force $-GM\rho\, dr/r^2$ to get the Eddington limit

$$L = \frac{4\pi GMc}{\kappa_T} = 1.38 \times 10^{38}\left(\frac{M}{M_\odot}\right) \text{ergs/sec}\ .$$

Microscopically, we could also get this result by noting that since the scattering cross section is forward-backward symmetric ($\propto 1 + \cos^2\theta$), a photon of momentum p deposits on the average a momentum p to the electron per collision, and a radial component p_r. The energy of the photon is pc, but its contribution to the outward luminosity is reduced by the factor $v_r/c = p_r/p$. Thus the radial momentum deposited to an electron at radius r per unit time is just proportional to the net radial luminosity $\mathcal{L}$ (constant at all radii)

$$\frac{(\text{momentum})}{(\text{time})} = \underbrace{\frac{L}{4\pi r^2(\hbar\omega)}}_{\substack{\text{equivalent radial}\\ \text{photons per area}\\ \text{per time}}} \times \underbrace{\sigma_T}_{\text{cross section}} \times \underbrace{\left(\frac{\hbar\omega}{c}\right)}_{\substack{\text{momentum}\\ \text{per photon}}} = \frac{L\sigma_T}{4\pi r^2 c}\ .$$

Momentum per unit time is a force, and we equate this to the inward gravitational force on the proton which is associated with the electron:

$$\frac{L\sigma_T}{4\pi r^2 c} = \frac{GM\,m_p}{r^2}$$

which gives the same result as above.

Solution 12.6. Let ϕ^g and ϕ^e be respectively the gravitational and electrostatic potentials. The potential energy of an electron is then

$$\Phi = m\phi^g + e\phi^e \ ,$$

(m, e = electron mass, charge) and the force on an electron is $\underset{\sim}{F} = -\nabla\Phi$. In a static situation there can be no force component tangential to the conducting surface, and hence Φ must be a constant on the inside surface. Now there is neither mass $(\nabla^2\phi^g = 0)$ nor charge $(\nabla^2\phi^e = 0)$ inside the container so $\nabla^2\Phi = 0$. The solution to Laplace's equation $\nabla^2\Phi = 0$ inside the conductor equipotential is then Φ = constant. If an electron is then introduced into this field, it will feel no force. [Note: Since the electron is introduced at the center of the container, image-charge forces can be ignored by symmetry.]

Solution 12.7. As the contents are heated, mass-energy is added according to $c^2 dM = \frac{7}{2}\,kN\,dT$ ($\frac{7}{2}$ since air is diatomic; N is the total number of molecules in the cylinder). However the center of gravity of the gas rises which acts to decrease the weight (distance to earth's center increases). The position of the center of gravity is:

$$z_{cg} = \int_0^h \rho z\,dz \Big/ \int_0^h \rho\,dz \ , \qquad \rho = \rho_0 \exp\left(-\frac{mg}{kT} z\right) \ ,$$

so the change of z_{cg} with temperature is given by

$$\frac{dz_{cg}}{dT} = \frac{k}{mg}\left[1-\left(\frac{\eta}{\sinh\eta}\right)^2\right], \qquad \eta \equiv \frac{mgh}{2kT}$$

where m is the average mass of an air molecule.

The weight change is thus

$$dW = \left(g\,\frac{dM}{dT} + M\,\frac{dg}{dz_{cg}}\,\frac{dz_{cg}}{dT}\right)dT = \left(\frac{7}{2}\,\frac{gkN}{c^2} - mN\,\frac{2g}{R_\oplus}\,\frac{k}{mg}\left[1-\frac{\eta^2}{\sinh^2\eta}\right]\right)dT .$$

The right-hand side is negative when

$$1 - \frac{\eta^2}{\sinh^2\eta} > \frac{7}{4}\,\frac{gR_\oplus}{c^2} = .122\times 10^{-8} .$$

This implies that $\eta > 6.05\times 10^{-5}$ and that $h = 2kT\eta/mg > 218$ cm.

Solution 12.8. Since Poisson's equation is $U_{,kk} = 4\pi\rho_0$ we have

$$T_{jk,k} = \frac{1}{4\pi}\,(U_{,jk}U_{,k} + U_{,j}U_{,kk} - U_{,nj}U_{,n}) = \rho_0 U_{,j} .$$

Thus the Newtonian equation of motion can be written as

$$\rho_0\,\frac{dv_j}{dt} = -\,\frac{\partial t_{jk}}{\partial x^k} - \rho_0\,\frac{\partial U}{\partial x^j} = \frac{-\partial}{\partial x^k}\,(T_{jk}+t_{jk})_{,k} . \qquad (1)$$

Now use the Newtonian equation of mass conservation

$$\partial\rho_0/\partial t + (\rho_0 v^k)_{,k} = 0 . \qquad (2)$$

From Equation (2) and the expression for the convective derivative $d/dt = \partial/\partial t + v_k\partial/\partial x^k$, one can write

$$\rho_0\,\frac{dv_j}{dt} = \rho_0\left(\frac{\partial v_j}{dt} + v^k v^j_{,k}\right) = (\rho_0 v_j)_{,t} + (\rho_0 v_j v_k)_{,k}$$

and thus Equation (1) can be written as

$$(\rho_0 v_j)_{,t} + (T_{jk}+t_{jk}+\rho_0 v_j v_k)_{,k} = 0 . \qquad (3)$$

Solution 12.9. Consider a virtual displacement $\underset{\sim}{\delta}$; the energy must be stationary in $\underset{\sim}{\delta}$. If we include in the energy, the gravitational potential energy of the work done on or by forces, then the change in energy for any one force is

$$dE_i = -\underset{\sim}{F}_i \cdot \underset{\sim}{\delta} + (dE_i)(\underset{\sim}{g}/c^2) \cdot \underset{\sim}{x}_i$$

and the change in energy from raising the object mass is

$$dE = -M\underset{\sim}{g} \cdot \underset{\sim}{\delta} .$$

Thus

$$dE_{total} = -\sum \underset{\sim}{F}_i \cdot \underset{\sim}{\delta} \left(1 - \frac{\underset{\sim}{g} \cdot \underset{\sim}{x}_i}{c^2}\right) - M\underset{\sim}{g} \cdot \underset{\sim}{\delta}$$

for any $\underset{\sim}{\delta}$ and hence

$$\sum_i \underset{\sim}{F}_i \left(1 - \frac{\underset{\sim}{g} \cdot \underset{\sim}{x}_i}{c^2}\right) = -M\underset{\sim}{g} .$$

Note that by taking the gravitational potential relative to the center of gravity, we avoid having to add the gravitational potential energy of the gravitational potential energy. (This argument is due to K. Nordtvedt.)

Solution 12.10. From the previous problem

$$\sum \underset{\sim}{F}_i \, (1 - \underset{\sim}{g} \cdot \underset{\sim}{x}_i/c^2) = -M\underset{\sim}{g} ;$$

If p is pressure and $d\underset{\sim}{A}$ is directed surface area, then $\underset{\sim}{F} = p\,d\underset{\sim}{A}$ and the above result becomes

$$-\int p(x)(1 - \underset{\sim}{g} \cdot \underset{\sim}{x}/c^2)\, d\underset{\sim}{A} = -M\underset{\sim}{g}$$

or

$$-\int p\,d\underset{\sim}{A} + \int p\underset{\sim}{g} \cdot \frac{\underset{\sim}{x}}{c^2}\, d\underset{\sim}{A} = -M\underset{\sim}{g}$$

for $\underset{\sim}{g}$ constant. Now with $\underset{\sim}{g} = -\frac{GM}{r^2}\underset{\sim}{e}_r$, $\underset{\sim}{x} = dr\,\underset{\sim}{e}_r$, $d\underset{\sim}{A} = \pm\, dA\,\underset{\sim}{e}_r$, we have

$$-A[p(r+dr)-p(r)]\underset{\sim}{e}_r - dr\,\frac{GM}{c^2r^2}\underset{\sim}{e}_r\int p\,dA = -M\underset{\sim}{g} = A\,dr\rho\,\frac{GM}{r^2}\underset{\sim}{e}_r$$

and thus

$$\frac{dp}{dr} = -\frac{GM}{r^2}\left(\rho + \frac{p}{c^2}\right).$$

This result holds in any theory of gravity that satisfies $T^{\mu\nu}{}_{;\nu} = 0$ (such theories are called metric theories) and is independent of the field equations. Each theory, however, gives its own prescription for finding $M(r)$, which is defined such that the local acceleration of gravity is $g = -GM/r^2$.

Solution 12.11. The preferred coordinates of Newtonian theory are "universal time" t and Galilean space coordinates x^j. The equation of motion

$$\frac{d^2x^j}{dt^2} = -\frac{\partial\Phi}{\partial x^j}$$

can be written in 4-dimensional form:

$$\frac{d^2t}{d\lambda^2} = 0$$

$$\frac{d^2x^j}{d\lambda^2} + \frac{\partial\Phi}{\partial x^j}\left(\frac{dt}{d\lambda}\right)^2 = 0\,.$$

From this we read off the Christoffel symbols:

$$\Gamma^j{}_{00} = \frac{\partial\Phi}{\partial x^j}; \quad \text{all other } \Gamma^\alpha{}_{\beta\gamma} \text{ vanish}\,.$$

The standard formula for the components of the Riemann tensor gives

$$R^j{}_{0k0} = -R^j{}_{00k} = \frac{\partial^2\Phi}{\partial x^j\,\partial x^k}; \quad \text{all other } R^\alpha{}_{\beta\gamma\delta} \text{ vanish}\,.$$

Suppose these were derivable from a metric, then we would have

$$R_{j0k0} = g_{j\alpha} R^{\alpha}{}_{0k0} = g_{jm} \frac{\partial^2 \Phi}{\partial x^m \partial x^k} ,$$

but

$$R_{0jk0} = g_{0\alpha} R^{\alpha}{}_{jk0} = 0 \neq -R_{j0k0} .$$

This violates the antisymmetry property $R_{\alpha\beta\gamma\delta}$ has if it is derived from a metric.

Solution 12.12. Let the Newtonian test-particle trajectories be

$$x^j = x^j(t, n) .$$

Here n is a parameter telling which trajectory we are considering. The vector

$$\underset{\sim}{n} = \frac{\partial}{\partial n} = \frac{\partial x^k}{\partial n} \frac{\partial}{\partial x^k} = n^k \frac{\partial}{\partial x^k}$$

is a connecting vector between neighboring trajectories. The relative acceleration of neighboring trajectories is

$$\frac{\partial^2 n^j}{\partial t^2} = \frac{\partial^2}{\partial t^2}\left(\frac{\partial x^j}{\partial n}\right) = \frac{\partial}{\partial n}\left(\frac{\partial^2 x^j}{\partial t^2}\right) = \frac{\partial}{\partial n}\left(-\frac{\partial \Phi}{\partial x^j}\right) = -n^k \frac{\partial}{\partial x^k}\left(\frac{\partial \Phi}{\partial x^j}\right) = -n^k \frac{\partial^2 \Phi}{\partial x^k \partial x^j} ,$$

where we have used the equation of motion $\partial^2 x^j/\partial t^2 = -\partial\Phi/\partial x^j$. The equation of geodesic deviation is

$$\frac{D^2 n^\alpha}{d\tau^2} = -R^{\alpha}{}_{\beta\gamma\delta}\, n^\gamma u^\beta u^\delta .$$

In the Newtonian limit (velocities $\sim \epsilon$, gravitational fields $\sim \epsilon^2$)

$$u^0 = 1 + \mathcal{O}(\epsilon), \quad u^j = \mathcal{O}(\epsilon), \quad \Gamma^{\alpha}{}_{\beta\gamma} = \mathcal{O}(\epsilon^2)$$

and $\mathbf{n}$, which connects events of equal proper time, connects events of equal coordinate time to $\mathcal{O}(\epsilon)$. Thus, $\frac{D}{d\tau} = \frac{d}{dt} + \mathcal{O}(\epsilon)$ and so

$$\frac{d^2 n^j}{dt^2} = -R^j{}_{0k0}\, n^k + \mathcal{O}(\epsilon) \; .$$

The Riemann components are then

$$R^j{}_{0k0} = \frac{\partial^2 \Phi}{\partial x^k \partial x^j}$$

in the Newtonian limit. Note that, unless the velocities involved approach the speed of light, the other components of $R^\alpha{}_{\beta\gamma\delta}$ do not enter into the equation of relative motion of test particles.

Solution 12.13. Newtonian gravity is characterized by instant "action at a distance," where simultaneity is defined by the slices of a "universal time" field t. We want a relativistically invariant expression for the gravitational acceleration at a point in terms of the matter density $T^{\mu\nu}$ in the rest of the universe. Spacetime is taken to have the Minkowski metric everywhere. Only T^{00} should enter, so we first dot into two universal time vectors:

$$\nabla t \cdot \mathbf{T} \cdot \nabla t \quad (\text{or} \quad t_{,\alpha} T^{\alpha\beta} t_{,\beta}) \; .$$

Thus the inverse square law acceleration of a particle at position x becomes

$$a_{grav} = G \int_{t=\text{constant}} \mathbf{l} \, \frac{\nabla t \cdot \mathbf{T}(x') \cdot \nabla t}{(\mathbf{l} \cdot \mathbf{l})^{\frac{3}{2}}} \, d^3 x'$$

where $\mathbf{l}$ is the spacelike vector connecting x' to x and d^3x' is the proper volume element on the constant t slice. This is not yet consistent with special relativity, because a is not necessarily orthogonal to the particle's 4-velocity $\mathbf{u}$. We must apply a projection operator, giving a final answer

$$a_{grav} = G \int \frac{\nabla t \cdot \mathbf{T} \cdot \nabla t}{(\mathbf{l} \cdot \mathbf{l})^{\frac{3}{2}}} \, [\mathbf{l} + (\mathbf{u} \cdot \mathbf{l})\, \mathbf{u}] \, d^3 x' \; .$$

This is manifestly consistent with special relativity because it is (i) written in terms of geometrical objects, and (ii) defines a geometrical object ($\mathbf{a}$ with $\mathbf{a}\cdot\mathbf{u} = 0$). Signals can be sent faster than light by changing the distribution of masses at a point. The gravitational field then changes on surfaces of constant t. Since these surfaces are spacelike, they are outside the lightcone of every observer. Nevertheless the theory is not acausal; since every observer's worldline is timelike, t is always increasing in his future direction. Likewise, t increases along any light ray he emits. So any signal can at best connect events of constant t; no sequence of signals can return to an observer at an earlier t than they started. This is to be contrasted with the case of tachyons (Problem 1.6) which *are* acausal. The reason for the difference is that this problem has some "pregeometry" in the scalar field t, which slices up spacetime in a universally time-ordered fashion. Tachyons "slice" spacetime in a manner that varies from observer to observer.

Solution 12.14. Take the separation as $\underset{\sim}{\eta} = (\text{separation})\, \underset{\sim}{e}_z$; then due to gravitational forces

$$\frac{d^2\eta^z}{dt^2} = -R^z{}_{\beta\gamma\delta}\, u^\beta \eta^\gamma u^\delta \,.$$

Since the particles start from rest, if we confine our observations to short time, $u^j \approx 0$, so that

$$\frac{d^2\eta^z}{dt^2} = -R^z{}_{0z0}\, \eta^z \,.$$

The Riemann component for a weak field is easily calculated

$$R^z{}_{0z0} = \frac{\partial^2}{\partial z^2}\left(-\frac{M}{z}\right) = -\frac{2M}{z^3} \,.$$

Due to the E field, there is an acceleration of the lower particle

$$\underset{\sim}{a} = \frac{q}{m} E \underset{\sim}{e}_z \,.$$

Thus

$$\frac{d^2\eta^z}{dt^2} = \frac{2M}{z^3}\eta^z - \frac{q}{m}E .$$

It may seem disturbing at first that the two terms look as if they can be comparable. But suppose our laboratory is limited to a chunk of space-time $\Delta x^\mu \approx L$. Then at the end of the experiment the gravitational forces will have caused a change in η^z:

$$\delta\eta^z \sim \frac{1}{2}\frac{2M}{z^3}\eta^z L^2 .$$

But $\eta^z < L$ so

$$\delta\eta^z = \mathcal{O}(L^3) .$$

The change in η^z due to the electric field is

$$\delta\eta^z \sim -\frac{q}{m}\cdot E\cdot\frac{1}{2}L^2 = \mathcal{O}(L^2)$$

so, if we sufficiently limit the "size" of our experiment we can ignore gravitational effects.

Solution 12.15. The geometrical object which tells about the charge is a scalar, vector, 2-tensor, etc. for spins 0, 1, 2, etc. For example, the charge density of a particle of rest-charge q_0 moving along a worldline $x^\mu = z^\mu(\tau)$ with 4-velocity $\mathbf{u}$ is

$$\rho(x^\sigma) = q_0\int \delta^4[x^\mu - z^\mu(\tau)]\,d\tau \qquad \text{(scalar)}$$

$$J^\mu(x^\sigma) = q_0\int u^\mu\delta^4[x^\sigma - z^\sigma(\tau)]\,d\tau \qquad \text{(vector)}$$

$$\qquad (1)$$

$$T^{\mu\nu}(x^\sigma) = q_0\int u^\mu u^\nu\delta^4[x^\sigma - z^\sigma(\tau)]\,d\tau \qquad \text{(spin 2)}$$

$$T^{\mu\nu\cdots\rho}(x^\sigma) = q_0\int u^\mu u^\nu\cdots u^\rho\delta^4[x^\sigma - z^\sigma(\tau)]\,d\tau . \qquad \text{(spin s)}$$

The charge of a particle in a laboratory volume, measured by the "Coulomb" part of its interaction in the lab is

$$q = \int T^{00\cdots0}\, d^3x$$

(lab frame)

$$= (-1)^s \int T^{\alpha\beta\cdots\gamma} u_\alpha^{LAB} u_\beta^{LAB} \cdots u_\gamma^{LAB}\, d^3x\,. \qquad (2)$$

Substituting Equation (1) into Equation (2) gives the measured charge of a single particle

$$q = q_0 \int d^3x \int d\tau\, (-\mathbf{u}\cdot\mathbf{u}^{LAB})^s\, \delta^4[x^\sigma - z^\sigma(\tau)]$$

$$= q_0 \int d^3x\, dt \left(\frac{d\tau}{dt}\right) (-\mathbf{u}\cdot\mathbf{u}^{LAB})^s\, \delta^4[x^\sigma - z^\sigma(\tau)]\,.$$

Integrating out the delta function, and using $\mathbf{u}\cdot\mathbf{u}^{LAB} = -\gamma$ and $d\tau/dt = 1/\gamma$, we get $q = q_0\gamma^{s-1} \approx q_0\left(1+\frac{s-1}{2}v^2\right)$ for $v \ll c$. Now for a perfect gas we have $\langle v^2\rangle = 3kT/m_0$. So when we sum the charges of all the particles in the gas we get

$$Q = \sum_n q = nq_0\left[1 + \frac{s-1}{2}\left(\frac{3kT}{m_0}\right)\right],$$

and from the temperature dependence given in the problem we now read off $s = 5$.

Solution 12.16. Suppose, to the contrary, that there were another spin-2 field. Since it is massless, and has infinite range, it must admit weak, plane waves as solutions. Also since the field is spin-2 it must couple to a symmetric, traceless "charge" tensor $J^{\mu\nu}$. Moreover, for weak fields and slow velocities, "charged" particles which are much smaller than a wavelength must couple predominantly through their J^{00} (i.e. the "Coulomb" limit).

Consider a plane wave incident on two particles (initially located at the same point) of charge $J^{00} = q_1$, $J^{00} = q_2$. For a sinusoidal plane wave of a definite linear polarization the particles must execute linear sinusoidal motion whose amplitude is proportional to their charge and inversely proportional to their mass. Hence, at an instant in time, we can define a polarization vector (analog of the electric field E) for the field

$$\underset{\sim}{V} \equiv \left(\frac{q_1}{m_1} - \frac{q_2}{m_2}\right)^{-1} (\underset{\sim}{x}_1 - \underset{\sim}{x}_2)$$

which is independent of the numerical value of q_1, q_2, m_1, m_2. But this is a contradiction! Since the field has *spin-2* structure, a polarization state should be *unchanged* by 180° rotation, but under 180° rotation $\underset{\sim}{V} \to -\underset{\sim}{V}$ implying $\underset{\sim}{V} = 0$. What is wrong? We assumed that the particles responded *differently* to the field. We now see that $\underset{\sim}{x}_1 = \underset{\sim}{x}_2$ for all charges and masses is the only possibility. This is equivalent to saying that *all* particles have a fixed charge-to-mass ratio: q/m = constant. By choice of units we can make $q = m$. Hence, in the weak field, plane-wave limit $J^{00} = T^{00}$, the stress energy. But by Lorentz invariance we must have $J^{\mu\nu} = T^{\mu\nu}$ (e.g. we cannot introduce $T^{\mu}{}_{\mu}$ since this is not spin-2).

We have shown that our "arbitrary" spin-2 field is coupled to $T^{\mu\nu}$. Hence, it is just a piece of gravity, i.e. whatever we measure experimentally as gravity. (No claim that it is general relativity is made.) R. P. Feynman has pointed out that the assumption of a "point" particle, with no *intrinsic* polarization vector, has entered crucially. Thus it *is* possible to have the field couple to individual particles with spin; but for *unpolarized* bulk matter, there can be no net force.

Solution 12.17. In c.g.s. units we have

$$\begin{aligned} c &= 2.998 \times 10^{10} \text{ cm/sec} \\ c^2 &= 8.998 \times 10^{20} \text{ erg/gm} \\ G/c^2 &= 0.7425 \times 10^{-28} \text{ cm/gm} \\ G^{\frac{1}{2}} &= 2.582 \times 10^{-4} \text{ esu/gm} . \end{aligned}$$

(The value of $G^{\frac{1}{2}}$ comes from the fact that in Coulomb's or Newton's law, force = e^2/r or Gm^2/r^2.) Thus, in gravitational units, the values are

$$\text{(a)}\quad \hbar = \hbar\left(\frac{G}{c^2}\right) c\left(\frac{1}{c^2}\right) = \frac{\hbar G}{c^3} \simeq \frac{(\text{erg sec})(\text{cm/gm})(\text{cm/sec})}{(\text{erg/gm})} \simeq 2.611\times 10^{-66}\,\text{cm}^2$$

$$\text{(b)}\quad e = e\left(\frac{G}{c^2}\right)\left(\frac{1}{G^{\frac{1}{2}}}\right) \simeq \frac{(\text{esu})(\text{cm/gm})}{(\text{esu/gm})} \simeq \frac{eG^{\frac{1}{2}}}{c^2}\ 1.381\times 10^{-34}\,\text{cm}$$

$$\text{(c)}\quad \frac{e}{m} = \frac{e}{G^{\frac{1}{2}}m} \simeq \frac{(\text{esu})}{(\text{gm})(\text{esu/gm})} \simeq 2.042\times 10^{21}$$

$$\text{(d)}\quad M_\odot = \frac{GM_\odot}{c^2} \simeq (\text{gm})(\text{cm/gm}) \simeq 1.48\times 10^{5}\,\text{cm}$$

$$\text{(e)}\quad L_\odot = L_\odot\left(\frac{G}{c^2}\right)\left(\frac{1}{c^2}\right)\left(\frac{1}{c}\right) \simeq \frac{(\text{erg/sec})(\text{cm/gm})}{(\text{erg/gm})(\text{cm/sec})} \simeq 1.07\times 10^{-26}$$

$$\text{(f)}\quad 300^0\text{K} = \frac{(300^0\text{K})\,k\left(\frac{G}{c^2}\right)}{c^2} \simeq \frac{(^\circ\text{K})(\text{erg}/^\circ\text{K})(\text{cm/gm})}{(\text{erg/gm})} \simeq 3.42\times 10^{-63}\,\text{cm}$$

$$\text{(g)}\quad (1\text{ year}) = (1\text{ year})c = 9.460\times 10^{17}\,\text{cm}$$

$$\text{(h)}\quad (1\text{ volt}) = \left(\frac{1}{299.8}\,\frac{\text{erg}}{\text{esu}}\right) G^{\frac{1}{2}}\left(\frac{1}{c^2}\right) \simeq \frac{(\text{erg/esu})(\text{esu/gm})}{(\text{erg/gm})} \simeq 9.58\times 10^{-28}.$$

Solution 12.18.

$$L^* = (\hbar G/c^3)^{\frac{1}{2}} = 1.616\times 10^{-33}\,\text{cm}$$

$$T^* = (\hbar G/c^5)^{\frac{1}{2}} = 5.391\times 10^{-44}\,\text{sec}$$

$$M^* = (\hbar c/G)^{\frac{1}{2}} = 2.177\times 10^{-5}\ \text{gm}\ .$$

These combinations were noticed by Planck, immediately after he discovered his constant. They are now called the Planck length, mass, and time. They are uniquely defined up to factors of order unity (e.g. using h instead of $\hbar$), since by dimensional analysis there is only one way of mapping three dimensional quantities into three specified dimensional results (i.e. there are no free dimensionless combinations).

Solution 12.19. Force balance gives us

$$Gm_N^2/r^2 = m_N\omega^2\left(\frac{1}{2}r\right) .$$

According to Bohr quantization the separation radius must satisfy

$$2m\omega\left(\frac{1}{2}r\right)^2 = n\hbar .$$

From these two equations, the separation is

$$r = 2n^2\hbar^2/Gm_N^3 .$$

The lowest energy level is the $n=1$ level. In this level the separation is

$$r = 2\hbar^2/Gm_N^3 = 6\times10^{24}\,\text{cm} \approx 6\times10^6 \text{ light years!}$$

CHAPTER 13: SOLUTIONS

Solution 13.1. According to the Bianchi identities the Einstein tensor $R_{\mu\nu} - \frac{1}{2} g_{\mu\nu}R$ is divergenceless so that the divergence of our generalized field equation is

$$\left(\frac{1}{2} - a\right) R_{,\mu} = 8\pi T_{\mu\ ;\nu}^{\ \nu} .$$

On the other hand, if we contract our field equations first and then differentiate we find

$$(1-4a)R_{,\mu} = 8\pi T_{,\mu}$$

so that the equations of motion must be

$$T_{\mu\ ;\nu}^{\ \nu} = \kappa T_{,\mu}, \qquad \kappa \equiv \frac{\frac{1}{2} - a}{1-4a} .$$

In the Newtonian limit, and for a fluid with density ρ and negligible pressures, the $\mu = 0$ component of this equation is

$$\frac{\partial \rho}{\partial t} + \nabla \cdot (\rho \underset{\sim}{v}) = \kappa \frac{\partial \rho}{\partial t}$$

where $\underset{\sim}{v}$ is the velocity of fluid flow. If κ were not zero (i.e. if a were not $\frac{1}{2}$) this would differ from the Newtonian continuity equation, and violate the cõnservation of mass in Newtonian order.

Solution 13.2. The vanishing of the Weyl tensor allows us to write the metric in the conformally flat form

$$g_{\mu\nu} = e^{2\phi}\eta_{\mu\nu}$$

where $\phi << 1$ in the Newtonian limit (nearly flat spacetime). We have then (see Problem 9.19)

$$R \approx -6\nabla^2\phi \ .$$

For nonrelativistic stress energies $T^{\mu}{}_{\mu} \approx T^0{}_0 \approx -\rho$ and the field equations become

$$-6\phi_{,\alpha\beta}\eta^{\alpha\beta} = \kappa T = -\kappa\rho \ .$$

In the Newtonian limit time variations will be slow compared to spatial variations (in $c = 1$ units) so we have

$$6\phi_{,ij}\delta^{ij} = \kappa\rho \ .$$

For $\kappa = 24\pi$ this is the usual Newtonian equation for the Newtonian potential ϕ. The solution to Problem 12.11 implies that Newtonian trajectories are geodesics in a geometry with $g_{00} \approx -(1+2\phi)$. Thus in the Newtonian limit the proposed theory agrees with Newtonian theory and ϕ plays the role of the Newtonian potential.

For the metric of a massive object like the sun the function ϕ must fall off far from the object where spacetime approaches flatness. To see that light is not deflected we need only notice [see Problem 9.18] that the null geodesics of $g_{\mu\nu} = e^{2\phi}\eta_{\mu\nu}$ are identical to the null geodesics of $\eta_{\mu\nu}$, which are certainly not deflected by the massive object. Thus when we compare, in the asymptotically flat distant regions, the directions of a photon's motion before and after it interacts with the massive object, we will observe no deflection.

Near the earth's surface the metric is of the form

$$ds^2 = e^{2\phi(z)}(-dt^2 + dx^2 + dy^2 + dz^2)$$

where z corresponds to height above the surface. If particles are to fall with the proper acceleration we must have $\phi \approx -gz$. From the geodesic equation we find that for the energy of a photon moving vertically

$$\frac{dp^0}{dz} = -\Gamma^0{}_{0z}\, p^0 = -\phi_{,z}\, p^0 \ .$$

Thus a photon loses energy at the same rate as a particle and the proposed theory agrees with Pound-Rebka experiment, as any theory based on geodesic motion must.

Solution 13.3. The static field equation for the Brans-Dicke scalar field is

$$\nabla^2\phi = \frac{8\pi T}{(3+2\omega)} . \tag{1}$$

Now, for a spherical shell of mass M, radius R,

$$T = -\rho = \frac{M\delta(r-R)}{4\pi r^2} \tag{2}$$

so that the ϕ field inside and outside the shell is

$$\phi_I = \phi_1 , \quad r < R \tag{3a}$$

$$\phi_{II} = \phi_\infty + \frac{2}{3+2\omega}\frac{M}{r}, \quad r > R \tag{3b}$$

where ϕ_1 and ϕ_∞ are constants. Matching ϕ_I and ϕ_{II} at $r = R$ gives

$$\phi_1 = \phi_\infty + \frac{2}{3+2\omega}\frac{M}{R} . \tag{4}$$

Now, we also know that in the Brans-Dicke theory, the local gravitational constant, in terms of ϕ, is (Weinberg, Equation 9.9.11)

$$G = \phi^{-1}\left(\frac{4+2\omega}{3+2\omega}\right) . \tag{5}$$

Thus

$$\phi_\infty = \frac{4+2\omega}{3+2\omega}\frac{1}{G_\infty}$$

$$G_1 = \phi_1^{-1}\left(\frac{4+2\omega}{3+2\omega}\right) \approx \left(\frac{4+2\omega}{3+2\omega}\right)\phi_\infty^{-1}\left[1 - \frac{2M}{\phi_\infty R(3+2\omega)}\right]$$

and finally

$$G_1 = G_\infty\left[1 - \frac{G_\infty M}{R}\frac{1}{(2+\omega)}\right] . \tag{6}$$

This solution is correct only to lowest order in M/R, not only because of the terms neglected here, but also because in higher order one would have to account for the changed metric $g_{\mu\nu}$ due to the mass and use the curved space ∇ in Equation (1) rather than the flat space one.

Solution 13.4. If there is to be no preferred frame for the vacuum, the stress-energy of the vacuum must look the same in any Lorentz frame. The tensor $\eta_{\mu\nu}$ looks the same in any Lorentz frame so that the stress-energy $\rho_{vac}\eta_{\mu\nu}$, or in general coordinates $\rho_{vac}g_{\mu\nu}$, does not single out any frame. [The uniqueness of $g_{\mu\nu}$ can be seen as follows: if there is no preferred frame there can be no preferred vectors, hence no preferred eigenvectors. This is only possible if *all* vectors are eigenvectors, but if $S_\mu{}^\nu V_\nu = KV_\mu$ for all K, then $S_\mu{}^\nu$ must be proportional to $\delta_\mu{}^\nu$.]

The field equations with vacuum and matter sources are

$$R^{\mu\nu} - \frac{1}{2} g^{\mu\nu} R = 8\pi (T^{\mu\nu}_{matter} + g^{\mu\nu}\rho_{vac}) \ .$$

When we compare this with the field equations including the cosmological term

$$R^{\mu\nu} - \frac{1}{2} g^{\mu\nu} R + \Lambda g^{\mu\nu} = 8\pi T^{\mu\nu}_{matter}$$

we see that we can identify the cosmological constant with the vacuum energy density according to $\Lambda = 8\pi\rho_{vac}$.

If the vacuum contains particles of mass m at a separation of $\lambda \sim \hbar/mc$ then the energy density would be of order $m(mc/\hbar)^3$. This is of order 10^4 grams/cc for electrons and 10^{17} grams/cc for protons: both absurdly large.

If the vacuum energy is to be considered as arising from the gravitational interaction, of energy $\sim G\frac{m^2}{\lambda}$, of nearby particles, the equivalent mass density would be $c^{-2}(Gm^2/\lambda)/\lambda^3 = Gm^6c^2/\hbar^4$. For protons this is of order 10^{-22} grams/cc and for electrons 10^{-41} grams/cc. The density of our Galaxy is 10^{-23} grams/cc so the 10^{-22} grams/cc associated with protons would have measurable consequences even for Galactic dynamics. The 10^{-41} grams/cc associated with electrons, on the other hand, is small even compared with cosmological densities (10^{-31} grams/cc) and would have no measurable effects.

These arguments should be considered only suggestive and somewhat ad hoc. On dimensional grounds ρ_{vac} must be of form

$$\frac{m^4c^3}{\hbar^3}\left(\frac{Gm^2}{c\hbar}\right)^n .$$

For $n = 1$ we get somewhat reasonable mass densities. The Zel'dovich argument is an attempt to give a physical justification for $n = 1$.

Solution 13.5. If we contract the Einstein field equations we get a relation between the trace of $T_{\mu\nu}$ and the Ricci scalar: $R = -8\pi T_\mu^{\ \mu}$. In the local Lorentz frame in which $T_\mu^{\ \nu}$ is diagonal, $T_\mu^{\ \mu} = -\rho + p_x + p_y + p_z$, where p_x, p_y, p_z are the (principal) pressures. For all known equations of state $\rho \geq 3p$ so $T_\mu^{\ \mu}$ should always be negative and R should be positive.

The electromagnetic stress-energy is traceless (see Problem 4.16) so $R = 0$ if the only stress-energy is electromagnetic.

Solution 13.6. Given $T_\alpha^{\ \beta}$ in his own orthonormal frame, the observer can find its four eigenvalues and eigenvectors:

$$T_\alpha^{\ \beta} W_\beta = \lambda W_\alpha . \tag{1}$$

We are given that one eigenvector is timelike. Normalize it so that $W_\alpha W^\alpha = -1$. If its components are $W^\alpha = (\gamma, \gamma v^j)$, then perform a Lorentz transformation with velocity v^j to the rest frame of $\mathbf{W}$. In that frame $W^{\hat{\alpha}} = (1, \underset{\sim}{0})$ and Equation (1) gives $T^{\hat{0}\hat{0}} = -\lambda_{\text{timelike}} \equiv \rho$, and $T^{\hat{0}\hat{k}} = 0$. The 3×3 matrix $T_{\hat{j}\hat{k}}$ can be diagonalized by a spatial rotation, and from Equation (1) we see that the diagonal elements are the remaining eigenvalues $\lambda_i = p_i$ with $i = 1,2,3$. Now let $\mathbf{u}$ be an arbitrary 4-velocity, with components $u^{\hat{\alpha}} = (\bar{\gamma}, \bar{\gamma}\bar{v}^j)$ in the frame of the observer for whom $T^{\alpha\beta}$ is diagonal. The weak energy condition is that $T^{\alpha\beta}u_\alpha u_\beta \geq 0$, that is

$$\rho + \bar{v}_1^2 p_1 + \bar{v}_2^2 p_2 + \bar{v}_3^2 p_3 \geq 0 \qquad (2)$$

where $\bar{v}^j$ is arbitrary and $\bar{v}^2 \leq 1$. The necessary and sufficient condition that Equation (2) be satisfied is

$$\rho \geq 0 \qquad \rho + p_i \geq 0 \qquad (3)$$

a test which can be made by the original observer after he has solved Equation (1). For the case when $T^{\alpha\beta}$ has no timelike eigenvector, see Hawking and Ellis, p. 89.

Solution 13.7. For $n = 0$ the statement reads $\mathbf{u}\cdot\mathbf{u} \leq 0$ and is trivial. For $n = 1$ the statement is $\mathbf{u}\cdot\mathbf{T}\cdot\mathbf{u} \geq 0$ for all timelike $\mathbf{u}$. This is just the weak energy condition. For $n = 2$ the statement is that $(\mathbf{u}\cdot\mathbf{T})\cdot(\mathbf{u}\cdot\mathbf{T}) \leq 0$ or that $\mathbf{u}\cdot\mathbf{T}$ must be nonspacelike. But for an observer with 4-velocity $\mathbf{u}$, $\mathbf{u}\cdot\mathbf{T} = (-\rho, [\text{Energy flux}]^i)$ so, by the dominant energy condition, $|\rho| >$ $|\text{Energy flux}|$ and $\mathbf{u}\cdot\mathbf{T}$ is indeed nonspacelike.

Now consider any n and the statement

$$(-1)^n \mathbf{u}\cdot\mathbf{T}\cdot\mathbf{T} \cdots \mathbf{T}\cdot\mathbf{u} \leq 0 \ .$$

Since $\mathbf{u}\cdot\mathbf{T}$ is nonspacelike, the statement for any n is equivalent to the statement for $n-2$. But we have shown that the dominant energy condition implies the $n=1$ and $n=2$ cases so it follows that the dominant energy condition implies the statement for all n. The converse follows immediately, any stress energy satisfying the statement for $n=1$ and $n=2$ satisfies all the requirements of the dominant energy condition.

Solution 13.8. Yes. Write down a metric $g_{\mu\nu}$ such that $g_{\mu\nu} = \eta_{\mu\nu}$ for $t < 0$ (this is clearly empty space) and $g_{\mu\nu}$ = (arbitrary functions of your choice) for $t \geq 0$. The only constraint on the arbitrary functions is that they be twice differentiable and match smoothly to the flat-space values at the $t = 0$ surface. Now from this metric $g_{\mu\nu}$ we can calculate the Ricci tensor and scalar and we can define a tensor $T_{\mu\nu} \equiv (8\pi)^{-1}(R_{\mu\nu} - \frac{1}{2} g_{\mu\nu}R)$. This tensor is obviously symmetric, and by the contracted Bianchi identities is divergenceless. If we fill spacetime with a stress-energy equal to this, then $g_{\mu\nu}$ is a solution of the Einstein field equations. (*However*, for a physically meaningful solution there must be further requirements for $T_{\mu\nu}$, e.g. energy density should be everywhere nonnegative. This will not in general be true for the procedure above.)

Solution 13.9. We use Killing's equation $\xi_{(\alpha;\beta)} = 0$ and the "hypersurface-orthogonal" statement (see Problem 10.8) of staticity $\xi_{[\alpha;\beta}\xi_{\gamma]} = 0$, to derive

$$\begin{aligned} 0 &= (\xi_{[\alpha;\beta}\xi_{\gamma]})^{;\gamma} \\ &= \frac{1}{3}(\xi_{\alpha;\beta}\xi_{\gamma} + \xi_{\gamma;\alpha}\xi_{\beta} + \xi_{\beta;\gamma}\xi_{\alpha})^{;\gamma} \\ &= \frac{1}{3}(R_{\lambda\gamma\beta\alpha}\xi^{\lambda}\xi^{\gamma} + R_{\lambda\ \alpha\gamma}^{\ \gamma}\xi^{\lambda}\xi_{\beta} + R_{\lambda\ \gamma\beta}^{\ \gamma}\xi^{\lambda}\xi_{\alpha}) \end{aligned}$$

where we have used the result $\xi_{\mu;\alpha\beta} = R_{\gamma\beta\alpha\mu}\xi^{\gamma}$ from Problem 10.7. The first term above vanishes by symmetry and the remaining terms give Ricci components, so we have

$$\begin{aligned} 0 = \xi^{\lambda}R_{\lambda[\alpha}\xi_{\beta]} &= 8\pi\xi^{\lambda}\left(T_{\lambda[\alpha} - \frac{1}{2} g_{\lambda[\alpha}T\right)\xi_{\beta]} \\ &= 8\pi\xi^{\lambda}\left((\rho + p)u_{\lambda}u_{[\alpha} + \frac{1}{2}(\rho - p)g_{\lambda[\alpha}\right)\xi_{\beta]} \ . \end{aligned}$$

The last term vanishes trivially so we can conclude $u_{[\alpha}\xi_{\beta]} = 0$ and hence $\mathbf{u} \propto \boldsymbol{\xi}$.

A less rigorous proof is possible if the equivalent "stationary and time reversible" condition for staticity (see Problem 10.8) is used. In this case we see immediately that if $\mathbf{u}$ is not parallel to $\boldsymbol{\xi}$ then $T^{0i} \neq 0$ in the "time reversible frame." But this implies $G^{0i} \neq 0$ which is incompatible with time reversibility of the metric.

Solution 13.10. From the Bianchi identities $G^{\nu\mu}{}_{;\mu} \equiv 0$ we have

$$\frac{\partial}{\partial t} G^{\nu 0} \equiv - G^{\nu i}{}_{,i} - G^{\sigma\mu}\Gamma^{\nu}{}_{\sigma\mu} - G^{\nu\sigma}\Gamma^{\mu}{}_{\sigma\mu} .$$

Terms on the right cannot involve third derivatives of the metric with respect to time, hence $G^{\nu 0}$ cannot contain second time derivatives. It follows then that the four equations $G^{\nu 0} = 8\pi T^{\nu 0}$ serve as initial value equations, i.e. constraints on the data on the Cauchy hypersurface, while $G^{ij} = 8\pi T^{ij}$ are the dynamical equations.

An examination of the constraints on initial data is obscured by the arbitrariness in the metric associated with the arbitrariness in the coordinate system. To eliminate this confusion we can choose at the outset four constraints on the metric. A convenient choice is Gaussian normal coordinates: $g_{00} = -1$ and $g_{0i} = 0$. We now appear to have six field variables g_{ij}, six dynamical equations $G^{ij} = 8\pi T^{ij}$ which we can solve for $\partial^2 g_{ij}/\partial t^2$, and four initial value equations $G^{\nu 0} = 8\pi T^{\nu 0}$ giving relations among the initial data g_{ij} and $\partial g_{ij}/\partial t$ on the initial hypersurface.

We can now differentiate the four initial value equations with respect to time and use the dynamical equations to eliminate terms of the form $\partial^2 g_{ij}/\partial t^2$. This generates four new relations among the initial value data. [These new relations must be independent of the original four because $\partial G^{\nu 0}/\partial t = 8\pi\, \partial T^{\nu 0}/\partial t$, and we can specify $\partial T^{\nu 0}/\partial t$ independently of any previously considered constraints.]

We have now eight constraints on the twelve initial value functions g_{ij} and $\partial g_{ij}/\partial t$. It follows that we can independently specify four initial value functions on the hypersurface.

If we think of these four initial value functions as the values of two field variables and their time derivatives on the cauchy hypersurface, then we can picture the time evolution of the gravitational field as being described by the dynamics of these two fields. For this reason we usually speak of the gravitational field as having two dynamical degrees of freedom.

Solution 13.11. The components of the Landau-Lifschitz pseudotensor are given by the rather awesome expression

$$(-g)t^{\alpha\beta}_{\ \ L-L} = (16\pi)^{-1}\Big[\mathfrak{g}^{\alpha\beta}{}_{,\lambda}\mathfrak{g}^{\lambda\mu}{}_{,\mu} - \mathfrak{g}^{\alpha\lambda}{}_{,\lambda}\mathfrak{g}^{\beta\mu}{}_{,\mu} + \tfrac{1}{2}g^{\alpha\beta}g_{\lambda\mu}\mathfrak{g}^{\lambda\nu}{}_{,\rho}\mathfrak{g}^{\rho\mu}{}_{,\nu}$$

$$- g^{\alpha\lambda}g_{\mu\nu}\mathfrak{g}^{\beta\nu}{}_{,\rho}\mathfrak{g}^{\mu\rho}{}_{,\lambda} - g^{\beta\lambda}g_{\mu\nu}\mathfrak{g}^{\alpha\nu}{}_{,\rho}\mathfrak{g}^{\mu\rho}{}_{,\lambda} + \underline{g_{\lambda\mu}g^{\nu\rho}\mathfrak{g}^{\alpha\lambda}{}_{,\nu}\mathfrak{g}^{\beta\mu}{}_{,\rho}}$$

$$\underline{+ \tfrac{1}{8}(2g^{\alpha\lambda}g^{\beta\mu} - g^{\alpha\beta}g^{\lambda\mu})(2g_{\nu\rho}g_{\sigma\tau} - g_{\sigma\rho}g_{\nu\tau})\mathfrak{g}^{\nu\tau}{}_{,\lambda}\mathfrak{g}^{\rho\sigma}{}_{,\mu}}\Big]$$

where

$$\mathfrak{g}^{\alpha\beta} \equiv (-g)^{\frac{1}{2}}g^{\alpha\beta} .$$

For the metric we are considering

$$\mathfrak{g}^{00} = -(1-2\phi)^{\frac{3}{2}}(1+2\Phi)^{-\frac{1}{2}}, \quad \mathfrak{g}^{ij} = (1-4\Phi^2)^{\frac{1}{2}}\delta^{ij}, \quad \mathfrak{g}^{0i} = 0 .$$

Notice that all terms in the expression for $t^{\alpha\beta}_{\ \ L-L}$ involve the product of the derivatives of two $\mathfrak{g}$'s, so to lowest order $t^{\alpha\beta}$ will go as the product of the first derivative of Φ. We need to keep therefore, only the terms proportioned to $\Phi_{,i}$ when we calculate the derivatives of the $\mathfrak{g}$'s:

$$\mathfrak{g}^{00}{}_{,i} \approx 4\phi_{,i} \qquad g^{ij}{}_{,k} \approx -4\phi\phi_{,k}\delta^{ij} \approx 0 .$$

If we keep only the $\mathfrak{g}^{00}{}_{,i}$ type terms, only the underlined terms in the expression for $t^{\alpha\beta}_{\ \ L-L}$ survive and we find

$$t^{\alpha\beta}_{\ \ L-L} = (16\pi)^{-1}\Big[-g^{ij}\mathfrak{g}^{\alpha 0}{}_{,i}\mathfrak{g}^{\beta 0}{}_{,j} + \tfrac{1}{8}(2g^{\alpha l}g^{\beta m} - g^{\alpha\beta}g^{lm})\mathfrak{g}^{00}{}_{,l}\mathfrak{g}^{00}{}_{,m}\Big] .$$

From this, with $g^{\alpha\beta} \approx \eta^{\alpha\beta}$ and $\mathfrak{g}^{00}{}_{,i} = 4\phi_{,i}$ we easily find

$$t^{00}_{\ \ L-L} = (16\pi)^{-1}\Big[-\delta^{ij}\mathfrak{g}^{00}{}_{,i}\mathfrak{g}^{00}{}_{,j} + \tfrac{1}{8}\delta^{lm}\mathfrak{g}^{00}{}_{,l}\mathfrak{g}^{00}{}_{,m}\Big] = \frac{-7}{8\pi}\delta^{ij}\Phi_{,i}\Phi_{,j}$$

$$t^{0i}_{\ \ L-L} = 0$$

$$t^{ij}_{\ \ L-L} = (16\pi)^{-1}\Big[\tfrac{1}{8}(2\delta^{il}\delta^{jm} - \delta^{ij}\delta^{lm}\mathfrak{g}^{00}{}_{,l}\mathfrak{g}^{00}{}_{,m}\Big] = (4\pi)^{-1}\Big(\Phi_{,i}\Phi_{,j} - \tfrac{1}{2}\delta_{ij}\Phi_{,m}\Phi^{,m}\Big) .$$

Note that the spatial components t^{ij}_{L-L} agree with the stress tensor given in Problem 12.8.

Solution 13.12. At a given point the value of a scalar is unchanged so

$$\phi^{new}(x^{new}(P)) = \phi^{old}(x^{old}(P)) = \phi^{old}(x^{new}(P) - \boldsymbol{\xi}(P))$$

$$\phi^{new}(x^{new}) \approx \phi^{old}(x^{new}) - \xi^{a}\phi_{,a} .$$

For a vector field

$$\begin{aligned} V^{new}_{\mu} &= \frac{\partial(x^{old})^{\nu}}{\partial(x^{new})^{\mu}} V^{old}_{\nu}(x^{old}) \\ &= (\delta^{\nu}_{\mu} - \xi^{\nu}_{,\mu})(V^{old}_{\nu}(x^{new}) - V^{old}_{\nu,\sigma}\xi^{\sigma}) \\ &\approx V^{old}_{\mu}(x^{new}) - V_{\nu}\xi^{\nu}_{,\mu} - V_{\mu,\sigma}\xi^{\sigma} . \end{aligned}$$

For a tensor field, similarly

$$\begin{aligned} T^{new}_{\mu\nu}(x^{new}) &= \frac{\partial(x^{old})^{a}}{\partial(x^{new})^{\mu}} \frac{\partial(x^{old})^{\beta}}{\partial(x^{new})^{\nu}} T^{old}_{a\beta}(x^{old}) \\ &= (\delta^{a}_{\mu} - \xi^{a}_{,\mu})(\delta^{\beta}_{\nu} - \xi^{\beta}_{,\nu})(T^{old}_{a\beta}(x^{new}) - T_{a\beta,\sigma}\xi^{\sigma}) \\ &\approx T^{old}_{\mu\nu} - T_{\mu\beta}\xi^{\beta}_{,\nu} - T_{\beta\nu}\xi^{\beta}_{,\mu} - T_{\mu\nu,\sigma}\xi^{\sigma} . \end{aligned}$$

Notice that

$$\xi_{\mu;\nu} = g_{\mu\sigma}\xi^{\sigma}_{;\nu} = g_{\mu\sigma}(\xi^{\sigma}_{,\nu} + \xi^{a}\Gamma^{\sigma}_{a\nu}) = g_{\mu\sigma}\xi^{\sigma}_{,\nu} + \xi^{a}\Gamma_{\mu a\nu}$$

and that

$$\xi_{(\mu;\nu)} = g_{\sigma(\mu}\xi^{\sigma}_{,\nu)} + \xi^{a}\left(\frac{1}{2}\, g_{\mu\nu,a}\right) .$$

For the metric tensor then

$$\begin{aligned} g^{\text{new}}_{\mu\nu}(x^{\text{new}}) &= g^{\text{old}}_{\mu\nu}(x^{\text{new}}) - g_{\mu\nu,\sigma}\xi^{\sigma} - 2g_{\sigma(\mu}\xi^{\sigma}{}_{,\nu)} \\ &= g^{\text{old}}_{\mu\nu}(x^{\text{new}}) - 2\xi_{(\mu;\nu)} \\ &= \eta_{\mu\nu} + h^{\text{old}}_{\mu\nu}(x^{\text{new}}) - 2\xi_{(\mu;\nu)} \end{aligned}$$

and we have

$$h^{\text{new}}_{\mu\nu}(x^{\text{new}}) = h^{\text{old}}_{\mu\nu}(x^{\text{new}}) - 2\xi_{(\mu;\nu)} \ .$$

Since, to lowest order, the metric is Minkowskian we can take

$$\xi_{(\mu;\nu)} = \xi_{(\mu,\nu)} + \mathcal{O}(\xi h) \ .$$

Solution 13.13. Since the Christoffel symbols $\Gamma_{\alpha\mu\beta} = \frac{1}{2}(h_{\alpha\beta,\mu} + h_{\alpha\mu,\beta} - h_{\beta\mu,\alpha})$ are of order h we may ignore the terms that are products of the Γ's in the formula relating the Riemann components to the Γ's, thus we are left with

$$\begin{aligned} R_{\alpha\mu\beta\nu} &\approx g_{\alpha\lambda}(\Gamma^{\lambda}{}_{\mu\nu,\beta} - \Gamma^{\lambda}{}_{\mu\beta,\nu}) \approx 2\Gamma_{\alpha\mu[\nu,\beta]} \\ &= (h_{\alpha[\nu,\beta],\mu} - h_{\mu[\nu,\beta],\alpha}) \end{aligned}$$

and the formula is verified.

Under a gauge transformation $h_{\mu\nu} \to h_{\mu\nu} - 2\xi_{(\mu,\nu)}$ and

$$R_{\alpha\mu\beta\nu} \to R_{\alpha\mu\beta\nu} - \xi_{\alpha[,\nu,\beta],\mu} + \xi_{\mu[,\nu,\beta],\alpha} - \xi_{\nu[,\alpha,\beta],\mu} + \xi_{\nu[,\mu,\beta],\alpha} \cdot$$

Since partial derivatives commute the added terms cancel and the form of the Riemann tensor is unchanged.

Solution 13.14. Under a gauge transformation (see Problem 13.12) the metric perturbations become

$$h'_{\alpha\beta} = h_{\alpha\beta} - 2\xi_{(\alpha,\beta)} \ .$$

If we contract this we see that the trace $(h \equiv h^{\sigma}_{\sigma} \equiv \eta^{\sigma\lambda} h_{\sigma\lambda})$ transforms as $h' = h - 2\xi^{\sigma}{}_{,\sigma}$, (where $\xi^{\sigma} \equiv \eta^{\sigma\lambda}\xi_{\lambda}$). The transformation law for $\bar{h}_{\alpha\beta}$ is then

$$\bar{h}'_{\alpha\beta} = h'_{\alpha\beta} - \frac{1}{2}\eta_{\alpha\beta}h' = h_{\alpha\beta} - 2\xi_{(\alpha,\beta)} - \frac{1}{2}\eta_{\alpha\beta}(h - 2\xi^{\sigma}{}_{,\sigma})$$
$$= \bar{h}_{\alpha\beta} - \xi_{\alpha,\beta} - \xi_{\beta,\alpha} + \eta_{\alpha\beta}\xi^{\sigma}{}_{,\sigma}$$

and

$$\bar{h}'_{\alpha\beta}{}^{,\beta} = \eta^{\lambda\beta}\bar{h}'_{\alpha\beta,\lambda} = \bar{h}_{\alpha\beta}{}^{,\beta} - \xi_{\alpha,\beta}{}^{,\beta} - \xi_{\beta,\alpha}{}^{,\beta} + \xi_{\sigma}{}^{,\sigma}{}_{,\alpha}$$
$$= \bar{h}_{\alpha\beta}{}^{,\beta} - \xi_{\alpha,\beta}{}^{,\beta} = \bar{h}_{\alpha\beta}{}^{,\beta} - \Box\xi_{\alpha} \ .$$

So we need only choose the four functions ξ_{α} such that they satisfy the familiar wave equation $\Box\xi_{\alpha} = \bar{h}_{\alpha\beta}{}^{,\beta}$.

Such a solution can always be found (e.g. the retarded integral of classical electromagnetic wave theory). The solution, of course, is not unique. We can always add to ξ_{α} any solution of $\Box\xi_{\alpha} = 0$.

Solution 13.15. In Problem 13.13 we found an expression for the linearized Riemann tensor. By contracting this we can get the linearized Ricci tensor

$$R_{\mu\nu} = \frac{1}{2}(h_{\nu\alpha}{}^{,\alpha}{}_{,\mu} + h_{\mu\alpha}{}^{,\alpha}{}_{,\nu} - \Box h_{\mu\nu} - h_{,\mu\nu})$$

where $h \equiv h_{\sigma}^{\ \sigma}$. In the Lorentz gauge $\bar{h}_{\alpha\beta}{}^{,\beta} = 0$ implies $h_{\alpha\beta}{}^{,\beta} = \frac{1}{2}h_{,\alpha}$ so that

$$R_{\mu\nu} = \frac{1}{2}(h_{,\mu\nu} - \Box h_{\mu\nu} - h_{,\mu\nu}) = -\frac{1}{2}\Box h_{\mu\nu}$$

and by contracting again we have $R = -\frac{1}{2}\Box h$. The field equations then are

$$R_{\mu\nu} - \frac{1}{2}g_{\mu\nu}R = -\frac{1}{2}\left(\Box h_{\mu\nu} - \frac{1}{2}\eta_{\mu\nu}\Box h\right) = -\frac{1}{2}\Box\bar{h}_{\mu\nu}$$

so that

$$\Box\bar{h}_{\mu\nu} = -16\pi T_{\mu\nu} \ .$$

Solution 13.16. For $\bar{h}_{\mu\nu}$ to satisfy the Lorentz gauge condition (see Problem 13.14), and the linearized field equations, we must have

$$0 = \bar{h}_{\mu\nu}{}^{,\nu} = \mathrm{Re}\,[ik^{\nu}A_{\mu\nu}\,e^{ik_{\alpha}x^{\alpha}}]$$

$$0 = \bar{h}_{\mu\nu,\sigma}{}^{,\sigma} = \mathrm{Re}\,[-k_{\sigma}k^{\sigma}A_{\mu\nu}\,e^{ik_{\alpha}x^{\alpha}}]\ .$$

So the 4-vector **k** must be null and must be orthogonal to **A**.

The gauge transformation of $\bar{h}_{\mu\nu}$ is given in Problem 13.14 as

$$\bar{h}^{\mathrm{new}}_{\mu\nu} = \bar{h}^{\mathrm{old}}_{\mu\nu} - 2\xi_{(\mu,\nu)} + \eta_{\mu\nu}\xi^{\sigma}{}_{,\sigma}\ .$$

Clearly the gauge transformation must be of a plane wave $e^{ik_{\alpha}x^{\alpha}}$ form. With $\xi_{\mu} = -iC_{\mu}e^{ik_{\alpha}x^{\alpha}}$ we have

$$A^{\mathrm{new}}_{\mu\nu} = A^{\mathrm{old}}_{\mu\nu} - (C_{\mu}k_{\nu} + C_{\nu}k_{\mu}) + \eta_{\mu\nu}C^{\alpha}k_{\alpha}\ .$$

If this equation is contracted with k^{μ} we see that

$$A^{\mathrm{new}}_{\mu\nu}k^{\mu} = -k^{\nu}C_{\mu}k^{\mu} + k^{\nu}C^{\alpha}k_{\alpha} = 0$$

so **A** remains orthogonal to **k**. We can now apply the transverse-traceless condition, $(A^{\mathrm{new}})_{\mu}{}^{\mu} = 0$ and $A^{\mathrm{new}}_{\mu 0} = 0$, to our equation to get

$$0 = (A^{\mathrm{old}})_{\mu}{}^{\mu} + 2\,\mathbf{C}\cdot\mathbf{k}$$

$$0 = A^{\mathrm{old}}_{\mu 0} - (C_0k_{\mu} + C_{\mu}k_0) + \eta_{\mu 0}\,\mathbf{C}\cdot\mathbf{k}\ .$$

These can be combined into

$$0 = A^{\mathrm{old}}_{\mu 0} - (C_0k_{\mu} + C_{\mu}k_0) - \tfrac{1}{2}\,\eta_{\mu 0}\,A^{\mathrm{old}}{}_{\mu}{}^{\mu}\ .$$

Taking $\mu = 0$ we have $C_0 = (2k_0)^{-1}\left(A^{\mathrm{old}}_{00} + \tfrac{1}{2}\,(A^{\mathrm{old}})_{\mu}{}^{\mu}\right)$. The spatial components are then easily found from

$$0 = A^{\mathrm{old}}_{i0} - (C_0k_i + C_ik_0)\ .$$

Solution 13.17. For a beam of light moving in the x-direction nonzero components of the stress-energy tensor are $T_{00} = T_{xx} = -T_{0x}$. From the form of the linearized field equations

$$\Box \bar{h}_{\alpha\beta} \equiv \bar{h}_{\alpha\beta,\gamma}{}^{,\gamma} = -16\pi T_{\alpha\beta}$$

and from the fact that all components of $h_{\alpha\beta}$ should be retarded integrals over the source, the solutions must be related in the same way that the sources are:

$$\bar{h}_{00} = \bar{h}_{xx} = -\bar{h}_{0x} \ ,$$

with all other components zero. Since the trace of $\bar{h}_{\alpha\beta}$ is zero, $\bar{h}_{\alpha\beta} = h_{\alpha\beta}$ and the above relation is also true if we remove the bars on the h's.

For a photon moving in the x-direction with affine parameter λ,

$$\frac{d^2y}{d\lambda^2} = -\frac{dx^\alpha}{d\lambda}\frac{dx^\beta}{d\lambda}\Gamma^y{}_{\alpha\beta} = -\left(\frac{dt}{d\lambda}\right)^2 (\Gamma^y{}_{00} + \Gamma^y{}_{xx} + 2\Gamma^y{}_{0x}) \ .$$

But the right hand side vanishes

$$2\Gamma^y{}_{0x} + \Gamma^y{}_{00} + \Gamma^y{}_{xx} = -\tfrac{1}{2}(h_{00,y} + h_{xx,y} - 2h_{0x,y}) = 0$$

so that $d^2y/d\lambda^2 = 0$ and similarly $d^2z/d\lambda^2 = 0$. Thus the photon continues to move parallel to the beam of light and therefore two thin beams of light initially moving parallel will not attract each other to the lowest order in each of their mass-energies. (In fact, it turns out that there is no attraction even in higher order.)

Solution 13.18. From linearized theory we know that in the Lorentz gauge

$$\Box \bar{h}_{\alpha\beta} = -16\pi T_{\alpha\beta} \ .$$

Since the source, and hence the field, is stationary this becomes a Poisson equation $\nabla^2 \bar{h}_{\alpha\beta} = -16\pi T_{\alpha\beta}$. To calculate $\bar{h}_{0y}$ we need T_{0y}. For a spherical mass configuration of mass density ρ rotating (in the positive sense about the z-axis) with angular velocity Ω, we have $T_{0y} = r\Omega\rho \sin\theta \cos\phi$. Notice that T_{0y} is proportional to the real part of the spherical harmonic $Y_{11}(\theta,\phi)$ and hence $\bar{h}_{0y}$ will also be. We therefore write $\bar{h}_{0y} = f(r)\sin\theta \cos\phi$ and the Poisson equation becomes

$$\frac{1}{r^2}\frac{d}{dr}\left[r^2 \frac{d}{dr} f(r)\right] - \frac{2f(r)}{r^2} = 16\pi r\Omega\rho \ .$$

If the source is a spherical shell of mass M, radius R, then $\rho = M\delta(r-R)/4\pi R^2$ and the equation can easily be integrated to show

$$f(r) = -\frac{4M\Omega}{3} \times \begin{bmatrix} r/R & \text{for} \quad r < R \\ \left(\frac{R}{r}\right)^2 & \text{for} \quad r > R \end{bmatrix}$$

and hence inside the shell

$$g_{0y} \approx h_{0y} = \bar{h}_{0y} = -\frac{4\Omega}{3} M \frac{r}{R} \sin\theta \cos\phi \ .$$

By symmetry

$$g_{\hat{0}\hat{\phi}} = g_{0y}\Big|_{\phi=0} = -\frac{4}{3}\Omega M \left(\frac{r}{R}\right)\sin\theta$$

and

$$\omega = -\frac{g_{0\phi}}{g_{\phi\phi}} = -\frac{g_{\hat{0}\hat{\phi}}}{(g_{\phi\phi})^{\frac{1}{2}}} = -\frac{g_{\hat{0}\hat{\phi}}}{r\sin\theta} = \frac{4\Omega M}{3R} \ .$$

Inside the shell then locally inertial frames will rotate relative to inertial frames far away. Mach's principle states that the inertial properties of spacetime depend on the motion of distant matter. The dragging of inertial frames certainly is an example of the influence of matter on the inertial properties of spacetime. The constancy of ω suggests that the influence of distant matter on inertial properties might be viewed as a $\frac{1}{r}$ law, so that the inertial properties are not primarily determined by nearby matter, but rather the influence of all the matter in the universe must be considered. (But aside from a few idealized examples such as this one, no one has ever been very successful in "deriving" Mach's principal from the field equations of general relativity.)

Solution 13.19. The equations of motion $T^{\mu\nu}{}_{;\nu} = 0$ are clearly gauge invariant, but in the Lorentz gauge (see Problem 13.15) $T^{\mu\nu}{}_{,\nu} = -(16\pi)^{-1}\Box\bar{h}^{\mu\nu}{}_{,\nu} = 0$ rather than $T_{\mu\nu}{}^{;\nu} = 0$. The difference between $T^{\mu\nu}{}_{;\nu}$ and $T^{\mu\nu}{}_{,\nu}$ is of order $T\Gamma$. But both T and Γ are of the order of $h_{\alpha\beta}$ so the inconsistent terms are of second order in the deviation of the metric from flatness.

Solution 13.20. By the principle of equivalence, the acceleration of a body is independent of the magnitude of its mass, and depends only on other masses with which it gravitates. Thus a positive mass attracts both positive and negative masses, and a negative mass repels them both.

Here, then, the acceleration of the negative mass is toward the positive mass, and of magnitude GM/ℓ^2. (Since $\ell >> M$, we are in the Newtonian limit.) The acceleration of the positive mass is in the same direction and of the same magnitude: the two masses try to chase each other.

The problem is more complicated once the particles are moving, because each sees the retarded field of the other.

Since we are given $M << \ell$, we shall solve the problem using linearized theory which will be justified as long as their distance in a frame comoving with either remains much greater than M. The equation

$$\Box\bar{h}_{\mu\nu} = -16\pi T_{\mu\nu} \tag{1}$$

has the retarded integral solution

$$\bar{h}_{\mu\nu}(t,\underset{\sim}{x}) = 4\int \frac{T_{\mu\nu}(t',\underset{\sim}{x}')}{|\underset{\sim}{x}-\underset{\sim}{x}'|}\,\delta(t'+|\underset{\sim}{x}-\underset{\sim}{x}'|-t)\,d^3x'dt' . \tag{2}$$

In linearized theory we can superpose the $T_{\mu\nu}$'s (that is the $h_{\mu\nu}$'s) of each particle. For a particle of mass m moving along a world-line

$$x^\mu = z^\mu(\tau) \tag{3}$$

we have

$$T_{\mu\nu}(t,\underset{\sim}{x}) = m\int u_\mu u_\nu \delta^4(x - z(\tau))\, d\tau = m u_\mu u_\nu \delta^3(\underset{\sim}{x} - \underset{\sim}{z}(t))/u^0 \tag{4}$$

where we have used $d\tau = dt/u^0$ to do the integral. Substitute Equation (4) in Equation (2):

$$\begin{aligned}\overline{h}_{\mu\nu}(t,\underset{\sim}{x}) &= 4m\int \frac{u_\mu u_\nu \delta^3[\underset{\sim}{x}' - \underset{\sim}{z}(t')]\,\delta(t' + |\underset{\sim}{x} - \underset{\sim}{x}'| - t)\, d^3x'dt'}{u^0|\underset{\sim}{x} - \underset{\sim}{x}'|} \\ &= 4m\int \frac{u_\mu(t')u_\nu(t')\delta(t' + |\underset{\sim}{x} - \underset{\sim}{z}(t')| - t)\, dt'}{u^0(t')\,|\underset{\sim}{x} - \underset{\sim}{z}(t')|}\,.\end{aligned} \tag{5}$$

By symmetry everything depends only on one spatial coordinate, say x. Let

$$R(t') = |x - z(t')| = \epsilon[x - z(t')] \tag{6}$$

where $\epsilon = \pm 1$ for $x \gtrless z$. Then

$$\overline{h}_{\mu\nu}(t,x) = 4m\int \frac{u_\mu u_\nu \delta(t' + R - t)\, dt'}{u^0 R}\,. \tag{7}$$

The 4-velocity u is determined from the equation of motion

$$du^\mu/d\tau = -\Gamma^\mu_{\alpha\beta}u^\alpha u^\beta$$

i.e.

$$du^\mu/d\tau + \eta^{\mu\nu}\, u^\alpha u^\beta(h_{\nu\alpha,\beta} + h_{\nu\beta,\alpha} - h_{\alpha\beta,\nu})/2 = 0\,. \tag{8}$$

Here the $h_{\alpha\beta}$ of particle 2 evaluated at the position of particle 1 is used to find u^μ of particle 1 and vice versa. By symmetry $u^y = u^z = 0$ for both particles. Write out Equation (8) for $\mu = 0$ and $\mu = x$, using $h_{00} = h_{xx}$ (easily verified from Equation (7)):

$$du^0/d\tau = \tfrac{1}{2}h_{00,0}({u^0}^2 - {u^x}^2) + h_{00,x}u^0u^x + h_{0x,x}{u^x}^2 \tag{10}$$

$$du^x/d\tau = \tfrac{1}{2}h_{00,x}({u^0}^2 - {u^x}^2) - h_{0x,0}{u^0}^2 - h_{00,0}u^0u^x\,. \tag{11}$$

Now evaluate the derivatives of the $h_{\alpha\beta}$ from Equation (7):

$$\bar{h}_{00,0} = 4m \frac{\partial}{\partial t} \int \frac{u^0(t')\delta(t'+R(t')-t)\,dt'}{R(t')} = -4m \int \frac{u^0\delta'(t'+R-t)\,dt'}{R}$$

where δ' denotes differentiation with respect to the argument of the delta function. Let

$$f(t') = t' + R(t')$$

$$\frac{df}{dt'} = 1 - \epsilon v(t'), \qquad v = \frac{dz}{dt'} .$$

Then

$$\begin{aligned}
\bar{h}_{00,0} &= -4m \int \frac{u^0}{R}\left[\frac{d}{df}\delta(f-t)\right]\frac{dt'}{df}\,df \\
&= 4m \int \frac{d}{df}\left[\frac{u^0}{R}\frac{dt'}{df}\right]\delta(f-t)\,df \\
&= 4m \left[\frac{dt'}{df}\frac{d}{dt'}\left(\frac{u^0}{R}\frac{dt'}{df}\right)\right]_{f=t} \\
&= 4m \left[\frac{1}{1-\epsilon v}\frac{d}{dt'}\frac{u^0}{(1-\epsilon v)R}\right]_{t'=t-R(t')} . \qquad (12)
\end{aligned}$$

Since $du^0/dt' \sim m$, to linear order in m we need not differentiate u^0 and v in Equation (12). Thus

$$\bar{h}_{00,0} = 4m \left[\frac{\epsilon v u^0}{(1-\epsilon v)^2 R^2}\right]_{\text{ret.}} \qquad (13)$$

where ret. means evaluated at the retarded time $t' = t - R(t')$. But

$$\begin{aligned}
v(t') &= v(t) + (dv/dt)(t'-t) + \cdots \\
&= v(t) + \mathcal{O}(m)
\end{aligned}$$

and similarly for $u^0 = (1-v^2)^{-\frac{1}{2}}$, and

$$\begin{aligned}
R(t') &= \epsilon[x - z(t')] \\
&= \epsilon[x - z(t) - v(t)(t'-t)] + \mathcal{O}(m) \\
&= \epsilon[x - z(t) + v(t)R(t')] ,
\end{aligned}$$

thus

$$R(t')(1-\epsilon v) = \epsilon[x - z(t)] \ .$$

Since $\bar{h}_{00,0}$ is being evaluated at the position of the other particle, we have

$$\bar{h}_{00,0} = 4m \frac{\epsilon v u^0}{(z_1 - z_2)^2} \tag{14}$$

where the quantities are evaluated at the instantaneous time t. Similarly

$$\bar{h}_{0x,0} = \frac{-u^x}{u^0} \bar{h}_{00,0} \tag{15}$$

$$\bar{h}_{xx,0} = \left(\frac{u^x}{u^0}\right)^2 \bar{h}_{00,0} \ . \tag{16}$$

The spatial derivatives can be evaluated by noting that x occurs only in R in Equation (7). Thus

$$\begin{aligned}
\bar{h}_{00,x} &= \epsilon \frac{\partial}{\partial R} 4m \int \frac{u^0 \delta(t'+R-t)\,dt'}{R} \\
&= 4m\epsilon\left[-\int \frac{u^0\delta(t'+R-t)\,dt'}{R^2} + \int \frac{u^0\delta'(t'+R-t)\,dt'}{R}\right] \\
&= 4m\epsilon\left[-\int \frac{u^0}{R^2}\,\delta(f-t)\,\frac{dt'}{df}\,df + \int \frac{u^0}{R}\frac{dt'}{df}\frac{d}{df}\delta(f-t)\,df\right] \\
&= -4m\epsilon\left[\frac{u^0}{(1-\epsilon v)R^2} + \frac{1}{(1-\epsilon v)}\frac{d}{dt'}\left\{\frac{u^0}{(1-\epsilon v)R}\right\}\right]_{\text{ret.}} \\
&= -4m\epsilon\left[\frac{u^0}{(1-\epsilon v)R^2} + \frac{u^0\epsilon v}{(1-\epsilon v)^2R^2}\right]_{\text{ret.}} \\
&= -\frac{4m\epsilon u^0}{(z_1-z_2)^2} \ .
\end{aligned} \tag{17}$$

Similarly

$$\bar{h}_{0x,x} = -\frac{u^x}{u^0}\bar{h}_{00,x} \tag{18}$$

$$\bar{h}_{xx,x} = \left(\frac{u^x}{u^0}\right)^2 \bar{h}_{00,x} \; . \tag{19}$$

Note that the gauge condition $h_{\mu,\alpha}^{\ \alpha} = 0$ is satisfied by Equations (14)-(19). Write out the equations of motion, Equations (10) and (11), for particle 1 letting $u^0 = \gamma$, $u^x = \gamma v$, $g = m/(z_1 - z_2)^2$. Take particle 1 to be the positive mass particle at $z_1 > z_2$, so that $\epsilon = +1$ and the source mass in Equations (14)-(19) is negative.

$$d\gamma_1/d\tau = -g\gamma_2 v_2(1+v_2^2) + 2g\gamma_2(1+v_2^2)\gamma_1^2 v_1 - 4g\gamma_2 v_2 \gamma_1^2 v_1^2 \tag{20}$$

$$d(\gamma_1 v_1)/d\tau = g\gamma_2(1+v_2^2) - 4g\gamma_2 v_2^2\gamma_1^2 + 2g\gamma_2 v_2(1+v_2^2)\gamma_1^2 v_1 \; . \tag{21}$$

Now the equations for $d\gamma_2/d\tau_2$ and $d(\gamma_2 v_2)/d\tau_2$ have the same right-hand sides as Equations (20)-(21), with $v_2 \leftrightarrow v_1$. (εm keeps the same sign.) Since $v_1 = v_2$ at $t = 0$, then $v_1 = v_2$ for all t. Thus $z_1 - z_2$ is constant for all t, i.e. $z_1 - z_2 = \ell$. We can drop the subscripts 1 and 2 in Equations (20) and (21) and get, after some simplification

$$d\gamma/d\tau = gv/\gamma \tag{22}$$

$$d(\gamma v)/d\tau = g/\gamma \tag{23}$$

These equations are not independent, because of the relation $\gamma = (1-v^2)^{-\frac{1}{2}}$. We find then that the differential equation for the motion is

$$dv/d\tau = g(1-v^2)^2 \; , \tag{24}$$

and hence

$$2g\tau = v/(1-v^2) + \tanh^{-1} v \tag{25}$$

where $\tau = 0$ when $v = 0$. Thus

$$\frac{dz}{dv} = \frac{dz}{dt}\frac{dt}{d\tau}\frac{d\tau}{dv} = \frac{v}{g}(1-v^2)^{-5/2}$$

implies

$$z_2 = \frac{1}{3g}\left[(1-v^2)^{-\frac{3}{2}} - 1\right]$$
$$z_1 = z_2 + \ell \tag{26}$$

and similarly

$$\frac{dt}{dv} = \frac{dt}{d\tau}\frac{d\tau}{dv} = \frac{1}{g}(1-v^2)^{-5/2} \tag{27}$$

implies

$$t = \frac{v}{g}\left[(1-v^2)^{-\frac{1}{2}} + \frac{v^2}{3}(1-v^2)^{-\frac{3}{2}}\right]. \tag{28}$$

Equations (26) and (28) are parametric equations for the trajectories with v as parameter. Note that while the coordinate separation of the particles remains constant, the proper separation as measured by an observer co-moving with one of the particles increases, approximately as $\gamma\ell$. Proof: Suppose particle 1 has velocity v at some instant t_1 at position $z_1(t_1)$. The Lorentz transformation to the locally comoving inertial frame is

$$x' = \gamma[x - z_1(t_1) - v(t-t_1)]$$
$$t' = \gamma[t - t_1 - v\{x - z_1(t_1)\}] .$$

Thus the trajectory of particle 2 in these coordinates is $z'_2(t'_2)$ where

$$z'_2 = \gamma[z_2(t_2) - z_1(t_1) - v(t_2 - t_1)]$$
$$t'_2 = \gamma[t_2 - t_1 - v\{z_2(t_2) - z_1(t_1)\}] .$$

The distance to particle 2 is $-z'_2(t'_2 = 0)$. Let

$$\beta = z_2(t_2) - z_1(t_1) .$$

Now $t'_2 = 0$ implies

$$t_2 = t_1 + v\beta$$

so

$$\begin{aligned}\beta &= z_2(t_1 + v\beta) - z_1(t_1) \\ &\approx z_2(t_1) + v\beta\, dz_2/dt\Big|_{t=t_1} + \frac{v^2\beta^2}{z}\, \frac{d^2 z_2}{dt^2}\Big|_{t=t_1} - z_1(t_1) \\ &= -\ell + v^2\beta + \mathcal{O}(m/L)\,.\end{aligned}$$

Solve for β:

$$\beta = \frac{-\ell}{(1 - v^2)}\,.$$

Thus

$$-z'_2(t'_2 = 0) = -\gamma(\beta - v^2\beta) \approx \gamma\ell\,.$$

CHAPTER 14: SOLUTIONS

Solution 14.1. First we determine what is the appropriate $T^{\mu\nu}$ for a point particle, say of mass m. In the momentarily comoving frame it should have only a T^{00} component, and this should be a δ-function of position, so

$$T^{\mu\nu} \propto \int \delta^4(x^\alpha - x^\alpha(\tau))\, u^\mu u^\nu \, d\tau$$

where $x^\alpha(\tau)$ is the particle's trajectory in spacetime as a function of its proper time τ. Now we want $T^{\mu\nu}$ to transform like a tensor. The product $u^\mu u^\nu$ already transforms as a tensor, but $\delta^4(x^\alpha - x^\alpha(\tau))$ is not a scalar; $\delta^4(x^\alpha - x^\alpha(\tau))(-g)^{-\frac{1}{2}}$, however, is a scalar:

$$1 = \text{scalar} = \int \delta^4(x^\alpha - x^\alpha(\tau))\, d^4x = \int \frac{\delta^4}{(-g)^{\frac{1}{2}}} (-g)^{\frac{1}{2}} d^4x \; . \tag{1}$$

Thus, normalizing to mass m, we have

$$T^{\mu\nu} = m \int \delta^4(x^\alpha - x^\alpha(\tau)) \frac{u^\mu u^\nu}{(-g)^{\frac{1}{2}}} d\tau \equiv \int \rho\, u^\mu u^\nu \, d\tau \tag{2}$$

where ρ contains all the terms except the u's. Now

$$0 = T^{\mu\nu}{}_{;\nu} = \int [(\rho u^\nu)_{;\nu} u^\mu + (\rho u^\nu) u^\mu{}_{;\nu}]\, d\tau \; . \tag{3}$$

Dotting with u^μ gives

$$0 = \int [-(\rho u^\nu)_{;\nu} + \rho(\nabla_{\mathbf{u}}\mathbf{u})\cdot\mathbf{u}]\, d\tau \; . \tag{4}$$

The second term vanishes because 4-acceleration $\nabla_{\mathbf{u}}\mathbf{u}$ is orthogonal to 4-velocity. We conclude therefore that the first term vanishes and hence, from Equation (3)

$$0 = \int (\rho u^\nu) u^\mu{}_{;\nu} \, d\tau \ .$$

This implies that $u^\nu u^\mu{}_{;\nu} = 0$ wherever the (delta function) ρ is nonzero, that is, wherever the particle is! But $u^\nu u^\mu{}_{;\nu} = 0$ is just the geodesic equation.

Solution 14.2. If a system is in thermal equilibrium, then no net energy will flow between two locations A and B if they are allowed to exchange energy. In particular, imagine a "light pipe" between the two locations which transmits a thermal flux of photons. Since the pipe and the system are static, there is no change in a photon's energy when it reflects internally in the pipe. There *is* a change in energy between A and B due to the gravitational redshift: $(h\nu)_A/(h\nu)_B = (-g_{00}{}^B)^{\frac{1}{2}}/(-g_{00}{}^A)^{\frac{1}{2}}$. Now use the fact that a black-body intensity $B_\nu(T_A)$ at A is changed by any kind of redshift to precisely another black-body spectrum with a temperature $T_B = [(h\nu)_B/(h\nu)_A]\,T_A$. If thermal equilibrium holds, this must be the ambient temperature at B, so that the fluxes up the pipe and down the pipe are identical. Thus $T_B = [(-g_{00}{}^A)^{\frac{1}{2}}/(-g_{00}{}^B)^{\frac{1}{2}}]\,T_A$, and

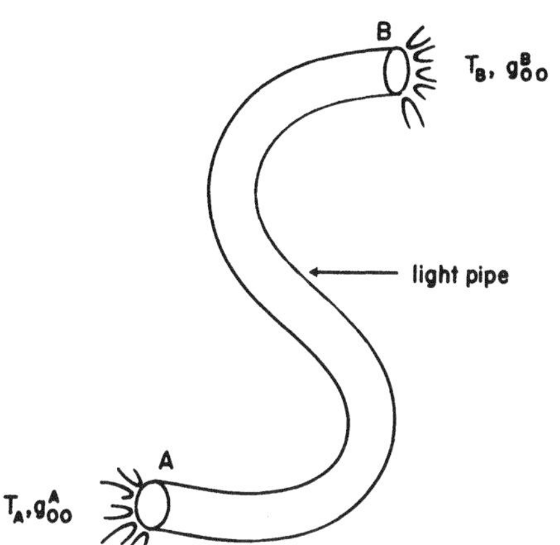

$$T(-g_{00})^{\frac{1}{2}} = \text{constant}$$

throughout the system.

Solution 14.3. For a perfect fluid the stress-energy tensor is $T^{\mu\nu} = (\rho + p) u^\mu u^\nu + p g^{\mu\nu}$, thus the equations of motions read

$$0 = T^{\mu\nu}{}_{;\nu} = (\rho + p)_{;\nu} u^\mu u^\nu + (\rho + p)(u^\mu{}_{;\nu} u^\nu + u^\mu u^\nu{}_{;\nu}) + p_{,\nu} g^{\mu\nu} \ .$$

To get the Euler equation, project this equation perpendicular to **u** using the projection tensor

$$P_{\alpha\beta} = u_\alpha u_\beta + g_{\alpha\beta} \ .$$

The result is

$$\begin{aligned} 0 &= P_{\alpha\mu} T^{\mu\nu}{}_{;\nu} \\ &= 0 + P_{\alpha\mu}(\rho+p)\, u^\mu{}_{;\nu} u^\nu + 0 + p_{,\nu}\, g^{\mu\nu} P_{\alpha\mu} \\ &= (\rho+p)\, u_{\alpha;\nu} u^\nu + p_{,\alpha} + p_{,\nu} u^\nu u_\alpha \end{aligned}$$

which is the Euler equation.

To take the Newtonian limit (see Solution 14.8), write the Euler equation as

$$\rho_0(1+\pi+p/\rho_0)(u_{j,\nu}u^\nu - \Gamma^\alpha{}_{\nu j} u_\alpha u^\nu) = -p_{,j} - u_j\, dp/d\tau \ .$$

Set $u^0 = 1$, $u_0 = -1$, $u_j = v_j$ and get

$$\rho_0(dv_j/d\tau + \Gamma^0{}_{0j}) \approx -p_{,j} \ .$$

But

$$\Gamma^0{}_{0j} \approx -\Gamma_{00j} = -\tfrac{1}{2}\, g_{00,j} \approx \phi_{,j} \ .$$

Thus

$$\frac{dv_j}{d\tau} = -\phi_{,j} - \frac{1}{\rho_0}\, p_{,j} \ .$$

Solution 14.4. The Euler equation for a perfect fluid is

$$(\rho+p)\nabla_{\mathbf{u}}\mathbf{u} = -\nabla p - \mathbf{u}\nabla_{\mathbf{u}} p \ .$$

Hydrostatic equilibrium implies the existence of a time Killing vector $\boldsymbol{\xi}$. The fluid 4-velocity, according to Problem 13.9, must be parallel to this Killing vector, so

$$\mathbf{u} = \boldsymbol{\xi}/|\boldsymbol{\xi}| \ .$$

(In components, this means that only $u^i = 0$, where $\boldsymbol{\xi} = \partial/\partial t$.) From Problem 10.14 we know

$$\nabla_{\mathbf{u}}\mathbf{u} = \frac{1}{2}\nabla \log|\boldsymbol{\xi}\cdot\boldsymbol{\xi}| .$$

We also know that $\partial p/\partial t = 0$, or $\nabla_{\mathbf{u}}p \propto \nabla_{\boldsymbol{\xi}}p = 0$, and therefore

$$\nabla p = -(\rho+p)\nabla \log|\boldsymbol{\xi}\cdot\boldsymbol{\xi}|^{\frac{1}{2}} .$$

But $\boldsymbol{\xi}\cdot\boldsymbol{\xi} = (\partial/\partial t)\cdot(\partial/\partial t) = g_{00}$, so $|\boldsymbol{\xi}\cdot\boldsymbol{\xi}|^{\frac{1}{2}} = (-g_{00})^{\frac{1}{2}}$, from which the result follows.

In the Newtonian limit, $p \ll \rho$ and $g_{00} \approx -(1+2\Phi)$ where the Newtonian potential $\Phi \ll 1$. Thus

$$\frac{\partial p}{\partial x^\nu} \approx -\rho\frac{\partial}{\partial x^\nu}\frac{1}{2}\log(1+2\Phi) \approx -\rho\frac{\partial\Phi}{\partial x^\nu}..$$

Solution 14.5. Using $p = \frac{1}{3}\rho$ in the equation of hydrostatic equilibrium (Problem 14.4) $p_{,\lambda} = (\rho+p)[\log(-g_{00})^{\frac{1}{2}}]_{,\lambda}$ we get

$$-\log(-g_{00})^{\frac{1}{2}} = \frac{1}{4}\int\frac{d\rho}{\rho} .$$

This integrates to $\rho = \text{constant}\times(-g_{00})^{-2}$. As long as g_{00} is finite, therefore, ρ cannot vanish and thus there is no free surface.

Solution 14.6. A static uniform gravitational field has a line element

$$ds^2 = g_{tt}\,dt^2 + g_{zz}\,dz^2 + dx^2 + dy^2 \qquad (1)$$

where g_{tt} and g_{zz} depend only on z. The nonzero Christoffel symbols are easily computed (Problem 7.6). The weight as measured by a scale at $z = 0$ is $W(0)$, where

$$W(z) = \int T^{\hat{z}\hat{z}}\,dx\,dy = \int g_{zz}T^{zz}\,dx\,dy . \qquad (2)$$

We wish to show that $W(0)$ depends only on $T^{\hat{t}\hat{t}} = g_{tt}T^{tt}$ and on no other $T^{\hat{\alpha}\hat{\beta}}$. We have

$$0 = T^{\alpha\beta}{}_{;\beta} = |g|^{-\frac{1}{2}} (|g|^{\frac{1}{2}} T^{\alpha\beta})_{,\beta} + \Gamma^{\alpha}{}_{\beta\gamma} T^{\beta\gamma} . \tag{3}$$

Take $\alpha = t$ and integrate over x, y, using the fact that $g_{\alpha\beta}$ is independent of x, y and t:

$$0 = \int T^{tt}{}_{,t} \, dx\, dy + \int T^{tx}{}_{,x} \, dx\, dy + \int T^{ty}{}_{,y} \, dx\, dy + |g|^{-\frac{1}{2}} (|g|^{\frac{1}{2}} \int T^{tz} \, dx\, dy)_{,z}$$
$$+ 2\Gamma^{t}{}_{tz} \int T^{tz} \, dx\, dy . \tag{4}$$

The 1st term vanishes because $T^{\hat{t}\hat{t}}$ is time-independent. The 2nd and 3rd terms give terms proportional to T^{tx} and T^{ty} on the boundary of the integration region. These vanish for a boundary outside the container. Equation (4) can be put in the form

$$0 = |g|^{-\frac{1}{2}} g_{tt}{}^{-1} (|g|^{\frac{1}{2}} g_{tt} \int T^{tz} \, dx\, dy)_{,z} . \tag{5}$$

Thus $|g|^{\frac{1}{2}} g_{tt} \int T^{tz} \, dx\, dy$ is independent of z. Since this term vanishes above the top of the container, it vanishes everywhere.

Now put $\alpha = z$ in Equation (3) and integrate over x and y:

$$0 = \int T^{zt}{}_{,t} \, dx\, dy + \int T^{zx}{}_{,x} \, dx\, dy + \int T^{zy}{}_{,y} \, dx\, dy + |g|^{-\frac{1}{2}} (|g|^{\frac{1}{2}} \int T^{zz} \, dx\, dy)_{,z}$$
$$+ \Gamma^{z}{}_{tt} \int T^{tt} \, dx\, dy + \Gamma^{z}{}_{zz} \int T^{zz} \, dx\, dy . \tag{6}$$

The 1st term vanishes as above, while the 2nd and 3rd give boundary contributions which vanish. Using Equation (2), we find

$$(|g_{tt}|^{\frac{1}{2}} W)_{,z} = |g_{tt}|^{\frac{1}{2}}{}_{,z} \int T^{\hat{t}\hat{t}} \, dx\, dy . \tag{7}$$

Integrate Equation (7) from $z = 0$ to the top of the container, and use the fact that $W(\text{top}) = 0$ to get

$$W(0) = |g_{tt}|^{-\frac{1}{2}}_{z=0} \int_0^{\text{top}} dz \int dx\, dy \, |g_{tt}|^{\frac{1}{2}}{}_{,z} T^{\hat{t}\hat{t}} , \tag{8}$$

which depends only on the component $T^{\hat{t}\hat{t}}$. (Solution due to W. Unruh.)

Solution 14.7. Since the gravitational field is stationary, there is a time-Killing vector $\boldsymbol{\xi} = \partial/\partial t$. Dot this into the Euler equation (Problem 14.3) for the flow of a perfect gas:

$$(\rho + p)\boldsymbol{\xi} \cdot \nabla_{\mathbf{u}}\mathbf{u} = -\boldsymbol{\xi} \cdot \nabla p - \boldsymbol{\xi} \cdot \mathbf{u}\,\nabla_{\mathbf{u}} p \ . \qquad (1)$$

But

$$\boldsymbol{\xi} \cdot \nabla_{\mathbf{u}}\mathbf{u} = \nabla_{\mathbf{u}}(\boldsymbol{\xi} \cdot \mathbf{u}) - (\nabla_{\mathbf{u}}\boldsymbol{\xi}) \cdot \mathbf{u} = \nabla_{\mathbf{u}}(\boldsymbol{\xi} \cdot \mathbf{u})$$

since by Killing's equation $\nabla\boldsymbol{\xi}$ is antisymmetric. Also, $\boldsymbol{\xi} \cdot \nabla p = \partial p/\partial t = 0$ for stationary flow. Putting $u_0 = \mathbf{u} \cdot \boldsymbol{\xi}$, we find $(\rho + p)\, du_0/d\tau = u_0\, dp/d\tau$ [Equation 2]. For adiabatic flow of a perfect fluid, the first law of thermodynamics (Problem 5.19) is $d\rho = (\rho + p)\, dn/n$. [Equation 3]. From Equations (2) and (3)

$$\frac{du_0}{u_0} = \frac{dn}{n} - \frac{d(\rho + p)}{\rho + p}$$

and hence $u_0 = \text{constant} \times \dfrac{n}{\rho + p}$.

Solution 14.8. The Newtonian limit of relativistic hydrodynamics is obtained by choosing a global, nearly-Lorentz, frame in which

$$\rho = \rho_0(1 + \pi), \qquad \pi \ll 1$$

$$g_{00} = -(1 + 2\phi), \qquad |\phi| \ll 1$$

$$p/\rho_0 \ll 1, \qquad v^2 \ll 1 \ .$$

Here $\rho_0 = nm_B$ is the rest-mass density (m_B = mean baryon rest mass), π is the specific internal energy, ϕ is the Newtonian potential and v is 3-velocity of the fluid. Since

$$u^{\hat{0}} = (1 - v^2)^{-\frac{1}{2}} \approx 1 + \frac{1}{2} v^2 \ ,$$

and

$$u_0 = (-g_{00})^{\frac{1}{2}} u^{\hat{0}} \approx 1 + \frac{1}{2} v^2 + \phi ,$$

the Bernoulli equation becomes

$$\left(1 + \frac{1}{2} v^2 + \phi\right) = \frac{\text{constant} \times n}{\rho_0(1+\pi+p/\rho_0)}$$

that is

$$\left(1 + \frac{1}{2} v^2 + \phi\right) (1+\pi+p/\rho_0) = \text{constant}$$

and finally

$$\frac{1}{2} v^2 + \phi + \pi + p/\rho_0 = \text{constant} .$$

Solution 14.9.

(a) The condition that $\boldsymbol{\xi}$ be a connecting vector between the world lines of neighboring particles can be written

$$\mathcal{L}_{\mathbf{u}} \boldsymbol{\xi} = \alpha \mathbf{u} \tag{1}$$

where α is a scalar function. In Problem 8.14, α was zero since $\boldsymbol{\xi}$ connected events of equal proper time; here we find α from the condition $\xi^{\hat{0}} = \boldsymbol{\xi} \cdot \mathbf{u} = 0$ along the world lines:

$$0 = \nabla_{\mathbf{u}}(\boldsymbol{\xi} \cdot \mathbf{u}) = \boldsymbol{\xi} \cdot \mathbf{a} + \mathbf{u} \cdot \nabla_{\mathbf{u}} \boldsymbol{\xi} . \tag{2}$$

Compare this with Equation (1) dotted with $\mathbf{u}$:

$$\mathbf{u} \cdot (\nabla_{\mathbf{u}} \boldsymbol{\xi} - \nabla_{\boldsymbol{\xi}} \mathbf{u}) = -\alpha$$

and find $\alpha = \boldsymbol{\xi} \cdot \mathbf{a}$. (Recall that $\mathbf{u} \cdot \nabla_{\boldsymbol{\xi}} \mathbf{u} = \frac{1}{2} \nabla_{\boldsymbol{\xi}} (\mathbf{u} \cdot \mathbf{u}) = 0$.) Thus we have

$$\nabla_{\mathbf{u}} \boldsymbol{\xi} = \nabla_{\boldsymbol{\xi}} \mathbf{u} + (\boldsymbol{\xi} \cdot \mathbf{a}) \mathbf{u} . \tag{3}$$

The condition for rigid body motion, $\nabla_{\mathbf{u}}(\boldsymbol{\xi} \cdot \boldsymbol{\xi}) = 0$, is equivalent to $\boldsymbol{\xi} \cdot \nabla_{\mathbf{u}} \boldsymbol{\xi} = 0$, or by Equation (3) to

$$\xi^\alpha \xi^\beta (u_{\alpha;\beta} + u_\alpha a_\beta) = 0 .$$

Since this must be true for any ξ it follows that

$$u_{(\alpha;\beta)} + u_{(\alpha}a_{\beta)} = 0$$

and hence (see Problem 5.18 for definitions) that

$$\sigma_{\alpha\beta} + \frac{1}{3}\theta P_{\alpha\beta} = u_{(\alpha;\beta)} + u_{(\alpha}a_{\beta)} = 0 \ .$$

Rigid body motion therefore can occur if and only if $\sigma_{\alpha\beta} = \theta = 0$.

(b) The symmetry $\sigma^{\alpha\beta} = \sigma^{\beta\alpha}$ reduces the condition $\sigma^{\alpha\beta} = 0$ to 10 equations. The 5 identities $u^\alpha\sigma_{\alpha\beta}$ and $P^{\alpha\beta}\sigma_{\alpha\beta} = \sigma^\alpha{}_\alpha = 0$ leave 5 independent equations. With the condition $\theta = 0$, we have 6 independent constraints altogether.

The rigid-body conditions do not constrain a^α or $\omega^{\alpha\beta}$. Each of these has 3 independent components ($a^\alpha u_\alpha = 0$, $\omega^{\alpha\beta} = -\omega^{\beta\alpha}$) so there are 6 degrees of freedom as in nonrelativistic rigid-body motion.

Solution 14.10. From the definition of $\theta = \nabla \cdot \mathbf{u}$ we have

$$\begin{aligned}\frac{d\theta}{d\tau} &= u^\beta(u^\alpha{}_{;\alpha})_{;\beta} = u^\beta u^\alpha{}_{;\alpha\beta} = u^\beta(u^\alpha{}_{;\beta\alpha} - R^\alpha{}_{\beta\alpha\gamma}u^\gamma) \\ &= u^\beta u^\alpha{}_{;\beta\alpha} - R_{\beta\gamma}u^\beta u^\gamma \ .\end{aligned} \tag{1}$$

But

$$\begin{aligned}u^\alpha{}_{;\beta\alpha}u^\beta &= (u^\alpha{}_{;\beta}u^\beta)_{;\alpha} - u^\alpha{}_{;\beta}u^\beta{}_{;\alpha} \\ &= a^\alpha{}_{;\alpha} - \left(\omega^\alpha{}_\beta + \sigma^\alpha{}_\beta + \tfrac{1}{3}\theta P^\alpha{}_\beta - a^\alpha u_\beta\right) \\ &\quad \times\left(\omega^\beta{}_\alpha + \sigma^\beta{}_\alpha + \tfrac{1}{3}\theta P^\beta{}_\alpha - a^\beta u_\alpha\right)\end{aligned} \tag{2}$$

where we have used Problem 5.18. Using the symmetry properties of $\omega_{\alpha\beta}$, $\sigma_{\alpha\beta}$, and $P_{\alpha\beta}$ and the fact that they are orthogonal to $\mathbf{u}$, Equations (1) and (2) imply

$$\begin{aligned}\frac{d\theta}{d\tau} &= a^\alpha{}_{;\alpha} - \omega^\alpha{}_\beta\omega^\beta{}_\alpha - \sigma^\alpha{}_\beta\sigma^\beta{}_\alpha - \tfrac{1}{3}\theta^2 - R_{\alpha\beta}u^\alpha u^\beta \\ &= a^\alpha{}_{;\alpha} + 2\omega^2 - 2\sigma^2 - \tfrac{1}{3}\theta^2 - R_{\alpha\beta}u^\alpha u^\beta \ .\end{aligned}$$

Solution 14.11. Since the fluid flows on geodesics $a_\alpha \equiv u^\beta u_{\alpha;\beta} = 0$, and since the flow is shear and expansion free $\sigma_{\alpha\beta} = \theta = 0$. The decomposition of $\nabla \mathbf{u}$ in Problem 5.18 then gives

$$u_{\alpha;\beta} = \omega_{\alpha\beta} \ .$$

But $\boldsymbol{\omega}$ is antisymmetric, so $u_{(\alpha;\beta)} = 0$ and hence $\mathbf{u}$ satisfies Killing's equation.

Solution 14.12. In geometrical language the observer sets up his local coordinates x^α as follows: If $\mathbf{u}$ is the 4-velocity of the center of his box along its worldline, he chooses $\mathbf{e}_0 = \mathbf{u}$ and $\mathbf{e}_\alpha \cdot \mathbf{e}_\beta = \eta_{\alpha\beta}$ along his world line. From each point $P(\tau)$ on his world-line (τ = proper time), he sends out spatial geodesics orthogonal to $\mathbf{u}$, with affine parameter equal to proper length s. If $\mathbf{n}$ is the tangent vector to the spatial geodesic through the point P near his world-line, then he assigns to P the coordinates $x^0 = \tau$, $x^k = sn^k$.

The equation of motion is now

$$\frac{d^2x^\alpha}{d\lambda^2} + \Gamma^\alpha{}_{\beta\gamma} \frac{dx^\beta}{d\lambda} \frac{dx^\gamma}{d\lambda} = 0 \tag{1}$$

where we can take λ to be the proper time of the particle. The 4-velocity of the particle is $dx^\alpha/d\lambda = (\gamma, \gamma \underset{\sim}{v})$, where $\gamma = (1-v^2)^{-\frac{1}{2}}$. First change $d/d\lambda$ to d/dt in Equation (1) using $d/d\lambda = \gamma\, d/dt$:

$$\frac{d^2x^\alpha}{dt^2} + \frac{1}{\gamma}\frac{d\gamma}{dt}\frac{dx^\alpha}{dt} + \Gamma^\alpha{}_{\beta\gamma} \frac{dx^\beta}{dt} \frac{dx^\gamma}{dt} = 0 \ . \tag{2}$$

Putting $\alpha = 0$, we get

$$\frac{1}{\gamma}\frac{d\gamma}{dt} + \Gamma^0{}_{\beta\gamma} \frac{dx^\beta}{dt} \frac{dx^\gamma}{dt} = 0 \ .$$

Substitute this in Equation (2) and get, for $\alpha = j$,

$$\frac{dv^j}{dt} + (-v^j\Gamma^0{}_{\beta\gamma} + \Gamma^j{}_{\beta\gamma})\frac{dx^\beta}{dt}\frac{dx^\gamma}{dt} = 0 . \tag{3}$$

To first order in v, Equation (3) is

$$\frac{dv^j}{dt} - v^j\Gamma^0{}_{00} + \Gamma^j{}_{00} + 2\Gamma^j{}_{k0}v^k = 0 . \tag{4}$$

Thus, to first order in $\underset{\sim}{x}$ and $\underset{\sim}{v}$, we have

$$\frac{dv^j}{dt} = v^j\Gamma^0{}_{00}\big|_{\underset{\sim}{x}=0} - \Gamma^j{}_{00}\big|_{\underset{\sim}{x}=0} - x^k\Gamma^j{}_{00,k}\big|_{\underset{\sim}{x}=0} - 2v^k\Gamma^j{}_{k0}\big|_{\underset{\sim}{x}=0} . \tag{5}$$

The Christoffel symbols at $\underset{\sim}{x} = 0$ are found using a result of Problem 7.17:

$$\Gamma^\beta{}_{\alpha 0}e_\beta = \nabla_{\mathbf{u}}e_\alpha = -\Omega^\beta{}_\alpha e_\beta$$

$$\Gamma_{\beta\alpha 0} = -\Omega_{\beta\alpha} \equiv a_\alpha u_\beta - a_\beta u_\alpha + \varepsilon_{\alpha\beta\lambda\sigma}u^\lambda\omega^\sigma .$$

Since $u^\alpha = (1,\underset{\sim}{0}), u_\alpha = (-1,\underset{\sim}{0}), a_\alpha = (0,\underset{\sim}{a}), \omega^\sigma = (0,\underset{\sim}{\omega})$ we get

$$\Gamma_{\beta\alpha 0} = 0 \quad \text{if} \quad \beta = \alpha$$

$$\Gamma_{0j0} = -\Gamma_{j00} = -a_j$$

$$\Gamma_{kj0} = -\Gamma_{jk0} = \varepsilon_{jk0m}\omega^m = \varepsilon_{jkm}\omega^m \tag{6}$$

The Γ's with no 0-index are found from the condition that the coordinate lines $x^\alpha = (\tau, sn^k)$, where τ and n^k are independent of s, are geodesics. Thus,

$$0 = \frac{d^2x^\alpha}{ds^2} + \Gamma^\alpha{}_{\beta\gamma}\frac{dx^\beta}{ds}\frac{dx^\gamma}{ds}$$

$$= 0 + \Gamma^\alpha{}_{jk}n^jn^k .$$

This is satisfied at $\underset{\sim}{x} = 0$ for arbitrary $\underset{\sim}{n}$, so

$$\Gamma_{\alpha jk} = 0 . \tag{7}$$

We find $\Gamma^j{}_{00,k}$ from the expression for the Riemann tensor:

$$R^\alpha{}_{\beta\gamma\delta} = \Gamma^\alpha{}_{\beta\delta,\gamma} - \Gamma^\alpha{}_{\beta\gamma,\delta} + \Gamma^\alpha{}_{\gamma\mu}\Gamma^\mu{}_{\beta\delta} - \Gamma^\alpha{}_{\delta\mu}\Gamma^\mu{}_{\beta\gamma} .$$

This gives

$$\Gamma^j{}_{00,k} = R^j{}_{0k0} + \Gamma^j{}_{0k,0} - \Gamma^j{}_{k\mu}\Gamma^\mu{}_{00} + \Gamma^j{}_{0\mu}\Gamma^\mu{}_{0k} .$$

At $\underset{\sim}{x} = 0$:

$$\Gamma^j{}_{0k,0} = -\varepsilon_{jkm}\omega^m{}_{,0}$$

$$\Gamma^j{}_{k\mu}\Gamma^\mu{}_{00} = 0$$

$$\begin{aligned}\Gamma^j{}_{0\mu}\Gamma^\mu{}_{0k} &= \Gamma^j{}_{00}\Gamma^0{}_{0k} + \Gamma^j{}_{0m}\Gamma^m{}_{0k} \\ &= a_j a_k + \varepsilon_{mjn}\omega^n \varepsilon_{kml}\omega^l ,\end{aligned}$$

and thus

$$\begin{aligned}\frac{dv^j}{dt} &= -a_j - x^k(R^j{}_{0k0} - \varepsilon_{jkm}\omega^m{}_{,0} + a_j a_k + \varepsilon_{mjn}\omega^n \varepsilon_{kml}\omega^l) + 2\varepsilon_{jkm}\omega^m v^k \\ &= -a_j(1+\underset{\sim}{a}\cdot\underset{\sim}{x}) - R^j{}_{0k0}x^k + (\underset{\sim}{x}\times\underset{\sim}{\omega}_{,0})^j - [(\underset{\sim}{x}\times\underset{\sim}{\omega})\times\underset{\sim}{\omega}]^j + 2(\underset{\sim}{v}\times\underset{\sim}{\omega})^j .\end{aligned}$$

The first term is the "inertial force" due to the acceleration of the reference frame. The $1+\underset{\sim}{a}\cdot\underset{\sim}{x}$ is a relativistic correction factor (see MTW Exercise 37-4). The second term is the "true" gravitational force. In the weak-field limit, a Newtonian physicist would separate $\underset{\sim}{a}$ into a contribution from the local acceleration of gravity, $\partial\Phi/\partial x_j$, plus an "absolute" acceleration $\underset{\sim}{a}_{abs}$. Thus,

$$\begin{aligned}a_j + R^j{}_{0k0}x^k &= (a_j)_{abs} + \partial\Phi/\partial x^j|_{\underset{\sim}{x}=0} + \partial^2\Phi/\partial x^j\partial x^k|_{\underset{\sim}{x}=0}\, x^k \\ &= (a_j)_{abs} + \partial\Phi/\partial x^j|_{\underset{\sim}{x}=\underset{\sim}{x}_{particle}} .\end{aligned}$$

The terms involving $\underset{\sim}{\omega}$ are the same as in nonrelativistic mechanics; the second and third are the centrifugal and Coriolis forces respectively.

Solution 14.13. Setting the divergence of the stress-energy tensor to zero, we have

$$0 = 4\pi T_{\mu\ ;\nu}^{\ \nu} = \Phi_{,\mu;\nu}\Phi^{,\nu} + \Phi_{,\mu}\Phi^{,\nu}_{\ ;\nu} - \frac{1}{2}\delta_\mu^{\ \nu}(2\Phi_{\alpha;\nu}\Phi^{,\alpha})$$

$$= (\Phi_{,\mu;\nu} - \Phi_{,\nu;\mu})\Phi^{,\nu} + \Phi_{,\mu}\Phi^{,\nu}_{\ ;\nu} \ .$$

But second covariant derivatives of a scalar commute $(\Phi_{[,\mu;\nu]} = \Phi_{[,\mu,\nu]} + \Phi_{,\sigma}\Gamma^\sigma_{\ [\mu\nu]} = 0)$ so only the last term in our equation survives and the equation of motion must be

$$\Box\Phi \equiv \Phi^{,\nu}_{\ ;\nu} = 0 \ .$$

Solution 14.14. In the proposed Equation (2) the term $\frac{1}{6}R\Phi$ is independent of the size of the laboratory in the same way that the ρ term is. In principle then, if we measure the Φ field we can measure the Ricci curvature; the conformally invariant Equation (2) then violates the spirit of the strong equivalence principle.

To find the anomalous R-term forces between two particles, let us go to the local Lorentz frame in which one of the particles is stationary. The Φ field due to this stationary particle of charge μ_1 satisfies

$$\Phi_{,j}^{\ ,j} - \frac{1}{6}R\Phi = \mu_1\,\delta(\mathbf{r}) \ .$$

In writing this equation we have assumed a global (or at least laboratory sized) inertial coordinate system. That is, we have ignored aspects of curvature other than the Ricci scalar. Presumably (and in fact) these other aspects of curvature do not violate strong equivalence.

The equation for Φ has the solution

$$\Phi = -\frac{\mu_1}{r}\exp\{-r/a\sqrt{6}\}, \qquad a = R^{-\frac{1}{2}}$$

so that the force on a particle of charge μ_2 is

$$F^r = \mu_2 \Phi_{,r} = \mu_1 \mu_2 / r^2 \, [1 + r/\sqrt{6}a] \exp\{-r/a\sqrt{6}\}$$

$$\approx \mu_1 \mu_2 / r^2 \, [1 + r^2/12a^2 + \mathcal{O}(r^3/a^3)] \, .$$

The anomalous R-term force $\approx \mu_1\mu_2/12a^2$ is then independent of particle separation, clearly violating the strong equivalence principle.

From the Einstein equations it is clear that the magnitude of the Ricci scalar is of the order of the density of mass-energy, so that the ratio of anomalous forces to "ordinary" scalar forces is

$$\frac{r^2}{12a^2} \sim r^2 \rho_{\text{mass-energy}}$$

$$\sim [r\,(\text{cm.})]^2 \, [\rho/\rho_{\text{nuclear}}] \cdot 10^{-14} \, .$$

Thus, even if the scalar forces are being measured inside matter at nuclear densities, $\rho_{\text{nuclear}} = 10^{14}\text{g/cm}^3$, a *very* large ($r \sim 100$ km) matter filled "laboratory" is necessary before the R-term forces become important. In any practical experiment the R-term forces would be miniscule.

Solution 14.15. According to Problem 7.7 (i) and Maxwell's equations

$$4\pi J^\mu = F^{\mu\nu}{}_{;\nu} = \frac{1}{|g|^{\frac{1}{2}}} \left(|g|^{\frac{1}{2}} F^{\mu\nu}\right)_{,\nu}$$

$$4\pi J^\mu{}_{;\mu} = 4\pi \frac{1}{|g|^{\frac{1}{2}}} \left(J^\mu |g|^{\frac{1}{2}}\right)_{,\mu} = \frac{1}{|g|^{\frac{1}{2}}} \left(|g|^{\frac{1}{2}} F^{\mu\nu}\right)_{,\nu\mu} \, .$$

Since F is antisymmetric this must vanish.

Solution 14.16. In flat spacetime we have

$$F^{\mu\nu}{}_{,\nu} = 4\pi J^\mu$$

$$F_{\mu\nu} = A_{\nu,\mu} - A_{\mu,\nu}$$

and hence

$$-A^{\mu,\nu}{}_{,\nu} + A^{\nu,\mu}{}_{,\nu} = 4\pi J^{\mu} \ . \qquad (1)$$

If we interchange μ and ν differentiation on the second term, as is allowed in flat spacetime, and then apply the "comma to semicolon" rule we get

$$-A^{\mu;\nu}{}_{;\nu} + A^{\nu}{}_{;\nu}{}^{;\mu} = 4\pi J^{\mu} \ . \qquad (2)$$

If, on the other hand, we apply the rule directly to Equation (1) we get

$$-A^{\mu;\nu}{}_{;\nu} + A^{\nu;\mu}{}_{;\nu} = 4\pi J^{\mu} \ .$$

The derivates on the second term can be exchanged by introducing a curvature term (see Problem 9.8)

$$-A^{\mu;\nu}{}_{;\nu} + A^{\nu}{}_{;\nu}{}^{;\mu} + R^{\mu}{}_{\sigma}A^{\sigma} = 4\pi J^{\mu} \ . \qquad (3)$$

Equations (2) and (3) have, in principle, measurably different consequences. As of 1975, there is *no* experimental evidence to tell which is correct.

Solution 14.17. If the curvature coupling term is of the sort found in Problem 14.16, we have (in a Lorentz gauge i.e. $A^{\mu}{}_{;\mu} = 0$) $\Box A^{\mu} + (R^{\mu}{}_{\nu}A^{\nu}) = 4\pi J^{\mu}$, where the term in parentheses may or may not be present. Since the Ricci tensor vanishes in vacuum, the only hope of detecting its term would be by experiments in matter (e.g. in glass). The first term is of order $A^{\mu}[\text{minimum}\,(\frac{c}{\nu},\ell)]^{-2}$. The magnitude of the Ricci tensor is (by the field equations) of order $G\rho/c^2 \approx .74 \times 10^{-28}\,\text{cm}^{-2} \times \left(\frac{\rho}{\text{gm.}\cdot\text{cm}^{-3}}\right)$. The fractional correction of the second term is thus

$$\frac{\delta A^{\mu}}{A^{\mu}} \sim \left(\frac{\text{minimum}\,(\frac{c}{\nu},\ell)(\rho/\text{gm.}\cdot\text{cm}^{-3})^{\frac{1}{2}}}{1.16 \times 10^{14}\ \text{cm}}\right)^2 .$$

Since an earthbound processes is presumably no larger than the size of the earth $\sim 6 \times 10^{8}$ cm, this is bounded by $(\delta A^{\mu}/A^{\mu}) \lesssim (5 \times 10^{-6})^2$ for

$\rho = 1$. One can also dream up fancier curvature coupling terms like $KR^{\alpha\beta\gamma\delta}R_{\alpha\beta\gamma\delta}A^{\mu}$ where K is a *dimensional* constant. Because there is no natural value to assume for K, there is no *a priori* way to estimate the magnitude of the effect resulting from such terms. Since the equations are similar to the Proca equation $\Box A^{\mu} + mA^{\mu} = 0$, experiments to measure the photon rest mass do put some limits on K, but so far they are not useful ones.

Solution 14.18. For the electromagnetic stress-energy

$$4\pi T^{\mu\nu} = -\left(F^{\mu}{}_{\alpha}F^{\alpha\nu} + \frac{1}{4} g^{\mu\nu}F_{\alpha\beta}F^{\alpha\beta}\right) \tag{1}$$

$$4\pi T^{\mu\nu}{}_{;\nu} = -\left(F^{\mu}{}_{\alpha;\nu}F^{\alpha\nu} + F^{\mu}{}_{\alpha}F^{\alpha\nu}{}_{;\nu} + \frac{1}{2} F_{\alpha\beta}F^{\alpha\beta;\mu}\right) = 0\ . \tag{2}$$

Now, the first and last terms above equal

$$\begin{aligned} -F^{\alpha\beta}g^{\mu\tau}\left(\frac{1}{2} F_{\alpha\beta;\tau} + F_{\tau\alpha;\beta}\right) &= -\frac{1}{2} F^{\alpha\beta}g^{\mu\tau}(F_{\alpha\beta;\tau} + F_{\tau\alpha;\beta} - F_{\tau\beta;\alpha}) \\ &= -\frac{1}{2} F^{\alpha\beta}g^{\mu\tau}(F_{\alpha\beta;\tau} + F_{\tau\alpha;\beta} + F_{\beta\tau;\alpha}) \\ &\propto F^{\alpha\beta}g^{\mu\tau}F_{[\alpha\beta;\tau]} = 0\ . \end{aligned}$$

If F is derivable from a potential A in the usual way, then $F_{[\alpha\beta;\tau]} = 0$ and we are left with

$$0 = 4\pi T^{\mu\nu}{}_{;\nu} = -F^{\mu}{}_{\alpha}F^{\alpha\nu}{}_{;\nu}\ .$$

This implies that $F^{\alpha\nu}{}_{;\nu} = 0$ except in the case that the determinant of the coefficients $F^{\mu}{}_{\alpha}$ vanishes. But $\det(F^{\mu}{}_{\alpha}) = -(\underset{\sim}{E}\cdot\underset{\sim}{B})^2$.

Solution 14.19. First note that H has been normalized to have dimensions of $(\text{mass})^2$; this means that Hamilton's equations are

$$dx^{\mu}/d\lambda = \partial H/\partial\pi_{\mu} \tag{1}$$

$$\partial\pi_{\mu}/d\lambda = -\partial H/\partial x^{\mu} \tag{2}$$

where λ is an affine parameter ($=\tau/m$ for a particle of mass m where τ is proper time). If H were divided by m, one would replace $d/d\lambda$ by $d/d\tau$ in Equations (1) and (2); $d/d\lambda$ has the advantage of being valid for massless particles as well.

Since the particle's 4-momentum is $p^\mu = dx^\mu/d\lambda$, and since $g^{\mu\nu}$ and A_μ are functions of x^α not π_α, Equation (1) gives

$$p^\mu = g^{\mu\nu}(\pi_\nu - eA_\nu) . \tag{3}$$

Equation (2) gives

$$\frac{d\pi_\alpha}{d\lambda} = -\frac{1}{2} g^{\mu\nu}{}_{,\alpha}(\pi_\mu - eA_\mu)(\pi_\nu - eA_\nu) + g^{\mu\nu} eA_{\mu,\alpha}(\pi_\nu - eA_\nu) .$$

But from Equation (3) we get

$$\begin{aligned} \frac{d\pi_\alpha}{d\lambda} &= \frac{d}{d\lambda}(g_{\alpha\mu}p^\mu + eA_\alpha) \\ &= g_{\alpha\mu}\frac{dp^\mu}{d\lambda} + g_{\alpha\mu,\beta}p^\beta p^\mu + eA_{\alpha,\beta}p^\beta \end{aligned}$$

(where we have used $d/d\lambda = p^\beta \partial/\partial x^\beta$). Multiplying by $g^{\alpha\gamma}$, we find

$$\frac{dp^\gamma}{d\lambda} + g^{\alpha\gamma}g_{\alpha\mu,\beta}p^\beta p^\mu + \frac{1}{2} g^{\alpha\gamma}g^{\mu\nu}{}_{,\alpha}p_\mu p_\nu = g^{\alpha\gamma}e(A_{\mu,\alpha} - A_{\alpha,\mu})p^\mu .$$

Now rearrange the derivatives of $g_{\alpha\beta}$ to form a Christoffel symbol. Since

$$0 = (g^{\mu\beta}g_{\beta\gamma})_{,\alpha} = g^{\mu\beta}{}_{,\alpha}g_{\beta\gamma} + g^{\mu\beta}g_{\beta\gamma,\alpha}$$

we have on multiplying by $g^{\gamma\nu}$

$$g^{\mu\nu}{}_{,\alpha} = -g^{\gamma\nu}g^{\mu\beta}g_{\beta\gamma,\alpha} .$$

Thus

$$\frac{dp^\gamma}{d\lambda} + g^{\alpha\gamma}\left(g_{\alpha\mu,\beta} - \frac{1}{2} g_{\beta\mu,\alpha}\right) p^\beta p^\mu = g^{\alpha\gamma}eF_{\alpha\mu}p^\mu .$$

Since

$$g_{\alpha\mu,\beta}p^{\beta}p^{\mu} = \frac{1}{2}(g_{\alpha\mu,\beta} + g_{\alpha\beta,\mu})p^{\beta}p^{\mu} ,$$

we get, finally

$$\frac{dp^{\gamma}}{d\lambda} + \Gamma^{\gamma}{}_{\beta\mu}p^{\beta}p^{\mu} = eF^{\gamma}{}_{\mu}p^{\mu}$$

which is the equation of motion.

Solution 14.20. For an uncharged test particle, $\mathbf{p}\cdot\boldsymbol{\xi}$ is conserved. We can guess that for a charged particle, $\boldsymbol{\pi}\cdot\boldsymbol{\xi}$ is conserved, where $\boldsymbol{\pi}$ is the canonical momentum $\mathbf{p} + e\mathbf{A}$ (see Problem 14.19). One way of seeing this is to note that corresponding to $\boldsymbol{\xi}$ there is an ignorable coordinate, i.e. we can choose coordinates so that the Hamiltonian H is independent of one coordinate. The conjugate canonical momentum is then a conserved quantity.

Alternatively, we can verify directly from the equation of motion

$$\nabla_{\mathbf{p}}\mathbf{p} = e\mathbf{F}\cdot\mathbf{p}$$

that

$$\begin{aligned}\nabla_{\mathbf{p}}(\boldsymbol{\pi}\cdot\boldsymbol{\xi}) &= (\nabla_{\mathbf{p}}\mathbf{p})\cdot\boldsymbol{\xi} + \mathbf{p}\cdot\nabla_{\mathbf{p}}\boldsymbol{\xi} + e(\nabla_{\mathbf{p}}\mathbf{A})\cdot\boldsymbol{\xi} + e\mathbf{A}\cdot\nabla_{\mathbf{p}}\boldsymbol{\xi}\\ &= e\boldsymbol{\xi}\cdot\mathbf{F}\cdot\mathbf{p} + 0 + e(\nabla_{\mathbf{p}}\mathbf{A})\cdot\boldsymbol{\xi} - e\mathbf{p}\cdot\nabla_{\mathbf{A}}\boldsymbol{\xi} .\end{aligned}$$

Here we have used Killing's equation to set the second term to zero and to rewrite the last term. Note that the order of vectors in the dot products in the first line is immaterial, but F has two "slots" and we must keep the vectors in the correct positions. Now the fact that the electromagnetic field admits an ignorable coordinate is expressed by the relation

$$0 = \mathcal{L}_{\boldsymbol{\xi}}\mathbf{A} = \nabla_{\boldsymbol{\xi}}\mathbf{A} - \nabla_{\mathbf{A}}\boldsymbol{\xi} .$$

Using this in the last term, we get

$$\begin{aligned}\nabla_{\mathbf{p}}(\boldsymbol{\pi}\cdot\boldsymbol{\xi}) &= e\boldsymbol{\xi}\cdot\mathbf{F}\cdot\mathbf{p} + e\boldsymbol{\xi}\cdot(\nabla\mathbf{A})\cdot\mathbf{p} - e\mathbf{p}\cdot(\nabla\mathbf{A})\cdot\boldsymbol{\xi} \\ &= e\boldsymbol{\xi}\cdot\mathbf{F}\cdot\mathbf{p} - e\boldsymbol{\xi}\cdot\mathbf{F}\cdot\mathbf{p} \\ &= 0\end{aligned}$$

where we have used the fact that $\mathbf{F}$ is $\nabla\mathbf{A}$ antisymmetrized. Thus $\boldsymbol{\pi}\cdot\boldsymbol{\xi}$ is a conserved quantity along the particle's trajectory.

Solution 14.21. Under the conformal transformation the four Maxwell equations

$$F_{[\alpha\beta,\nu]} = 0 = \tilde{F}_{[\alpha\beta,\nu]}$$

are clearly invariant. The other four Maxwell equations are

$$F^{\mu\nu}{}_{;\nu} = |g|^{-\frac{1}{2}}[F^{\mu\nu}|g|^{\frac{1}{2}}]_{,\nu} = 4\pi J^{\mu} \; .$$

Since

$$F^{\mu\nu} = g^{\mu\alpha}g^{\nu\beta}F_{\alpha\beta} = f^2\tilde{g}^{\mu\alpha}\tilde{g}^{\nu\beta}\tilde{F}_{\alpha\beta} = f^2\tilde{F}^{\mu\nu}$$

$$\tilde{g} \equiv \mathrm{Det}(\tilde{g}_{\alpha\beta}) = f^4 g$$

it follows that $|g|^{\frac{1}{2}}F^{\mu\nu} = |\tilde{g}|^{\frac{1}{2}}\tilde{F}^{\mu\nu}$ so that

$$\begin{aligned}F^{\mu\nu}{}_{;\nu} &= \frac{f^2}{|\tilde{g}|^{\frac{1}{2}}}[|\tilde{g}|^{\frac{1}{2}}\tilde{F}^{\mu\nu}]_{,\nu} = 4\pi J^{\mu} = 4\pi g^{\mu\alpha}J_{\alpha} \\ &= 4\pi(f\tilde{g}^{\mu\alpha})f\tilde{J}_{\alpha} = 4\pi f^2\tilde{J}^{\mu} \; ,\end{aligned}$$

and therefore

$$\tilde{F}^{\mu\nu}{}_{;\nu} = 4\pi\tilde{J}^{\mu} \; .$$

CHAPTER 15: SOLUTIONS

Solution 15.1. When the particle moves in the equatorial plane, we have $L^2 = p_\phi^2$, which is obviously a conserved constant since p_ϕ itself is. ($\boldsymbol{\xi} = \partial/\partial\phi$ is a Killing vector so $\boldsymbol{\xi}\cdot p$ is conserved; see Problem 10.10.) But by spherical symmetry the motion is always in the equatorial plane of *some* rotated coordinate system. If we could write this p_ϕ^2 as an invariant quantity (under rotations) and then evaluated the invariant in the original system, we would be done.

At some instant when the particle is at radius r the covariant 4-velocity $\tilde{p}$ has components $(p_t, p_r, p_\theta, p_\phi)$. Consider the "reduced" 4-velocity $\tilde{p}_\alpha^{red} = (0, 0, rp_\theta, rp_\phi)$ whose construction from $\tilde{p}_\alpha$ (e.g. by projection) is independent of θ and ϕ. Now when the motion is equatorial $p_\theta = 0$ and $\theta = \pi/2$ we have

$$L^2 = g^{\alpha\beta}\tilde{p}_\alpha^{red}\,\tilde{p}_\beta^{red} = g^{\phi\phi}\,r^2 p_\phi^2 = p_\phi^2 = \text{conserved}\;.$$

But in general

$$L^2 = g^{\theta\theta}\,r^2 p_\theta^2 + g^{\phi\phi}\,r^2 p_\phi^2 = p_\theta^2 + \frac{p_\phi^2}{\sin^2\theta}\;.$$

So this must be conserved in general, q.e.d.

Solution 15.2.

(a) Using the spherical symmetry of the metric we orient the coordinate axes so that the particle is at $\theta = \pi/2$ with $\dot{\theta} = 0$ at $\tau = 0$; then the unique solution of the geodesic equation

$$\frac{d}{d\tau}(r^2\dot{\theta}) = r^2 \sin\theta\,\cos\theta\,\dot{\phi}^2$$

is $\theta = \pi/2$ for all τ.

(b) Using the constant of motion L^2 of Problem 15.1, we have

$$\left(\frac{d\theta}{d\lambda}\right)^2 = (g^{\theta\theta}p_\theta)^2 = \frac{1}{r^4}\left(L^2 - \frac{p_\phi^2}{\sin^2\theta}\right) . \tag{1}$$

Suppose that the unperturbed orbit is $\theta = \pi/2$, $L = p_\phi = K$. Let the particle be perturbed out of the plane of the orbit i.e. $\theta = \pi/2 + \delta\theta$, $L = K + \delta L$, $p_\phi = K + \delta p_\phi$. To first order in δL and δp_ϕ and second order in $\delta\theta$, Equation (1) gives

$$\left[\frac{d(\delta\theta)}{d\lambda}\right]^2 = \frac{1}{r^4}\left[2K\delta L - 2K\delta p_\phi - K^2(\delta\theta)^2\right]$$

and hence

$$\frac{d^2(\delta\theta)}{d\lambda^2} = -\frac{K^2}{r^4}\delta\theta .$$

This shows that $\delta\theta$ oscillates about $\theta = \pi/2$ and does not grow in time. Hence the orbits are stably planar.

Solution 15.3. We have two constants for radial fall: $\mathbf{u}\cdot\mathbf{u} = -1$ and $\mathbf{u}\cdot\partial/\partial t = u_0 = -(1-2M/r)u^0$. From these we get

$$\begin{aligned}
\mathbf{u}\cdot\mathbf{u} = -1 &= -(1-2M/r)(u^0)^2 + (1-2M/r)^{-1}(u^r)^2 \\
&= [-(1-2M/r) + (1-2M/r)^{-1}(dr/dt)^2](u^0)^2 \\
&= [-(1-2M/r) + (1-2M/r)^{-1}(dr/dt)^2](u_0)^2(1-2M/r)^2 .
\end{aligned}$$

Solving for $(dr/dt)^2$ we find

$$(dr/dt)^2 = (1-2M/r)^2[1-(1-2M/r)(u_0)^{-2}] .$$

A stationary observer measures time intervals $d\hat{t} = (1-2M/r)^{\frac{1}{2}}dt$ and radial distances $d\hat{r} = (1-2M/r)^{-\frac{1}{2}}dr$ so he measures velocity

$$\frac{d\hat{r}}{d\hat{t}} = (1-2M/r)^{-1}\frac{dr}{dt} .$$

Notice that independent of the value of u_0 this locally measured velocity approaches the velocity of light as r approaches 2M.

Solution 15.4. From the first integrals of the geodesic equation (see Problem 15.3)

$$u_0 = -\tilde{E} = \text{constant}$$

$$g^{00}u_0^2 + g_{rr}(u^r)^2 = -1$$

we get

$$u^0 = \frac{dt}{d\tau} = \frac{\tilde{E}}{1-2M/r} \tag{1}$$

$$u^r = \frac{dr}{d\tau} = -(\tilde{E}^2 - 1 + 2M/r)^{\frac{1}{2}} . \tag{2}$$

The minus sign for the square root in Equation (2) corresponds to an in-falling particle.

Case (i). For this case $dr/d\tau = 0$ at $r=R$ implies $2M/R = 1-\tilde{E}^2$, i.e. $\tilde{E} < 1$. Thus Equation (2) can be rewritten

$$d\tau = \frac{dr}{\left(\frac{2M}{r} - \frac{2M}{R}\right)^{\frac{1}{2}}} . \tag{3}$$

This can be integrated in closed form to give

$$\tau = \left(\frac{R^3}{8M}\right)^{\frac{1}{2}} \left[2\left(\frac{r}{R} - \frac{r^2}{R^2}\right)^{\frac{1}{2}} + \cos^{-1}\left(\frac{2r}{R} - 1\right)\right] \tag{4}$$

where the constant of integration has been chosen so that $\tau = 0$ at $r = R$. It is convenient to rewrite Equation (4) in parametric form by introducing the "cycloid parameter"

$$\eta = \cos^{-1}\left(\frac{2r}{R} - 1\right)$$

so that we have

$$r = \frac{1}{2} R(1+\cos\eta) \qquad (\eta = 0 \text{ at } r = R) \tag{5}$$

$$\tau = \left(\frac{R^3}{8M}\right)^{\frac{1}{2}} (\eta + \sin\eta) . \tag{6}$$

Equation (1) gives

$$t = \int \frac{\tilde{E}\,d\tau}{\left(1-\frac{2M}{r}\right)} = \tilde{E}\int \frac{dr}{d\eta}\frac{1}{1-\frac{2M}{r}}\,d\eta$$

$$= \left(1-\frac{2M}{R}\right)^{\frac{1}{2}} \int \frac{\left(\frac{R^3}{8M}\right)^{\frac{1}{2}}(1+\cos\eta)\,d\eta}{1-4M\,[R(1+\cos\eta)]^{-1}} \ .$$

From a table of integrals, we find

$$\frac{t}{2M} = \log\left|\frac{\left(\frac{R}{2M}-1\right)^{\frac{1}{2}} + \tan\left(\frac{\eta}{2}\right)}{\left(\frac{R}{2M}-1\right)^{\frac{1}{2}} - \tan\left(\frac{\eta}{2}\right)}\right| + \left(\frac{R}{2M}-1\right)^{\frac{1}{2}}\left[\eta + \frac{R}{4M}(\eta+\sin\eta)\right] \qquad (7)$$

where the constant of integration has been chosen so that $t = 0$ at $\eta = 0$ i.e. at $r = R$. Note that $t \to \infty$ when $\tan(\eta/2) \to (R/2M - 1)^{\frac{1}{2}}$ i.e. when $r \to 2M$.

Case (ii). Here $\tilde{E} = 1$ and Equation (2) gives

$$\tau = -\frac{2}{3}\left(\frac{r^3}{2M}\right)^{\frac{1}{2}} + \text{constant} \qquad (8)$$

and since

$$\frac{dt}{dr} = \frac{dt/d\tau}{dr/d\tau}$$

we get

$$t = -\frac{2}{3}\left(\frac{r^3}{2M}\right)^{\frac{1}{2}} - 4M\left(\frac{r}{2M}\right)^{\frac{1}{2}} + 2M\log\left|\frac{\left(\frac{r}{2M}\right)^{\frac{1}{2}}+1}{\left(\frac{r}{2M}\right)^{\frac{1}{2}}-1}\right| + \text{constant} \ . \qquad (9)$$

Case (iii). By analogy with case (i), choose R such that

$$\frac{2M}{R} = \tilde{E}^2 - 1 = (1-v_\infty^2)^{-1} - 1 = \frac{v_\infty^2}{1-v_\infty^2} \ .$$

Then changing the sign of R in Equation (4) gives

$$\tau = -\left(\frac{R^3}{8M}\right)^{\frac{1}{2}}\left[2\left(\frac{r}{R}+\frac{r^2}{R^2}\right)^{\frac{1}{2}} - \cosh^{-1}\left(\frac{2r}{R}+1\right)\right] \ . \qquad (10)$$

Note that $\tau = 0$ at $r = 0$, and $\tau = -\infty$ at $r = \infty$. Put

$$\eta = \cosh^{-1}\left(\frac{2r}{R} + 1\right)$$

and get

$$r = \frac{R}{2}(\cosh\eta - 1) \qquad (r = 0 \text{ at } \eta = 0) \tag{11}$$

$$\tau = -\left(\frac{R^3}{8M}\right)^{\frac{1}{2}}(\sinh\eta - \eta) \ . \tag{12}$$

The equation corresponding to Equation (7) is

$$\frac{t}{2M} = \log\left|\frac{\left(\frac{R}{2M}+1\right)^{\frac{1}{2}} + \coth\frac{\eta}{2}}{\left(\frac{R}{2M}+1\right)^{\frac{1}{2}} - \coth\frac{\eta}{2}}\right| - \left(\frac{R}{2M}+1\right)^{\frac{1}{2}}\left[\eta + \frac{R}{4M}(\sinh\eta - \eta)\right]. \tag{13}$$

Note that t goes from $-\infty$ to $+\infty$ as r goes from ∞ to $2M$.

Solution 15.5. Choose the equatorial plane to be $\theta = \pi/2$. Then $u^\theta = 0$. One first integral of the motion is the normalization of the 4-velocity which implies

$$g^{00}u_0^2 + g^{\phi\phi}u_\phi^2 + g_{rr}{u^r}^2 = -1 \ . \tag{1}$$

From the Killing vectors $\partial/\partial t$ and $\partial/\partial\phi$ two more first integrals are

$$u_\phi = \tilde{L} = \text{constant}$$

$$u_0 = -\tilde{E} = \text{constant}$$

where $\tilde{L}$ and $\tilde{E}$ are respectively the angular momentum per unit rest mass and energy per unit rest mass. Equation (1) now implies

$$\left(\frac{dr}{d\tau}\right)^2 = \frac{-1 - g^{00}\tilde{E}^2 - g^{\phi\phi}\tilde{L}^2}{g_{rr}} \tag{2}$$

or with the explicit metric functions:

$$dr/d\tau = \pm\left\{\left[-1+\tilde{E}^2\left(1-\frac{2m}{r}\right)^{-1}-\frac{\tilde{L}^2}{r^2}\right]\left(1-\frac{2m}{r}\right)\right\}^{\frac{1}{2}} . \qquad (3)$$

From $u^\phi = g^{\phi\phi}\tilde{L}$ we have $d\phi/d\tau = \tilde{L}/r^2$. Combining this with Equation (3) gives us finally

$$\frac{dr}{d\phi} = \pm\frac{r^2}{\tilde{L}}\left\{\left[-1+\tilde{E}^2(1-2M/r)^{-1}-\frac{\tilde{L}^2}{r^2}\right](1-2M/r)\right\}^{\frac{1}{2}} .$$

Solution 15.6. Choose $\theta = \pi/2$ and $p^\theta = 0$. If λ is an affine parameter such that $p^r = dr/d\lambda$ and $p^\phi = d\phi/d\lambda = r^{-2}p_\phi$, then

$$\frac{dr}{d\phi} = \frac{p^r}{p^\phi} .$$

From $p\cdot p = 0$ we get a relation amongst p^r, p^ϕ, and p^0 which can be solved for $dr/d\phi$:

$$(dr/d\phi)^2 = r^4(1-2M/r)[\gamma(1-2M/r)^{-1} - 1/r^2] \qquad (1)$$

where $\gamma \equiv p_0^2/p_\phi^2$ = constant. If we now introduce $u \equiv M/r$, Equation (1) becomes

$$(u')^2 = (1-2u)[\gamma M^2(1-2u)^{-1} - u^2] = \gamma M^2 - u^2 + 2u^3 \qquad (2)$$

where $'$ denotes $d/d\phi$. If we differentiate this equation we get the very simple second-order equation

$$u'' + u = 3u^2 . \qquad (3)$$

In the limit $M/b \ll 1$ the zeroth order solution can be taken to be the "straight line" solution $r\sin\phi = b$ or

$$u_0 = (M/b)\sin\phi .$$

Now write $u = u_0 + u_1 + \cdots$ with $u_1 \ll 1$ so that Equation (3) is approximately

$$u_1'' + u_1 \approx 3u_0^2 = 3\left(\frac{M}{b}\right)^2\sin^2\phi = \frac{3}{2}\left(\frac{M}{b}\right)^2(1-\cos 2\phi) .$$

This can be solved by inspection, and we get

$$u \approx \left(\frac{M}{b}\right)\sin\phi + \frac{1}{2}\left(\frac{M}{b}\right)^2 (3+\cos 2\phi) \ .$$

Now we can find the total deflection angle by calculating the two angles at which $r = \infty\,(u=0)$. These angles must satisfy:

$$2\sin\phi \approx -(M/b)(3+\cos 2\phi)$$

$$\phi \approx -2(M/b) \qquad \text{and} \qquad \phi \approx \pi + 2(M/b) \ .$$

Thus the total deflection angle is $4M/b$.

Solution 15.7.

(a) The orbital equation was derived in Problem 15.5. If we introduce $u \equiv M/r$ and two new parameters ϵ and u_0 in place of $\tilde{E}$, $\tilde{L}$, we can rewrite the orbital equation as

$$(du/d\phi)^2 + (u-u_0)^2 - \epsilon^2 u_0^2 = 6u_0(u-u_0)^2 + 2(u-u_0)^3 \ .$$

(See also MTW, Equation 25.47 or Weinberg, Equation 8.4.29.) The terms on the right are the relativistic terms; in Newtonian gravitation theory, the right side would be zero.

The lowest order solution, the Newtonian solution, is

$$u = u_0(1+\epsilon\cos\phi)$$

where ϵ is the eccentricity. Now note that the first correction term is

$$6u_0(u-u_0)^2 = \mathcal{O}(u_0^3\epsilon^2) = \mathcal{O}\left(\frac{m^3}{r_0^2}\epsilon^2\right)$$

and the second is $(u-u_0)^3 = \mathcal{O}\left(\frac{m^3}{r_0^3}\epsilon^3\right)$ so we can ignore it! Thus, first order corrections can be found from

$$(du/d\phi)^2 + (1-6u_0)(u-u_0)^2 = u_0^2\epsilon^2 \ .$$

Now define $\psi \equiv (1-6u_0)^{\frac{1}{2}}\phi$, $\mu = u - u_0$ so that

$$(du/d\psi)^2 + \mu^2 = u_0^2\epsilon^2/(1-6u_0) \ .$$

Note that the solution is periodic in ψ so that r is periodic in $\psi = (1-6u_0)^{\frac{1}{2}}\phi$. One orbit corresponds therefore to $\psi = 2\pi$ or $\phi = 2\pi(1-6M/r_0)^{-\frac{1}{2}}$ so that the change in the periastron is given by

$$\left.\frac{\delta\phi}{2\pi}\right|_{\text{per orbit}} = \frac{3M}{r_0} \ .$$

[For another way of doing this, better suited to large eccentricity, see A. S. Eddington, *The Mathematical Theory of Relativity* (Cambridge University Press (1922), Section 40).]

(b) From the Newtonian orbital equations, derive

$$\left(\frac{du}{d\phi}\right)^2 + u^2 + \frac{2\Phi(u)}{L^2} = \frac{2E_\infty}{L^2} \ .$$

With $\Phi = -M/r - AM/r^3$, this becomes

$$\left(\frac{du}{d\phi}\right)^2 + u^2 - \frac{2mu}{L^2} - \frac{2amu^3}{L^2} = \frac{2E_\infty}{L^2} \ .$$

Put this in the form

$$\left(\frac{du}{d\phi}\right)^2 + (1-c)(u-u_0)^2 - \frac{2am}{L^2}(u-u_0)^3 = \text{constant.}$$

By comparing terms of order u^2, conclude that $c = 6au_0m/L^2$. For nearly circular orbits $L^2 \approx mr$ so that $c = 6a/r_0^2$. From our experience with part (a) we know that the solution is periodic in the variable $(1+c)^{\frac{1}{2}}\phi$ so that the angular spread from periastron to periastron is

$$\Delta\phi = \frac{2\pi}{(1-c)^{\frac{1}{2}}} = \frac{2\pi}{\left(\dfrac{1-6a}{r_0^2}\right)^{\frac{1}{2}}}$$

and therefore

$$\left.\frac{\delta\phi}{2\pi}\right|_{\text{per orbit}} = \frac{3a}{r_0^2} .$$

(The periastron advances if $a > 0$ (oblate) and regresses if $a < 0$ (prolate).)

(c) In $c = G = 1$ units $M_\odot = 1.5$ km. For Mercury $r_0 = .58 \times 10^8$ km. So the relativistic advance is

$$\begin{aligned}\delta\phi &= .105 \text{ seconds of arc per orbit}\\ &= 42 \text{ seconds of arc per century}\end{aligned}$$

since the orbital period is .241 years. If this advance were due to the oblateness of the sun, then for the other planets we would have,

$$\begin{aligned}\delta\phi &= (42\ \text{sec/century})(r_{\text{Merc.}}/r)^2\ (\text{Period of Merc./Period})\\ &= (42\ \text{sec/century})(r_{\text{Merc.}}/r)^{\frac{7}{2}} .\end{aligned}$$

Numerical answers:

	$\frac{r_{\text{Merc}}}{r}$	general relativity sec/century	oblateness sec/century
Mercury	1	42	42
Venus	.536	8.8	4.7
Earth	.386	3.9	1.51
Mars	.245	1.25	.3

Solution 15.8. We solve the problem in two stages: First, compute the laser frequency as measured by a static observer at r ("SOR"). Second, compute the redshift between this observer and a static observer at infinity (SOI).

Let u_R be the 4-velocity of the rocket and p_R the photon momentum at r.

$$\frac{\nu_{\text{SOR}}}{\nu_0} = \frac{u_{\text{SOR}} \cdot p_R}{u_R \cdot p_R} . \qquad (1)$$

If v is the proper relative velocity between the rocket (R) and SOR observer, then in the rocket frame Equation (1) gives us

$$\frac{\nu_{SOR}}{\nu_0} = \frac{\gamma\nu_0(1+v\cos\alpha)}{\nu_0}$$

or

$$\nu_{SOR} = \gamma\nu_0(1+v\cos\alpha) \tag{2}$$

where $\gamma \equiv (1-v^2)^{-\frac{1}{2}}$. To compute v, recall that circular orbits have an angular velocity

$$\Omega \equiv \frac{d\phi}{dt}$$

exactly the same as Kepler's Third Law in Newtonian gravity:

$$\Omega = \left(\frac{M}{r^3}\right)^{\frac{1}{2}} . \tag{3}$$

(See Problem 17.4.) Thus

$$v = \frac{d\hat{\phi}}{d\hat{t}} = \frac{r\,d\phi}{(1-2M/r)^{\frac{1}{2}}dt} = \left[\frac{M}{r(1-2M/r)}\right]^{\frac{1}{2}} . \tag{4}$$

The observer at infinity then measures a frequency

$$\nu_{SOI} = \frac{\nu_{SOR}}{u^0_{SOR}} = \nu_{SOR}(1-2M/r)^{\frac{1}{2}}$$

$$\nu_{SOI} = \gamma\nu_0(1+v\cos\alpha)(1-2M/r)^{\frac{1}{2}}$$

where v is given by Equation (4).

Solution 15.9.

(a) From the first integral of the motion, one has the equation

$$(dr/dt)^2 + V^2(r) = \tilde{E}^2 \tag{1}$$

where V is as defined in Problem 15.11. If one now uses

$$d\phi/d\tau = p^{\phi} = \tilde{L}/r^2 \tag{2}$$

and defines the variable $u \equiv M/r$, then an equation relating u and ϕ may be obtained:

$$\left(\frac{du}{d\phi}\right)^2 = \frac{\tilde{E}^2 - (1-2u)(1+\bar{L}^2u^2)}{\bar{L}^2} \tag{3}$$

where $\bar{L} \equiv \tilde{L}/M$. Differentiate Equation (3) with respect to ϕ:

$$2u'u'' = -(2u\bar{L}^2 - 2 - 6\bar{L}^2u^2)u'\bar{L}^{-2}$$

or

$$u'' + u = \frac{1}{\bar{L}^2} + 3u^2 . \tag{4}$$

For large impact parameter, the right side of Equation (4) is small. Let $u = u_0 + \epsilon v$, with the right hand side of Equation (4) $= \mathcal{O}(\epsilon)$, then

$$u_0 = A\cos\phi , \quad A = \text{constant} , \tag{5}$$

for suitable choice of x, y, axes.

Substitution of u_0 into Equation (4) yields

$$v'' + v = \bar{L}^{-2} + 3A^2\cos^2\phi = \bar{L}^{-2} + \frac{3}{2}A^2(1+\cos 2\phi) . \tag{6}$$

The solution to Equation (6) is

$$v = \bar{L}^{-2} + \frac{3}{2}A^2 - \frac{A^2}{2}\cos 2\phi = \bar{L}^{-2} + 2A^2 - A^2\cos^2\phi$$

and hence

$$u = A\cos\phi - A^2\cos^2\phi + \bar{L}^{-2} + 2A^2 . \tag{7}$$

The asymptotes of the orbit are given by $u = 0$ i.e.

$$A^2\cos^2\phi - A\cos\phi - B = 0$$

$$B \equiv \frac{1}{\bar{L}^2} + 2A^2$$

$$\cos\phi = \frac{A - (A^2 + 4A^2B)^{\frac{1}{2}}}{2A^2} \approx -\frac{B}{A} . \tag{8}$$

Since B/A is a small quantity, the solution to Equation (8) is

$$\phi \approx \pi/2 - B/A \ . \tag{9}$$

Since there is equal deflection at both asymptotes, $(\phi \approx \pi/2, -\pi/2)$ the total deflection angle is

$$\Delta\phi = 2B/A \ . \tag{10}$$

If b is the distance of closest approach, then by the definition of A in Equation (5)

$$A = M/b \ . \tag{11}$$

Also,

$$\bar{L}^2 = \frac{L^2}{M^2 m^2} = \frac{b^2(E^2 - m^2)}{M^2 m^2} = \frac{b^2\beta^2}{M^2(1-\beta^2)}$$

$$B = \frac{M^2(1-\beta^2)}{b^2\beta^2} + \frac{2M^2}{b^2} = \frac{M^2(1+\beta^2)}{b^2\beta^2} \ . \tag{12}$$

Combining Equations (10), (11), (12) one has

$$\Delta\phi_{grav} = \frac{2M(1+\beta^2)}{b\beta^2} \ . \tag{13}$$

(b) The differential equation describing the trajectory is

$$\frac{d}{dt}(m\gamma\dot{r}) = \gamma m r \omega^2 - \frac{Ze^2}{r^2} \tag{14a}$$

$$m\gamma r^2\omega \equiv L = \text{constant} \tag{14b}$$

$$m\gamma - \frac{Ze^2}{r} \equiv E = \text{constant} \ . \tag{14c}$$

From Equation (14b), $d/dt = (L/m\gamma r^2)d/d\phi$ so Equation (14a) becomes

$$\frac{d^2}{d\phi^2}u + \left(1 - \frac{Z^2e^4}{L^2}\right)u = \frac{Ze^2E}{L^2} \tag{15}$$

where $u \equiv 1/r$. The solution to Equation (15) is

$$u = \frac{1}{b} \cos \tau\phi + \frac{Ze^2E}{L^2 - Z^2e^4}, \tag{16a}$$

$$\tau \equiv \left(1 - \frac{Z^2e^4}{L^2}\right)^{\frac{1}{2}}, \tag{16b}$$

which reduces, for large L, to

$$u = \frac{1}{b} \cos \phi + \frac{Ze^2E}{L^2}. \tag{17}$$

Now, the asymptotes are given by the equation

$$0 = \frac{1}{b} \cos \phi + \frac{Ze^2E}{L^2},$$

or $\phi \approx \frac{bZe^2E}{L}$, so that the total deflection angle is

$$\Delta\phi_{EM} = \frac{2bZe^2E}{L}. \tag{18}$$

From the relations,

$$E = m(1-\beta^2)^{-\frac{1}{2}}$$

$$L^2 = \frac{b^2m^2\beta^2}{1-\beta^2}$$

Equation (18) becomes

$$\Delta\phi_{EM} = \frac{2Ze^2(1-\beta^2)^{\frac{1}{2}}}{mb\beta^2}. \tag{19}$$

Equations (13) and (19) differ because of the difference in the tensor nature of gravity and vector nature of electromagnetism: the fields Lorentz transform differently and this induces a different dependence of the deflection on β when β is not small.

Solution 15.10. It is convenient to work in "outgoing Eddington-Finkelstein coordinates": Define r^* by $dr^*/dr = (1-2M/r)^{-1}$, that is

$$r^* = r + 2M \log(r/2M - 1) + \text{constant}$$

and define a retarded time coordinate

$$u = t - r^* .$$

The Schwarzschild metric in these coordinates is

$$ds^2 = -(1-2M/r)\,du^2 - 2du\,dr + r^2 d\Omega^2 . \tag{1}$$

Putting $ds = 0$, we see that outgoing radial photons travel along lines $u = \text{constant}$. The redshift is calculated by comparing the time between reception of photons and emission of photons:

$$\frac{\lambda_\infty}{\lambda_{\text{emitter}}} = \frac{(\Delta t)_\infty}{(\Delta\tau)_{\text{em}}} = \frac{(\Delta u)_\infty}{(\Delta\tau)_{\text{em}}} = \frac{(\Delta u)_{\text{em}}}{(\Delta\tau)_{\text{em}}} = \left.\frac{du}{d\tau}\right|_{\text{em}} . \tag{2}$$

Note how we can relate times at infinity to times at the emitter by using $u = \text{constant}$ along the rays. We must therefore find the component $du/d\tau = U^u$ of the emitter's 4-velocity $\mathbf{U}$ as a function of t.

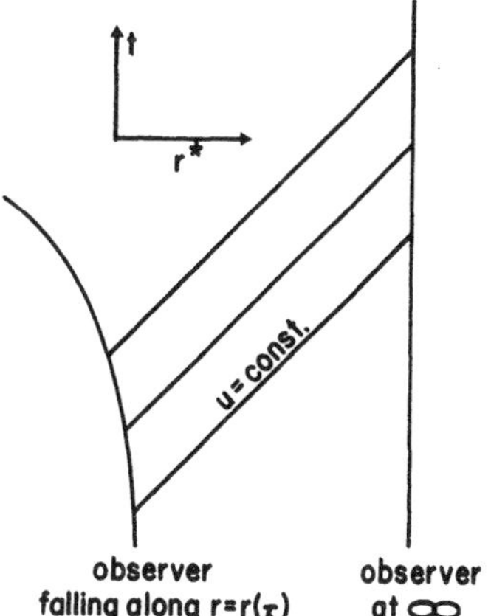

Now u and t are both ignorable coordinates and

$$U_t = U_u = \text{constant} = -\tilde{E} \tag{3}$$

and $\mathbf{U}\cdot\mathbf{U} = -1$ implies

$$U_r = \frac{-\tilde{E} - (\tilde{E}^2 - 1 + 2M/r)^{\frac{1}{2}}}{1 - 2M/r} , \tag{4}$$

therefore

$$U^u = g^{uu}U_u + g^{ur}U_r = 0 + [\tilde{E} + (\tilde{E}^2 - 1 + 2M/r)^{\frac{1}{2}}]\,(1-2M/r)^{-1} \tag{5}$$

and

$$\begin{aligned} U^r = g^{ur}U_u + g^{rr}U_r &= \tilde{E} + [-\tilde{E} - (\tilde{E}^2 - 1 + 2M/r)^{\frac{1}{2}}] \\ &= -(\tilde{E}^2 - 1 + 2M/r)^{\frac{1}{2}} \ . \end{aligned} \tag{6}$$

Thus

$$\frac{dr}{du} = \frac{U^r}{U^u} = \frac{-(\tilde{E}^2 - 1 + 2M/r)^{\frac{1}{2}}(1 - 2M/r)}{\tilde{E} + (\tilde{E}^2 - 1 + 2M/r)^{\frac{1}{2}}} \ . \tag{7}$$

Near $r = 2M$, Equation (7) gives

$$du \approx -2(1 - 2M/r)^{-1}dr \approx -2(r/2M - 1)^{-1}dr$$

$$u \approx -4M \log(r/2M - 1) + \text{constant}$$

$$1 - 2M/r \sim \exp(-u/4M) \ .$$

Thus Equation (5) implies that, as $r \to 2M$

$$U^u \sim e^{+u/4M} \ .$$

But $u = t + \text{constant}$ for an observer at fixed (large) r, so finally

$$\lambda_{\text{emitter}}/\lambda_\infty \sim \exp(-t/4M) \ .$$

Solution 15.11. For particle motion $u_t \equiv \tilde{E}$ and $u_\phi \equiv \tilde{L}$ are both constants of motion, so $\mathbf{u} \cdot \mathbf{u} = -1$ gives us

$$(dr/d\tau)^2 + V^2(r) = \tilde{E}^2 \tag{1a}$$

$$V \equiv [(1 - 2M/r)(1 + \tilde{L}^2/r^2)]^{\frac{1}{2}} \ . \tag{1b}$$

(See also MTW, Equation 25.15 or Weinberg, Equation 8.4.13.) By setting $\partial V^2/\partial r = 0$ we can find the maximum of V^2

$$V^2_{max} = \frac{\bar{L}^2 + 36 + (\bar{L}^2 - 12)(1 - 12/\bar{L}^2)^{\frac{1}{2}}}{54} \tag{2a}$$

$$\bar{L} \equiv \tilde{L}/M \ . \tag{2b}$$

If a particle has $\tilde{E} > V_{max}$, it is captured. Thus, the limiting $\bar{L}$ for capture is found by equating

$$\tilde{E}^2 = V_{max}^2 \ . \qquad (3)$$

(a) For large E (large $\bar{L}$), Equation (2a) reduces to

$$V_{max}^2 \approx \frac{\bar{L}^2 + 36 + (\bar{L}^2 - 12)(1 - 6/\bar{L}^2)}{54} = \frac{\bar{L}^2 + 9}{27} \ ,$$

and Equation (3) becomes

$$\bar{L}_{crit}^2 = 27\tilde{E}^2 - 9 \ . \qquad (4)$$

Corresponding to L_{crit} is a critical impact parameter, b_{crit}, given by

$$b_{crit} = \frac{L_{crit}}{p} = \frac{L_{crit}}{(E^2 - m^2)^{\frac{1}{2}}} \ ,$$

where m is the particle mass. Capture takes place for impact parameters $b < b_{crit}$. Thus the cross section for capture then becomes

$$\sigma_{cap} = \pi b_{crit}^2 = \frac{\pi L_{crit}^2}{(E^2 - m^2)} = \frac{\pi M^2 \bar{L}_{crit}^2}{(\tilde{E}^2 - 1)}$$

$$\approx \frac{\pi M^2}{\tilde{E}^2}\left(1 + \frac{1}{\tilde{E}^2}\right)(27\tilde{E}^2 - 9)$$

or

$$\pi b_{crit}^2 \approx 27\pi M^2 \left(1 + \frac{2}{3\tilde{E}^2}\right) \ . \qquad (5)$$

(b) For $\tilde{E} \approx 1$, (small β) we have $\tilde{E}^2 \approx 1 + \beta^2$ and Equation (3) can be approximated by

$$18 + 54\beta^2 \approx \bar{L}^2 + (\bar{L}^2 - 12)^{\frac{3}{2}}/\bar{L} \ .$$

We can solve for $\bar{L}^2$ to first order in β^2 to find

$$\bar{L}^2_{crit} = 16(1+2\beta^2) + \mathcal{O}(\beta^4)$$

and

$$\sigma_{capt} = \pi b^2_{crit} = \frac{\pi M^2 \bar{L}^2_{crit}}{(\tilde{E}^2 - 1)} \approx \frac{16\pi M^2}{\beta^2} . \tag{6}$$

Solution 15.12. Choose Paul's orbit to be in the $\theta = \pi/2$ plane. Since the orbit is circular $u^r = 0$ and from the r-component of the geodesic equation, $Du^r/d\tau = 0$, we get

$$\Gamma^r{}_{tt}(u^t)^2 + \Gamma^r{}_{\phi\phi}(u^\phi)^2 = 0$$

and hence

$$\omega^2 \equiv \left(\frac{d\phi}{dt}\right)^2 = \left(\frac{u^\phi}{u^t}\right)^2 = -\frac{\Gamma^r{}_{tt}}{\Gamma^r{}_{\phi\phi}} = \frac{M}{r^3} = \frac{1}{64M^2}$$

for Paul's orbit. Since $d\phi/dt = 1/8M$ and he traverses 20π radians between meetings, the coordinate time elapsed between meetings is

$$\Delta t = 160\pi M .$$

From $g_{tt}(u^t)^2 + g_{\phi\phi}(u^\phi)^2 = -1$ we get

$$u^t \equiv \frac{dt}{d\tau} = \left(1 - \frac{3M}{r}\right)^{-\frac{1}{2}} = 2$$

for Paul's orbit. Thus Paul measures an elapsed proper time

$$\Delta\tau_{Paul} = \frac{1}{2}\Delta t = 80\pi M \approx 251.5M .$$

For Peter's orbit we use the equations describing radial fall (Problem 15.4). The time for Peter to fall from R to $r = 4M$ must be $\frac{1}{2}\Delta t = 80\pi M$. We must find the values of R and η corresponding to a lapse in t of $80\pi M$. We can then find Peter's proper time from Solution 15.4, Equation (6):

$$\Delta\tau_{Peter} = 2\left(\frac{R^3}{8M}\right)^{\frac{1}{2}} (\eta + \sin\eta) . \tag{1}$$

The equations for R and η are Equations (7) and (5) of Solution 15.4

$$40\pi = \log\left[\frac{(X-1)^{\frac{1}{2}} + \tan\frac{1}{2}\eta}{(X-1)^{\frac{1}{2}} - \tan\frac{1}{2}\eta}\right] + (X-1)^{\frac{1}{2}}\left[\eta + \frac{1}{2}X(\eta+\sin\eta)\right] \tag{2}$$

$$4 = X(1+\cos\eta) \tag{3}$$

where $X \equiv R/2M$.

Now use physical intuition: Peter travels outward for a considerable time while Paul orbits. Therefore X should be quite large and $\eta \approx \pi$. Since logarithms are slowly varying functions, ignore the logarithm term and approximate

$$40\pi \approx X^{\frac{1}{2}}\left(\pi + \frac{1}{2}X\pi\right) .$$

This gives $X^{\frac{3}{2}} \approx 80\pi$ and $X \approx 18.5$. From Equation (3) then, $1 + \cos\eta \approx .216$ implies $\eta = 2.47$. More accurately,

$$\eta = 2.46029 \qquad X = \frac{R}{2M} = 17.91737$$

$$\Delta\tau_{\text{Peter}} = 468.72M .$$

Paul measures a much smaller (about half) proper time interval than Peter. This makes sense because Paul remains in a highly relativistic orbit deep within a gravitational well – both factors make his clocks "go slow." Peter spends most of his time at fairly small velocities and not very deep in a gravitational well (remember $R/2M \sim 18$).

Solution 15.13. We clearly need a transformation such that

$$r^2 = e^{2\mu}\bar{r}^2 \tag{1a}$$

$$e^{2\Lambda}dr^2 = e^{2\mu}d\bar{r}^2 . \tag{1b}$$

When these are combined, we get a differential equation for the transformation $\bar{r} = \bar{r}(r)$ in terms of the known function Λ:

$$\frac{d\bar{r}}{r} = e^{\Lambda}\frac{dr}{r} .$$

This can immediately be integrated, and our solution is

$$\bar{r} = \text{constant} \times \exp\left(\int \frac{e^{\Lambda}}{r} dr\right) \tag{2a}$$

$$e^{2\mu} = r^2/\bar{r}^2 . \tag{2b}$$

For the Schwarzschild metric $e^{\Lambda} = (1-2M/r)^{-\frac{1}{2}}$ and we have from Equation (2a)

$$\bar{r} = \text{constant} \times \exp\left(\int \frac{dr}{(r^2-2Mr)^{\frac{1}{2}}}\right)$$

which gives

$$r = \bar{r}(1+M/2\bar{r})^2 \tag{3}$$

if we choose the integration constant so that $r \to \bar{r}$ as $r \to \infty$. Equation (3) can also be expressed as

$$\bar{r} = \frac{1}{2}\left[r - M \pm (r(r-2M))^{\frac{1}{2}}\right] . \tag{4}$$

From Equation (2b) we have

$$e^{2\mu} = (1+M/2\bar{r})^4 . \tag{5}$$

From this it follows that the area of a sphere with r and t constant is

$$A = (1+M/2\bar{r})^4\bar{r}^2 \int d\theta \sin\theta\, d\phi = 4\pi\bar{r}^2(1+M/2\bar{r})^4. \tag{6}$$

In constructing the embedding diagram note that (i) according to Equation (4) the mapping from r to $\bar{r}$ is double valued and (ii) the coordinate $\bar{r}$ only describes the region of the Schwarzschild geometry with $r \geq 2M$ (In Equation (4), $\bar{r}$ is complex for $r < 2M$.)

The embedding diagram for a surface of constant t and r requires a function $z(r)$ such that

$$ds^2 = dz^2 + dr^2 + r^2 d\phi^2 = [1 + (dz/dr)^2]\, dr^2 + r^2 d\phi^2$$
$$= (1 - 2M/r)^{-1} dr^2 + r^2 d\phi^2 \ .$$

The solution is

$$z = [8M(r - 2M)]^{\frac{1}{2}} \ . \tag{7}$$

This parabola is plotted in the accompanying figure. The embedding surface is the paraboloid generated by rotating this curve about the r axis.

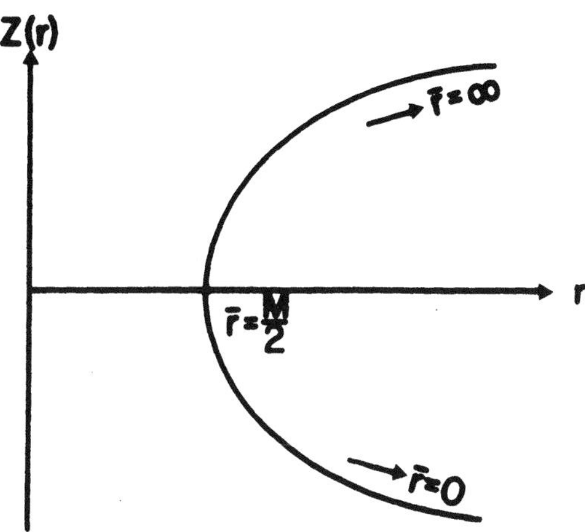

Solution 15.14.

(a) Let the boost be in the $e_{\hat{x}}$ direction. For a velocity parameter $\psi = \tanh^{-1}\beta$ the transformation is given by

$$\Lambda^{\hat{t}}{}_{\hat{t}'} = \Lambda^{\hat{x}}{}_{\hat{x}'} = \cosh\psi \qquad \Lambda^{\hat{t}}{}_{\hat{x}'} = \Lambda^{\hat{x}}{}_{\hat{t}'} = \sinh\psi$$

so that

$$\begin{aligned}
R_{\hat{t}'\hat{x}'\hat{t}'\hat{x}'} &= \Lambda^{\alpha}{}_{\hat{t}'}\Lambda^{\beta}{}_{\hat{x}'}\Lambda^{\gamma}{}_{\hat{t}'}\Lambda^{\delta}{}_{\hat{x}'}R_{\hat{\alpha}\hat{\beta}\hat{\gamma}\hat{\delta}} \\
&= \cosh^4\psi\, R_{\hat{t}\hat{x}\hat{t}\hat{x}} + \cosh^2\psi\,\sinh^2\psi\, R_{\hat{t}\hat{x}\hat{x}\hat{t}} \\
&\quad + \sinh^2\psi\,\cosh^2\psi\, R_{\hat{x}\hat{t}\hat{t}\hat{x}} + \sinh^4\psi\, R_{\hat{x}\hat{t}\hat{x}\hat{t}} \\
&= (\cosh^4\psi - 2\sinh^2\psi\,\cosh^2\psi + \sinh^4\psi)\, R_{\hat{t}\hat{x}\hat{t}\hat{x}} \\
&= (\cosh^2\psi - \sinh^2\psi)^2\, R_{\hat{t}\hat{x}\hat{t}\hat{x}} \\
&= R_{\hat{t}\hat{x}\hat{t}\hat{x}} \ .
\end{aligned}$$

(b) In the Schwarzschild geometry the nonvanishing physical components of the Riemann tensor are

$$R_{\hat{t}\hat{r}\hat{t}\hat{r}} = -R_{\hat{\theta}\hat{\phi}\hat{\theta}\hat{\phi}} = -2M/r^3$$

$$R_{\hat{t}\hat{\theta}\hat{t}\hat{\theta}} = R_{\hat{t}\hat{\phi}\hat{t}\hat{\phi}} = -R_{\hat{r}\hat{\theta}\hat{r}\hat{\theta}} = -R_{\hat{r}\hat{\phi}\hat{r}\hat{\phi}} = M/r^3$$

and the components related to these by symmetry. To show that all physical components are invariant for a radial boost we can explicitly calculate 20 independent, boosted components in a straightforward manner. Another way is to define, in the Schwarzschild geometry, the vectors

$$\mathbf{l} \equiv \mathbf{e}_{\hat{t}} + \mathbf{e}_{\hat{r}} \qquad \mathbf{n} \equiv \mathbf{e}_{\hat{t}} - \mathbf{e}_{\hat{r}} \qquad \mathbf{m} \equiv \mathbf{e}_{\hat{\theta}} + i\mathbf{e}_{\hat{\phi}}$$

and to observe that the Riemann tensor can be written as a sum of products of the vectors:

$$\mathbf{R} = \frac{1}{2}\frac{M}{r^3}\{-(\mathbf{n}\wedge\mathbf{l})(\mathbf{n}\wedge\mathbf{l}) + (\mathbf{m}\wedge\mathbf{m}^*)(\mathbf{m}\wedge\mathbf{m}^*)$$
$$+ \text{Real}\,[(\mathbf{n}\wedge\mathbf{m})(\mathbf{l}\wedge\mathbf{m}^*) + (\mathbf{l}\wedge\mathbf{m}^*)(\mathbf{n}\wedge\mathbf{m})]\} \,.$$

For a boost in the r-direction with velocity parameter ψ, clearly the components of $\mathbf{m}$ are invariant. From $\mathbf{e}_{\hat{t}} = \cosh\psi\, \mathbf{e}_{\hat{t}'} - \sinh\psi\, \mathbf{e}_{\hat{r}'}$ and $\mathbf{e}_{\hat{r}} = \cosh\psi\, \mathbf{e}_{\hat{r}'} - \sinh\psi\, \mathbf{e}_{\hat{t}'}$ it follows that

$$\mathbf{l} = \mathbf{e}_{\hat{t}} + \mathbf{e}_{\hat{r}} = e^{-\psi}(\mathbf{e}_{\hat{t}'} + \mathbf{e}_{\hat{r}'})$$

$$\mathbf{n} = \mathbf{e}_{\hat{t}} - \mathbf{e}_{\hat{r}} = e^{\psi}(\mathbf{e}_{\hat{t}'} + \mathbf{e}_{\hat{r}'})$$

so that

$$l_{\hat{\mu}'} n_{\hat{\nu}'} = l_{\hat{\mu}} n_{\hat{\nu}} \,.$$

From the form of $\mathbf{R}$ then, it is clear that all components are invariant.

To see that this same conclusion does not apply to boosts in other directions, notice that for a boost in the ϕ direction, with velocity parameter ψ

$$
\begin{aligned}
R_{\hat{r}'\hat{t}'\hat{r}'\hat{\phi}'} &= \Lambda^{\alpha}{}_{\hat{t}'}\Lambda^{\beta}{}_{\hat{\phi}'}R_{\hat{r}\alpha\hat{r}\beta} \\
&= \sinh\psi \cosh\psi \,(R_{\hat{r}\hat{t}\hat{r}\hat{t}} + R_{\hat{r}\hat{\phi}\hat{r}\hat{\phi}}) \\
&= -(M/r^3) \sinh\psi \cosh\psi \\
&\neq R_{\hat{r}\hat{t}\hat{r}\hat{\phi}} \, .
\end{aligned}
$$

Solution 15.15. In Kruskal coordinates, the Schwarzschild geometry is given by

$$ds^2 = \frac{32M^3}{r} e^{-r/2M} (-dv^2 + du^2) + r^2 d\Omega^2 \tag{1a}$$

where

$$\left(\frac{r}{2M} - 1\right) e^{r/2M} = u^2 - v^2 \, . \tag{1b}$$

Setting $v =$ constant, and $\theta = \pi/2$ in Equation (1a) and equating the line element to a Euclidean surface of revolution gives

$$\frac{32M^3}{r} e^{-r/2M} du^2 + r^2 d\phi^2 = [1 + (dz/dr)^2] dr^2 + r^2 d\phi^2 \tag{2}$$

or

$$1 + \left(\frac{dz}{dr}\right)^2 = \frac{32M^3}{r} e^{-r/2M} \left(\frac{du}{dr}\right)^2 .$$

We can calculate du/dr from Equation (1b) to arrive at

$$1 + \left(\frac{dz}{dr}\right)^2 = \frac{r}{2M} \frac{e^{r/2M}}{(r/2M - 1) e^{r/2M} + v^2} \, . \tag{3}$$

For embedding, z must be a real function, so we must have

$$\frac{r/2M \; e^{r/2M}}{(r/2M - 1) e^{r/2M} + v^2} \geq 1 \, . \tag{4}$$

From Equation (4), we conclude

$$r/2M \geq \log v^2 \, .$$

Thus, for $|v| > 1$, there will be a minimum value of r inside which the embedding breaks down.

Geometrically, the necessary condition for embedding is that

$$\frac{d\,(\text{circumference})}{d\,(\text{proper radius})} \leq 2\pi \ .$$

We now consider the general condition on dv/du which allows Euclidean embedding: Suppose we have a slice described by $v(u)$. Then

$$\begin{aligned} ds^2 &= \frac{32M^3}{r}\, e^{-r/2M}\left[1-\left(\frac{dv}{du}\right)^2\right] du^2 + r^2\, d\phi^2 \\ &= \frac{32M^3}{r}\, e^{-r/2M}\left[1-\left(\frac{dv}{du}\right)^2\right]\left[\frac{(r/8M^2)\,e^{r/2M}}{\left(u - v\,\frac{dv}{du}\right)}\right]^2 dr^2 + r^2\, d\phi^2 \end{aligned} \tag{6}$$

and we require

$$\frac{r}{2M}\, e^{r/2M}\left[1-\left(\frac{dv}{du}\right)^2\right] \geq \left(u - v\,\frac{dv}{du}\right)^2 \tag{7}$$

at each r.

Solution 15.16. To avoid the suggestive connotations of coordinate names, relabel r and t as follows: $r \to z$, $t \to w$. The metric then takes the form

$$ds^2 = -dw^2 + \frac{4}{9}\left[\frac{(9/2)M}{z-w}\right]^{\frac{2}{3}} dz^2 + [(9/2)M\,(z-w)^2]^{\frac{2}{3}}\, d\Omega^2 . \tag{1}$$

Now, define a new coordinate, r, by

$$r = \left[\frac{9}{2}M\,(z-w)^2\right]^{\frac{1}{3}} . \tag{2}$$

(We are led to do this because Equation (1) is manifestly spherically symmetric and therefore, the coefficient of $d\Omega^2$ is a geometrically defined curvature radius.) With r thus defined we have

$$-\left(\frac{2r^3}{9M}\right)^{\frac{1}{2}} + z = w \tag{3a}$$

$$dw = dz - \left(\frac{r}{2M}\right)^{\frac{1}{2}} dr \ . \tag{3b}$$

Substituting Equations (2), (3) into Equation (1) gives:

$$ds^2 = -\left(1-\frac{2M}{r}\right)dz^2 + 2\left(\frac{r}{2M}\right)^{\frac{1}{2}} dz\,dr - \frac{r}{2M}\,dr^2 + r^2\,d\Omega^2 . \tag{4}$$

Next we want to try to diagonalize the dz^2, $dz\,dr$, and dr^2 terms: Define a coordinate t and a function $F(r)$ by

$$z = t + F(r) , \tag{5}$$

substitute this into Equation (4):

$$\begin{aligned} ds^2 &= -\left(1-\frac{2M}{r}\right)(dt^2 + 2F'dt\,dr + F'^2 dr^2) \\ &\quad + 2\left(\frac{r}{2M}\right)^{\frac{1}{2}} dr\,(dt + F'dr) - \frac{r}{2M}\,dr^2 + r^2 d\Omega^2 , \end{aligned} \tag{6}$$

and choose F so that metric is diagonal, i.e.

$$\left(1-\frac{2M}{r}\right)F' = \left(\frac{r}{2M}\right)^{\frac{1}{2}} . \tag{7}$$

With this choice for F', Equation (6) becomes

$$\begin{aligned} ds^2 &= -\left(1-\frac{2M}{r}\right)dt^2 - \left(1-\frac{2M}{r}\right)^{-1}\left(\frac{r}{2M}\right)dr^2 \\ &\quad + 2\left(\frac{r}{2M}\right)\left(1-\frac{2M}{r}\right)^{-1} dr^2 - \frac{r}{2M}\,dr^2 + r^2\,d\Omega^2 \end{aligned}$$

or

$$ds^2 = -\left(1-\frac{2M}{r}\right)dt^2 + \left(1-\frac{2M}{r}\right)^{-1} dr^2 + r^2\,d\Omega^2 \tag{8}$$

which is the Schwarzschild metric in curvature coordinates and is, of course, a static spacetime. The original coordinates of Equation (1) are called "Lemaitre coordinates." Its time coordinate w measures the proper time of freely-infalling observers; each observer moves along a line $z = \text{constant}$.

Solution 15.17. A coordinate-stationary observer in Problem 15.16 satisfies $z = \text{constant}$ (see solution to 15.16). By Equation (5) of Solution 15.16, this implies

$$dt = -F'dr = -\left(\frac{r}{2M}\right)^{\frac{1}{2}} \left(1 - \frac{2M}{r}\right)^{-1} dr \tag{1}$$

according to Equation (7) of 15.16. From this equation we get

$$(dr/dt)^2 = (2M/r)\left(1 - \frac{2M}{r}\right)^2 . \tag{2}$$

Comparison of Equation (2) with Equation (3) of Solution 15.3 (with $u_0 = 1$ for a particle falling from rest at infinity) shows that Equation (2) represents a radially falling particle, with unit energy-to-rest-mass ratio at infinity.

Solution 15.18. The relativistic Bernoulli equation (Problem 14.7)

$$u_0 = \left(\frac{n}{\rho + p}\right)\left(\frac{n_\infty}{p_\infty + \rho_\infty}\right)^{-1} \tag{1}$$

can be expressed in terms of the fluid's proper radial velocity $v = (1 - 2M/r)\,dr/dt$, by using the relationship between u_0 and dr/dt

$$g_{00}(u^0)^2 + g_{rr}(u^r)^2 = -1 ,$$

to yield

$$v^2 = 1 - \frac{\left(1 - \frac{2M}{r}\right)\left(\frac{\rho + p}{n}\right)^2}{\left(\frac{\rho_\infty + p_\infty}{n_\infty}\right)^2} . \tag{2}$$

One can now use rest mass conservation $(u^\alpha n |g|^{\frac{1}{2}})_{,\alpha} = 0$ to obtain (after again expressing u^r in terms of v)

$$\frac{v(1 - 2M/r)^{\frac{1}{2}}\, nr^2}{(1 - v^2)^{\frac{1}{2}}} = \text{constant} \equiv \frac{\dot{M}}{4\pi} . \tag{3}$$

Incidentally, in this solution n is a rest mass density, not a number density: $n_{here} \equiv m_p n_{usual}$.

Now, we would like to express n in terms of the sound speed a (using $p = Kn^\gamma$; compare Solution 5.25):

$$a^2 = \frac{dp}{d\rho} = \frac{\gamma K n^{\gamma-1}}{1 + \gamma K n^{\gamma-1}/(\gamma-1)} .$$

This implies

$$\gamma K n^{\gamma-1} = \frac{a^2}{1 - a^2/(\gamma-1)} = \frac{a^2(\rho+p)}{n} \tag{4}$$

or

$$n = \left[\frac{a^2}{\gamma K(1 - a^2/(\gamma-1))}\right]^{\frac{1}{\gamma}-1} . \tag{5}$$

For the purposes of finding a self-consistent solution for some constant $\dot{M}$ and some v(r), Equations (2), (3) and (5) can be thought of as defining $\dot{M}$ as a function of r and v. We can then find the radius at which the flow becomes supersonic by setting $v = a$ and setting $d\dot{M} = 0$. First we must express r in terms of a. Using Equations (3) and (4), (with $v = a$), we get

$$r = \frac{2M}{1 - (1 - a^2/(\gamma-1))^2 (1 - a^2)\tau_\infty} \tag{6}$$

where

$$\tau_\infty \equiv \left(\frac{\rho+p}{n}\right)_\infty .$$

Substituting Equations (6), (5) into Equation (3), we get

$$\frac{\dot{M}}{4\pi} = \frac{4M^2 a\left(1 - \frac{a^2}{\gamma-1}\right)\left[\frac{a^2}{\gamma K(1 - a^2/(\gamma-1))}\right]^{\frac{1}{\gamma-1}}}{[1 - (1-a^2)(1 - a^2/(\gamma-1))^2 \tau_\infty]^2} . \tag{7}$$

Now, we expand Equation (4)

$$\tau_\infty \approx 1 + \frac{a_\infty^2}{\gamma-1}$$

and work to lowest order in a. From $d \log \dot{M}/da = 0$, we have

$$a^2 = \frac{2a_\infty^2}{5 - 3\gamma}, \quad \gamma \neq 5/3 \tag{8a}$$

$$a^2 = 2\cdot 3^{-\frac{1}{2}} a_\infty, \quad \gamma = 5/3 . \tag{8b}$$

Using Equations (8) and Equation (6), we conclude finally that the radius at which the flow becomes supersonic is

$$r_s = \left(\frac{5-3\gamma}{8}\right)\frac{2M}{a_\infty^2} \qquad \gamma \neq 5/3 \tag{9a}$$

$$r_s = \frac{3^{\frac{1}{2}} M}{4a_\infty} \qquad \gamma = 5/3 . \tag{9b}$$

Solution 15.19. The form of $\Box\Phi$ for a diagonal metric is

$$\Box\Phi = \nabla\cdot(\nabla\Phi) = (-g)^{-\frac{1}{2}}[(-g)^{\frac{1}{2}} g^{\alpha\beta}\Phi_{,\beta}]_{,\alpha} \tag{1}$$

$$\begin{aligned}\Box\Phi = (-g)^{-\frac{1}{2}}[&\{(-g)^{\frac{1}{2}} g^{00}\Phi_{,0}\}_{,0} + \{(-g)^{\frac{1}{2}} g^{rr}\Phi_{,r}\}_{,r} \\ &+ \{(-g)^{\frac{1}{2}} g^{\theta\theta}\Phi_{,\theta}\}_{,\theta} + \{(-g)^{\frac{1}{2}} g^{\phi\phi}\Phi_{,\phi}\}_{,\phi}] .\end{aligned} \tag{2}$$

In the Schwarzschild metric, curvature coordinates,

$$g^{00} = -\left(1-\frac{2M}{r}\right)^{-1}, \quad g^{rr} = \left(1-\frac{2M}{r}\right) \tag{3}$$

and $g^{\theta\theta}$ and $g^{\phi\phi}$ have their flat space form. It is clear that the third and fourth terms of Equation (2) have their flat space form (with flat space angular operator L^2) and thus, Equation (2) has the form

$$\Box\Phi = -\left(1-\frac{2M}{r}\right)^{-1}\Phi_{,00} + \frac{1}{r^2}\left[r^2\left(1-\frac{2M}{r}\right)\Phi_{,r}\right]_{,r} + \frac{L^2\Phi}{r^2} . \tag{4}$$

The field equations $\Box\Phi = 0$ then imply that

$$0 = \Phi_{,00} - \frac{1}{r^2}\left(1-\frac{2M}{r}\right)\left[r^2\left(1-\frac{2M}{r}\right)\Phi_{,r}\right]_{,r} - \frac{L^2}{r^2}\left(1-\frac{2M}{r}\right)\Phi . \tag{5}$$

Let Φ have the form

$$\Phi = \frac{1}{r}\,\psi(r,t)\,Y_{\ell m}(\theta,\phi)\,. \tag{6}$$

Substituting Equation (6) into Equation (5), we have

$$\psi_{,00} - \left(1-\frac{2M}{r}\right)\left[\left(1-\frac{2M}{r}\right)\psi_{,r}\right]_{,r} + V_\ell(r)\psi = 0 \tag{7a}$$

where

$$V_\ell(r) = \left(1-\frac{2M}{r}\right)\left[\frac{2M}{r^3}+\frac{\ell(\ell+1)}{r^2}\right] \,. \tag{7b}$$

Solution 15.20. The Brans-Dicke field equations can be put into the form

$$G^{\mu\nu} + F^{\mu\nu}(\Phi_{;\alpha},\Phi_{;\alpha\beta}) = T^{\mu\nu}\,, \tag{1a}$$

$$\Box\Phi = T^{\mu}_{\ \mu} \tag{1b}$$

where every term of $F^{\mu\nu}$ is proportional to derivatives of the scalar field Φ. In vacuum, a solution of Equation (1b) is

$$\Phi = \text{constant}\,. \tag{2}$$

Substituting Equation (2) into (1a), one obtains

$$G^{\mu\nu} = 0\,,$$

which are the vacuum Einstein equation. Consequently, the Schwarzschild metric is a solution of the vacuum Brans-Dicke equations. (The most general static, spherically symmetric vacuum metric in Brans-Dicke theory has two arbitrary parameters corresponding to mass and scalar charge. A spherical black hole has zero scalar charge and is the Schwarzschild solution.)

CHAPTER 16: SOLUTIONS

Solution 16.1. Since an observer's time axis points along his 4-velocity, if he is stationary in the (t, r, θ, ϕ) coordinate system then his time axis must be along $\partial/\partial t$ (and normalized properly). Accordingly, take $e_{\hat{0}} = f\,\partial/\partial t$, where f is to be determined by normalization:

$$-1 = e_{\hat{0}} \cdot e_{\hat{0}} = f^2 \frac{\partial}{\partial t} \cdot \frac{\partial}{\partial t} = f^2 g_{00} = -f^2 e^{2\Phi} .$$

Thus $f = e^{-\Phi}$ and

$$e_{\hat{0}} = e^{-\Phi} \frac{\partial}{\partial t} . \tag{1}$$

The three directions, $\partial/\partial\theta$, $\partial/\partial\phi$, $\partial/\partial r$ are all orthogonal to $\partial/\partial t$ and choosing basis vectors along them, with unit normalization (same technique as above) yields

$$e_{\hat{r}} = e^{-\mu} \frac{\partial}{\partial r}, \quad e_{\hat{\theta}} = e^{-\mu} r^{-1} \frac{\partial}{\partial \theta}, \quad e_{\hat{\phi}} = e^{-\mu}(r \sin\theta)^{-1} \frac{\partial}{\partial \phi} . \tag{2}$$

The basis vectors in Equations (1) and (2) constitute an orthonormal proper reference frame for a stationary observer. Dual 1-forms, $\tilde{\omega}^{\hat{\alpha}}$, are found by the requirement

$$<\tilde{\omega}^{\hat{\alpha}}, e_{\hat{\beta}}> = \delta^{\alpha}{}_{\beta} \tag{3}$$

and are easily seen to be

$$\tilde{\omega}^{\hat{0}} = e^{\phi}\widetilde{dt}, \quad \tilde{\omega}^{\hat{r}} = e^{\mu}\widetilde{dr}, \quad \tilde{\omega}^{\hat{\theta}} = e^{\mu} r\, \widetilde{d\theta}, \quad \tilde{\omega}^{\hat{\phi}} = e^{\mu} r \sin\theta\, \widetilde{d\phi}. \tag{4}$$

Solution 16.2. The observer makes measurements in his local orthonormal frame

$$e_{\hat{t}} = e^{-\Phi} e_t, \quad e_{\hat{r}} = e^{-\lambda} e_r, \quad e_{\hat{\theta}} = r^{-1} e_{\theta}, \quad e_{\hat{\phi}} = (r \sin\theta)^{-1} e_{\phi};$$

$$\tilde{\omega}^{\hat{t}} = e^{\Phi}\widetilde{dt}, \quad \tilde{\omega}^{\hat{r}} = e^{\lambda}\widetilde{dr}, \quad \tilde{\omega}^{\hat{\theta}} = r\widetilde{d\theta}, \quad \tilde{\omega}^{\hat{\phi}} = r \sin\theta\, \widetilde{d\phi} .$$

Here the metric is as given in the introduction, except $(1-2m/r)^{-1} \equiv e^{2\lambda}$. The volume of a fluid element which he measures is

$$\begin{aligned} V &= \tilde{\omega}^{\hat{r}} \wedge \tilde{\omega}^{\hat{\theta}} \wedge \tilde{\omega}^{\hat{\phi}} \\ &= e^{\lambda} r^2 \sin\theta \, \widetilde{dr} \wedge \widetilde{d\theta} \wedge \widetilde{d\phi} \\ &= e^{\lambda} r^2 \sin\theta \, dr \, d\theta \, d\phi \, . \end{aligned}$$

The pressure at opposite points on the vertical walls of the fluid element are the same, but because of the radial pressure gradient, the top and bottom pressures are not:

$$|p_{top} - p_{bottom}| = |p_{,\hat{r}} \tilde{\omega}^{\hat{r}}| = |p_{,r} e^{-\lambda} \tilde{\omega}^{\hat{r}}| \, .$$

Thus,

$$\begin{aligned} |F_{buoy.}| &= |(p_{top} - p_{bottom}) \times \text{area}| \\ &= |p_{,r} e^{-\lambda} \omega^{\hat{r}} \omega^{\hat{\theta}} \omega^{\hat{\phi}}| \\ &= |p_{,r} e^{-\lambda} V| \, . \end{aligned}$$

Since the direction of the buoyant force is radial,

$$F_{buoy.} = e^{-\lambda} p_{,r} V e_{\hat{r}} = \left(1 - \frac{2m}{r}\right)^{\frac{1}{2}} p_{,r} V e_{\hat{r}} \, .$$

Using the OV equation for $p_{,r}$, we find

$$F_{grav.} = -F_{buoy.} = -\frac{(\rho + p)(m + 4\pi r^3 p) V e_{\hat{r}}}{r^2 \left(1 - \frac{2m}{r}\right)^{\frac{1}{2}}} \, .$$

The Newtonian result is

$$p_{,r} V e_{\hat{r}} = -\frac{\rho m V}{r^2} e_{\hat{r}} \, .$$

Solution 16.3. A spherically symmetric metric can depend only on t, dt, r, dr, and $d\theta^2 + \sin^2\theta \, d\phi^2$:

$$ds^2 = -A(r,t)\,dt^2 + B(r,t)\,dr^2 + 2C(r,t)\,dr\,dt + D(r,t)\,(d\theta^2 + \sin^2\theta \, d\phi^2) \, .$$

Choose a new radial variable $r' = D^{\frac{1}{2}}(r,t)$. Dropping the prime, we have

$$ds^2 = -E(r,t)dt^2 + F(r,t)dr^2 + 2G(r,t)dr\,dt + r^2(d\theta^2 + \sin^2\theta\, d\phi^2) .$$

Eliminate G by defining a new time variable

$$dt' = H(r,t)[E(r,t)dt - G(r,t)dr]$$

where H is an integrating factor chosen to make the right-hand side a perfect differential. The metric then takes the standard form (with the prime dropped)

$$ds^2 = -e^{2\Phi(r,t)}dt^2 + e^{2\lambda(r,t)}dr^2 + r^2(d\theta^2 + \sin^2\theta\, d\phi^2) .$$

From Problem 9.20, the Einstein tensor has nontrivial components

$$G_{\hat{t}\hat{t}} = 2e^{-2\lambda}\lambda_{,r}/r + (1-e^{-2\lambda})/r^2 \tag{1}$$

$$G_{\hat{t}\hat{r}} = 2e^{-(\Phi+\lambda)}\lambda_{,t}/r \tag{2}$$

$$G_{\hat{r}\hat{r}} = (e^{-2\lambda}-1)/r^2 + 2e^{-2\lambda}\Phi_{,r}/r \tag{3}$$

$$\begin{aligned}G_{\hat{\theta}\hat{\theta}} = G_{\hat{\phi}\hat{\phi}} &= e^{-2\lambda}(\Phi_{,rr} + \Phi_{,r}{}^2 - \Phi_{,r}\lambda_{,r} + \Phi_{,r}/r - \lambda_{,r}/r) \\ &\quad - e^{-2\Phi}(\lambda_{,tt} + \lambda_{,t}{}^2 - \Phi_{,t}\lambda_{,t}) .\end{aligned} \tag{4}$$

When the vacuum field equations $G_{\hat{\alpha}\hat{\beta}} = 0$ are imposed, Equation (2) shows that λ is independent of t and Equation (1) becomes

$$\frac{2d\lambda}{1-e^{2\lambda}} = \frac{dr}{r}$$

from which we find

$$e^{2\lambda} = \left(1 - \frac{2M}{r}\right)^{-1} . \tag{5}$$

The constant of integration has been chosen as $2M$ for convenience. Equations (3) and (4) are now equivalent equations determining Φ. (They had to be equivalent because of the identity $G^{\hat\beta}{}_{\hat\alpha;\hat\beta} = 0$.) The quickest way to find Φ is to add Equation (1) to Equation (3). This gives $\Phi_{,r} = -\lambda_{,r}$ so $e^{2\Phi} = f(t)e^{-2\lambda}$. By defining a new time variable $dt' = f^{\frac{1}{2}}(t)dt$, f can be eliminated, so the final form of the metric is

$$ds^2 = -\left(1-\frac{2M}{r}\right)dt^2 + \left(1-\frac{2M}{r}\right)^{-1}dr^2 + r^2(d\theta^2+\sin^2\theta\, d\phi^2).$$

This is the Schwarzschild metric which is static.

Note: The r coordinate used here is defined by $4\pi r^2$ = surface area of spheres defined by the spherical symmetry. If r were not a monotonic function, it could not be used as a coordinate. However, a detailed investigation of this case (MTW, p. 845) shows there are no other solutions.

Solution 16.4. From Birkhoff's Theorem (see Problem 16.3), we know that the line element in vacuum for a spherically symmetric system is the Schwarzschild solution. Inside a self-gravitating hollow sphere we therefore have a Schwarzschild solution, but here the point $r = 0$ is not a singularity; therefore, the mass M (which is actually an integration constant in the solution of Einstein's equations) must be set to zero to avoid infinite values of the M/r terms in the metric. The metric inside the hollow sphere is therefore a flat space metric and consequently a test particle there experiences no forces.

Solution 16.5. The vacuum field equations for the scalar field Φ and metric $g_{\mu\nu}$ in the Brans-Dicke theory are

$$\begin{aligned} R_{\alpha\beta} - \frac{1}{2}g_{\alpha\beta}R = \frac{\omega}{\Phi^2}&(\Phi_{,\alpha}\Phi_{,\beta} - \frac{1}{2}g_{\alpha\beta}\Phi_{,\gamma}\Phi^{,\gamma}) \\ &+ \Phi^{-1}(\Phi_{,\alpha;\beta} - g_{\alpha\beta}\Box\Phi) \end{aligned} \tag{1}$$

$$\Box\Phi = 0 . \tag{2}$$

If the standard form of the spherically symmetric static metric $(-e^{2U}dt^2 + e^{2\lambda}dr^2 + r^2d\Omega^2)$ is used, $\Box\Phi = 0$ becomes

$$(e^{U-\lambda} r^2 \Phi_{,r})_{,r} = 0 \ . \tag{3}$$

The general solution is

$$\Phi = a \int_r^\infty \frac{e^{\lambda - U}}{r^2} dr + b \ . \tag{4}$$

Near $r = 0$ the metric functions e^{2U} and $e^{2\lambda}$ approach unity. (The metric is Lorentzian near $r = 0$.) The integral in Equation (4) therefore blows up near $r = 0$ and Φ takes the form

$$\Phi \approx a/r + b \tag{5}$$

near the origin. But if a is nonzero, the right hand side of Equation (1) is of order a^2/r^2. (This is not a coordinate effect; the trace of the right hand side is of order a^2/r^2 so by Equation (1) the scalar curvature blows up as a^2/r^2.) For a solution regular at the origin, then, a must be zero, i.e. Φ must be constant. For Φ constant, Equation (1) reduces to the vacuum static spherically symmetric field equations of general relativity. From Birkhoff's theorem we know that the only solution to these equations which is well behaved at $r = 0$ is $g_{\mu\nu} = \eta_{\mu\nu}$.

Solution 16.6. Examine the components at any point in an orthonormal tetrad $[e_{\hat{t}}, e_{\hat{r}}, e_{\hat{\theta}}, e_{\hat{\phi}}]$. Spherical symmetry implies the components are invariant under rotations:

$$e_{\hat{\theta}} \to \cos\alpha \ e_{\hat{\theta}} + \sin\alpha \ e_{\hat{\phi}}$$

$$e_{\hat{\phi}} \to -\sin\alpha \ e_{\hat{\theta}} + \cos\alpha \ e_{\hat{\phi}} \ .$$

The components $T_{\hat{t}\hat{t}}$, $T_{\hat{t}\hat{r}}$ and $T_{\hat{r}\hat{r}}$ are invariant under rotations. The pair $(T_{\hat{t}\hat{\theta}}, T_{\hat{t}\hat{\phi}})$ transforms like a 2-dimensional vector under rotations and so is not invariant. Thus $T_{\hat{t}\hat{\theta}} = T_{\hat{t}\hat{\phi}} = 0$ and similarly $T_{\hat{r}\hat{\theta}} = T_{\hat{r}\hat{\phi}} = 0$. The only 2-dimensional matrix invariant under rotations is a multiple of the identity matrix, so

$$\begin{pmatrix} T_{\hat\theta\hat\theta} & T_{\hat\theta\hat\phi} \\ T_{\hat\phi\hat\theta} & T_{\hat\phi\hat\phi} \end{pmatrix} = T_{\hat\theta\hat\theta}\begin{pmatrix} 1 & 0 \\ 0 & 1 \end{pmatrix}.$$

Thus, there are four independent components: $T_{\hat t\hat t}$, $T_{\hat t\hat r}$, $T_{\hat r\hat r}$ and $T_{\hat\theta\hat\theta} = T_{\hat\phi\hat\phi}$.

Solution 16.7. The standard form for the metric of a static, spherically symmetric star is

$$ds^2 = -e^{2\Phi}dt^2 + e^{2\Lambda}dr^2 + r^2 d\Omega^2 . \tag{1}$$

The equation $T^{\alpha\beta}{}_{;\beta} = 0$ is most easily evaluated by recalling that it is equivalent to the first law of thermodynamics

$$d\rho/d\tau = \frac{\rho+p}{n}\frac{dn}{d\tau} \tag{2}$$

(see Problem 5.19) and the Euler equation

$$(\rho+p)u_{\alpha;\beta}u^\beta = -p_{,\alpha} - p_{,\beta}u^\beta u_\alpha \tag{3}$$

(see Problem 14.3). For a static star, both sides of Equation (2) vanish identically. The only nonzero component of $\mathbf{u}$ is u^t and $\mathbf{u}\cdot\mathbf{u} = -1$ implies $u^t = e^{-\Phi}$. Since only $p_{,r}$ is nonzero, the only nontrivial component of Equation (3) is

$$\begin{aligned} -p_{,r} &= (\rho+p)u_{r;\beta}u^\beta = -(\rho+p)\Gamma^\alpha{}_{r\beta}u_\alpha u^\beta \\ &= -(\rho+p)\Gamma^t{}_{rt}u_t u^t = (\rho+p)\Phi_{,r} . \end{aligned}$$

Solution 16.8. Let $m(r)$ be a solution to the Newtonian equations for a polytropic star, namely

$$dm/dr = 4\pi r^2\rho \tag{1}$$

$$dp/dr = -\frac{Gm\rho}{r^2} \tag{2}$$

$$p = K\rho^\gamma . \tag{3}$$

When $\gamma = 4/3$ the star is neutrally stable; this is seen in the fact that

$$\tilde{m}(r) = m(\alpha\, r)$$
$$\tilde{\rho}(r) = \alpha^3 \rho(\alpha\, r)$$
$$\tilde{p}(r) = \alpha^4 p(\alpha\, r)$$

is also a solution to Equations (1), (2), (3) and has the same total mass $m(\infty)$ and adiabat K.

The relativistic version of Equation (1) is identical; Equation (3), the given equation of state, is also identical by hypothesis. The relativistic form of Equation (2) is [Weinberg, Equation 11.1.13 or MTW, Equation 23.22]

$$\frac{dp}{dr} = -\frac{Gm\rho}{r^2}\left(1+\frac{p}{\rho}\right)\left(1+\frac{4\pi r^3 p}{m}\right)\left(1-\frac{m}{r}\right)^{-1} . \tag{4}$$

Since the star we are considering is nearly Newtonian, the terms in brackets are all very close to unity. Hence, one can find an exact solution to Equation (4) which is nearly equal to the Newtonian solution to Equation (2) [e.g. by substituting into Equation (4) to find a small correction to p and by iterating]. Call this solution $\hat{m}(r)$, $\hat{\rho}(r)$, $\hat{p}(r)$. From this generate

$$\tilde{m}(r) = \hat{m}(\alpha\, r)$$
$$\tilde{\rho}(r) = \alpha^3 \hat{\rho}(\alpha\, r)$$
$$\tilde{p}(r) = \alpha^4 \hat{p}(\alpha\, r) .$$

Equations (1) and (3) remain equalities, but the ratio of the right hand side of Equation (4) to the left hand side is, to lowest order,

$$\frac{\text{right}}{\text{left}} = 1 + (\alpha - 1)\left[\frac{\hat{p}}{\hat{\rho}} + 4\pi(\alpha r)^3 \frac{\hat{p}}{\hat{m}} + \frac{2\hat{m}}{\alpha r}\right] . \tag{5}$$

For $\alpha > 1$, which corresponds to shrinking the star to smaller radius, the right side is greater than the left side, hence the pressure gradient is everywhere insufficient to support the star and it must go to a smaller radius yet. This cannot be an equilibrium because it would still be almost Newtonian and the same argument would apply. Therefore, $\gamma = 4/3$ is unstable, and the range of stability must begin at $4/3 + \varepsilon$.

Solution 16.9. Increasing the mass of a degenerate star makes the material more compact until the Fermi energy of the gas (electron gas in case of white dwarf; neutron gas in the case of neutron star) rises to a relativistic level. If A is the number of baryons in the star, and R is the radius of the star, then the Fermi energy is approximately (in the relativistic regime)

$$E_F \sim \frac{\hbar c}{R} A^{\frac{1}{3}} \tag{1}$$

(In this equation we have ignored such details as the fact that there are twice as many pressure supporting fermions per baryon in a neutron star as in a white dwarf.) The gravitational mass-energy per fermion is approximately

$$E_G \sim -\frac{GAm_B^2}{R} \tag{2}$$

where m_B is the mass of a baryon.

Note that both the compressional and gravitational energy depend in the same way upon R. The decisive question is therefore the sign of the total coefficient of R^{-1}. When it is *positive*, R will increase, the Fermi energy will become nonrelativistic, the energy of compression will then fall faster than R^{-1} and a stable equilibrium will be achieved at some finite R. When the coefficient is *negative*, collapse will set in and continue. The critical number of baryons at which the star becomes unstable is thus obtained by equating E_G and E_F:

$$A_{\text{crit.}} = \left(\frac{\hbar c}{Gm_B^2}\right)^{\frac{3}{2}} \approx 10^{57}\,. \tag{3}$$

Equation (3) corresponds to a critical limiting mass of

$$M_{crit.} = m_B A_{crit.} \approx 2M_\odot \tag{4}$$

and limiting radii of

$$R_{crit.} \sim \hbar c\, A_{crit.}^{\frac{1}{3}} \times \begin{cases} (m_e c^2)^{-1} & \text{for white dwarf} \\ (m_B c^2)^{-1} & \text{for neutron star} \end{cases}$$

$$\approx \begin{cases} 5 \times 10^8 \text{ cm} & \text{for white dwarf} \\ 2.7 \times 10^5 \text{ cm} & \text{for neutron star} \end{cases}$$

Solution 16.10. At a particular point P let an observer construct a sphere S passing through P and defined such that the geometry is the same at all points on S. Then, from the line element, the *measured* area of this sphere will be

$$A(r) = 4\pi r^2 . \tag{1}$$

The gradient of the scalar function A(r) is a 1-form, i.e.

$$\nabla A(r) \equiv \tilde{d}A = 8\pi r\, \tilde{d}r . \tag{2}$$

Thus,

$$\tilde{d}A \cdot \tilde{d}A = 16\pi^2 r^2\, \tilde{d}r \cdot \tilde{d}r = 16\pi^2 r^2 (1 - 2m/r) \tag{3}$$

and

$$m(r) = \frac{(A/\pi)^{\frac{1}{2}}}{4} \left(1 - \frac{\tilde{d}A \cdot \tilde{d}A}{16\pi A}\right) . \tag{4}$$

Solution 16.11.

(a) Under the transformation, the metric

$$ds^2 = -\left(1 - \frac{2M}{r}\right) dt^2 + \left(1 - \frac{2M}{r}\right)^{-1} dr^2 + r^2 d\Omega^2 \tag{1}$$

becomes

$$ds^2 = -\left(1 - \frac{2M}{r}\right) du^2 - 2du\, dr + r^2 d\Omega^2 . \tag{2}$$

(b) We wish to compute the Einstein tensor $G_{\alpha\beta}$ when $M = M(u)$. The Christoffel symbols are most easily found from the Lagrangian for geodesics (see Problem 7.25):

$$L = -\left(1-\frac{2M}{r}\right)\dot{u}^2 - 2\dot{u}\dot{r} + r^2\dot{\theta}^2 + r^2\sin^2\theta\,\dot{\phi}^2 . \tag{3}$$

The Euler-Lagrange equation

$$\frac{d}{ds}\left(\frac{\partial L}{\partial \dot{r}}\right) - \frac{\partial L}{\partial r} = 0 \tag{4}$$

gives the geodesic equation

$$\ddot{u} - (M/r^2)\dot{u}^2 + r\dot{\theta}^2 + r\sin^2\theta\,\dot{\phi}^2 = 0 \tag{5}$$

from which we read off the nonzero Christoffel symbols

$$\Gamma^u{}_{uu} = -\frac{M}{r^2}, \qquad \Gamma^u{}_{\theta\theta} = r, \qquad \Gamma^u{}_{\phi\phi} = r\sin^2\theta .$$

Similarly, Equation (4) with r replaced by u, θ and ϕ gives the remaining Christoffel symbols

$$\Gamma^r{}_{uu} = \frac{M}{r^2}\left(1-\frac{2M}{r}\right) - \frac{M'}{r}, \qquad \Gamma^r{}_{\theta\theta} = -r\left(1-\frac{2M}{r}\right),$$

$$\Gamma^r{}_{\phi\phi} = -\left(1-\frac{2M}{r}\right)r\sin^2\theta, \qquad \Gamma^r{}_{ur} = \frac{M}{r^2}, \qquad \Gamma^\theta{}_{r\theta} = \frac{1}{r},$$

$$\Gamma^\theta{}_{\phi\phi} = -\sin\theta\cos\theta, \qquad \Gamma^\phi{}_{r\phi} = \frac{1}{r}, \qquad \Gamma^\phi{}_{\theta\phi} = \cot\theta .$$

These all have their Schwarzschild values except $\Gamma^r{}_{uu}$. From the formula for the Ricci tensor (MTW, Equation 8.51b)

$$R_{\alpha\beta} = \Gamma^\gamma{}_{\alpha\beta,\gamma} - (\log|g|^{\frac{1}{2}})_{,\alpha\beta} + (\log|g|^{\frac{1}{2}})_{,\gamma}\Gamma^\gamma{}_{\alpha\beta} - \Gamma^\gamma{}_{\beta\delta}\Gamma^\delta{}_{\alpha\gamma}$$

we see that all the $R_{\alpha\beta}$ will have their Schwarzschild values (i.e. zero) except

$$R_{uu} = -2M'/r^2 .$$

The Ricci scalar is

$$R = g^{uu}R_{uu} = 0$$

so the only nonzero component of the Einstein tensor is

$$G_{uu} = -2M'/r^2$$

which implies that

$$T_{uu} = -M'/4\pi r^2 ,$$

and that all other components of T are zero. Since the vector $k = \nabla u$, that is $k_\alpha = (1,0,0,0)$, is null, the stress-energy tensor

$$T = -\frac{M'}{4\pi r^2} k \otimes k$$

corresponds to a pure radiation field. Thus a physical interpretation of this solution (called the Vaidya metric) is the exterior metric of a spherical star whose mass is decreasing because of the energy carried off in radiation.

Solution 16.12. The equations of stellar structure are (MTW, Equation 23.5 or Weinberg, Chapter 11.1)

$$ds^2 = -e^{2\Phi}dt^2 + \left(1-\frac{2m}{r}\right)^{-1} dr^2 + r^2 d\Omega^2 \tag{1}$$

$$m = \int_0^r 4\pi r^2 \rho \, dr \tag{2}$$

$$\frac{dp}{dr} = -\frac{(\rho+p)(m+4\pi r^3 p)}{r(r-2m)} \tag{3}$$

$$\frac{d\Phi}{dr} = \frac{m+4\pi r^3 p}{r(r-2m)} . \tag{4}$$

Since $\rho = \rho_0$, a constant, Equation (2) gives

$$m = 4\pi r^3 \rho_0/3 \tag{5}$$

and

$$M = 4\pi R^3 \rho_0/3 , \tag{6}$$

where R is the radius and M the total mass for the star. Substitute Equation (5) in Equation (3) to get

$$\frac{dp}{4\pi(\rho_0+p)(\frac{1}{3}\rho_0+p)} = -\frac{rdr}{1-(8\pi/3)r^2\rho_0} .$$

Integrate and solve for p:

$$\frac{p}{\rho_0} = \frac{(1-2Mr^2/R^3)^{\frac{1}{2}} - (1-2M/R)^{\frac{1}{2}}}{3(1-2M/R)^{\frac{1}{2}} - (1-2Mr^2/R^3)^{\frac{1}{2}}} . \tag{7}$$

The constant of integration has been determined by demanding $p = 0$ at $r = R$. The quantity Φ is most easily found from the equation:

$$\frac{d\Phi}{dp} = \frac{d\Phi/dr}{dp/dr} = -\frac{1}{\rho_0+p} .$$

Integrate to find

$$\begin{aligned} e^{\Phi} &= \text{constant}/(\rho_0+p) \\ &= \frac{3}{2}(1-2M/R)^{\frac{1}{2}} - \frac{1}{2}(1-2Mr^2/R^3)^{\frac{1}{2}} \end{aligned} \tag{8}$$

where Equation (7) has been used to fix the constant of integration so that

$$e^{\Phi(r=R)} = (1-2M/R)^{\frac{1}{2}} .$$

(That is, so that the metric will match smoothly to the Schwarzschild metric outside the star.)

We thus have a 1-parameter family of stellar models, conveniently parametrized by the central pressure

$$\frac{p_c}{\rho_0} = \frac{1-(1-2M/R)^{\frac{1}{2}}}{3(1-2M/R)^{\frac{1}{2}}-1} . \tag{9}$$

The central pressure becomes infinite when

$$3\left(1-\frac{2M}{R}\right)^{\frac{1}{2}} - 1 = 0$$

i.e. the limiting value of $R/2M$ is $9/8$.

The dominant energy condition demands $p_c < \rho_0$, so that the right side of Equation (9) is < 1. This implies $R/2M > 4/3$.

Solution 16.13. A relativistic zero-temperature Fermi gas has the equation of state $p = 1/3\,\rho$, so the equations of structure (see Problem 16.12) read

$$d\rho/dr = -\frac{4\rho(m + 4\pi r^3\rho/3)}{r(r - 2m)} \tag{1}$$

$$dm/dr = 4\pi r^2 \rho \,. \tag{2}$$

Substitution of the suggested solution $m(r) = 3r/14$ into Equation (2) gives $\rho(r) = (3/14)(4\pi r^2)^{-1}$; substitution of this ρ and m into Equation (1) verifies that they are a solution. From the equation of state $p(r) = (1/14)(4\pi r^2)^{-1}$. To find $n(r)$ we use the relations for a relativistic Fermi gas (see Problem 5.24)

$$n = \int_0^{p_F} \left(\frac{2}{h^3}\right) 4\pi p^2 dp = \frac{8\pi}{3h^3}\, p_F^3 \tag{3}$$

$$\rho = \int_0^{p_F} p\left(\frac{2}{h^3}\right) 4\pi p^2 dp = \frac{8\pi}{4h^3}\, p_F^3 \,. \tag{4}$$

Eliminating p_F (the Fermi momentum) from Equations (3) and (4) gives

$$n(r) = \frac{8\pi}{3h^3}\left(\frac{4h^3\rho}{8\pi}\right)^{\frac{3}{4}}$$

and substitution of $\rho(r)$ gives

$$n(r) = \frac{K}{r^{\frac{3}{2}}}\,, \qquad K = \frac{8\pi}{3h^3}\left(\frac{4h^3}{8\pi}\cdot\frac{3}{14}\cdot\frac{1}{4\pi}\right)^{\frac{3}{4}} . \tag{5}$$

The total number of particles inside radius r is

$$N(r) = \int_0^r n(r)\,d\,(\text{proper volume})$$

$$= \int_0^r n(r)\,e^{\Lambda} 4\pi r^2 dr$$

$$= \frac{14\pi}{3} K r^{\frac{3}{2}} \tag{6}$$

which is finite for all r. [Note: In deriving Equation (6) we have used

$$e^{2\Lambda} \equiv g_{rr} = \left(1 - \frac{2m(r)}{r}\right)^{-1} = 7/4\Big].$$

The 3-geometry of the $t =$ constant hypersurface has the metric

$${}^{(3)}ds^2 = g_{rr} dr^2 + r^2 d\Omega^2 = (7/4)\,dr^2 + r^2 d\Omega^2 \ .$$

This obviously has a *conical* singularity at the origin, because a 2-sphere of circumference $2\pi r$ will have a radius of $(7/4)^{\frac{1}{2}} r$ which is greater than r.

The embedding equation for a radial cut is

$${}^{(3)}ds^2 = (7/4)\,dr^2 = dr^2 + dz^2$$

which implies

$$z = \pm\,(3/4)^{\frac{1}{2}} r$$

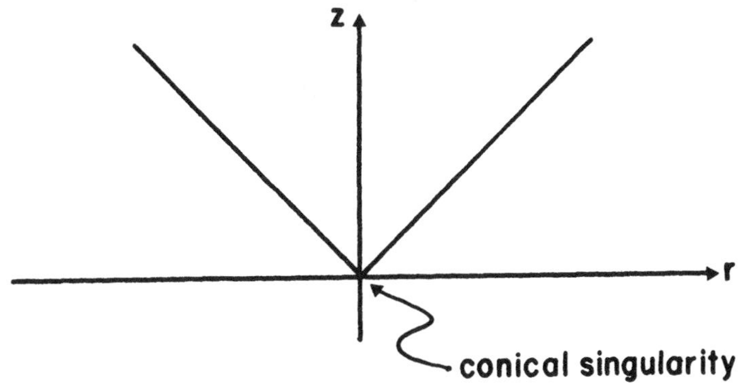

Solution 16.14. The metric cannot be continuous across the shell in curvature coordinates, because $g_{rr} = (1-2m(r)/r)^{-1}$ and $m(r)$ changes across the shell. Therefore we use isotropic coordinates, in which the metric *will* be continuous and (matching flat space inside to mass M outside) has the form

$$ds^2 = -e^{\gamma}dt^2 + e^{\alpha}(dr^2 + r^2 d\Omega^2) \tag{1}$$

$$e^{\gamma} = \begin{cases} \left(1-\frac{M}{2r}\right)^2\left(1+\frac{M}{2r}\right)^{-2} & r > R \\ \left(1-\frac{M}{2R}\right)^2\left(1+\frac{M}{2R}\right)^{-2} & r < R \end{cases} \tag{2a}$$

$$e^{\alpha} = \begin{cases} \left(1+\frac{M}{2r}\right)^4 & r > R \\ \left(1+\frac{M}{2R}\right)^4 & r < R\,. \end{cases} \tag{2b}$$

In Equations (1) and (2) we have used the fact that the geometry is Schwarzschild outside the shell and flat in the interior (Birkhoff's Theorem – see Problem 16.3). Also, the surface of the shell is located at $r = R$ (in isotropic radial coordinate distance).

Next we define the integrated stress, $\Lambda^{\hat{\alpha}}{}_{\hat{\beta}}$ in the following invariant manner

$$\Lambda^{\hat{\alpha}}{}_{\hat{\beta}} \equiv \lim_{\epsilon\to 0}\int_{R-\epsilon}^{R+\epsilon} T^{\hat{\alpha}}{}_{\hat{\beta}}\, d\hat{r} = \lim_{\epsilon\to 0}\int_{R-\epsilon}^{R+\epsilon} T^{\hat{\alpha}}{}_{\hat{\beta}}\, e^{\frac{1}{2}\alpha}\, dr\,. \tag{3}$$

The components of Λ may be evaluated by using the Einstein field equations and by noting that, according to Equations (2) and (3), only terms in $G^{\mu}{}_{\nu}$ with second derivatives of the metric potentials can contribute.

The components of $G^{\mu}{}_{\nu}$ for the metric of Equation (1) are (see Problem 9.20)

$$G^0{}_0 \sim \alpha'' e^{-\alpha} \sim -\frac{d}{dr^2}(e^{-\alpha}) \tag{4a}$$

$$G^r{}_r \sim 0 \tag{4b}$$

$$G^{\theta}{}_{\theta} = G^{\phi}{}_{\phi} \sim \frac{1}{2} e^{-\alpha}(\alpha'' + \gamma'') \sim -\frac{1}{2}\left[\frac{d^2}{dr^2}(e^{-\alpha}) - \frac{d}{dr}(e^{-\alpha}\gamma')\right]$$

where $\sim$ denotes the process of discarding all terms not involving second derivatives. Using Equations (3) and (4), and the Einstein field equations, we get

$$\Lambda^{\hat{0}}{}_{\hat{0}} = \Lambda^{0}{}_{0} = -(8\pi)^{-1} \lim_{\epsilon\to 0} \int_{R-\epsilon}^{R+\epsilon} e^{\alpha/2} \frac{d^2}{dr^2}(e^{-\alpha})\, dr \sim -(8\pi)^{-1} \lim_{\epsilon\to 0} e^{\alpha/2} \frac{d}{dr}(e^{-\alpha})\Big|_{R-\epsilon}^{R+\epsilon}$$

or

$$\Lambda^{\hat{0}}{}_{\hat{0}} = \frac{-M}{4\pi R^2(1+M/2R)^3} \tag{4c}$$

$$\Lambda^{\hat{r}}{}_{\hat{r}} = \Lambda^{r}{}_{r} = 0 \tag{5}$$

$$\Lambda^{\hat{\theta}}{}_{\hat{\theta}} = \Lambda^{\hat{\phi}}{}_{\hat{\phi}} = -(8\pi)^{-1} \lim_{\epsilon\to 0} \int_{R-\epsilon}^{R+\epsilon} \frac{1}{2} e^{\alpha/2} \left[\frac{d^2}{dr^2}(e^{-\alpha}) - \frac{d}{dr}(e^{-\alpha}\gamma')\right] dr$$

$$= -(8\pi)^{-1} \lim_{\epsilon\to 0} \frac{1}{2} e^{\alpha/2} \left[\frac{d}{dr}(e^{-\alpha}) - e^{-\alpha}\gamma'\right]\Big|_{R-\epsilon}^{R+\epsilon}$$

or

$$\Lambda^{\hat{\theta}}{}_{\hat{\theta}} = \frac{M}{8\pi R^2}\left(\frac{M/2R}{1-M/2R}\right)\left(1+\frac{M}{2R}\right)^{-3} . \tag{6}$$

To have proper interpretation of Equations (4) to (6) we must remember that R denotes the radius of the shell in isotropic coordinates. In curvature coordinates, the radius $\mathcal{R}$ is

$$\mathcal{R} = R(1+M/2R)^2 . \tag{7}$$

The proper surface mass density is just

$$\frac{M}{4\pi\mathcal{R}^2} = \frac{M}{4\pi R^2(1+M/2R)^4} . \tag{8}$$

The Newtonian limit of Equations (4) to (6) is just

$$\Lambda^{\hat{0}}{}_{\hat{0}} \to -\frac{M}{4\pi R^2}$$

$$\Lambda^{\hat{r}}{}_{\hat{r}} = 0$$

$$\Lambda^{\hat{\theta}}{}_{\hat{\theta}} = \Lambda^{\hat{\phi}}{}_{\hat{\phi}} \to \frac{M^2}{16\pi R^3} .$$

Solution 16.15. The dominant energy condition requires

$$|T^{\hat{0}\hat{0}}| > |T^{\hat{k}\hat{k}}| . \tag{1}$$

Using the results of Problem 16.14, and integrating Equation (1) through the thin shell, we have the inequality

$$\left(1+\frac{M}{2R}\right)^{-3} > \frac{1}{2}\,\frac{M/2R}{1-M/2R}\left(1+\frac{M}{2R}\right)^{-3} \tag{2}$$

where R is the isotropic radial coordinate. Letting $x = M/2R$ we have $2-2x > x$ or $x < 2/3$, giving $R > \frac{3}{4}M$, or for the curvature radial coordinate $\mathfrak{R}$

$$\mathfrak{R} = R\left(1+\frac{M}{2R}\right)^2 > \frac{25}{12}M , \tag{3}$$

about 1.04 Schwarzschild radii.

Solution 16.16. In isotropic coordinates (see Problem 16.14), the redshift to radial infinity is

$$z = \frac{\lambda_{\text{infinity}}}{\lambda_{\text{surface}}} - 1 = \frac{(-g_{00})_{\infty}^{\frac{1}{2}}}{(-g_{00})_{S}^{\frac{1}{2}}} - 1 = (e^{-\gamma/2})_S - 1$$

$$= \frac{1+M/2R}{1-M/2R} - 1 = \frac{M/R}{1-M/2R} .$$

The integrated energy density and transverse pressure are respectively

$$-\Lambda^{\hat{0}}{}_{\hat{0}} = \frac{M}{4\pi R^2}\left(1+\frac{M}{2R}\right)^{-3},$$

$$\Lambda^{\hat{\theta}}{}_{\hat{\theta}} = \Lambda^{\hat{\phi}}{}_{\hat{\phi}} = \frac{M}{8\pi R^2}\left(\frac{M/2R}{1-M/2R}\right)\left(1+\frac{M}{2R}\right)^{-3}.$$

Hence the ratio of integrated pressure to energy is

$$\frac{\Lambda^{\hat{\theta}}{}_{\hat{\theta}}}{-\Lambda^{\hat{0}}{}_{\hat{0}}} = \frac{M/4R}{1-M/2R} = \frac{1}{4}z$$

so the redshift to radial infinity is

$$z = -4\Lambda^{\hat{\theta}}{}_{\hat{\theta}}/\Lambda^{\hat{0}}{}_{\hat{0}} .$$

If the dominant energy condition is satisfied, $|\Lambda^{\hat{\theta}}{}_{\hat{\theta}}| \leq |\Lambda^{\hat{0}}{}_{\hat{0}}|$, so $z \leq 4$.

Solution 16.17. The non-vanishing components of the fluid 4-velocity are u^t and $u^{\phi} = \Omega u^t$, i.e.

$$\mathbf{u} = u^t\left(\frac{\partial}{\partial t}+\Omega\frac{\partial}{\partial\phi}\right).$$

Let

$$\boldsymbol{\xi} = \frac{\partial}{\partial t} + \Omega\frac{\partial}{\partial\phi} = \boldsymbol{\xi}_{(t)} + \Omega\boldsymbol{\xi}_{(\phi)} .$$

Then $\boldsymbol{\xi}$ is a Killing vector, since Ω is a constant. By Problem 10.14, we have

$$\nabla_{\mathbf{u}}\mathbf{u} = \nabla \log|\boldsymbol{\xi}\cdot\boldsymbol{\xi}|^{\frac{1}{2}}$$

where, since $\mathbf{u}\cdot\mathbf{u} = -1$, $|\boldsymbol{\xi}\cdot\boldsymbol{\xi}|^{\frac{1}{2}} = 1/u^t$. Thus, the Euler equation (see Problem 14.3) for the fluid becomes

$$(\rho+p)\nabla\log(u^t)^{-1} = -\nabla p - (\nabla_{(u^t\boldsymbol{\xi})}p)\mathbf{u} .$$

Since p is independent of t and ϕ, it follows that $\nabla_{\boldsymbol{\xi}}p = 0$. Thus,

$$(\rho+p)\nabla\log u^t = \nabla p .$$

Solution 16.18. First, note that $dp = (\rho + p) d \log u^t$ (hydrostatic equilibrium; see Problem 16.17). Next take the exterior derivative of this equation, and use $ddp = 0$ to get

$$0 = d(\rho + p) \wedge d \log u^t$$
$$= d(\rho + p) \wedge (\rho + p)^{-1} d\rho$$

which implies that

$$d\rho \wedge dp = 0 ,$$

and therefore that the surface $\rho = \text{constant}$ coincides with the surface $p = \text{constant}$.

Solution 16.19. The surface is a surface of constant p (that is $p = 0$). By Problem 16.18, it is a surface of constant ρ, and also of constant u^t, since

$$dp \propto du^t .$$

But, in the notation of Solution 16.17,

$$(u^t)^{-1} = |\boldsymbol{\xi} \cdot \boldsymbol{\xi}|^{\frac{1}{2}} = |g_{tt} + 2g_{t\phi}\Omega + g_{\phi\phi}\Omega^2|^{\frac{1}{2}}$$

so on the surface

$$g_{tt} + 2g_{t\phi}\Omega + g_{\phi\phi}\Omega^2 = \text{constant} .$$

Problem 16.20. If $\mathbf{p}$ is the 4-momentum of a photon and $\mathbf{u}_{em}$, $\mathbf{u}_\infty$ are the 4-velocities respectively of an emitting point on the star and of the observer at ∞, then the ratio of emitted and observed frequencies is

$$\frac{\nu_{em}}{\nu_\infty} = \frac{\mathbf{p} \cdot \mathbf{u}_{em}}{\mathbf{p} \cdot \mathbf{u}_\infty} .$$

But $\mathbf{u}_{em}$ has only t and ϕ components, so

$$\mathbf{p} \cdot \mathbf{u}_{em} = p_t u^t + p_\phi u^\phi = p_t u^t (1 + \Omega \ell)$$

where $\Omega \equiv u^\phi/u^t$ is the angular velocity of the star and $\ell \equiv p_\phi/p_t$ is the impact parameter of the photon relative to the axis of rotation. Note that p_ϕ and p_t are conserved along the photon's trajectory. Also, $p\cdot u_\infty = p_t$ since $u^t = 1$ for an observer at infinity and thus

$$z = u^t(1+\Omega\ell) - 1 = \frac{1+\Omega\ell}{|g_{tt} + 2\Omega g_{t\phi} + \Omega^2 g_{\phi\phi}|_{em}^{\frac{1}{2}}} - 1 .$$

Since $g_{tt} + 2\Omega g_{t\phi} + \Omega^2 g_{\phi\phi}$ is constant on the surface of a rigidly rotating star (see Problem 16.19), the variation in z across the surface of the star is

$$\Delta z = \frac{\Omega\Delta\ell}{|g_{tt} + 2\Omega g_{t\phi} + \Omega^2 g_{\phi\phi}|_{surface}^{\frac{1}{2}}} .$$

All photons reaching an observer on the axis must have $p_\phi = 0$, i.e. $\ell = 0$, i.e. $\Delta\ell = 0$. Thus $\Delta z = 0$ for such an observer.

Solution 16.21. The configuration of perfect fluid is convectively stable if its mass (i.e. rest mass + energy) does not change under a redistribution of baryons. We first calculate its change in mass, δM, due to an addition of δA baryons from infinity.

A distant astrophysicist drops a total mass-energy $\mu_B\delta A + W_0$ onto the fluid, where μ_B is the average rest mass of the baryons and W_0 is additional energy to be used for injecting the baryons into the fluid. A static observer at radius r catches the baryons and energy and measures a total mass-energy of

$$W = u^0 p_0 = e^{-\Phi}(\mu_B\delta A + W_0) \tag{1}$$

where $e^{-\Phi}$ is the redshift factor and comes from the metric

$$ds^2 = -e^{2\Phi}dt^2 + (1-2m/r)^{-1}dr^2 + r^2 d\Omega^2 . \tag{2}$$

With energy W, the local observer must heat and compress the baryons to local thermodynamic conditions and then open a space for them in the star

$$W_{\delta A\,(\text{local conditions})} = (\rho/n)\delta A \tag{3a}$$

$$W_{\text{open}} = p\delta V = (p/n)\delta A \tag{3b}$$

where n is baryon number density. (Note that, since we have assumed that the star is already at an extremum of total mass-energy, the change in energy of the star due to readjustment of its structure as mass is displaced for the δA baryons can be neglected.) The excess energy left over is

$$\begin{aligned} W_{ex}(r) &= W - (W_{\delta A} + W_{\text{open}}) \\ &= e^{-\Phi}(\mu_B \delta A + W_0) - \frac{(\rho+p)}{n}\delta A \,. \end{aligned} \tag{4}$$

The local observer then converts a fraction $(1-e^{\Phi})$ of W_{excess} into kinetic energy so that he can throw the rest to the observer at infinity, who then catches energy

$$W_{ex}(\infty) = W_{ex}(r)e^{\Phi} = \mu_B \delta A + W_0 - e^{\Phi}\frac{(\rho+p)}{n}\delta A \,. \tag{5}$$

The observer at infinity, therefore, measures a mass increase of the star, δM, of

$$\delta M = e^{\Phi}\frac{(\rho+p)}{n}\delta A = e^{\Phi}\left(\frac{\partial \rho}{\partial n}\right)_S \delta A \tag{6}$$

where the second relation follows from the first law of thermodynamics.

For convective stability we require δM to be independent of r, i.e.

$$e^{\Phi}(\rho+p)/n = \text{constant} \,. \tag{7}$$

We can easily show, using the Euler equation,

$$\frac{dp}{dr} = -(\rho+p)\frac{d\Phi}{dr} \tag{8}$$

that Equation (7) is equivalent to adiabatic gradients of thermodynamic variables. Taking the derivative of Equation (7) with respect to r gives

$$-\frac{(\rho+p)}{n}\frac{dn}{dr}+\frac{d\rho}{dr}+\frac{dp}{dr}+(\rho+p)\frac{d\Phi}{dr}=0\,. \tag{9}$$

Using Equation (8), we can put Equation (9) in the form

$$\frac{d\rho}{dr}=\frac{(\rho+p)}{n}\frac{dn}{dr}=\left(\frac{\partial\rho}{\partial n}\right)_s\frac{dn}{dr}\,. \tag{10}$$

From this equation, it is obvious that the fluid must be isentropic.

Solution 16.22. Begin with the Euler equation for a rigidly rotating star:

$$(\rho+p)^{-1}\,\partial p/\partial x^\mu=\partial\log u^0/\partial x^\mu \tag{1}$$

and compare it with the equation for constant injection energy

$$(\rho+p)/nu^0=\text{constant}\,. \tag{2}$$

Take the derivative of the logarithm of Equation (2) and substitute in from Equation (1), to obtain

$$(\rho+p)^{-1}\,\partial p/\partial x^\mu=n^{-1}\,\partial n/\partial x^\mu\,. \tag{3}$$

Equation (3) is the first law of thermodynamics specialized to isentropic systems.

Solution 16.23. Since $T^\mu{}_\nu$ vanishes outside the star, we can take the volume integral to be over all space at time t. The field equations imply

$$8\pi\left(T^\mu{}_\nu-\frac{1}{2}\delta^\mu{}_\nu T\right)=R^\mu{}_\nu$$

so the mass integral is

$$\begin{aligned}I&=-\int(2T^\mu{}_\nu-\delta^\mu{}_\nu T)\xi^\nu d^3\Sigma_\mu\\&=-(4\pi)^{-1}\int R^\mu{}_\nu\xi^\nu d^3\Sigma_\mu\\&=(4\pi)^{-1}\int\xi^{\mu;\nu}{}_{;\nu}d^3\Sigma_\mu\\&=(8\pi)^{-1}\oint\xi^{\mu;\nu}d^2\Sigma_{\mu\nu}\,.\end{aligned}$$

[The third equality follows from Problem 10.6; the last equality follows from Stokes' theorem, Problem 8.10(c).] If we choose coordinates (t,r,θ,ϕ) which reduce to spherical polar coordinates at infinity, then

$$d^2\Sigma_{\mu\nu} = \frac{1}{2!}\,\varepsilon_{\mu\nu\alpha\beta}\,\frac{\partial(x^\alpha, x^\beta)}{\partial(\theta,\phi)}\,d\theta\,d\phi = r^2\sin\theta\,d\theta\,d\phi$$

and thus

$$\xi^{\mu;\nu}d^2\Sigma_{\mu\nu} = 2\xi^{t;r}r^2\sin\theta\,d\theta\,d\phi\ .$$

The asymptotic form of the metric is

$$ds^2 = -\left(1-\frac{2M}{r}\right)dt^2 - \frac{4J\sin^2\theta}{r}\,dt\,d\phi + \left(1+\frac{2M}{r}\right)(dr^2 + r^2d\theta^2 + r^2\sin^2\theta\,d\phi^2)\,.$$

Since the only nonzero component of $\boldsymbol{\xi}_{(t)}$ is $\xi^t = 1$

$$\begin{aligned}
\xi^{t;r} &\approx \xi^t{}_{;r} \\
&= \xi^t{}_{,r} + \Gamma^t{}_{\alpha r}\xi^\alpha = 0 + \Gamma^t{}_{tr} \\
&\approx -\frac{1}{2}\,g_{tt,r} = M/r^2\ .
\end{aligned}$$

For the mass integral then

$$I = (8\pi)^{-1}\int \frac{2M}{r^2}\,r^2\sin\theta\,d\theta\,d\phi = M$$

as required. Similarly, we have

$$\begin{aligned}
S &= \int T^\mu{}_\nu\,\xi^\nu{}_{(\phi)}d^3\Sigma_\mu \\
&= (8\pi)^{-1}\int R^\mu{}_\nu\,\xi^\nu{}_{(\phi)}d^3\Sigma_\mu - (16\pi)^{-1}\int R\,\xi^\mu{}_{(\phi)}d^3\Sigma_\mu\,.
\end{aligned}$$

Since $d^3\Sigma_\phi = 0$ the second term vanishes and

$$\begin{aligned}
S &= -(8\pi)^{-1}\int \xi^{\mu;\nu}{}_{;\nu}d^3\Sigma_\mu - 0 = -(16\pi)^{-1}\oint \xi^{\mu;\nu}d^2\Sigma_{\mu\nu} \\
&= -(16\pi)^{-1}\oint 2\xi^{t;r}r^2\sin\theta\,d\theta\,d\phi\ .
\end{aligned}$$

But

$$\begin{aligned}\xi^{t;r} &\approx \xi^{t}{}_{,r} + \Gamma^{t}{}_{\alpha r}\xi^{\alpha} = 0 + \Gamma^{t}{}_{\phi r} \\ &= g^{tt}\Gamma_{t\phi r} + g^{t\phi}\Gamma_{\phi\phi r} \\ &\approx -\frac{1}{2}\, g_{t\phi,r} + \left(\frac{-2J}{r^3}\right)\frac{1}{2}\, g_{\phi\phi,r} \\ &= -3J\sin^2\theta/r^2 ,\end{aligned}$$

and thus

$$S = (16\pi)^{-1}\int 6J\sin^3\theta\, d\theta\, d\phi = J .$$

Solution 16.24. In curvature coordinates, $\xi^{\nu}{}_{(t)} = \delta^{\nu}{}_{t}$ and $d^3\Sigma_{\mu} = \delta^{t}{}_{\mu}|g|^{\frac{1}{2}}\, dr\, d\theta\, d\phi$, thus the integral is

$$\begin{aligned}I &= -\int (2T^{\mu}{}_{\nu} - \delta^{\mu}{}_{\nu}T)\xi^{\nu}{}_{(t)} d^3\Sigma_{\mu} \\ &= -\int (2T^{t}_{t} - T)\, e^{\Phi+\lambda}\, r^2 \sin\theta\, d\theta\, d\phi\, dr .\end{aligned}$$

But

$$\begin{aligned}T^{t}_{t} &= (\rho + p)\, u^{t}u_{t} + p\delta^{t}_{t} = -\rho \\ T &= -(\rho + p) + 4p = 3p - \rho .\end{aligned}$$

Since everything is independent of θ and ϕ, the integration over angles gives 4π and therefore

$$I = \int_0^R (\rho + 3p)\, e^{\Phi+\lambda} 4\pi r^2 dr . \tag{1}$$

We can transform this by integration by parts. Since

$$m(r) = \int_0^r 4\pi r^2 \rho\, dr \tag{2}$$

we have

$$\int_0^R \rho\, e^{\Phi+\lambda} 4\pi r^2 dr = [m\, e^{\Phi+\lambda}]_0^R - \int_0^R m(e^{\Phi+\lambda})'\, dr \tag{3}$$

where a prime denotes d/dr. The stellar structure equations tell us p' so we write

$$\int_0^R 3p\,e^{\Phi+\lambda}4\pi r^2 dr = [p\,e^{\Phi+\lambda}4\pi r^3]_0^R - \int_0^R 4\pi r^3 (p\,e^{\Phi+\lambda})'\,dr\,. \tag{4}$$

The first term on the right in Equation (3) is $m(R) = M$, since $m = 0$ at $r = 0$ and $\Phi = -\lambda$ at $r = R$. The first term on the right in Equation (4) is zero since $p = 0$ at $r = R$. Thus, $I = M$ provided the sum of the remaining terms in Equations (3) and (4) vanish. This sum is

$$\int_0^R m(e^{\Phi+\lambda})'\,dr + \int_0^R 4\pi r^3 (p\,e^{\Phi+\lambda})'\,dr$$

$$= \int_0^R dr\,e^{\Phi+\lambda}[m(\Phi'+\lambda') + 4\pi r^3\{p' + p(\Phi'+\lambda')\}]\,. \tag{5}$$

Now

$$e^{2\lambda} = (1-2m/r)^{-1}$$

so we get

$$\lambda' = \frac{(4\pi r^3\rho - m)}{r(r-2m)}\,.$$

If this result and the relations

$$\Phi' = \frac{m + 4\pi r^3 p}{r(r-2m)}$$

$$p' = \frac{-(\rho+p)(m+4\pi r^3 p)}{r(r-2m)}$$

are substituted in the expression in Equation (5), all terms cancel and it reduces to zero as required.

Solution 16.25. Let θ, ϕ and the radial coordinate R be comoving with the fluid of the star. We can then write the metric as

$$\begin{aligned} ds^2 = {} & g_{tt}dt^2 + g_{tR}dR\,dt + g_{RR}dR^2 + g_{ti}\,dt\,dx^i \\ & + g_{Ri}\,dR\,dx^i + g_{ij}\,dx^i dx^j \quad (i,j = \theta,\phi) \end{aligned}$$

where $i, j = \theta, \phi$. If the metric is spherically symmetric, then for $dt = dR = 0$ we must have

$$ds^2 = g_{ij}\, dx^i dx^j = r^2(R,t)(d\theta^2 + \sin^2\theta\, d\phi^2) .$$

Furthermore, g_{ti} and g_{Ri} define vectors in the 2-dimensional θ, ϕ space, so by the isotropy of spherical shells g_{ti} and g_{Ri} must vanish. Finally, we can choose a new time coordinate $T = T(t, r)$ to remove the g_{tr} term, without affecting the comoving status of the radial coordinate, and we have

$$ds^2 = g_{TT} dT^2 + g_{RR} dR^2 + r^2(R,T)\, d\Omega^2 .$$

Thus in general we can have both comoving coordinates and a diagonal metric.

Now consider the motion of a fluid element. From the fluid 4-velocity

$$\mathbf{u} = (-g_{TT})^{-\frac{1}{2}}\, \mathbf{e}_T$$

we have the fluid acceleration:

$$\nabla_{\mathbf{u}}\mathbf{u} = \mathbf{e}_R \left[\frac{du^R}{d\tau} + (u^T)^2 \Gamma^R{}_{TT}\right] = -\frac{1}{2}\, \mathbf{e}_R (u^T)^2 g^{RR} g_{TT,R} \cdot$$

It follows then that the fluid acceleration is zero (and hence the pressure gradient vanishes) if and only g_{TT} is a function of T only. But if that is so, we can define a new time coordinate $\tau(T)$ from the equation

$$d\tau/dT = (-g_{TT})^{\frac{1}{2}}$$

representing proper time for the fluid. Thus the three nice properties are all satisfied if and only if the pressure gradient is zero; since the pressure must be zero on the surface of the star, a vanishing pressure gradient is equivalent to zero pressure.

Solution 16.26.

(a) The first law of thermodynamics (see Problem 5.19) in the comoving frame is

$$\frac{\dot{\rho}}{\rho + p} = \frac{\dot{n}}{n}$$

where n is baryon density. The number of baryons in a shell of thickness dR is $(4\pi n r^2 e^{\Lambda})dR$, so the law of conservation of baryons in this shell

$$\partial(nr^2 e^{\Lambda})/\partial t = 0$$

and thus

$$\frac{\dot{\rho}}{\rho + p} = -\frac{2\dot{r}}{r} - \dot{\Lambda} .$$

We can now combine this with

$$G^{\hat{t}}{}_{\hat{R}} = 0 = \frac{2}{r} e^{-\Phi-\lambda}(\dot{r}' - \dot{r}\Phi' - r'\dot{\Lambda})$$

(see Problem 9.20) and with

$$T^{\beta}{}_{R;\beta} = 0 = p' + (p+\rho)\Phi'$$

to arrive at

$$-\dot{\rho} = (\rho + p)\left(\frac{2\dot{r}}{r} + \frac{\dot{r}'}{r'}\right) + \frac{\dot{r}p'}{r'} .$$

This relation can now be substituted in the expression for $\dot{m}$,

$$\dot{m} = 4\pi \int_0^R (2r\dot{r}r'\rho + r^2\dot{r}'\rho + r^2 r'\dot{\rho})dR$$

$$= -4\pi \int_0^R (2rr'\dot{r}p + r^2\dot{r}'p + r^2\dot{r}p')dR$$

$$= -4\pi \int_0^R (r^2\dot{r}p)'dR .$$

Since $r^2\dot{r}p = 0$ at $R = 0$ it follows that

$$\dot{m} = -4\pi r^2 p\dot{r} .$$

(b) From $G^{\hat{t}}_{\hat{t}} = -8\pi\rho$ (see Problem 9.20) we have

$$8\pi\rho r^2 = 1 - r'^2 e^{-2\Lambda} + \dot{r}^2 e^{-2\Phi} + 2r\dot{r}\, e^{-2\Phi}\dot{\Lambda} + 2r\, e^{-2\Lambda}(r'\Lambda' - r'') .$$

Now use $G^{\hat{t}}_{\hat{R}} = 0$ to eliminate $\dot{\Lambda}$, and note that $8\pi r^2 \rho r'$ is an exact differential with respect to R

$$\begin{aligned} 8\pi\rho r^2 r' &= r' - (r')^3 e^{-2\Lambda} + \dot{r}^2 r' e^{-2\Phi} + 2r\dot{r}\, e^{-2\Phi}(\dot{r}' - \dot{r}\Phi') \\ &\quad + 2rr' e^{-2\Lambda}(r'\Lambda' - r'') \\ &= [r - e^{-2\Lambda} r(r')^2 + r(\dot{r})^2 e^{-2\Phi}]' . \end{aligned}$$

Integration gives the desired relation

$$\begin{aligned} 2m &= r(1 - e^{-2\Lambda}(r')^2 + \dot{r}^2 e^{-2\Phi}) \\ &= r(1 - \Gamma^2 + U^2) . \end{aligned}$$

Solution 16.27. From Problem 16.26 we have

$$dr/d\tau = U \tag{1}$$

$$dm/d\tau = -4\pi r^2 p U \tag{2}$$

$$U = \frac{dr}{d\tau} = \pm\left[\Gamma^2 + \left(\frac{2m}{r} - 1\right)\right]^{\frac{1}{2}} . \tag{3}$$

The first two equations show that $dr/d\tau$ and $dm/d\tau$ have opposite signs, so for the shell to pass from $2m/r < 1$ to $2m/r > 1$ it must be that

$$\dot{r} \leq 0 , \quad \dot{m} \geq 0$$

upon passage through $r = 2m$. From Equation (3) then

$$\frac{dr}{d\tau} = -\left[\Gamma^2 + \left(\frac{2m}{r} - 1\right)\right]^{\frac{1}{2}} .$$

The radius therefore will continue to fall as long as $\Gamma^2 + \left(\frac{2m}{r} - 1\right) > 0$, but as long as the radius is decreasing, the mass must be increasing and the factor $\frac{2m}{r} - 1$ continues to increase, so the collapse cannot stop.

Suppose the shell has reached the point $2m/r - 1 = \varepsilon > 0$. Since this factor continues to increase, then

$$dr/d\tau \leq \varepsilon^{\frac{1}{2}}$$

and the shell will reach $r = 0$ in a proper time $\leq 2m/\varepsilon^{\frac{1}{2}}$. [This solution is due to J. M. Bardeen.]

Solution 16.28. In Problem 16.25 we saw that we can choose $\Phi = 0$ for pressure-free collapse. From $G^R{}_R = 0$ (see Problem 9.20) with $\Phi = 0$ we have then

$$\begin{aligned} 0 &= 2r\ddot{r} + 1 - r'^2 e^{-2\Lambda} + \dot{r}^2 \\ &= 2r\ddot{r} + 1 - \Gamma^2 + U^2 \end{aligned}$$

and therefore (see Problem 16.26)

$$\ddot{r} = \frac{d^2 r}{d\tau^2} = -\frac{M}{r^2} .$$

Solution 16.29. From the dynamical equation for mass (see Problem 16.26)

$$\dot{m} = -4\pi p r^2 \dot{r} = 0$$

mass is clearly time independent, as is physically obvious. By differentiating the Γ^2 equation in Problem 16.26 we get

$$2\Gamma\dot{\Gamma} = 2U\dot{U} + \frac{2m}{r^2}\dot{r} = 2\dot{r}\left(\ddot{r} + \frac{m}{r^2}\right)$$

and using the result of Problem 16.28 we see that Γ, too, is time independent.

The dynamical equation for $r(R, \tau)$ is now a simple first order equation and can be solved parametrically:

<u>Case (i)</u> $\Gamma^2 - 1 < 0$:

$$r = \frac{m(b)}{1 - \Gamma^2}(1 + \cos\eta) \tag{1a}$$

$$\tau = \frac{m(b)}{(1-\Gamma^2)^{\frac{3}{2}}}\,(\eta + \sin\eta) + F(b) \tag{1b}$$

where F(b) is an arbitrary function.

Case (ii) $\Gamma^2 - 1 = 0$:

$$r = \left\{\frac{3}{2}\,(m(b))^{\frac{1}{2}}\,[g(b) - \tau]\right\}^{\frac{2}{3}} \tag{2}$$

where g(b) is an arbitrary function.

Case (iii) $\Gamma^2 - 1 > 0$:

$$r = \frac{m(b)}{\Gamma^2 - 1}\,(\cosh\eta - 1) \tag{3a}$$

$$\tau = \frac{m(b)}{(\Gamma^2 - 1)^{\frac{3}{2}}}\,(\sinh\eta - \eta) + H(b) \tag{3b}$$

where H(b) is an arbitrary function.

Note that there are three free functions – i.e. functions that we can choose by choosing the physical situation. By choosing m(b) we can choose the distribution of mass at some initial time and forever after. The integration "constants" F(b), g(b), H(b) correspond to the choice of the r value of each fluid element on an initial hypersurface, i.e. the choice of $r(b, t=0)$.

The choice of $\Gamma^2(b)$ corresponds to the choice of the velocity of a fluid element on an initial hypersurface. By comparison with the equation for a radial geodesic, $\Gamma^2 - 1$ can be thought of as the conserved "energy at ∞" for the fluid shell. Therefore, it should not be surprising that there are three solution regimes: we can choose to give a fluid shell less than the escape velocity (a), precisely the escape velocity (b), or greater than the escape velocity (c). Note that the sign of t can be reversed in the above solutions so that, for example, solution (c) can correspond to shells falling inward. Note that in general if the three free functions are not chosen carefully, mass shells are going to cross!

Solution 16.30.

(i) The stress tensor is $T_{\alpha\beta} = \rho u_\alpha u_\beta$. The assumptions of spherical symmetry and uniform density are equivalent to assuming isotropy and homogeneity, so the metric must be a Friedmann solution:

$$ds^2 = -d\tau^2 + a^2(\tau)[d\chi^2 + \Sigma^2(\chi)(d\theta^2 + \sin^2\theta\, d\phi^2)] \tag{1}$$

where

$$\Sigma = \begin{cases} \sin\chi\ , & k = 1 \\ \chi\ , & k = 0 \\ \sinh\chi\ , & k = -1\ . \end{cases}$$

The surface of the star is at some constant value of the "radial" coordinate, $\chi = \chi_0$ say. From Equation (1), the proper circumferential radius of the star is $R = a(\tau)\Sigma(\chi_0)$. The Einstein field equations (see Problems 19.16, 19.18) for zero pressure are

$$\left(\frac{a_{,\tau}}{a}\right)^2 = -\frac{k}{a^2} + \frac{8\pi\rho}{3} \tag{2}$$

$$\rho a^3 = \text{constant}\ . \tag{3}$$

Equation (2) can be rewritten as

$$\rho = \frac{3}{8\pi}\left[\left(\frac{R_{,\tau}}{R}\right)^2 + \frac{k\Sigma^2(\chi_0)}{R^2}\right]. \tag{4}$$

Thus, if $R_{,\tau} = 0$ at any finite value of R, k must be $+1$ so that ρ is positive. Conversely, if $k = -1$, $R_{,\tau}$ is never zero. The case $k = 0$ corresponds to $R_{,\tau} \to 0$ as $R \to \infty$.

(ii) The Euler equation for zero pressure implies $\nabla_{\mathbf{u}}\mathbf{u} = 0$, i.e. each fluid element moves along a geodesic. By spherical symmetry the geodesics are radial.

(iii) We shall do the matching for $k = +1$; the cases $k = 0$ or $k = -1$ are similar. For $k = +1$, Equations (2) and (3) give

$$(a_{,\tau})^2 = a_m/a - 1 \tag{5}$$

where the constant in Equation (3) has been fixed by setting $a = a_m$ = maximum of a, when $a_{,\tau} = 0$. Equation (5) can be integrated as it stands; it is convenient, however, to introduce a new time parameter by

$$d\tau = a d\eta . \tag{6}$$

Equations (5) and (6) then give

$$a = \frac{1}{2} a_m(1+\cos\eta) \tag{7}$$

$$\tau = \frac{1}{2} a_m(\eta+\sin\eta) . \tag{8}$$

The constants of integration have been chosen to make $a = a_m$ and $\tau = 0$ at $\eta = 0$, and to make $a = 0$ and $\tau = \pi a_m/2$ at $\eta = \pi$.

The intrinsic 3-geometry of the surface of the star, as measured from the interior, is found by putting $\chi = \chi_0$ in Equation (1):

$$\begin{aligned} {}^{(3)}ds^2 &= -d\tau^2 + a^2(\tau)\sin^2\chi_0\, d\Omega^2 \\ &= a^2(\eta)(-d\eta^2 + \sin^2\chi_0\, d\Omega^2) . \end{aligned} \tag{9}$$

The exterior metric is

$$ds^2 = -(1-2M/r)dt^2 + (1-2M/r)^{-1}dr^2 + r^2\, d\Omega^2 . \tag{10}$$

The surface of the star is at $r = R(\tau)$, and from the equations for radial geodesics (see Problem 15.4), $R(\tau)$ is given by

$$R = \frac{1}{2} R_i(1+\cos\eta) \tag{11}$$

$$\tau = \left(\frac{R_i^3}{8M}\right)^{\frac{1}{2}} (\eta + \sin\eta) \tag{12}$$

$$u^t = \frac{dt}{d\tau} = \frac{(1-2M/R_i)^{\frac{1}{2}}}{1-2M/R} . \tag{13}$$

Thus the geometry of the surface is

$$\begin{aligned} {}^{(3)}ds_+^2 &= -d\tau^2 + R^2(\tau)\,d\Omega^2 \\ &= -\left(\frac{R_i^3}{8M}\right)(1+\cos\eta)^2\,d\eta^2 + \frac{R_i^2}{4}(1+\cos\eta)^2\,d\Omega^2 . \end{aligned} \tag{14}$$

Comparison with Equation (9) shows that the 3-geometries match provided we identify

$$R_i = a_m \sin\chi_0 , \quad 2M = a_m \sin^3\chi_0 . \tag{15}$$

Now compute the extrinsic curvature $K_{ij}^{(-)}$ in the interior. The normal to the surface is

$$\mathbf{n} = a^{-1}\,\partial/\partial\chi \tag{16}$$

(recall that $\mathbf{n}\cdot\mathbf{n} = 1$), while the vectors $\mathbf{u} = \partial/\partial\tau$, $\partial/\partial\theta$ and $\partial/\partial\phi$ lie in the surface. Let the indices i, j range over τ, θ, ϕ. Then

$$\begin{aligned} K_{ij} &\equiv -\mathbf{e}_i\cdot\nabla_j\mathbf{n} = -\mathbf{e}_i\cdot\Gamma^\alpha{}_{nj}\,\mathbf{e}_\alpha = -g_{i\alpha}\Gamma^\alpha{}_{nj} \\ &= -\Gamma_{inj} = -\frac{1}{2}(g_{in,j} + g_{ij,n} - g_{nj,i}) = -\frac{1}{2}g_{ij,n} \end{aligned} \tag{17}$$

since $g_{in} = a^{-1}g_{i\chi} = 0$. From the metric we see that

$$K_{\tau\tau}^{(-)} = K_{\tau\theta}^{(-)} = K_{\tau\phi}^{(-)} = K_{\theta\phi}^{(-)} = 0 \tag{18}$$

$$K_{\theta\theta}^{(-)} = K_{\phi\phi}^{(-)}\sin^{-2}\theta = -\frac{1}{2}a_m(1+\cos\eta)\sin\chi_0\cos\chi_0 . \tag{19}$$

In the exterior metric, the 4-velocity is

$$\mathbf{u} = u^t \mathbf{e}_t + u^r \mathbf{e}_r \ .$$

The normal vector

$$\mathbf{n} = n^t \mathbf{e}_t + n^r \mathbf{e}_r$$

satisfies

$$\mathbf{n}\cdot\mathbf{n} = 1 = g_{tt}(n^t)^2 + g_{rr}(n^r)^2 \tag{20}$$

$$\mathbf{n}\cdot\mathbf{u} = 0 = n^t u_t + n^r u_r \ . \tag{21}$$

Since

$$\mathbf{u}\cdot\mathbf{u} = -1 = g^{tt}(u_t)^2 + g^{rr}(u_r)^2 \tag{22}$$

and

$$g^{rr} = (g_{rr})^{-1} = -(g^{tt})^{-1} = -g_{tt} = 1 - 2M/r$$

the equations above imply that

$$n^t = u_r \, , \qquad n^r = -u_t \ . \tag{23}$$

Let i, j range over τ, θ, ϕ as before. Then Equation (17) holds for the exterior metric, since $g_{in} = \mathbf{n}\cdot\mathbf{e}_i = 0$. Equation (18) holds for the exterior metric because $\mathbf{e}_\tau\cdot\mathbf{e}_\tau = \mathbf{u}\cdot\mathbf{u} = -1$, $\mathbf{e}_\tau\cdot\mathbf{e}_\theta = \mathbf{e}_\tau\cdot\mathbf{e}_\phi = \mathbf{e}_\theta\cdot\mathbf{e}_\phi = 0$. Also,

$$\begin{aligned} K^{(+)}_{\theta\theta} &= K^{(+)}_{\phi\phi} \sin^{-2}\theta = -\tfrac{1}{2}\,(r^2)_{,n} = -\tfrac{1}{2}\,(r^2)_{,r} n^r \\ &= r u_t = -R\,(1-2M/R_i)^{\frac{1}{2}} \\ &= -\frac{1}{2} R_i\ (1+\cos\eta)(1-2M/R_i)^{\frac{1}{2}} \\ &= -\frac{1}{2} a_m \sin\chi_0 \cos\chi_0\,(1+\cos\eta) \end{aligned}$$

where we have used Equations (23), (13), (11) and (15). Thus $K^{(+)}_{ij} = K^{(-)}_{ij}$ and the proof is complete.

CHAPTER 17: SOLUTIONS

Solution 17.1. One finds the mass and angular momentum of an asymptotically Minkowskian (flat) metric by finding a coordinate system in which it takes the form

$$g_{00} = -\left(1-\frac{2\tilde{M}}{r}+\mathcal{O}(r^{-2})\right) \tag{1}$$

$$g_{0j} = -\left(4\varepsilon_{jkl}\,\tilde{S}^k\frac{x^l}{r^3}+\mathcal{O}(r^{-3})\right) . \tag{2}$$

The constants $\tilde{M}$ and $\tilde{S}^k$ which occur in the expansion are then the mass and intrinsic angular momentum (see Weinberg, Section 9.4 or MTW, Section 19.3).

Expanding the Kerr metric in Boyer-Lindquist coordinates in powers of r^{-1} gives the leading terms

$$ds^2 = -\left(1-\frac{2M}{r}+\cdots\right)dt^2 - \left(\frac{4aM}{r}\sin^2\theta+\cdots\right)dt\,d\phi$$
$$+ (1+\cdots)[dr^2+r^2(d\theta^2+\sin^2\theta\,d\phi^2)] .$$

Transforming to cartesian coordinates by $x \equiv r\sin\theta\cos\phi$, $y \equiv r\sin\theta\sin\phi$, $z \equiv r\cos\theta$ gives

$$ds^2 = -\left(1-\frac{2M}{r}+\cdots\right)dt^2 - \left(\frac{4aM}{r^3}+\cdots\right)(x\,dy-y\,dx)dt$$
$$+ (1+\cdots)(dx^2+dy^2+dz^2) \tag{3}$$

and comparison with Equations (1) and (2) gives at once

$$\tilde{M} = M$$

$$\tilde{S} = aM\,e_{\hat{z}} .$$

Solution 17.2. Suppose the car has mass m and length L before crushing and length L′ and "lumpiness" h afterward, and the hole has mass M. Suppose the internal stress per mass that steel can support is $\epsilon(\sim 0.1$ electron volt per nuclear mass; see Problem 5.6). Comparing gravitational pressure forces to internal stresses, we see that the conditions for crushing are

$$\left(\frac{GM}{L^2}\right)L > \epsilon \qquad \text{(condition for crushing to begin)} \tag{1}$$

$$\left(\frac{GM}{L'^2}\right)h > \epsilon \qquad \text{(condition for lumps no larger than h after crushing ends).} \tag{2}$$

Since cars typically crush to .1 of their original dimension, we see that the two inequalities are equivalent for $h/L' = .1$, which is a reasonable value. Taking $L' \sim 100\,\text{cm}$ and $\epsilon \sim 10^9\,\text{ergs/gm}$ gives $M \gtrsim 10^{18}\,\text{gm}$.

The time to crush a wreck is set by the free-fall time $(L^3/GM)^{\frac{1}{2}} \sim 10^{-5}\,\text{sec}$, so on the order of 10^8 can be processed per hour.

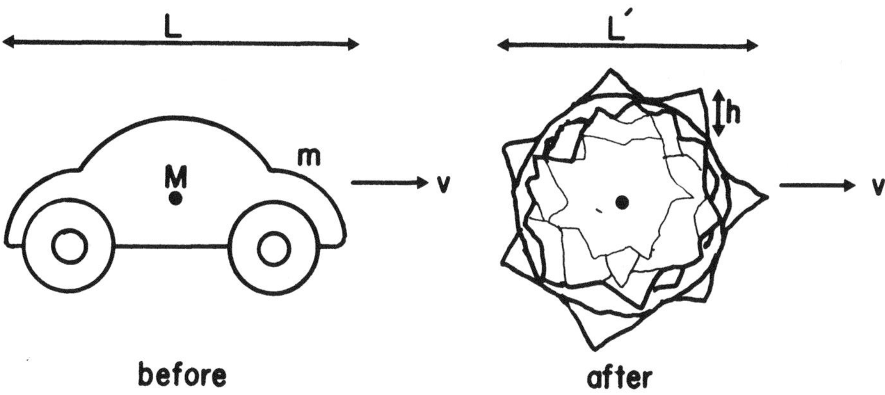

Solution 17.3. For the rocket ship to be moving along a timelike world-line, its 4-velocity must satisfy (from the Schwarzschild metric)

$$1 = -\mathbf{u}\cdot\mathbf{u} = \left(1-\frac{2M}{r}\right)\left(\frac{dt}{d\tau}\right)^2 - \left(1-\frac{2M}{r}\right)^{-1}\left(\frac{dr}{d\tau}\right)^2 - r^2\left(\frac{d\theta}{d\tau}\right)^2 - r^2\sin^2\theta\left(\frac{d\phi}{d\tau}\right)^2 .$$

Inside the horizon all of these terms are negative, except the one in $(dr/d\tau)^2$, so

$$\left(\frac{2M}{r}-1\right)^{-1}\left(\frac{dr}{d\tau}\right)^2 \geq 1 .$$

We also know (e.g. from the Eddington-Finkelstein or Kruskal picture) that the sign of $dr/d\tau$ must be negative for a "future directed" (i.e. physical) observer and thus,

$$dr < -\left(\frac{2M}{r}-1\right)^{\frac{1}{2}} d\tau$$

and

$$\tau_{max} = -\int_{2M}^{0}\left(\frac{2M}{r}-1\right)^{-\frac{1}{2}} dr$$

$$= \left[r^{\frac{1}{2}}(2M-r)^{\frac{1}{2}} + M\cos^{-1}\left(\frac{r}{M}-1\right)\right]_{2M}^{0} = \pi M .$$

Solution 17.4. We derive the case for a Kerr black hole and then obtain the Schwarzschild case by setting $a = 0$. The Kerr metric is (in the usual Boyer-Lindquist coordinates)

$$ds^2 = -\left(1-\frac{2Mr}{\Sigma}\right)dt^2 - \left(4Mar\,\frac{\sin^2\theta}{\Sigma}\right)dt\,d\phi + \frac{\Sigma}{\Delta}\,dr^2 + \Sigma\,d\theta^2 + \left(r^2+a^2+2Ma^2\,r\,\frac{\sin^2\theta}{\Sigma}\right)\sin^2\theta\,d\phi^2 \qquad (1)$$

where

$$\Delta \equiv r^2 - 2Mr + a^2 , \qquad \Sigma \equiv r^2 + a^2\cos^2\theta .$$

The geodesic equation is

$$d^2x^\mu/d\tau^2 + \Gamma^\mu{}_{\alpha\beta}u^\alpha u^\beta = 0 \qquad (2)$$

but the first term vanishes since u^t and u^ϕ, the nonvanishing components of $\mathbf{u}$, are both constant. The r component of Equation (2) is

$$0 = \Gamma_{r\alpha\beta}u^\alpha u^\beta = (\Gamma_{rtt}dt^2 + 2\Gamma_{rt\phi}dt\,d\phi + \Gamma_{r\phi\phi}d\phi^2)(d\tau)^{-2} . \qquad (3)$$

Since the metric coefficients depend only on r and θ, the relevant Γ's in the equatorial $(\theta=\pi/2)$ plane are

$$\begin{aligned}
\Gamma_{rtt} &= -\frac{1}{2}\, g_{tt,r} &&= M/r^2 \\
2\Gamma_{rt\phi} &= -g_{t\phi,r} &&= -2Ma/r^2 \\
\Gamma_{r\phi\phi} &= -\frac{1}{2}\, g_{\phi\phi,r} &&= \frac{Ma^2}{r^2} - r\ .
\end{aligned} \tag{4}$$

So if $\Omega \equiv d\phi/dt$, Equations (3) and (4) give

$$0 = \Omega^2\left(\frac{Ma^2}{r^2} - r\right) - \frac{2Ma}{r^2}\,\Omega + \frac{M}{r^2}\ , \tag{5}$$

a quadratic whose two roots (corresponding to direct and retrograde orbits) are

$$\Omega = M^{\frac{1}{2}}/(\pm r^{\frac{3}{2}} + aM^{\frac{1}{2}})\ . \tag{6}$$

For $a = 0$ this becomes $\Omega^2 = M/r^3$ which (by coincidence in these coordinates) is exactly the Newtonian Kepler's law.

Solution 17.5. The Reissner-Nordström metric is

$$ds^2 = -A dt^2 + A^{-1} dr^2 + r^2(d\theta^2 + \sin^2\theta\, d\phi^2)$$

where $A \equiv 1 - 2M/r + Q^2/r^2$. We first find the Keplerian frequency $\Omega \equiv d\phi/dt$ of circular orbits, in the same manner as in Solution 17.4. Here

$$\Gamma_{rtt} = -\frac{1}{2}\, g_{tt,r} = \frac{M}{r^2} - \frac{Q^2}{r^3}$$

$$2\Gamma_{rt\phi} = -g_{t\phi,r} = 0$$

$$\Gamma_{r\phi\phi} = -\frac{1}{2}\, g_{\phi\phi,r} = -r$$

so the quadratic equation for Ω becomes

$$\Omega^2 r - \left(\frac{M}{r^2} - \frac{Q^2}{r^3}\right) = 0$$

and the orbital frequency is

$$\Omega = \pm\left(\frac{M}{r^3} - \frac{Q^2}{r^4}\right)^{\frac{1}{2}} .$$

The proper velocity of the orbiting observer relative to a coordinate stationary observer is

$$\hat{v} = \frac{d\hat{\phi}}{d\hat{t}} = \frac{r\, d\phi}{A^{\frac{1}{2}} dt} = \left(\frac{Mr - Q^2}{r^2 - 2Mr + Q^2}\right)^{\frac{1}{2}} .$$

(Carets indicate orthonormal components in the stationary frame.) In the stationary frame the local electromagnetic field has only the component

$$E_{\hat{r}} = \frac{Q}{r^2} .$$

Lorentz transforming this with the standard relations for E-M fields gives, in the orbiting frame, all components zero except

$$E_{\hat{r}} = (1-\hat{v}^2)^{-\frac{1}{2}} \frac{Q}{r^2} = \frac{Q}{r^2}\left(\frac{r^2 - 2Mr + Q^2}{r^2 - 3Mr + 2Q^2}\right)^{\frac{1}{2}}$$

$$B_{\hat{\theta}} = (1-\hat{v}^2)^{-\frac{1}{2}}\, \hat{v} \frac{Q}{r^2} = \frac{Q}{r^2}\left(\frac{Mr - Q^2}{r^2 - 3Mr + 2Q^2}\right)^{\frac{1}{2}} .$$

Solution 17.6. The specific angular momentum a and charge e of a Kerr-Newman black hole of mass m cannot be arbitrarily large, but must satisfy the inequality

$$\frac{a^2}{c^2} + \frac{Ge^2}{c^4} \leq \frac{G^2}{c^4} M^2 . \tag{1}$$

(Here the G's and c's have been put in for clarity.) Equation (1) follows from the requirement that there exist a horizon, located at

$$r_H = M + (M^2 - e^2 - a^2)^{\frac{1}{2}} .$$

If Equation (1) is violated, one has a "naked singularity," which also turns out to be acausal. Measurements of the electron indicate that its spin, charge and mass give the result that the a^2 term is of order $10^{-22}\,\text{cm}^2$, the e^2 term of order $10^{-68}\,\text{cm}^2$ and the m^2 term of order $10^{-110}\,\text{cm}^2$. Thus, the inequality is rather violently violated and the electron is not a black hole.

Solution 17.7. Since we have three relations for the components of the 4-momentum ($p\cdot p = -m^2$, $p_t = E$, $p_\phi = L$) the orbital equations can be reduced to the form

$$(dr/d\lambda)^2 = V(E, L, r)$$

where λ is an affine parameter and V is some effective potential. The condition for a circular orbit is that $dr/d\lambda$ remain zero, i.e.

$$V(E, L, r) = 0 \tag{1}$$

$$V'(E, L, r) = 0 \tag{2}$$

where $V' \equiv \partial V/\partial r$. By the implicit function theorem, Equations (1) and (2) can be solved for

$$E = E(r), \qquad L = L(r)$$

provided

$$\begin{vmatrix} \dfrac{\partial V}{\partial E} & \dfrac{\partial V}{\partial L} \\ \dfrac{\partial V'}{\partial E} & \dfrac{\partial V'}{\partial L} \end{vmatrix} \neq 0 . \tag{3}$$

(This condition is in fact satisfied.) Then dE/dr and dL/dr can be found by differentiating Equations (1) and (2):

$$0 = \frac{dV}{dr} = \frac{\partial V}{\partial E}\frac{dE}{dr} + \frac{\partial V}{\partial L}\frac{dL}{dr} + V' \tag{4}$$

$$0 = \frac{dV'}{dr} = \frac{\partial V'}{\partial E}\frac{dE}{dr} + \frac{\partial V'}{\partial L}\frac{\partial L}{\partial r} + V'' . \tag{5}$$

Now consider an orbit at $r = r_0$ perturbed to $r = r_0 + \epsilon$. The perturbed orbit equation is

$$(dr/d\lambda)^2 = V(r_0) + \epsilon V'(r_0) + \frac{1}{2}\epsilon^2 V''(r_0) + \cdots$$

where $V(r_0) = 0$ and $V'(r_0) = 0$ (condition for unperturbed circular orbit). The V'' term then governs stability and V'' must be negative if the orbit is stable. For a *marginally* stable orbit we have $V'' = 0$ in addition to $V = 0$ and $V' = 0$, so the only solution to Equations (4) and (5) is $dE/dr = 0$ and $dL/dr = 0$. It is physically clear that these extrema are minima.

Solution 17.8.

(a) From $\mathbf{u}\cdot\mathbf{u} = -1$ and $\Omega \equiv u^\phi/u^0$ we have

$$\begin{aligned} -1 &= g_{00}(u^0)^2 + 2g_{0\phi}u^0u^\phi + g_{\phi\phi}(u^\phi)^2 \\ &= (u^0)^2(g_{00} + 2\Omega\, g_{0\phi} + \Omega^2 g_{\phi\phi}) . \end{aligned} \tag{1}$$

and hence

$$u^0 = (-g_{00} - 2\Omega\, g_{0\phi} - \Omega^2 g_{\phi\phi})^{-\frac{1}{2}}$$

where, in the $\theta = \pi/2$ plane,

$$-g_{00} = 1 - 2M/r , \quad -g_{0\phi} = 2Ma/r , \quad -g_{\phi\phi} = -(r^3 + a^2 r + 2Ma^2)/r .$$

The other components are easily found in terms of u^0:

$$u^\phi = \Omega u^0$$

$$u_0 = g_{00}\,u^0 + g_{0\phi}\,u^\phi = u^0(g_{00} + \Omega\, g_{0\phi})$$

$$u_\phi = g_{0\phi}u^0 + g_{\phi\phi}u^\phi = u^0(g_{0\phi} + \Omega\, g_{\phi\phi}) .$$

(b) From Equation (1) we have

$$Y \equiv g_{00} + 2\Omega\, g_{0\phi} + \Omega^2 g_{\phi\phi} < 0 . \tag{2}$$

The observer can be coordinate stationary ($\Omega = 0$) only where $g_{00} < 0$, i.e. only for $r > r_0 = 2M$ in the equatorial plane.

(c) The discriminant of Y defined in Equation (2) is

$$(g_{0\phi})^2 - g_{00}g_{\phi\phi} = r^2 - 2Mr + a^2 = (r - r_+)(r - r_-) ,$$

and is negative if $r_- < r < r_+$. If the discriminant of Y is negative, then Y has the same sign for any Ω, i.e. the same sign as $g_{\phi\phi}$ which is positive. This violates Equation (2), which assumed only timelike motion and constant r. The observer therefore cannot remain at constant r.

Note that as $r \to r_+$, then Ω goes to $-g_{0\phi}/g_{00} = a/2Mr_+$.

Solution 17.9. The conserved energy of a particle is the dot product of its 4-momentum with the time Killing vector $\boldsymbol{\xi}_{(t)} = \partial/\partial t$:

$$E = -p \cdot \left(\frac{\partial}{\partial t}\right) = -p_t \, . \tag{1}$$

The Kerr metric is of the form

$$ds^2 = -e^{2\nu}dt^2 + e^{2\psi}(d\phi - \omega dt)^2 + e^{2\mu_1}dr^2 + e^{2\mu_2}d\theta^2 \, . \tag{2}$$

The contravariant components of $g^{\alpha\beta}$ are

$$\begin{gathered} g^{tt} = -e^{-2\nu} \qquad g^{t\phi} = -\omega e^{-2\nu} \qquad g^{\phi\phi} = e^{-2\psi} - \omega^2 e^{-2\nu} \\ g^{rr} = e^{-2\mu_1} \qquad g^{\theta\theta} = e^{-2\mu_2} \, . \end{gathered} \tag{3}$$

If μ is the mass of the particle, then

$$\begin{aligned} -\mu^2 = p\cdot p = & -e^{-2\nu}p_t^2 - e^{-2\nu}\, 2\omega\, p_t\, p_\phi \\ & + (e^{-2\psi} - e^{-2\nu}\omega^2)\, p_\phi^2 + e^{-2\mu_1}p_r^2 + e^{-2\mu_2}p_\theta^2 \, . \end{aligned} \tag{4}$$

Solving the quadratic equation (4) for $E = p_t$, we get

$$E = \omega p_\phi + [e^{2\nu - 2\psi}p_\phi^2 + e^{2\nu}(e^{-2\mu_1}p_r^2 + e^{-2\mu_2}p_\theta^2 + \mu^2)]^{\frac{1}{2}} \, . \tag{5}$$

The sign of the square root in Equation (5) is positive so that $E = +\mu$ for a particle at rest at infinity. If E is to be negative, we must have p_ϕ negative and

$$[e^{2\nu-2\psi}p_\phi^{\,2} + e^{2\nu}(e^{-2\mu_1}p_r^{\,2} + e^{-2\mu_2}p_\theta^{\,2} + \mu^2)]^{\frac{1}{2}} < -\omega p_\phi \,. \tag{6}$$

The boundary of the region of negative E can be found by letting $p_r = p_\theta = 0$, and $\mu \to 0$ (i.e. highly relativistic particle). This gives $e^{2\nu-2\psi} < \omega^2$ which is equivalent to $g_{tt} > 0$. That is, the orbit must be inside the ergosphere. (This result also follows from Problem 10.15.)

When a rocket ship fires a bullet we have conservation of 4-momentum

$$\mathbf{p}_{\text{before}} = \mathbf{p}_{\text{after}} + \mathbf{p}_{\text{bullet}} \,. \tag{7}$$

Dotting this into the time Killing vector gives

$$E_{\text{before}} = E_{\text{after}} + E_{\text{bullet}} \,. \tag{8}$$

Since the rocket ship can fall into the ergosphere from infinity, we have $E_{\text{before}} > \mu$. In the ergosphere it can fire a bullet sufficiently fast with negative p_ϕ so that $E_{\text{bullet}} < 0$. Then, by Equation (7), $E_{\text{after}} > E_{\text{before}}$ and the rocket coasts off to infinity with increased total energy. Note that since both the rocket's trajectory and the bullet's negative energy trajectory are timelike, the bullet can be fired from the rocket with a locally measured velocity less than c.

Solution 17.10. In Problem 17.8 we saw that $\Omega \to a/2Mr_+$ when $\theta = \pi/2$. To prove this is so for any θ we write, in the notation of Problem 17.9:

$$\Omega = \frac{p^\phi}{p^t} = \frac{g^{\phi\phi}p_\phi + g^{\phi t}p_t}{g^{tt}p_t + g^{\phi t}p_\phi} = \omega - e^{2\nu}\,\frac{e^{-2\psi}p_\phi}{p_t + \omega p_\phi} \,.$$

At $\Delta = 0$ (that is, on the horizon) the metric function $e^{-2\psi}$ is finite but $e^{2\nu}$ vanishes, hence $\Omega = \omega$ and at $\Delta = 0$

$$\omega = \frac{-g_{\phi t}}{g_{\phi\phi}} = \frac{a}{2Mr_+} .$$

Solution 17.11. Start with the r and θ equations for orbital motion in the Kerr geometry (see e.g. MTW, Equation 33.32):

$$(r^2 + a^2\cos\theta)\,dr/d\tau = \pm V_r^{\frac{1}{2}} \tag{1}$$

$$(r^2 + a^2\cos\theta)\,d\theta/d\tau = \pm V_\theta^{\frac{1}{2}} \tag{2}$$

where

$$V_r = [E(r^2+a^2) - La]^2 - \Delta\,[r^2 + (L-aE)^2 + Q] \tag{3}$$

$$V_\theta = Q - \cos^2\theta\,[a^2(1-E^2) + L^2/\sin^2\theta] \tag{4}$$

$$\Delta \equiv r^2 - 2Mr + a^2$$

and where E, L, Q are constants of the motion. To have θ motion across $\theta = 0$ and $\theta = \pi$ we must have V_θ positive at both $\theta = 0$ and $\theta = \pi$. By Equation (4) this implies $L = 0$ and $Q > a^2(1-E^2)$. Equation (3) then is

$$V_r = E^2(r^2+a^2)^2 - \Delta\,[r^2 + a^2 + I] \tag{5}$$

where

$$I \equiv Q - a^2(1-E^2) > 0 .$$

The condition that the orbit be at constant coordinate r is $V_r = 0$ and $dV_r/dr = 0$ (condition for a turning point and a *perpetual* turning point). The condition on the derivative gives

$$0 = 2E^2r(r^2+a^2) - (r-M)E^2(r^2+a^2)^2/\Delta - r\Delta \tag{6}$$

where $V_r = 0$ and Equation (5) have been used to eliminate the constant I. This result can be rewritten

$$E^2 = r\Delta^2\,[(r^2+a^2)(r^3 - 3Mr^2 + a^2r + a^2M)]^{-1} . \tag{7}$$

For very large r we have $E \approx \left(1 - \frac{M}{2r} + \cdots\right)$ which is the Newtonian binding energy of a circular orbit. As we decrease r, Equation (7) tells how E changes. Clearly, since the numerator is positive, the only way Equation (7) can fail to provide a solution is for the denominator to go to zero (its going to ∞ is obviously impossible outside the horizon). Thus there exist circular polar orbits down to a radius r given by

$$r^3 - 3Mr^2 + a^2 r + a^2 M = 0 \; . \tag{8}$$

This is easily solved with the cubic formula to give the minimum radius of a polar orbit

$$r = M\left[1 + \frac{2(3-\tilde{a}^2)^{\frac{1}{2}}}{3^{\frac{1}{2}}} \cos\left\{\frac{1}{3}\cos^{-1}\left(\frac{3^{\frac{3}{2}}(1-\tilde{a}^2)}{(3-\tilde{a}^2)^{\frac{3}{2}}}\right)\right\}\right]$$

where $\tilde{a} \equiv a/M$. Note that $E^2 \to \infty$ in the limiting case, so the limiting case is a photon orbit. The limiting radius decreases from $r = 3M$ for $a = 0$ to $r = (1+2^{\frac{1}{2}})M$ for $a = M$.

Solution 17.12. Let $\boldsymbol{\xi}$ be the timelike Killing vector present because the black hole is static. Then Killing's equation tells us

$$\xi_{\alpha;\beta} = -\xi_{\beta;\alpha} \tag{1}$$

and the requirement for a static geometry (see Problem 10.8) is

$$\xi_{[\alpha;\beta}\xi_{\gamma]} = 0 \; . \tag{2}$$

A static observer has 4-velocity parallel to $\boldsymbol{\xi}$, i.e.

$$\mathbf{u} = \boldsymbol{\xi}/v^{\frac{1}{2}} \tag{3}$$

$$v = -\xi^\alpha \xi_\alpha \; . \tag{4}$$

For a photon of momentum p, the infinite red-shift surface occurs when

$$0 = (\mathbf{p}\cdot\mathbf{u})_\infty/(\mathbf{p}\cdot\mathbf{u})_{\text{emitter}} = v^{\frac{1}{2}}_{\text{emitter}}$$

where we have used the fact that $\mathbf{p}\cdot\boldsymbol{\xi}$ is constant along the geodesic trajectory of the photon and v is normalized to unity at infinity. Thus, the ergosurface is the surface at which $\boldsymbol{\xi}$ becomes null. We now prove that the ergosurface, $v = 0$, is a null hypersurface, i.e. that its normal vector is null: We use Equation (1) to write Equation (2) in the form

$$\xi_{\alpha;\beta}\xi_\gamma + \xi_{\gamma;\alpha}\xi_\beta - \xi_{\gamma;\beta}\xi_\alpha = 0\,, \tag{5}$$

then we dot ξ^γ into Equation (5):

$$v\xi_{\alpha;\beta} + v_{,[\alpha}\xi_{\beta]} = 0\,. \tag{6}$$

Equation (6) shows that $v_{,\alpha}$ is parallel to ξ_α wherever $v = 0$. But $v_{,\alpha}$ is the normal to $v = 0$, i.e. the normal is a null vector.

The proof has to be modified in the degenerate case when $v_{,\alpha} = 0$, since it is possible that $v = 0$ in a finite region, not just on a hypersurface. However, a detailed investigation [B. Carter, J. Math. Phys. *10*, 70, (1969)] shows that this is not so.

The theorem proven here corresponds to a Schwarzschild black hole, where $r = 2M$ is both the horizon and infinite red-shift surface for static observers (in fact, for all observers). A Kerr black hole is an example of the case where the ergosurface does not coincide with the horizon.

Solution 17.13. At the outer horizon $(r = r_+)$ the metric with $dt = dr = 0$ becomes

$$ds^2 = (r_+^2 + a^2\cos^2\theta)\,d\theta^2 + \frac{(r_+^2 + a^2)^2\sin^2\theta\,d\phi^2}{(r_+^2 + a^2\cos^2\theta)}\,.$$

The area of the horizon is then

$$A = \iint g^{\frac{1}{2}}\,d\theta\,d\phi = \iint (r_+^2 + a^2)\sin\theta\,d\theta\,d\phi = 4\pi(r_+^2 + a^2)\,.$$

Since r_+ is the larger solution of $\Delta = 0$ we have $r_+ = M + (M^2 + Q^2 - a^2)^{\frac{1}{2}}$ from which the result follows.

Solution 17.14. The two initial black holes have the same area, so the total initial surface area $(Q=0$ in Problem 17.13) is

$$A_i = 16\pi M_1 [M_1 + (M_1^2 - a^2)^{\frac{1}{2}}] .$$

The final black hole is a Schwarzschild black hole, so

$$A_f = 4\pi(2M_2)^2 = 16\pi M_2^2 .$$

The inequality $A_f \geq A_i$ becomes

$$M_2^2 \geq M_1^2 + M_1 (M_1^2 - a^2)^{\frac{1}{2}} .$$

If $a = M_1$, this means $M_2 \geq M_1$. The total initial mass was $2M_1$, so the final mass must be at least half of this. According to Hawking's theorem then, up to 50% of the original mass can be radiated away!

This is the most efficient of the collisions without charge. Suppose two black holes of masses M_1 and M_2, and angular momentum parameters a_1 and a_2 coalesce to form a black hole of mass M_3, and parameter a_3; then $A_f \geq A_i$ implies that

$$M_1 [M_1 + (M_1^2 - a_1^2)^{\frac{1}{2}}] + M_2 [M_2 + (M_2^2 - a_2^2)^{\frac{1}{2}}] \leq M_3 [M_3 + (M_3^2 - a_3^2)^{\frac{1}{2}}] .$$

One gets the most energy out when the equality obtains. Moreover, for given M_1, M_2, a_1, a_2 the smallest M_3 (hence the largest radiation of energy) will occur for $a_3 = 0$. Similarly, for given M_3 (with $a_3 = 0$), the largest $M_1 + M_2$ (hence the largest radiation of energy) will occur for $|a_1| = |a_2| = M_1 = M_2$. Since $a_3 = 0$, conservation of angular momentum implies that $a_1 = -a_2$. This is just the situation we worked in detail above. The conclusion, therefore, is that when two Kerr holes collide and coalesce, no more than 50% of their original mass can be radiated away.

Solution 17.15. The area A and "reduced area" $\tilde{A}$ of a Kerr black hole are

$$A \equiv 8\pi \tilde{A} = 8\pi [M^2 + (M^4 - J^2)^{\frac{1}{2}}] \tag{1}$$

or equivalently,

$$\tilde{A}^2 - 2\tilde{A}M^2 + J^2 = 0 . \tag{2}$$

Taking the first variation, we have

$$(\tilde{A} - M^2)\delta\tilde{A} = 2\tilde{A}M\delta M - J\delta J . \tag{3}$$

The left-hand side is nonnegative by the second law. Let a particular wave mode have the t and ϕ dependence $\exp(-i\omega t + im\phi)$. All scalar, electromagnetic and gravitational waves satisfy

$$\delta M = \frac{\omega}{m}\delta J \tag{4}$$

so we have

$$\left(2AM - J\frac{m}{\omega}\right)\delta M \geq 0 . \tag{5}$$

For an amplified wave conservation of energy requires that the mass of the hole decrease, so we want $\delta M < 0$; thus the term in parentheses in Equation (5) must be negative. This condition can be rewritten as

$$\frac{2Mr_+}{a} - \frac{m}{\omega} < 0 \tag{6}$$

where r_+ is the coordinate radius of the event horizon, or (if we take into account that m and ω can have either sign) the condition is

$$0 \lessgtr \omega \lessgtr \frac{ma}{2Mr_+} \tag{7}$$

where $\lessgtr$ means "lies between." Note that $a/2Mr_+$ is just Ω the "angular velocity of the black hole" (see Problem 17.10).

Solution 17.16.

(a) Using the results of Problem 7.7,

$$\Box\Phi = (-g)^{-\frac{1}{2}}[(-g)^{\frac{1}{2}}g^{\alpha\beta}\Phi_{,\alpha}]_{,\beta} ,$$

and $(-g)^{\frac{1}{2}} = (r^2 + a^2\cos^2\theta)\sin\theta$, we have

$$0 = \left[-\frac{(r^2+a^2)^2}{\Delta} + a^2\sin^2\theta\right]\frac{\partial^2\Phi}{\partial t^2} - \frac{Mar}{\Delta}\frac{\partial^2\Phi}{\partial t\,\partial\phi} + \left(\frac{1}{\sin^2\theta} - \frac{a^2}{\Delta}\right)\frac{\partial^2\Phi}{\partial\phi^2}$$
$$+ \frac{\partial}{\partial r}\left(\Delta\frac{\partial}{\partial r}\Phi\right) + \frac{1}{\sin\theta}\frac{\partial}{\partial\theta}\left(\sin\theta\frac{\partial\Phi}{\partial\theta}\right). \qquad (1)$$

(b) In Equation (1) we take the t and ϕ dependence to be

$$\Phi(t, r, \theta, \phi) = e^{-i\omega t}e^{im\phi}R(r)S(\theta)$$

and we divide through by Φ. This gives us

$$\frac{1}{R}\frac{d}{dr}\left(\Delta\frac{dR}{dr}\right) + \omega^2\frac{(r^2+a^2)^2}{\Delta} - \frac{4Mar\omega m}{\Delta} + \frac{a^2m^2}{\Delta}$$
$$= \frac{-1}{S\sin\theta}\frac{d}{d\theta}\left(\sin\theta\frac{dS}{d\theta}\right) + a^2\omega^2\sin^2\theta + \frac{m^2}{\sin^2\theta}. \qquad (2)$$

Since the left-hand side of Equation (2) is a function of r alone and the right-hand side must be a function of θ alone; each side must be a constant A, thus

$$\frac{1}{\sin\theta}\frac{d}{d\theta}\left(\sin\theta\frac{dS}{d\theta}\right) - \left(a^2\omega^2\sin^2\theta + \frac{m^2}{\sin^2\theta} - A\right)S = 0 \qquad (3)$$

and

$$\frac{d}{dr}\left(\Delta\frac{dR}{dr}\right) + \left[\frac{\omega^2(r^2+a^2)^2 - 4Mar\omega m + a^2m^2}{\Delta} - A\right]R = 0. \qquad (4)$$

Since S must be regular at $\theta = 0$ and π, Equation (3) is an eigenvalue equation for A. The function S is in fact a "spheroidal wave function"; in the simplest case when $a\omega = 0$, we have $S = P_{\ell m}(\cos\theta)$ and $A = \ell(\ell+1)$

(c) It is convenient (but not essential) to introduce a "tortoise" coordinate r^* which satisfies $dr^*/dr = (r^2+a^2)/\Delta$. Note that the interval (r_+, ∞) in the r-coordinate is stretched to $(-\infty, \infty)$ in r^*. Equation (4) now becomes

$$\frac{d^2R}{dr^{*2}} + \frac{2r\Delta}{(r^2+a^2)^2}\frac{dR}{dr^*} + \left[\omega^2 + \frac{a^2m^2 - 4Marm\omega - \Delta A}{(r^2+a^2)^2}\right]R = 0 . \qquad (5)$$

As $r \to \infty$, we get

$$\frac{d^2R}{dr^{*2}} + \frac{2}{r}\frac{dR}{dr^*} + \omega^2 R \approx 0$$

which has solutions

$$R \sim r^{-1} e^{\pm i\omega r^*}$$

corresponding to ingoing and outgoing waves.

(d) As $\Delta \to 0$, Equation (5) gives

$$\frac{d^2R}{dr^{*2}} + \left[\omega^2 - \frac{2am\omega}{2Mr_+} + \frac{a^2m^2}{(2Mr_+)^2}\right]R = 0$$

and hence

$$R \sim e^{\pm i(\omega - m\omega_+)r^*}$$

where

$$\omega_+ \equiv a/2Mr_+ .$$

(e) Since all physical observers at the horizon are related by Lorentz transformations, they will all agree whether a particular solution represents ingoing or outgoing waves. Thus we can calculate with any convenient observer. We choose an observer at constant r just outside the horizon. Since he is inside the ergosphere, he is dragged around in the positive ϕ direction with some $\Omega = d\phi/dt > 0$ where, from Problem 17.10, $\Omega = \omega_+$. This observer sees the local (t,r) dependence of the solution

$$\Phi = e^{-i\omega t} e^{im\phi} e^{\pm i(\omega - m\omega_+)r^*} S(\theta) \qquad (6)$$

as

$$\Phi = e^{-i(\omega - m\omega_+)t} e^{\pm i(\omega - m\omega_+)r^*} e^{im\tilde{\phi}} S(\theta) .$$

where we have put $\tilde{\phi} = \phi - \omega_+ t$. Thus $e^{-i(\omega - m\omega_+)r^*}$ corresponds to in-going waves.

(f) The stress-energy tensor for a scalar field is

$$4\pi T_{\alpha\beta} = \Phi_{,(\alpha}\Phi^*_{,\beta)} - \frac{1}{2} g_{\alpha\beta} |\Phi_{,\gamma}\Phi^{,\gamma}| .$$

(The complex conjugation is necessary because the representation of Φ is complex.) The energy flux vector is $J_\beta = -T_{\alpha\beta}\xi^\alpha$, where $\boldsymbol{\xi}$ is the time Killing vector $\partial/\partial t$ (see Problem 10.11). The energy flux into the horizon is found by integrating the radial component of $\mathbf{J}$ over the 2-surface $r = r_+$:

$$\frac{dE}{dt} = \int T_t^r |g|^{\frac{1}{2}} d\theta \, d\phi .$$

Since

$$4\pi T_t^r = \mathrm{Re}(\Phi_{,t}\Phi^{*,r}) = \mathrm{Re}\left(\Phi_{,t}\Phi^*_{,r^*} \frac{r^2 + a^2}{\Sigma}\right)$$
$$= \omega(\omega - m\omega_+) S^2(\theta) \frac{2Mr_+}{\Sigma} ,$$

we have

$$\frac{dE}{dt} = \omega(\omega - m\omega_+) \frac{2Mr_+}{4\pi} \int S^2(\theta) \sin\theta \, d\theta \, d\phi$$

and dE/dt is negative, i.e. energy flows out of the horizon if $\omega - m\omega_+ < 0$, i.e. if $0 < \omega/m < \omega_+$ in agreement with Problem 17.15.

Solution 17.17. Suppose charge δQ and energy δE are dropped into the hole. We then have

$$\delta(Q^2 - M^2) = 2Q\delta Q - 2M\delta E . \qquad (1)$$

Not all values of δQ and δE are allowed, however. The radial "effective potential" equation (see e.g. MTW, Equation 33.32) for a charged particle falling in the Reissner-Nordström geometry is

$$r^2 \frac{dr}{d\tau} = -\{(r^2\delta E - Q\delta Q r)^2 - \Delta(\mu_0^2 r^2 + L_z^2 + \mathcal{Q})\}^{\frac{1}{2}} \tag{2}$$

where E, L_z and $\mathcal{Q}$ are conserved quantities, and μ_0 is the particle rest mass. A particle which crosses the event horizon into the black hole has $dr/d\tau \leq 0$ at the event horizon $r = r_+$. Since $\Delta = 0$ at $r = r_+$, (2) becomes

$$r^2 \frac{dr}{d\tau} = -(r_+^2\delta E - Q\delta Q r_+) \leq 0 \tag{3}$$

and hence $\delta E > Q\delta Q/r_+$. Combining this with Equation (1) we see that

$$\delta(Q^2 - M^2) \leq 2\delta E(r_+ - M) = 2\delta E(M^2 - Q^2)^{\frac{1}{2}} . \tag{4}$$

As $Q^2 \to M^2$ the right side goes to zero as a square root, so no sequence of adding δE's can make the integrated left side positive.

Solution 17.18.

(a) $$\tilde{\omega}^{\hat{t}} \cdot \tilde{\omega}^{\hat{t}} = |g_{tt} - \omega^2 g_{\phi\phi}|^{\frac{1}{2}} \widetilde{dt} \cdot |g_{tt} - \omega^2 g_{\phi\phi}|^{\frac{1}{2}} \widetilde{dt}$$
$$= -(g_{tt} - \omega^2 g_{\phi\phi}) g^{tt} .$$

[The minus sign is chosen because $g_{tt} - \omega^2 g_{\phi\phi} < 0$.] The Kerr metric has only the $g_{t\phi}$ term off the diagonal. Inverting $g_{\alpha\beta}$ we find $g^{tt} = g_{tt} - \omega^2 g_{\phi\phi}$ and $\tilde{\omega}^{\hat{t}} \cdot \tilde{\omega}^{\hat{t}} = -1$. The calculation of other inner products to verify $\tilde{\omega}^{\hat{\alpha}} \cdot \tilde{\omega}^{\hat{\beta}} = \eta^{\alpha\beta}$ is similar.

(b) The dual basis $e_{\hat{\alpha}}$ is defined by $\langle\tilde{\omega}^{\hat{\alpha}}, e_{\hat{\beta}}\rangle = \delta^{\alpha}{}_{\beta}$. If $\tilde{\omega}^{\hat{\alpha}} = A^{\hat{\alpha}}{}_{\beta}\tilde{\omega}^{\beta}$ and $e_{\hat{\beta}} = B_{\hat{\beta}}{}^{\gamma} e_{\gamma}$, then the matrix $B_{\hat{\beta}}{}^{\gamma}$ must be the transposed inverse of $A^{\hat{\alpha}}{}_{\beta}$. To find B from A we can separate the t-ϕ piece from the (diagonal) r-θ piece:

$$\begin{bmatrix} \tilde{\omega}^{\hat{t}} \\ \tilde{\omega}^{\hat{\phi}} \end{bmatrix} = \begin{bmatrix} |g_{tt} - \omega^2 g_{\phi\phi}|^{\frac{1}{2}} & 0 \\ -\omega(g_{\phi\phi})^{\frac{1}{2}} & (g_{\phi\phi})^{\frac{1}{2}} \end{bmatrix} \begin{bmatrix} \widetilde{dt} \\ \widetilde{d\phi} \end{bmatrix}$$

implies that

$$\begin{bmatrix} e_{\hat{t}} \\ \\ e_{\hat{\phi}} \end{bmatrix} = \begin{bmatrix} |g_{tt} - \omega^2 g_{\phi\phi}|^{-\frac{1}{2}} & \omega |g_{tt} - \omega^2 g_{\phi\phi}|^{-\frac{1}{2}} \\ \\ 0 & (g_{\phi\phi})^{-\frac{1}{2}} \end{bmatrix} \begin{bmatrix} e_t \\ \\ e_\phi \end{bmatrix} .$$

Since the r-θ piece is diagonal, it is trivially inverted:

$$e_{\hat{r}} = (\Delta/\Sigma)^{\frac{1}{2}} e_r \qquad e_{\hat{\theta}} = \Sigma^{-\frac{1}{2}} e_\theta .$$

(c) The 4-velocity has zero rotation if $\omega_{\alpha\beta} = 0$ (see Problem 5.18) or equivalently if $\mathbf{u}$ is hypersurface orthogonal:

$$u_{[\alpha;\beta} u_{\gamma]} = 0$$

(see Problem 7.23). The quickest way to check this condition is to use forms and the equivalence of $u_{[\hat{\alpha};\hat{\beta}]}$ and $\mathbf{d}\tilde{\boldsymbol{\omega}}^{\hat{t}}$. The condition $u_{[\hat{\alpha};\hat{\beta}}u_{\hat{\gamma}]} = 0$ is then equivalent to $\mathbf{d}\tilde{\boldsymbol{\omega}}^{\hat{t}} \wedge \tilde{\boldsymbol{\omega}}^{\hat{t}} = 0$. But $\tilde{\boldsymbol{\omega}}^{\hat{t}} = \alpha\, \mathrm{dt}$ where $\alpha = |g_{tt} - \omega^2 g_{\phi\phi}|^{\frac{1}{2}}$ $= \alpha(r, \theta)$ and thus

$$\mathbf{d}\tilde{\boldsymbol{\omega}}^{\hat{t}} = \alpha_{,r} \widetilde{dr} \wedge \widetilde{dt} + \alpha_{,\theta} \widetilde{d\theta} \wedge \widetilde{dt}$$

so $\mathbf{d}\tilde{\boldsymbol{\omega}}^{\hat{t}} \wedge \tilde{\boldsymbol{\omega}}^{\hat{t}} = 0$ since $\widetilde{dt} \wedge \widetilde{dt} = 0$.

(d) The 4-velocity of the ZAMO is

$$\mathbf{u} = e_{\hat{t}} = |g_{tt} - \omega^2 g_{\phi\phi}|^{-\frac{1}{2}} (e_t + \omega\, e_\phi) .$$

The vectors e_t and e_ϕ are Killing vectors. The vector field $e_t + \omega e_\phi$ is not a Killing vector because ω is not constant. *However* any particular ZAMO *is* moving along the Killing vector $\boldsymbol{\xi} = e_t + \omega_0 e_\phi$ (ω_0 a fixed constant, which is chosen to equal ω at the radius of the particular ZAMO under consideration). Thus the results of Problem 10.14 apply and we get

$$\mathbf{a} = \frac{1}{2} \nabla \log |\boldsymbol{\xi} \cdot \boldsymbol{\xi}| = \frac{1}{2} \nabla \log |g_{tt} + 2\omega_0 g_{t\phi} + \omega_0^2 g_{\phi\phi}| .$$

When we set $g_{t\phi} = -\omega g_{\phi\phi}$ and evaluate at $\omega = \omega_0$ this becomes

$$a = \frac{1}{2}\nabla \log |g_{tt} - \omega^2 g_{\phi\phi}| .$$

Solution 17.19. In Boyer-Lindquist coordinates, for $t = \text{constant}$ and $r = r_+$, the Kerr metric gives the metric of the horizon:

$$ds^2 = (r_+^2 + a^2\cos^2\theta)\, d\theta^2 + (2Mr_+)^2 \sin^2\theta (r_+^2 + a^2\cos^2\theta)^{-1} d\phi^2 . \qquad (1)$$

(We have used the relation $r_+^2 + a^2 = 2Mr_+$ in $g_{\phi\phi}$.) The Gaussian curvature K of a 2-surface is the same as the Riemannian curvature of Problem 9.23, where the Riemann tensor is computed from the 2-dimensional metric. It is convenient to use an orthonormal basis dyad. For a metric of the form

$$ds^2 = e^{2\alpha_1}(dx^1)^2 + e^{2\alpha_2}(dx^2)^2 \qquad (2)$$

we let

$$\tilde{\omega}^{\hat{1}} = e^{\alpha_1}\widetilde{dx}^1 , \qquad \tilde{\omega}^{\hat{2}} = e^{\alpha_2}\widetilde{dx}^2 \qquad (3)$$

and we have

$$K = R_{\hat{1}\hat{2}\hat{1}\hat{2}} = R_{1212}\, e^{-2(\alpha_1 + \alpha_2)} . \qquad (4)$$

The quickest way to find the Riemann tensor is with forms as in Problems 8.27 and 9.20. Since

$$d\tilde{\omega}^{\hat{1}} = e^{\alpha_1}\widetilde{d\alpha}_1 \wedge \widetilde{dx}^1 = e^{\alpha_1}\alpha_{1,2}\widetilde{dx}^2 \wedge \widetilde{dx}^1 = \alpha_{1,2}e^{-\alpha_2}\tilde{\omega}^{\hat{2}} \wedge \tilde{\omega}^{\hat{1}}$$

and a similar expression with 1 and 2 interchanged, we find

$$\tilde{\omega}^{\hat{1}}{}_{\hat{2}} = e^{-\alpha_2}\alpha_{1,2}\tilde{\omega}^{\hat{1}} - e^{-\alpha_1}\alpha_{2,1}\tilde{\omega}^{\hat{2}} . \qquad (5)$$

(Recall that $d\tilde{\omega}^{\hat{\alpha}} = -\tilde{\omega}^{\hat{\alpha}}{}_{\hat{\beta}} \wedge \tilde{\omega}^{\hat{\beta}}$ and $\tilde{\omega}_{\hat{\alpha}\hat{\beta}} = -\tilde{\omega}_{\hat{\beta}\hat{\alpha}}$.) The only nontrivial curvature form is

$$\begin{aligned}\mathcal{R}^{\hat{1}}{}_{\hat{2}} &= d\tilde{\omega}^{\hat{1}}{}_{\hat{2}} + \tilde{\omega}^{\hat{1}}{}_{\hat{a}} \wedge \tilde{\omega}^{\hat{a}}{}_{\hat{2}} \\ &= d(\alpha_{1,2} e^{\alpha_1-\alpha_2} \widetilde{dx}^1 - \alpha_{2,1} e^{\alpha_2-\alpha_1} \widetilde{dx}^2) + 0 \\ &= (\alpha_{1,2} e^{\alpha_1-\alpha_2})_{,2} \widetilde{dx}^2 \wedge \widetilde{dx}^1 - (\alpha_{2,1} e^{\alpha_2-\alpha_1})_{,1} \widetilde{dx}^1 \wedge \widetilde{dx}^2 \\ &= -e^{-(\alpha_1+\alpha_2)}[(e^{\alpha_2}{}_{,1} e^{-\alpha_1})_{,1} + (e^{\alpha_1}{}_{,2} e^{-\alpha_2})_{,2}]\tilde{\omega}^{\hat{1}} \wedge \tilde{\omega}^{\hat{2}} .\end{aligned}$$

Since $\mathcal{R}_{\hat{1}\hat{2}} = R_{\hat{1}\hat{2}\hat{1}\hat{2}}\, \tilde{\omega}^{\hat{1}} \wedge \tilde{\omega}^{\hat{2}}$ we have

$$K = R_{\hat{1}\hat{2}\hat{1}\hat{2}} = -e^{-(\alpha_1+\alpha_2)}[(e^{\alpha_2}{}_{,1} e^{-\alpha_1})_{,1} + (e^{\alpha_1}{}_{,2} e^{-\alpha_2})_{,2}] . \tag{6}$$

Using the explicit form of the metric in Equation (1), with $\theta = x^1$ and $\phi = x^2$, we find from Equation (6) that

$$K = 2Mr_+ \frac{r_+^2 - 3a^2\cos^2\theta}{(r_+^2 + a^2\cos^2\theta)^3} . \tag{7}$$

Note that if $r_+^2 - 3a^2 < 0$, then K is negative for a range of θ around the poles $(\theta = 0, \pi)$. But when K is negative, the surface cannot be globally embedded in a Euclidean 3-space. Since $r_+ = M + (M^2 - a^2)^{\frac{1}{2}}$, the condition that $r_+^2 - 3a^2$ be less than zero is equivalent to $a > 3^{\frac{1}{2}}M/2$.

The Gauss-Bonnet theorem says that

$$\int K d^2S = 2\pi\chi$$

where χ is the Euler characteristic of the surface ($\chi = 2$ for a sphere, 0 for a torus, etc.). Since

$$d^2S = (g)^{\frac{1}{2}} d\theta\, d\phi = 2Mr_+ \sin\theta\, d\theta\, d\phi$$

we have

$$\chi = (2\pi)^{-1}\int_0^{2\pi} d\phi \int_0^{\pi} (2Mr_+)^2 \frac{r_+^2 - 3a^2\cos^2\theta}{(r_+^2 + a^2\cos^2\theta)^3}\sin\theta\, d\theta$$

$$= (2Mr_+)^2\, 2\int_0^1 \frac{r_+^2 - 3a^2x^2}{(r_+^2 + a^2x^2)^3}\, dx$$

$$= (2Mr_+)^2\, 2\left[\frac{x}{(r_+^2 + a^2x^2)^2}\right]_0^1$$

$$= 2$$

so the surface is topologically a 2-sphere. (For further details see L. Smarr, Phys. Rev. D. *7*, 289, (1973).)

Solution 17.20. The energy amplification described in Problems 17.15 and 17.16 corresponds to stimulated emission in quantum-mechanical language. If we know the rate for stimulated emission, we can find the rate for spontaneous emission as follows:

Let p be the probability per unit of phase space of the spontaneous emission of a quantum. If we send N quanta in, the probability of getting $N+1$ out by stimulated emission is $(N+1)p$, since the quanta (gravitons, photons) are bosons. To simplify the argument, assume $p \ll 1$ so we do not have to consider the emission of more than one extra quantum; dimensionally the answer would not be changed if we were to include this refinement. The expectation value of the excess number of quanta emitted is

$$\langle \Delta N \rangle = (\text{excess number}) \times (\text{probability of that number})$$
$$= 1 \times (N+1)p\ .$$

The amplification is

$$A \equiv \frac{\langle \Delta N \rangle}{N} = \left(1 + \frac{1}{N}\right)p \approx p \qquad (1)$$

in the classical limit $(N \to \infty)$. Fermi's Golden Rule says that the rate of spontaneous emission is

$$\frac{dN}{dt} \sim \int_{\text{phase space}} p \,. \tag{2}$$

In Equation (2) replace p by A and recall that the classical effect is described in terms of modes with "quantum numbers" ℓ and m and a frequency ω. Thus

$$\frac{dN}{dt} \sim \sum_{\ell,m} \int A\, d\omega \,. \tag{3}$$

Now A goes rapidly to zero for large ℓ and m, as in all barrier-penetration problems. Also from Problems 17.15 and 17.16, A is non-zero for $\omega \sim 1/M$ over a range $\Delta\omega \sim 1/M$, where M is the mass of the black hole. Since A is a number of order unity for low ℓ and m, replace A by 1 in Equation (3) and drop the summation. Then

$$\frac{dN}{dt} \sim \Delta\omega \sim \frac{1}{M} \,.$$

The hole loses energy at a rate

$$\frac{dE}{dt} \sim \hbar\omega \frac{dN}{dt} \sim \frac{\hbar}{M^2} \,.$$

This implies a loss of angular momentum at a rate

$$\frac{dJ}{dt} \sim \frac{m}{\omega}\frac{dE}{dt} \sim \frac{\hbar}{M} \,.$$

Thus the time-scale for angular momentum loss is

$$\tau \sim J/(dJ/dt) \sim M^2(\hbar/M) \sim M^3/\hbar \sim (M/10^{15}\text{gm})^3\, 10^{10} \text{ years.}$$

Thus for $M \lesssim 10^{15}$gm, J will already have become essentially zero by this process. For $M \sim 10^{33}$gm, the fraction lost will be $\sim 10^{-54}$ in 10^{10} years.

Solution 17.21. For all static spacetimes, the "surface of infinite red-shift" ($g_{00} = 0$) coincides with the horizon (Problem 17.12). Therefore, if the above metric describes a black hole, its surface is at $g_{00} = 0$ or $\bar{\rho} = \frac{1}{2}$, where $\bar{\rho} \equiv \rho/m$. We will now show that, in fact, the surface $\bar{\rho} = \frac{1}{2}$ is not contained in the same manifold as the one which becomes asymptotically flat $(\bar{\rho} \to \infty)$ and thus cannot be reached by objects in our universe.

A glance at the $g_{\rho\rho}$ portion of the metric shows that $\bar{\rho} = 1.5$ is an infinite proper radial distance away from any $\bar{\rho} > 1.5$. To see that $\bar{\rho} = 1.5$ is also infinite proper time away and thus removed from the manifold it is sufficient to consider radial null geodesics. (The reader can verify that for nonradial or timelike geodesics, matters get even worse):

$$g_{\rho\rho}(p^{\rho})^2 + g^{00}(p_0)^2 = 0$$

or

$$\frac{d\rho}{d\lambda} = \left(\bar{\rho} - \frac{3}{2}\right) \frac{(\bar{\rho} + \frac{1}{2})}{(\bar{\rho} - \frac{1}{2})^2} p_0$$

where λ is an affine parameter. Remembering that p_0 is a constant of the motion (stationary metric), the above equations show that

$$p_0^{-1} \int_{\bar{\rho}}^{\bar{\rho} = \frac{3}{2}} \frac{(\bar{\rho} - \frac{1}{2})^2 \, d\rho}{(\bar{\rho} - \frac{3}{2})(\bar{\rho} + \frac{1}{2})} = \int_{\lambda}^{\lambda_0} d\lambda \; ,$$

Since the integral on the left diverges at $\bar{\rho} = \frac{3}{2}$, the $\bar{\rho} = \frac{3}{2}$ surface is an infinite affine distance away. There are no singularities for $\bar{\rho} \geq \frac{3}{2}$ and $\bar{\rho}$ is a monotonically decreasing coordinate for radially infalling photons and particles, so $\bar{\rho} = \frac{1}{2}$, as well as $\bar{\rho} = \frac{3}{2}$, must be off the physical manifold.

The above conclusions also indicate that the surface of a star collapsing to form the given vacuum metric can never reach $\bar{\rho} = \frac{3}{2}$; thus for realistic physical systems, the given metric is realizable only for $\bar{\rho} > \frac{3}{2}$.

CHAPTER 18: SOLUTIONS

Solution 18.1. To find the gravitational radiation we use the square of the third time derivative of the quadrupole moment

$$\text{Power} = \frac{G}{5c^5}\left(\dddot{I}\right)^2 \sim \frac{GM^2L^4}{5c^5t^6} .$$

The appropriate parameters for the mean Massachusetts motorist: M = mass of fist and forearm $\sim 2 \times 10^3$ gm, L = length of forearm $\sim$ 50 cm, t = period of average shake $\sim$ .2 sec. The calculation in cgs units then yields a radiative power on the order of 2×10^{-43} ergs/sec.

To calculate his total energy expended, use the fact that muscles are almost totally nonconservative, so every shake uses up an energy roughly equal to the arms maximum kinetic energy:

$$\text{Power} = \frac{1}{2}\, M\, \frac{L^2}{t^2} \cdot \frac{1}{t} \approx 3 \times 10^8 \text{ erg/sec} .$$

The fractional "efficiency" is thus $\sim 10^{-51}$.

Solution 18.2. The gravitational binding energy of the system is of order M^2/R and the kinetic energy is of order MR^2/T_{dyn}^2, where T_{dyn} characterizes motions of the system. For a system in equilibrium the virial theorem gives

$$T_{dyn}^2 \sim R^3/M .$$

The rate at which energy is radiated gravitationally is governed by the square of the third time derivative of the quadrupole moment or roughly

$$P^{GW} \sim M^2R^4/T_{dyn}^6 .$$

The time for radiation reaction to affect the system is the time it takes to radiate a significant fraction of its energy as gravitational waves:

$$T = \frac{\text{K.E.}}{P^{GW}} \sim \frac{T^4_{dyn}}{MR^2} \sim \frac{RT^2_{dyn}}{M^2} .$$

The ratio of this gravitational wave characteristic time to the dynamical time is then of order

$$\frac{T}{T_{dyn}} \sim \frac{RT_{dyn}}{M^2} \sim \left(\frac{R}{M}\right)^{\frac{5}{2}} .$$

Evidently gravitational radiation reaction forces play an appreciable role only in relativistically compact systems.

Solution 18.3. The three independent orientations of the dipole correspond to the three components of a three dimensional vector. Since the quadrupole is a second rank, three dimensional tensor (9 component) which is symmetric (3 constraints) and traceless (1 constraint) there are five independent "orientations." An example of five independent quadrupole tensors is the five tensors with all components zero except:

(i) $I_{zz} = -\frac{1}{2} I_{xx} = -\frac{1}{2} I_{yy} = 1$ (ii) $I_{yy} = -\frac{1}{2} I_{xx} = -\frac{1}{2} I_{zz} = 1$

(iii) $I_{xy} = I_{yx}$ (iv) $I_{xz} = I_{zx}$ (v) $I_{yz} = I_{zy}$.

Solution 18.4. Assume the rod to be in the xy plane rotating about the z axis. If μ is the mass per unit length of the rod the *time dependent* parts of the reduced quadrupole tensor are

$$\text{I\kern-0.5em-}_{xx} = \mu \cos^2(\omega t) \cdot 2 \int_0^{\ell/2} x^2 dx$$

$$= \frac{\mu \ell^3}{12} \cos^2 \omega t = \frac{M\ell^2}{24} \cos(2\omega t) + \text{constant}$$

$$\ddagger_{yy} = -\frac{M\ell^2}{24}\cos 2\omega t + \text{constant}$$

$$\ddagger_{xy} = \ddagger_{yx} = \frac{M\ell^2}{24}\sin 2\omega t \ ,$$

so that

$$L_{GW} = \frac{1}{5}\langle \dddot{\ddagger}_{jk}\dddot{\ddagger}_{jk}\rangle = \frac{(2\omega)^6}{5}\left(\frac{M\ell^2}{24}\right)^2 \langle 2\cos^2 2\omega t + 2\sin^2 2\omega t\rangle$$

$$= \frac{2}{45}\omega^6 M^2\ell^4 \,. \text{ (Note units with } G = c = 1!)$$

To calculate the charge centrifuged to the ends of the bar, balance electrostatic and centripetal forces:

$$|e\nabla\Phi| = rm\omega^2 \ ,$$

where Φ is the electrostatic potential and e and m are the charge and mass of an electron. Thus the charge density induced will be of order

$$\rho \sim -\nabla^2\Phi \sim -(m/e)\omega^2 \ .$$

This charge depletion in the main part of the rod will of course be balanced by electrons squashed against the ends with charge $(m/e)\omega^2\ell A$ where A = rod cross section. The rod thus becomes an electric quadrupole with quadrupole moment of order $(m/e)\omega^2\ell^3 A$. The electric quadrupole radiation due to this induced moment will be roughly ω^6 times the square of the quadrupole moment, or roughly $\omega^{10}(m/e)^2\ell^6 A^2$. The ratio of electromagnetic damping to gravitational damping is then

$$\frac{L_{EM}}{L_{GW}} \sim \frac{\omega^{10}(m/e)^2\ell^6 A^2}{\omega^6\rho^2 A^2\ell^6} = \left[\frac{(m/e)^2\omega^2}{\rho}\right]^2 \ .$$

In $G = c = 1$ units $(m/e) \approx \frac{1}{2}\times 10^{-21}$ and $\rho(10\text{g/cm}^3) \approx 10^{-27}$ and $\omega(1\text{kH}_z) \approx \frac{1}{3}\times 10^{-7}$ so that

$$L_{EM}/L_{GW} \sim 10^{-18} \ .$$

For once gravitational radiation wins out.

Solution 18.5. The radiation reaction force on an element of the source is

$$dF_i = -\Phi_{,i}\rho\, d^3x\ .$$

The rate at which the source element loses energy is the rate at which the reaction forces do work on the source element: $v^i dF_i$ where v is the velocity of the source element. For the whole source, then

$$\frac{dE}{dt} = -\int \Phi_{,i}\, v^i \rho\, d^3x = \int \Phi (v^i\rho)_{,i}\, d^3x\ .$$

Here we have used Gauss' divergence theorem and evaluated the surface integral on a surface outside the source to obtain the second equality. The equation of continuity tells us that $\underset{\sim}{\nabla}\cdot(\underset{\sim}{v}\rho) = -\partial\rho/\partial t$ so

$$\frac{dE}{dt} = -\int \Phi \frac{\partial\rho}{\partial t}\, d^3x = -\frac{1}{5}\, {\rlap{-}I}^{(5)}_{jk}\, \frac{d}{dt}\int x^j x^k \rho\, d^3x$$

$$= -\frac{1}{5}\, {\rlap{-}I}^{(5)}_{jk}\, \frac{d}{dt}\left({\rlap{-}I}_{jk} + \frac{1}{3}\delta_{jk}\int r^2\rho\, d^3x\right)\ .$$

Since ${\rlap{-}I}_{jk}\delta_{jk} = 0$, we drop the final term to arrive at

$$dE/dt = -\left(\frac{1}{5}\right) {\rlap{-}I}^{(5)}_{jk}\, \dot{\rlap{-}I}_{jk}\ .$$

Now we time average over several oscillations. Since secular changes in parameters of the source are assumed to be small over the several oscillations, we have

$$\frac{1}{T}\int_0^T {\rlap{-}I}^{(5)}_{jk}\, \dot{\rlap{-}I}_{jk}\, dt \approx \frac{1}{T}\int_0^T \dddot{\rlap{-}I}_{jk}\, \dddot{\rlap{-}I}_{jk}\, dt$$

and

$$\frac{dE}{dt} = -\frac{1}{5}\left\langle \dddot{\rlap{-}I}_{jk}\, \dddot{\rlap{-}I}_{jk}\right\rangle\ .$$

To find the rate of loss of angular momentum we use the fact that for an element of the source

$$\frac{dJ}{dt} = \text{radiation reaction torque} = \mathbf{r}\times d\mathbf{F} = -\mathbf{r}\times\nabla\Phi\rho\, d^3x$$

and for the whole source

$$\frac{dJ^i}{dt} = -\varepsilon^{ijk}\int x^j\Phi_{,k}\rho\, d^3x$$

$$= -\varepsilon^{ijk}\int x^j\left(\frac{2}{5}\mathcal{I}^{(5)}_{km}x^m\right)\rho\, d^3x$$

$$= -\frac{2}{5}\,\varepsilon^{ijk}\,\mathcal{I}^{(5)}_{km}\int x^j x^m \rho\, d^3x = -\frac{2}{5}\,\varepsilon^{ijk}\,\mathcal{I}^{(5)}_{km}\mathcal{I}_{jm}\,.$$

Time averaging, and integrating (in time) by parts as above gives us

$$\frac{dJ^i}{dt} = -\frac{2}{5}\,\varepsilon^{ijk} < \dddot{\mathcal{I}}_{km}\,\dddot{\mathcal{I}}_{jm} > .$$

Solution 18.6. Let m_1, m_2 and r_1, r_2 be respectively the masses, and distances from the center of mass, of the two stars rotating at angular frequency ω. From Newtonian mechanics we know that $r_1 m_1 = r_2 m_2 = R\mu$, where $R \equiv r_1 + r_2$ and μ = reduced mass = $m_1 m_2/(m_1 + m_2)$. The total energy of the stars can now be written as a function of M, R, and μ:

$$E = \frac{1}{2}\,r_1^2 m_1 + \frac{1}{2}\,r_2^2 m_2\;\;\omega^2 - \frac{m_1 m_2}{R} = \frac{1}{2}\,\omega^2 R\mu - \frac{\mu M}{R} = -\frac{1}{2}\,\frac{\mu M}{R}\,.$$

Here $M = m_1 + m_2$ and Kepler's laws have been used.

The power dissipated as gravitational waves is easily calculated once the reduced quadrupole moment,

$$\mathcal{I}_{jk} \equiv \int \rho\left(x_j x_k - \frac{1}{3}\,r^2\,\delta_{jk}\right)d^3x\,,$$

is evaluated. Note that the "reducing term" $\int \frac{1}{3}\rho r^2 \delta_{jk} d^3x$ will be constant in time and can be ignored. Suppose the z axis to be the axis of rotation and ϕ the azimuthal angle from the x-axis to the line joining the stars, then (aside from time independent terms) we have

$$I_{xx} = (r_1^2 m_1 + r_2^2 m_2)\cos^2\phi = \frac{1}{2} R^2\mu \cos 2\phi + \text{constant}$$

$$I_{yy} = -\frac{1}{2} R^2\mu \cos 2\phi + \text{constant}$$

$$I_{xy} = I_{yx} = \frac{1}{2} R^2\mu \sin 2\phi \ .$$

Since $\phi = \omega t$, it is straightforward to differentiate three times, perform the sum $\dddot{\mathcal{I}}_{jk}\dddot{\mathcal{I}}_{jk}$ and time average to find

$$\frac{1}{5} < \dddot{\mathcal{I}}_{jk}\dddot{\mathcal{I}}_{jk} > = \frac{1}{5}(2\omega)^6\left(\frac{1}{2}R^2\mu\right)^2 < \sin^2 2\omega t + \sin^2 2\omega t + 2\cos^2 2\omega t >$$

$$= \frac{32}{5}\omega^6 R^4\mu^2 = \frac{32}{5}\frac{M^3\mu^2}{R^5} = P_{GW} \ .$$

Since this power is radiated at the expense of orbital energy, it follows that

$$\frac{dE}{dt} = \frac{1}{2}\frac{\mu M}{R^2}\frac{dR}{dt} = -P_{GW} = -\frac{32}{5}\frac{M^3\mu^2}{R^5} \ .$$

The resulting differential equation

$$R^3\frac{dR}{dt} = -\frac{64}{5}M^2\mu$$

is easily integrated to give

$$R^4 = -(256/5)\, M^2\mu t + \text{constant} \ .$$

If we denote by t_0 the time at which the separation goes to zero according to this formula, then

$$R^4 = \frac{256}{5} M^2\mu (t_0 - t) \ .$$

We can find the time t_0 by noting that at $t = 0$, R must be the present separation R_{now}, thus

$$t_0 = \frac{5 R_{now}^4}{256\ M^2 \mu} .$$

Solution 18.7. In terms of the total orbital energy E and angular momentum L, the semimajor axis and eccentricity are

$$a = -m_1 m_1 / 2E \tag{1}$$

$$e^2 = 1 + \frac{2E\,L^2 (m_1 + m_2)}{m_1^3 m_2^3} \tag{2}$$

so that

$$\frac{da}{dt} = \frac{m_1 m_2}{2E^2} \frac{dE}{dt} \tag{3}$$

$$\frac{de}{dt} = \frac{m_1 + m_2}{m_1^3 m_2^3 e} \left(L^2 \frac{dE}{dt} + 2EL \frac{dL}{dt} \right) . \tag{4}$$

If the separation of the two particles is r, where

$$r = \frac{a(1-e^2)}{1 + e \cos\theta} \tag{5}$$

(see figure), then

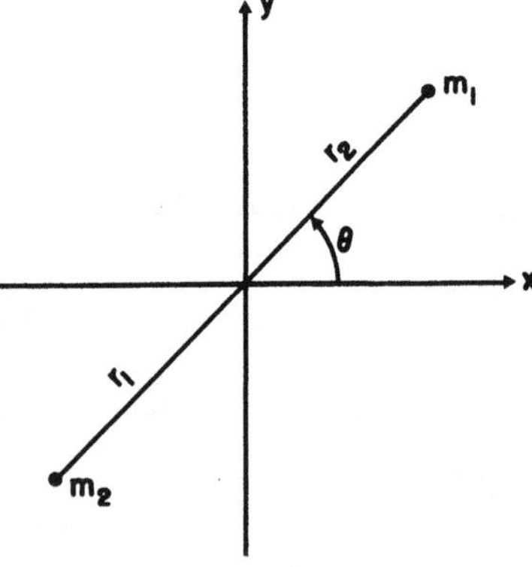

$$r_1 = \frac{m_2}{m_1 + m_2}\, r$$

$$r_2 = \frac{m_1}{m_1 + m_2}\, r$$

so the components of the quadrupole tensor are

$$I_{xx} = m_1 x_1^2 + m_2 x_2^2 = \frac{m_1 m_2}{m_1 + m_2}\, r^2 \cos^2\theta$$

$$I_{yy} = \frac{m_1 m_2}{m_1 + m_2}\, r^2 \sin^2\theta$$

$$I_{xy} = \frac{m_1 m_2}{m_1 + m_2} r^2 \sin\theta \cos\theta \tag{6}$$

$$I \equiv I_{xx} + I_{yy} = \frac{m_1 m_2}{m_1 + m_2} r^2 .$$

According to the Newtonian equations of motion

$$\dot{\theta} = \frac{[(m_1 + m_2) a(1 - e^2)]^{\frac{1}{2}}}{r^2} \tag{7}$$

so Equation (5) gives

$$\dot{r} = e \sin\theta \left[\frac{m_1 + m_2}{a(1 - e^2)}\right]^{\frac{1}{2}} . \tag{8}$$

Now we compute the successive time derivatives of I_{ij}, using Equations (5), (7) and (8) to simplify the expressions. The results are

$$\dot{I}_{xx} = \frac{-2m_1 m_2 r \cos\theta \sin\theta}{[(m_1 + m_2) a(1 - e^2)]^{\frac{1}{2}}} \tag{9}$$

$$\ddot{I}_{xx} = \frac{-2m_1 m_2}{a(1 - e^2)} (\cos 2\theta + e \cos^3\theta) \tag{10}$$

$$\dddot{I}_{xx} = \frac{2m_1 m_2}{a(1 - e^2)} (2 \sin 2\theta + 3e \cos^2\theta \sin\theta)\dot{\theta} \tag{11}$$

$$\dot{I}_{yy} = \frac{2m_1 m_2}{[(m_1 + m_2) a(1 - e^2)]^{\frac{1}{2}}} r(\sin\theta \cos\theta + e \sin\theta) \tag{12}$$

$$\ddot{I}_{yy} = \frac{2m_1 m_2}{a(1 - e^2)} (\cos 2\theta + e \cos\theta + e \cos^3\theta + e^2) \tag{13}$$

$$\dddot{I}_{yy} = \frac{-2m_1 m_2}{a(1 - e^2)} (2 \sin 2\theta + e \sin\theta + 3e \cos^2\theta \sin\theta)\dot{\theta} \tag{14}$$

$$\dot{I}_{xy} = \frac{m_1 m_2 r(\cos^2\theta - \sin^2\theta + e\cos\theta)}{[(m_1+m_2)a(1-e^2)]^{\frac{1}{2}}} \tag{15}$$

$$\ddot{I}_{xy} = \frac{-2m_1 m_2}{a(1-e^2)}(\sin 2\theta + e\sin\theta + e\sin\theta\cos^2\theta) \tag{16}$$

$$\dddot{I}_{xy} = \frac{-2m_1 m_2}{a(1-e^2)}(2\cos 2\theta - e\cos\theta + 3e\cos^3\theta)\dot{\theta} \tag{17}$$

$$\dddot{I} = \dddot{I}_{xx} + \dddot{I}_{yy} = \frac{-2m_1 m_2 e\sin\theta\dot{\theta}}{a(1-e^2)} . \tag{18}$$

The rate of loss of energy is then:

$$\frac{dE}{dt} = -\frac{1}{5}\dddot{\text{\sout{I}}}_{jk}\dddot{\text{\sout{I}}}_{jk} = -\frac{1}{5}\left(\dddot{I}_{jk}\dddot{I}_{jk} - \frac{1}{3}\dddot{I}^2\right)$$

$$= -\frac{1}{5}\left(\dddot{I}^2_{xx} + 2\dddot{I}^2_{xy} + \dddot{I}^2_{yy} - \frac{1}{3}\dddot{I}^2\right)$$

$$= \frac{-8m_1^2 m_2^2}{15a^2(1-e^2)^2}[12(1+e\cos\theta)^2 + e^2\sin^2\theta]\dot{\theta}^2$$

where we have used Equations (9)-(18) and straightforward algebraic simplifications. Now we must average over an orbit. The period of an orbit is (Kepler's third law)

$$T = \int_0^{2\pi}\frac{1}{\dot{\theta}}\,d\theta = \frac{2\pi a^{\frac{3}{2}}}{(m_1+m_2)^{\frac{1}{2}}}$$

so that

$$\left\langle\frac{dE}{dt}\right\rangle = \frac{1}{T}\int_0^T \frac{dE}{dt}\,dt = \frac{1}{T}\int_0^{2\pi}\frac{dE}{dt}\frac{1}{\dot{\theta}}\,d\theta$$

$$= \frac{-32}{5}\frac{m_1^2 m_2^2(m_1+m_2)}{a^5(1-e^2)^{7/2}}\left(1 + \frac{73}{24}e^2 + \frac{37}{96}e^4\right) . \tag{19}$$

Similarly, the rate of loss of angular momentum is

$$\begin{aligned}\frac{dL}{dt} &= \frac{-2}{5}\,\varepsilon_{zij}\ddot{\mathcal{I}}_{ik}\dddot{\mathcal{I}}_{kj} = \frac{-2}{5}\,\varepsilon_{zij}\ddot{I}_{ik}\dddot{I}_{kj} \\ &= \frac{-2}{5}\,[\ddot{I}_{xy}(\dddot{I}_{yy} - \dddot{I}_{xx}) + \dddot{I}_{xy}(\ddot{I}_{xx} - \ddot{I}_{yy})] \\ &= \frac{-8}{5}\,\frac{m_1^2 m_2^2}{a^2(1-e^2)^2}\,[4 + 10e\cos\theta + e^2(9\cos^2\theta - 1) \\ &\qquad + e^3(3\cos^3\theta - \cos\theta)]\,\dot{\theta}\end{aligned}$$

which when time averaged gives us

$$\begin{aligned}\left\langle\frac{dL}{dt}\right\rangle &= \frac{1}{T}\int_0^{2\pi}\frac{dL}{dt}\,\frac{1}{\dot{\theta}}\,d\theta \\ &= \frac{-32}{5}\,\frac{m_1^2 m_2^2 (m_1+m_2)^{\frac{1}{2}}}{a^{7/2}(1-e^2)^2}\left(1 + \frac{7}{8}\,e^2\right). \qquad (20)\end{aligned}$$

From Equations (3) and (4) we have, finally:

$$\begin{aligned}\left\langle\frac{da}{dt}\right\rangle &= \frac{2a^2}{m_1 m_2}\left\langle\frac{dE}{dt}\right\rangle \\ &= \frac{-64}{5}\,\frac{m_1 m_2 (m_1+m_2)}{a^3(1-e^2)^{7/2}}\left(1 + \frac{73}{24}\,e^2 + \frac{37}{96}\,e^4\right) \qquad (21)\end{aligned}$$

$$\begin{aligned}\left\langle\frac{de}{dt}\right\rangle &= \frac{m_1+m_2}{m_1 m_2 e}\left[\frac{a(1-e^2)}{m_1+m_2}\left\langle\frac{dE}{dt}\right\rangle - \frac{(1-e^2)^{\frac{1}{2}}}{a^{\frac{1}{2}}(m_1+m_2)^{\frac{1}{2}}}\left\langle\frac{dL}{dt}\right\rangle\right] \\ &= \frac{-304}{15}\,\frac{m_1 m_2 (m_1+m_2)e}{a^4(1-e^2)^{5/2}}\left(1 + \frac{121}{304}\,e^2\right). \qquad (22)\end{aligned}$$

From Equation (22) we see that de/dt is negative, so the eccentricity decreases because of radiation reaction.

If we put $e = 0$ in Equation (21), we recover the results of Problem 18.6. For the integration of Equations (21) and (22) when $e \neq 0$, see P. C. Peters, Phys. Rev. *136*, 1224 (1964).

Solution 18.8. The Ricci tensor for the wave must vanish. By contracting the result in Problem 13.13, we have then

$$0 = 2R_{\mu\nu} = h^{\alpha}{}_{\nu,\alpha\mu} + h^{\alpha}{}_{\mu,\alpha\nu} - h_{\mu\nu}{}^{,\alpha}{}_{\alpha} - h_{,\mu\nu} \ .$$

Since the perturbations are functions of $u = t - x$, the third term $\Box h_{\mu\nu}$ automatically vanishes and the components R_{22}, R_{23}, R_{33} automatically vanish. For $\mu,\nu = 0,2$ we find

$$\frac{d^2}{du^2}(h^0{}_2 - h^1{}_2) = 0 \ .$$

From this we conclude $h^0{}_2 = h^1{}_2$. [The integration constants are of no consequence; they are not wavelike; they carry no energy, and since they make no contribution to the Riemann tensor, they are gauge removable.] Similarly we have

$$h^0{}_3 = h^1{}_3 \quad \text{from} \quad \mu,\nu = 0,3$$

$$h^0{}_0 - h^1{}_0 = \frac{1}{2}h \quad \text{from} \quad \mu,\nu = 0,0$$

$$h^2{}_2 + h^3{}_3 = 0 \quad \text{from} \quad \mu,\nu = 1,0 \ .$$

Other choices of $\mu\nu$ give no further, independent relations.

These four constraints leave us with six independent components of $h_{\alpha\beta}$. A gauge transformation of the form $\xi_\mu(t-x)$ leaves the $h_{\alpha\beta}$ as functions only of $t - x$ and hence does not invalidate our four constraints above. We choose then a gauge transformation of this type which annihilates h_{00}, h_{11}, h_{02}, h_{03} and which leaves us with only the quantities $h_{23} = h_{32}$ and $h_{22} - h_{33}$ which are invariant for such a transformation. The explicit form of the gauge transformation is

$$\xi_0(u) = \frac{1}{2}\int^u h_{00}(\tilde{u})\,d\tilde{u} \qquad \xi_1(u) = -\frac{1}{2}\int^u h_{11}(\tilde{u})\,d\tilde{u}$$

$$\xi_2(u) = \int^u h_{02}(\tilde{u})\,d\tilde{u} \qquad \xi_3(u) = \int^u h_{03}(\tilde{u})\,d\tilde{u}\ .$$

Now we consider whether these are the same components as we would get by projecting out the transverse traceless part of the perturbation. Since we have $h = h^0{}_0 + h^1{}_1$ projecting out the transverse part is equivalent to projecting out the TT part. But projecting out the transverse part leaves h_{22}, h_{23}, h_{33} invariant, as did our gauge transformation. Thus we get the same components either way.

Solution 18.9. Let the z axis be the symmetry axis and let ϕ be the cyclic axial coordinate. At an instant of time, $t = \text{constant}$, in the asymptotically flat region far outside the sources through which the waves are propagating, the angular momentum inside a closed 2-surface $r = r_1$ is (by symmetry only an axial component)

$$J_1 = -(16\pi)^{-1} \oint_{r=r_1} \xi^{\mu;\nu}\, d^2\Sigma_{\mu\nu} \tag{1}$$

where $\boldsymbol{\xi} = \partial/\partial\phi$. (See Problem 16.23. The proof goes through even if there is no time Killing vector and even in the presence of waves since waves modify only the space-space parts of the metric — see MTW, Equation 19.5.) Similarly, the angular momentum inside a closed 2-surface $r = r_2 (r_2 > r_1)$ is

$$J_2 = -(16\pi)^{-1} \oint_{r=r_2} \xi^{\mu;\nu}\, d^2\Sigma_{\mu\nu}\ . \tag{2}$$

Thus the angular momentum of the waves between r_1 and r_2 is

$$J_2 - J_1 = -(16\pi)^{-1} \oint_{r=r_2} \xi^{\mu;\nu} d^2\Sigma_{\mu\nu} + (16\pi)^{-1} \oint_{r=r_1} \xi^{\mu;\nu} d^2\Sigma_{\mu\nu}$$

$$= \int_{r_1 \le r \le r_2} T^{\mu}{}_{\nu} \xi^{\nu} d^3\Sigma_{\mu}$$

where we have used Stokes' theorem as in Solution 16.23. Since $T^{\mu}{}_{\nu}$ vanishes for $r_1 \le r \le r_2$, $J_2 - J_1$ is zero and so the angular momentum flux of the waves vanishes.

Solution 18.10. A plane wave propagating in the z-direction can be described at the position of a detector by

$$h_{xx} = -h_{yy} = \mathrm{Re}\{A_+ e^{-i\omega t}\} \qquad h_{xy} = h_{yx} = \mathrm{Re}\{A_\times e^{-i\omega t}\} .$$

If the plane wave is monochromatic (and hence 100% polarized) A_+ and $A_\times$ will be constant. More generally if we are considering waves in a narrow bandwidth about ω, A_+ and $A_\times$ will be slow functions of time (i.e. slow compared to the wave oscillations). Since the intensity is proportional to the absolute squares of the A's we can define a polarization tensor

$$\rho_{ab} \equiv \frac{\langle A_a A_b^* \rangle}{|A_+|^2 + |A_\times|^2}$$

in analogy with the electromagnetic approach. [See for example Landau and Lifschitz, p. 122ff.] Following this analogy further we notice that this matrix is hermitean with a unit trace so we can write it as

$$\rho_{ab} = \frac{1}{2}\begin{bmatrix} 1+\xi_3 & \xi_1 - i\xi_2 \\ \xi_1 + i\xi_2 & 1-\xi_3 \end{bmatrix}.$$

As in electromagnetism ξ_1, ξ_2, ξ_3 will be called the *Stokes parameters* of the wave. [Note: Some authors call these the *normalized Stokes parameters*.]

These parameters characterize the polarization state of the wave. To get insight into the meaning of the parameters, let us consider a mixture of (1) unpolarized waves, (2) circularly polarized waves, and (3) linearly polarized waves with an orientation rotated about the x and y axis by an angle ψ. Mathematically these three cases correspond to superpositions of

(1) $A_+ = G_1(t) \quad A_\times = G_2(t)$ with $<|G_1(t)|^2> = <|G_2(t)|^2> \equiv <|G|^2>$
and with $<G_1(t)G_2^*(t)> = 0$

(2) $A_+ = H(t) \qquad A_\times = \pm\, iH(t)$

(3) $A_+ = F(t)\cos 2\psi \qquad A_\times = F(t)\sin 2\psi$.

We assume of course that G_1, G_2, F, and H are uncorrelated. The polarization matrix is then

$$\rho = \frac{1}{I}\begin{bmatrix} <|F|^2>\cos^2 2\psi + <|G|>^2 + <|H|>^2 & <|F|^2>\sin 2\psi\cos 2\psi \mp i<|H|^2> \\ <|F|^2>\sin 2\psi\cos 2\psi \pm i<|H|^2> & <|F|^2>\sin^2 2\psi + <|G|^2> + <|H|^2> \end{bmatrix}$$

$$I \equiv <|F|^2> + 2<|G|^2> + 2<|H|^2>$$

so that

$$\xi_3 = \frac{1}{I}<|F|^2>\cos 4\psi \qquad \xi_1 = \frac{1}{I}<|F|^2>\sin 4\psi \qquad \xi_2 = \pm\frac{2}{I}<|H|^2> .$$

We see finally that the polarization properties of the radiation can be found from the Stokes parameters in a manner very similar to that of electromagnetism.

Fraction of radiation linearly polarized $= (\xi_3^2 + \xi_1^2)^{\frac{1}{2}}$

Orientation of linear polarization: $\tan 4\psi = \xi_1/\xi_3$

Fraction of radiation circularly polarized $= |\xi_2|$
(ξ_2 positive for right circular polarization)

Fraction of radiation polarized $= (\xi_1^2 + \xi_2^2 + \xi_3^2)^{\frac{1}{2}}$.

Solution 18.11. The motion is most easily analyzed in a nearly Newtonian coordinate system. The motion of the particles (and the observer) then has the form

$$\ddot{x} \sim Rx \sim \ddot{h}x \ .$$

Here R denotes the magnitude of the Riemann tensor for the waves and h denotes the (dimensionless) metric perturbations for the waves. Since the waves are weak and oscillatory, the positions are fairly constant so we can integrate the equation of motion to get

$$x(t) - x_0 = x_0 h(t) + \alpha t + \beta + \mathcal{O}(h^2)$$

$$x_0 = \text{initial position} \ .$$

If we ignore the higher order corrections, we see that α and β must be zero since the particles were at rest in their initial positions before the wave burst arrived. After passage of the waves $h = 0$ and the particles must again be at rest in their initial positions.

Solution 18.12. The normal modes of the rod (taken to be oriented along the x-axis) are found from the equation

$$\frac{\partial^2 \xi}{\partial t^2} + \frac{1}{\tau}\frac{\partial \xi}{\partial t} - a^2 \frac{\partial^2 \xi}{\partial x^2} = 0 \ . \tag{1}$$

Here $\xi(x,t)$ is the displacement of an element of the rod, τ is the damping time and a is the speed of sound. The modes are of the form

$$\xi = e^{-i\omega t}\, e^{-t/2\tau}\, u(x) \tag{2}$$

where

$$a^2 \frac{d^2 u}{dx^2} + \left(\omega^2 + \frac{1}{4\tau^2}\right) u = 0 \ . \tag{3}$$

Since the ends of the rod are free,

$$\left.\frac{\partial\xi}{\partial x}\right|_{x=\pm L/2} = 0 \tag{4}$$

where L is the length of rod. The solutions of Equation (3) satisfying Equation (4) are

$$u_n = \begin{cases} \sin n\pi x/L, & n \text{ odd} \\ \cos n\pi x/L, & n \text{ even} \end{cases} \tag{5}$$

with frequencies $\omega_n = n\pi a/L$ if we assume $\omega_n \gg 1/\tau$.

Consider a gravitational wave propagating in the z-direction. The force field which it induces is

$$F_j = -x^k R_{j0k0}(t-z). \tag{6}$$

If we take the rod to be at $z = 0$, we must add a driving term

$$F(t,x) = -xR_{x0x0}(t) \tag{7}$$

to the right-hand side of Equation (1). We can express the resulting displacement as a superposition of normal modes:

$$\xi = \sum_n B_n(t)\,u_n(x). \tag{8}$$

The u_n are orthogonal, so Equation (1) gives

$$\ddot{B}_n + \frac{1}{\tau}\dot{B}_n + \omega_n^2 B_n = \frac{\int_{-L/2}^{L/2} F\,u_n\,dx}{\int_{-L/2}^{L/2} u_n^2\,dx}. \tag{9}$$

Since F is antisymmetric about $x = 0$, only the modes with n odd contribute. For n odd the right hand side of Equation (9) is

$$-R_{x0x0}(t)\,L/(2\pi^2 n^2).$$

If the rod is driven on resonance, i.e. $R_{x0x0} \sim e^{-i\omega_n t}$, Equation (9) with $B_n \sim e^{-i\omega_n t}$ gives

$$|B_n| \propto \left|\frac{\tau}{\omega_n n^2}\right| \propto \frac{1}{n^3}\,. \tag{10}$$

The energy flux F in the gravitational wave scales as $|R_{x0x0}/\omega_n|^2 \propto 1/n^2$, so

$$\text{sensitivity} \propto \frac{|B_n|^2}{1/n^2} \propto \frac{1}{n^4}\,. \tag{11}$$

Solution 18.13. Inside the material we have

$$T^{0\mu}_{GW,\mu} = T^{0i}_{GW,i} = F^0$$

where F^0 is the rate, per unit volume, at which external forces are doing work on the gravitational waves. Isolate a chunk of the cement of cross section A and put a Gaussian "pill box" around it. From Gauss' theorem in three dimensions we have

$$\int T^{0i}_{GW}\, d\Sigma_i = \int T^{0i}{}_{,i}\; d\,\text{Vol}.$$

$$\left(T^{0x}_{GW}\Big|_{out} - T^{0x}_{GW}\Big|_{in}\right)A = \int F^0\, d\,\text{Vol}.$$

= rate at which wave energy increases inside

= − rate at wave energy is absorbed by cement

and hence:

(rate at which momentum is absorbed per unit area per unit time) = (rate at which energy is absorbed per unit area per unit time) .

Solution 18.14. Take the wave to be propagating in the z-direction and the molecular displacement to be in the x-direction. If ξ is the displacement of the molecule, the equation of geodesic deviation gives us a dynamical equation for the molecule's motion. In a complex representation, the equation is the real part of

$$\ddot{\xi} + \Gamma\dot{\xi} + \omega_0^2\xi = -R^x{}_{0x0}X_0 = \frac{1}{2}\ddot{h}_{xx}X_0$$

or

$$\xi(-\omega^2 - i\omega\Gamma + \omega_0^2) = -\frac{\omega^2}{2}h_{xx}X_0 = -\frac{\omega^2}{2}A_+X_0e^{-i\omega t}\ .$$

Here ω is the frequency of the wave, $\Gamma \equiv (\text{damping time})^{-1}$, and X_0 is the distance in the x-direction of the equilibrium position of the molecule from the center of mass. We can easily solve for ξ,

$$\xi = \frac{1}{2}\omega^2A_+X_0e^{-i\omega t}[(\omega^2-\omega_0^2) - i\omega\Gamma]/\Delta$$

$$\Delta \equiv (\omega^2-\omega_0^2)^2 + \omega^2\Gamma^2$$

and for $\dot{\xi}$

$$\dot{\xi} = -\frac{1}{2}\omega^3A_+X_0e^{-i\omega t}[\omega\Gamma + i(\omega^2-\omega_0^2)]/\Delta\ .$$

From this we can easily evaluate the rate at which the molecule of mass m absorbs energy:

$$\begin{aligned}
<\dot{x}F^x> &= \frac{1}{2}\operatorname{Re}\{(\dot{\xi})(F^x)^*\} \\
&= \frac{1}{2}\operatorname{Re}\left\{(\dot{\xi})\left(-\frac{1}{2}m\omega^2A_+^*X_0e^{i\omega t}\right)\right\} \\
&= \frac{m\omega^6}{8}\Gamma|A_+|^2X_0^2\ .
\end{aligned}$$

To find the momentum absorption we must recall that there *is* a longitudinal geodesic deviation, though it is smaller by a factor of order v/c

$$\begin{aligned}
F^z &= -mR^z{}_{xx0}X_0\dot{\xi} = \frac{m}{2}\ddot{h}_{xx}X_0\dot{\xi} \\
&= -\frac{1}{2}m\omega^2X_0\operatorname{Re}\{A_+e^{-i\omega t}\}\operatorname{Re}\{\dot{\xi}\}\ .
\end{aligned}$$

The time averaged force is then

$$\begin{aligned}
<F^z> &= \frac{1}{4}m\omega^6X_0^2\Gamma<\operatorname{Re}(A_+e^{-i\omega t})\cdot\operatorname{Re}(A_+e^{-i\omega t})> \\
&= \frac{1}{8}m\omega^6\Gamma|A_+|^2X_0^2
\end{aligned}$$

which demonstrates on a microscopic basis the reason for the result in Problem 18.13.

Solution 18.15. Consider two particles a distance d apart. Their relative velocity fluctuates under the influence of the wave by an amount $\Delta v \approx h\omega d$. If the two particles collide during this cycle of the wave, their distance must have been of order $d \sim v/\omega$, so $\Delta v \sim hv$. On the average then, particles collide with energies fractionally larger by $(\Delta v/v)^2$ than they would in the absence of the wave. Since the collision randomizes their direction of motion, this energy is *not* subtracted out half a cycle later when the wave is reversed. Another effect also adds of order $(\Delta v/v)^2$ fractional energy per collision: the fact that particles moving toward each other at $v + \Delta v$ have a greater chance of colliding than do particles with $v - \Delta v$. We conclude that at temperature T every collision is, on the average, increased in energy by order $(\Delta v/v)^2 kT \sim h^2 kT$. The number of collisions per volume per time is $n_0 v/\ell \sim (n_0/\ell)(kT/m)^{\frac{1}{2}}$, where n_0 is the number density and m the mass of the particles. Thus the internal energy absorbed per volume per time is approximately

$$\frac{dE_{int}}{dt} \sim \frac{h^2}{n_0^{\frac{1}{2}} m^{\frac{1}{2}} \ell} (n_0 kT)^{\frac{3}{2}} \sim \frac{h^2}{n_0^{\frac{1}{2}} m^{\frac{1}{2}} \ell} E_{int}^{\frac{3}{2}} . \tag{1}$$

This is integrated to give

$$E_{int}^{\frac{1}{2}}(t) \sim \frac{t_0 E_{int}^{\frac{1}{2}}(t=0)}{t_0 - t} \tag{2}$$

where $t_0 \sim (\ell/h^2)(m/kT)^{\frac{1}{2}}$. Thus, $E_{int} \to \infty$ in a finite time t_0; of course this does not happen: our Newtonian treatment breaks down at a time $t \sim t_0$ when the gas becomes relativistic.

The "front" of the wave loses energy at the rate given by Equation (1). Since its internal energy density is of order $\omega^2 h^2 c^2/G$, it is damped in a distance

$$\ell_{\text{damping}} \sim c\tau_{\text{damping}} \sim \frac{\omega^2 h^2 c^3/G}{h^2 (n_0 kT)^{\frac{3}{2}}/(n_0^{\frac{1}{2}} m^{\frac{1}{2}} \ell)}$$

$$\sim \ell \left(\frac{c}{v_{\text{thermal}}}\right)^3 \left(\frac{\omega^2}{G m n_0}\right) .$$

(The first term in parenthesis approaches unity for a relativistic gas; the second approaches unity for a wave at the "gravitational plasma frequency" of the gas.)

Solution 18.16. The characteristic quadrupole gravitational-wave power of a physical process is $(G/c^5)\,|\dddot{Q}|^2 \sim (G/c^5) M^2 v^4/T^2$ where M, v, T are relevant masses, velocities and times for the process. This can be conveniently written as

$$P_{GW} \sim \frac{G}{c^5}\left(\frac{Mv^2}{T}\right)^2 \sim \frac{P^2}{3 \times 10^{59}\,\text{ergs/sec}} .$$

Here $P \sim Mv^2/T$ represents the characteristic "power" of internal energy flows.

For an explosion of energy E, characteristic time τ the "internal power flow" is E/τ. If we assume that this internal power flow couples efficiently to gravitational waves, then in $c = G = 1$ units

$$P_{GW} \sim (E/\tau)^2$$

and

$$E_{GW} \sim P_{GW}\tau \sim E^2/\tau .$$

The typical graviton emerging from this explosion will have energy $\hbar\omega \sim \hbar/\tau$ so the number of gravitons emerging will be of order

$$N \sim \frac{E^2/\tau}{\hbar/\tau} = \frac{E^2}{\hbar} \sim \left(\frac{E}{10^{16}\,\text{ergs}}\right)^2 .$$

Solution 18.17. The mass motions relevant to gravitational wave generation are the collisions of electrons with lattice points, photon scattering, etc. Since the thermal velocities of the electrons greatly exceed their drift velocity, the energy involved in a collision is of order kT. According to the analysis of Problem 18.16, the gravitational wave power generated by one electron should be $(kT/\tau)^2$ where τ is the mean time between collisions of the electron with the lattice. The total energy, in gravitational waves, emitted during the bulb lifetime must then be

$$E_{GW} \sim N\left(\frac{kT}{\tau}\right)^2 T_L/(10^{59}\,\text{ergs/sec.})$$

$$N = \text{number of conduction electrons in filament}$$

$$T_L = \text{bulb life time} \sim 4 \times 10^6 \text{ seconds}.$$

The typical graviton generated should have a frequency of roughly τ^{-1} so that n, the number of gravitons is of order

$$n \sim \frac{E_{GW}\tau}{\hbar} \sim \frac{N}{\tau\hbar}\,\frac{(kT)^2\,T_L}{10^{59}\,\text{ergs/sec.}}\;.$$

For the light bulb kT must be on the order of an electron volt, since the thermal emission is primarily in the visible, and rough values of N and τ are 10^{17} electrons and 10^{-13} seconds. For these values the number of gravitons emitted during the bulb's lifetime is of order 10^{-19}.

To estimate the number of gravitons given out by the lightbulb when it is dropped, we assume that the energy of the falling light bulb is converted to kinetic energy of shattering; and we calculate the resulting production of gravitational waves according to the formula in Problem 18.16. If the light bulb weighs 20 grams and is dropped from a height of 1 meter, it gains $\sim 10^6$ ergs. Let us assume that the shattering is characterized by a time of .1 sec, so that the total power in shattering is 10^7 ergs/sec. If 10% of this is relevant to quadrupole oscillations, then

$$P_{GW} \sim \frac{(10^6 \text{ ergs/sec})^2}{10^{59} \text{ ergs/sec}} \approx 10^{-47} \text{ ergs/sec}$$

$$E_{GW} \sim P_{GW} \times .1 \text{ sec} \approx 10^{-48} \text{ ergs} .$$

The number of $\hbar\omega \sim \hbar/.1 \text{ sec} \sim 10^{-26}$ erg gravitons is then on the order of 10^{-22}.

Solution 18.18. To linearized order in the gravitational field, the interaction Hamiltonian with a system whose stress-energy tensor is $T_{\mu\nu}$ is

$$H = (8\pi G)^{\frac{1}{2}} T_{\mu\nu} h^{\mu\nu} . \qquad (1)$$

We shall use units with $c = \hbar = 1$, so $G^{\frac{1}{2}} = 1.616 \times 10^{-33}$ cm. For the electron in the hydrogen atom,

$$T_{\mu\nu} = m_e u_\mu u_\nu = \frac{p_\mu p_\nu}{m_e} ; \qquad (2)$$

we work in the transverse-traceless gauge for $h^{\mu\nu}$, so

$$h_{0\mu} = 0, \qquad h^{ij}{}_{,j} = 0 \qquad (3)$$

and thus

$$H = (8\pi G)^{\frac{1}{2}} \frac{p_i p_j h^{ij}}{m_e} . \qquad (4)$$

Since a graviton has spin 2, the decay must be to an s state. The transition matrix element from a 3d state to a 1s state with the emission of a graviton with wave-vector $\underset{\sim}{k}$ and polarization λ (for right-hand circular polarization $\lambda = 1$ and for left-hand circular polarization $\lambda = -1$) is

$$T = \langle 1s; \underset{\sim}{k}, \lambda | H | 3d; 0 \rangle . \qquad (5)$$

We now expand

$$h^{ij} = \frac{1}{(2\pi)^{\frac{3}{2}}} \int \frac{d^3k}{(2\omega)^{\frac{1}{2}}} \sum_\lambda (e^{-i\underset{\sim}{k}\cdot\underset{\sim}{r}} e^{ij}_{k,\lambda} a_{k,\lambda} + e^{i\underset{\sim}{k}\cdot\underset{\sim}{r}} e^{ij*}_{k,\lambda} a^\dagger_{k,\lambda}) . \qquad (6)$$

Here ω is the frequency of the wave and e^{ij} the polarization tensor. The creation and annihilation operators satisfy the equal-time commutation relations

$$[a_{k,\lambda}, a^{\dagger}_{k',\lambda'}] = \delta^3(\underset{\sim}{k}-\underset{\sim}{k}')\delta_{\lambda\lambda'}$$

and so

$$a_{k,\lambda}|0\rangle = 0$$

$$\langle \underset{\sim}{k},\lambda|a^{\dagger}_{k',\lambda'}|0\rangle = \delta^3(\underset{\sim}{k}-\underset{\sim}{k}')\delta_{\lambda\lambda'} . \tag{7}$$

Using Equations (4), (6), and (7) in Equation (5), we get

$$T = \frac{(8\pi G)^{\frac{1}{2}}}{m_e}\frac{1}{(2\pi)^{\frac{3}{2}}}\frac{1}{(2\omega)^{\frac{1}{2}}} \langle 1s\,|e^{i\underset{\sim}{k}\cdot\underset{\sim}{r}}p_i p_j\, e^{ij^*}_{k,\lambda}|3d\rangle \ . \tag{8}$$

Since $kr \sim (1\mathrm{eV}) \times 10^{-8}\,\mathrm{cm} \sim 10^{-4}$, we can make the approximation of putting $e^{i\underset{\sim}{k}\cdot\underset{\sim}{r}} = 1$ (cf. the "dipole" approximation for electromagnetic transitions). The transition rate is

$$\begin{aligned} d\Gamma &= 2\pi\,|T|^2 \times (\text{density of final states for graviton}) \\ &= 2\pi\,|T|^2\,\omega^2\,d\Omega \end{aligned} \tag{9}$$

where $d\Omega$ is the solid angle into which the graviton is emitted. Hence

$$\frac{d\Gamma}{d\Omega} = \frac{G\omega}{\pi\, m_e^2}\,|\langle 1s|p_i p_j\, e^{ij^*}_{k,\lambda}|3d\rangle|^2 \ . \tag{10}$$

If we use the explicit form of the wave functions, the matrix element in this equation can be evaluated. Since we do not observe the spins of the initial or final states of the atom, we will sum over the final spin states and average over the initial spin states. This is most easily done by introducing the pure spin-2 operator

$$Q_{ij} = p_i p_j - \frac{1}{3}\delta_{ij}\,p^2 \ . \tag{11}$$

Since $e^i_{\ i} = 0$, we can replace $p_i p_j$ by Q_{ij} in Equation (10). The spherical components of Q_{ij} are

$$\begin{aligned}
Q_2 &\equiv \frac{1}{2}(Q_{xx} - Q_{yy} + 2i\,Q_{xy}) \\
Q_1 &\equiv Q_{zx} + i\,Q_{zy} \\
Q_0 &\equiv \left(\frac{2}{3}\right)^{\frac{1}{2}}\left(Q_{zz} - \frac{1}{2}Q_{xx} - \frac{1}{2}Q_{yy}\right) \\
Q_{-1} &\equiv Q_{zx} - i\,Q_{zy} \\
Q_{-2} &\equiv \frac{1}{2}(Q_{xx} - Q_{yy} - 2i\,Q_{xy})\,.
\end{aligned} \tag{12}$$

Since e_λ^{ij} has x-y components (for $\lambda = \pm 1$):

$$2^{-\frac{1}{2}}\begin{bmatrix} 1 & \pm i \\ \pm i & -1 \end{bmatrix}$$

we find

$$Q_{ij}\, e_\lambda^{ij*} = 2^{\frac{1}{2}}\, Q_{-2\lambda}\,. \tag{13}$$

Since Q_σ has spin two, the Wigner-Eckart theorem says

$$<j_j m_j|Q_\sigma|j_i m_i> = <j_j\|Q\|j_i><2\sigma\, j_i m_i|j_j m_j> \tag{14}$$

where the "reduced" matrix is defined as

$$<j_j\|Q\|j_i> = \sum_{m_i m_j \sigma} \frac{1}{2j_j+1} <2\sigma\, j_i m_i|j_j m_j><j_j m_j|Q_\sigma|j_i m_i> \tag{15}$$

and where $<jm\, j_i m_i|j_j m_j>$ denotes a Clebsch-Gordon coefficient. Thus Equations (13) and (14) (with the n quantum number added) give

$$|<n_j j_j m_j|Q_{ij}e^{ij*}|n_i j_i m_i>|^2 = 2|<n_j j_j\|Q\|n_i j_i>|^2|<2\ -2\lambda\ j_i m_i|j_j m_j>|^2\,.$$

Now we sum over m_j and average over m_i, using

$$\frac{1}{2j_i+1}\sum_{m_i m_j}|<2\ -2\lambda\ j_i m_i|j_i m_j>|^2 = \frac{1}{2j_i+1}\sum_{m_i m_j}\frac{(2j_j+1)}{5}|<j_j - m_j j_i m_i|2{-}2\lambda>|^2$$

$$= \frac{1}{5}\,\frac{2j_j+1}{2j_i+1}$$

and thus

$$\frac{d\Gamma}{d\Omega} = \frac{G\omega}{\pi m_e^2} \frac{2}{5} \frac{2j_j+1}{2j_i+1} |<n_j j_j \| Q \| n_i j_i>|^2 . \tag{16}$$

This is independent of the polarization of the graviton and of the angle of emission. The total rate is obtained by multiplying by 2 (for the two polarization states) and by 4π (solid angle), giving

$$\Gamma = \frac{G\omega}{m_e^2} \frac{16}{5} \frac{2j_j+1}{2j_i+1} |<n_j j_j \| Q \| n_i j_i>|^2. \tag{17}$$

The reduced matrix element can be evaluated from Equation (14) and any convenient choice of m_i and m_j. For example,

$$<1s\|Q\|3d> = \frac{<100|Q_0|320>}{<2020|00>} . \tag{18}$$

The Clebsch-Gordon coefficient in Equation (18) is unity. From Equations (11) and (12),

$$Q_0 = \left(\frac{3}{2}\right)^{\frac{1}{2}} \left(p_z p_z - \frac{1}{3} p^2\right) .$$

The p^2 term does not contribute to the matrix element in Equation (18) because it cannot couple $j = 0$ to $j = 2$. Thus

$$<1s\|Q\|3d> = \left(\frac{3}{2}\right)^{\frac{1}{2}} <100|p_z p_z|320> . \tag{19}$$

In general one would introduce a complete set of states to evaluate the matrix element in Equation (19) as a sum of terms of the form $<100|p_z|n><n|p_z|320>$. Since the $|100>$ wave function is so simple, however, we can use

$$p_z \to -i\partial/\partial z$$

so

$$\begin{aligned} p_z^2 \to -\partial^2/\partial z^2 &= -\left(\cos\theta \frac{\partial}{\partial r} - \frac{\sin\theta}{r} \frac{\partial}{\partial\theta}\right)\left(\cos\theta \frac{\partial}{\partial r} - \frac{\sin\theta}{r} \frac{\partial}{\partial\theta}\right) \\ &= -\left(\cos^2\theta \frac{\partial^2}{\partial r^2} + \frac{\sin^2\theta}{r} \frac{\partial}{\partial r}\right) \end{aligned}$$

when acting on the spherically-symmetric $|100\rangle$ state. Thus

$$\langle 320|p_z^2|100\rangle = -\int_0^\infty r^2 dr \int_0^\pi \sin\theta\, d\theta \int_0^{2\pi} d\phi \left(\frac{5}{16\pi}\right)^{\frac{1}{2}} \frac{(3\cos^2\theta - 1)}{a^{\frac{3}{2}}} \frac{2(30)^{\frac{1}{2}}}{955} \left(\frac{r}{a}\right)^3$$

$$\times\, e^{-r/3a} \left(\cos^2\theta \frac{\partial^2}{\partial r^2} + \frac{\sin^2\theta}{r}\frac{\partial}{\partial r}\right) \frac{1}{(4\pi)^{\frac{1}{2}}} \frac{2}{a^{\frac{3}{2}}} e^{-r/a}$$

$$= \frac{-6^{\frac{1}{2}}}{191\, a^6} \int_0^\infty dr\, r^5 e^{-\frac{4r}{3a}} \int_0^\pi \sin\theta\, d\theta\, (3\cos^2\theta - 1)\left(\frac{\cos^2\theta}{a^2} - \frac{\sin^2\theta}{ra}\right)$$

$$= \frac{6^{\frac{1}{2}}}{191\, a^6} \frac{8}{15} \int_0^\infty dr\, e^{-\frac{4r}{3a}} \left(\frac{r^5}{a^2} + \frac{r^4}{a}\right)$$

$$= -\frac{1}{a^2} \frac{6^{\frac{1}{2}} \times 8 \times 243 \times 57}{191 \times 15 \times 2 \times 256}$$

$$= \frac{-.19}{a^2}$$

where $a = (m_e \alpha)^{-1}$ is the Bohr radius and $\alpha = 1/137$ is the fine structure constant.

$$\Gamma = \frac{G\omega}{m_e^2} \cdot \frac{16}{5} \cdot \frac{1}{5} \cdot \frac{3}{2} \cdot \frac{(.19)^2}{a^4}$$

$$= .36\, G\, m_e^2\, \omega \alpha^4 \,.$$

Now $G m_e^2 = 1.75 \times 10^{-45}$ and $\omega = 12\text{eV}$ for $3p \rightarrow 1s$. Thus the lifetime is $\Gamma^{-1} = 1.9 \times 10^{38}$ sec.

Solution 18.19. A star, and its thermal graviton flux are spherically symmetric only in the time averaged sense. Since a star's interior is composed of subatomic particles it is clear that, at a given time the star may be spherically symmetric to an astronomer, but to a mathematician, with his much more stringent requirements for symmetry, the star has a fine grain (atomic scale) graininess which makes all symmetry theorems inapplicable. It is, of course, the time dependence of these fine scale particle inhomogenieties which generate gravitational (and electromagnetic!) radiation.

The multipolarity of the flux is related to the scale of angular asymmetries of the flux. (This should not be confused with the fact that the generation of the waves is *locally* quadrupole, nor with the fact that the time averaged flux is monopole – i.e. isotropic.) For a star like our Sun most of the graviton flux will be generated in the high temperature (10^7°K), dense (50 gm/cm^3) core. If we take as a characteristic radius of this core one quarter of the sun's radius $\sim 2 \times 10^{10}$ cm and the interparticle distance of 10^{-8} cm implied by the density, we see that the angular scale of asymmetry is $\sim 10^{-18}$ radians. The spherical harmonic expansion of graininess of 10^{-18} radians requires ℓ values of order 10^{18}. In this fine-grain sense then, the flux has characteristically a $2^{10^{18}}$-pole distribution.

Solution 18.20. In a nearly inertial coordinate system centered at the center of mass, the effect of the Riemann tensor, according to the equation of geodesic deviation, is to accelerate free particles in the following way

$$\frac{d^2x^j}{dt^2} = -R_{j0k0}\,x^k .$$

We have here ignored the velocity dependent "magnetic" type terms which are smaller by a factor v/c. (See solution to Problem 18.14.) This equation makes it clear that the Riemann tensor has a longitudinal effect if and only if one of the spatial indices of R_{j0k0} is z, the direction of wave propagation. It follows that Ψ_2, Ψ_3, $\overline{\Psi}_3$ have longitudinal effects but that Ψ_4, $\overline{\Psi}_4$, Φ_{22} are purely transverse.

To investigate the spin of the waves, transform from x,y coordinates to new coordinates x', y' rotated about the z-axis in the positive sense by an angle ϕ. Under such a transformation the transformation for the Riemann components is, for instance

$$R_{x'0'z'0'} = R_{x0z0}\cos\phi + R_{y0z0}\sin\phi$$

$$R_{x'0'x'0'} = R_{x0x0}\cos^2\phi + R_{y0y0}\sin^2\phi + 2\sin\phi\cos\phi\, R_{x0y0}\,.$$

From the form of the transformation for the Riemann components, the transformation laws for the symbols defined in the problem are easily calculated:

$$\Psi_{2'} \equiv -\frac{1}{6}R_{z'0'z'0'} = -\frac{1}{6}R_{z0z0} = \Psi_2$$

$$\Psi_{3'} = e^{i\phi}\Psi_3$$

$$\bar{\Psi}_{3'} = e^{-i\phi}\Psi_3$$

$$\Psi_{4'} = e^{2i\phi}\Psi_4\,, \qquad \bar{\Psi}_{4'} = e^{-2i\phi}\bar{\Psi}_4$$

$$\Phi_{2'2'} = \Phi_{22}\,.$$

Since Ψ_2 and Φ_{22} are unaffected by rotation about z, the waves are scalar (spin 0) waves. For Ψ_3 and $\bar{\Psi}_3$, a rotation of 180° returns the wave to the same state of polarization (e.g. pure real), so these symbols must correspond to spin 1 waves. The symbols Ψ_4 and $\bar{\Psi}_4$ need only be rotated by 90° to be in the same polarization state so they represent spin 2 waves. These symbols in fact correspond to the circular polarization states of the waves of general relativity.

Solution 18.21. In the figure are the six polarization modes of a weak, plane, null gravitational wave permitted in the generic metric theory of gravity. The figure shows the displacement that each mode induces on a sphere of test particles. The wave propagates in the $+z$-direction (arrow at upper right) and has time dependence $\cos(\omega t)$. Solid line, snapshot at $\omega t = 0$, the broken line, one at $\omega t = \pi$. There is no displacement perpendicular to the plane of the figure. The above modes are obtained by using the equation of geodesic deviation

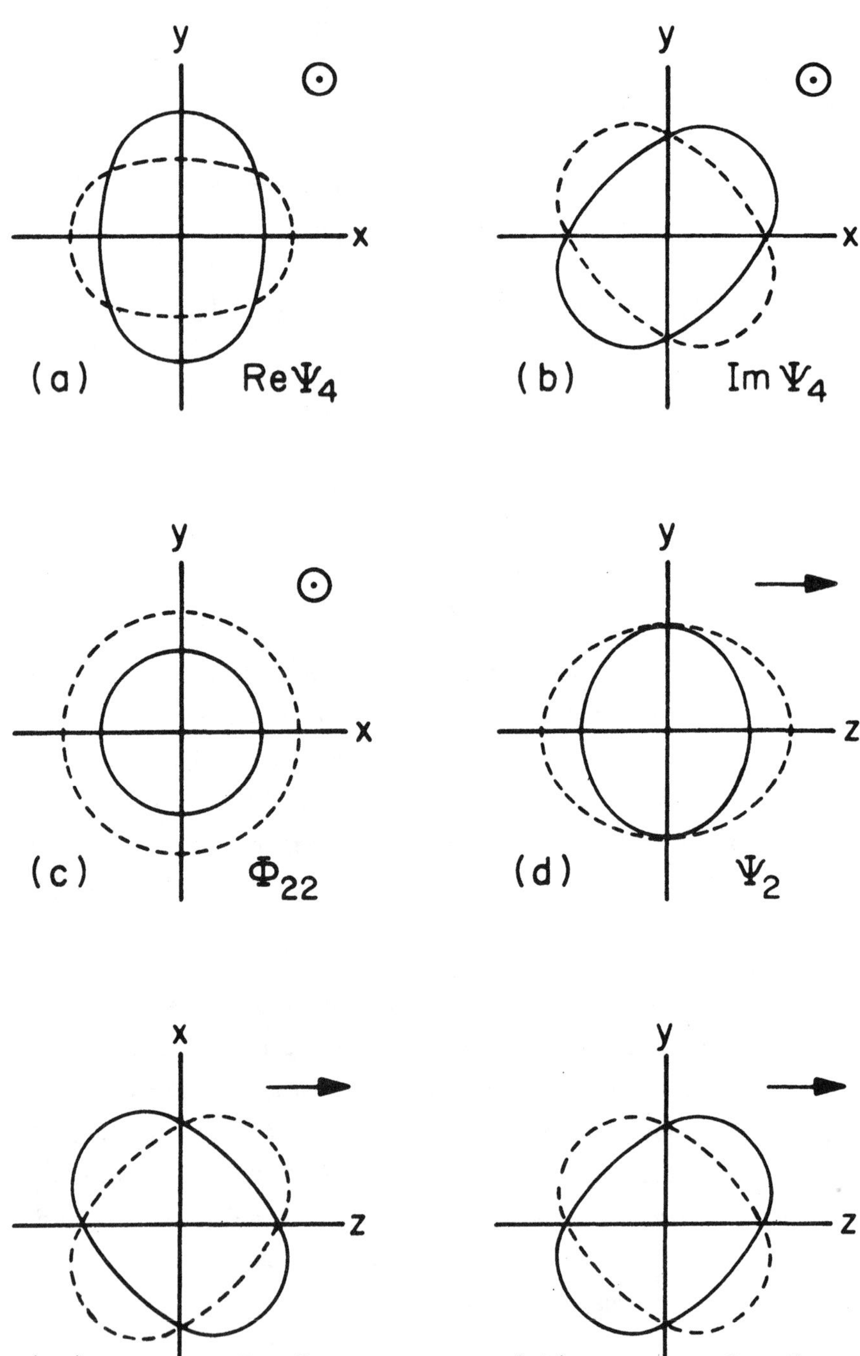
y
x
(a)
ReΨ_4
y
x
(b)
Im Ψ_4
y
x
(c)
Φ_{22}
y
z
(d)
Ψ_2
x
z
(e)
Re Ψ_3
y
z
(f)
Im Ψ_3

$$\frac{d^2x^k}{dt^2} \approx x^j R^k{}_{0j0} \, .$$

Solution 18.22. The function H is determined by the vacuum field equations $R_{\alpha\beta} = 0$. We first compute the Christoffel symbols from the Lagrangian for geodesics (see Problem 7.25)

$$L = \dot{x}^2 + \dot{y}^2 - \dot{u}\dot{v} + 2H\dot{u}^2 \, .$$

From the Euler-Lagrange equation

$$\frac{d}{ds}\left(\frac{\partial L}{\partial \dot{x}}\right) - \frac{\partial L}{\partial x} = 0$$

we get

$$\ddot{x} + H_{,x}\dot{u}^2 = 0$$

and therefore $\Gamma^x{}_{uu} = H_{,x}$. Similarly $\ddot{y} + H_{,y}\dot{u}^2 = 0$ implies $\Gamma^y{}_{uu} = H_{,y}$ and $\ddot{u} = 0$ implies $\Gamma^u{}_{\alpha\beta} = 0$. From

$$\ddot{v} - 2H_{,u}\dot{u}^2 - 4H_{,x}\dot{x}\dot{u} - 4H_{,y}\dot{y}\dot{u} = 0$$

we get

$$\Gamma^v{}_{uu} = -2H_{,u} \qquad \Gamma^v{}_{xu} = -2H_{,x} \qquad \Gamma^v{}_{yu} = -2H_{,y} \, .$$

All other Γ's are zero. In the expression (MTW, Equation 8.51b)

$$R_{\alpha\beta} = \Gamma^\mu{}_{\alpha\beta,\mu} - (\log(-g)^{\frac{1}{2}})_{,\alpha\beta} + (\log(-g)^{\frac{1}{2}})_{,\mu}\Gamma^\mu{}_{\alpha\beta} - \Gamma^\mu{}_{\sigma\beta}\Gamma^\sigma{}_{\alpha\mu}$$

only the first term contributes (note that $(-g)^{\frac{1}{2}} = 1$). We find the only nonzero term to be

$$R_{uu} = H_{,xx} + H_{,yy} \, .$$

Any harmonic function in x and y will therefore satisfy the field equations. For a plane wave, H is a quadratic function in x and y, so that (exercise!) the Riemann tensor is a function of u alone and has no singularities in the x-y plane.

CHAPTER 19: SOLUTIONS

Solution 19.1. The relevant Newtonian equations are

$$\nabla^2\Phi = 4\pi G\rho \quad \text{(gravitational potential)} \tag{1}$$

$$\partial\rho/\partial t + \nabla\cdot(\rho\underset{\sim}{v}) = 0 \quad \text{(continuity equation)} \tag{2}$$

$$\frac{\partial \underset{\sim}{v}}{\partial t} + (\underset{\sim}{v}\cdot\nabla)\underset{\sim}{v} = -\frac{\nabla p}{\rho} - \nabla\Phi \quad \text{(equation of motion)}\ . \tag{3}$$

If the universe is static and homogeneous, then $\underset{\sim}{v} = 0$, and ρ and p are constant in position and time. The solution of Equation (1) is then

$$\Phi = \frac{2}{3}\pi G\rho\,(\underset{\sim}{r}\cdot\underset{\sim}{r}) + \underset{\sim}{C}\cdot\underset{\sim}{r} + K$$

where $\underset{\sim}{C}$ and K are arbitrary constants of integration and $\underset{\sim}{r}$ is the position vector from the (arbitrary) origin. Equation (2) is identically zero. Equation (3), however is contradictory: the left side vanishes since $\underset{\sim}{v}=0$; on the right, ∇p vanishes by homogeneity, but $\nabla\Phi$ is not identically zero for any choice of $\underset{\sim}{C}$, K. Thus, there is no solution to the equations.

Solution 19.2. Since the spacetime is isotropic everywhere, it is in particular spherically symmetric around some nonsingular observer. But by Birkhoff's theorem (Problem 16.3) the only vacuum spherical solutions are the Schwarzschild metrics, and the only one which is regular at the origin – where our observer is – is the $M = 0$ case, i.e. flat Minkowski space.

Solution 19.3. By Liouville's theorem the number density of photons in phase space I_ν/ν^3 is a constant along a light ray, and (see Problem 5.10) it is also a Lorentz invariant. Using $(\nu_{\text{emitted}}/\nu_{\text{observed}}) = 1+z$, We have then

$$\text{Flux observed} = \Omega \int I_\nu^{\text{obs}}\, d\nu_{\text{obs}} = \Omega \int \left(\frac{I_\nu^{\text{obs}}}{\nu^3_{\text{obs}}}\right) \nu^3_{\text{obs}}\, d\nu_{\text{obs}}$$

$$= \Omega \int \frac{I_\nu^{\text{emit}}}{\nu^3_{\text{emit}}} \frac{\nu^3_{\text{emit}}\, d\nu_{\text{emit}}}{(1+z)^4} = \frac{\Omega}{(1+z)^4} \int I_\nu^{\text{emit}}\, d\nu_{\text{emit}}$$

$$= \frac{\Omega}{(1+z)^4} \frac{1}{\pi} \sigma T^4$$

where σ is the Stefan-Boltzmann constant. Note that since a redshifted black body spectrum is also a black body spectrum, the flux and spectrum observed is just that which would come from a nearby stationary object of solid angle Ω and temperature $T/(1+z)$.

In all the above it doesn't matter whether the redshift is due to doppler motion, gravitational redshift, or cosmological redshift, because Liouville's theorem holds in general.

Solution 19.4. Homogeneity implies that the scalar curvature R of the hypersurface must be constant. R can be found by specializing the formulas of Problem 9.20, or from the formula

$$R_{ij} = \Gamma^k{}_{ij,k} - (\log|g|^{\frac{1}{2}})_{,ij} + \Gamma^k{}_{ij}(\log|g|^{\frac{1}{2}})_{,k} - \Gamma^k{}_{im}\Gamma^m{}_{jk}$$

where $|g|^{\frac{1}{2}} = a^3 f r^2 \sin\theta$ and the Γ's are found from the Lagrangian for geodesics or as in Problem 7.6. This gives

$$R_{\theta\theta} = \Gamma^r{}_{\theta\theta,r} - (\log(\sin\theta))_{,\theta\theta} + \Gamma^r{}_{\theta\theta}(\log fr^2)_{,r} - \Gamma^\phi{}_{\theta\phi}\Gamma^\phi{}_{\theta\phi} - 2\Gamma^\theta{}_{\theta r}\Gamma^r{}_{\theta\theta}$$

$$= -(rf^{-2})_{,r} + \operatorname{cosec}^2\theta - r^{-1}f^{-3}(r^2 f)_{,r} - \cot^2\theta + 2f^{-2}$$

$$= [r^2(1-f^{-2})]_{,r}(2r)^{-1}\ .$$

Since

$$R_{\hat{\theta}\hat{\theta}} = R_{\theta\theta}(a^2 r^2)^{-1} ,$$

and

$$R_{\hat{r}\hat{r}} = R_{\hat{\phi}\hat{\phi}} = R_{\hat{\theta}\hat{\theta}}$$

by isotropy, we get

$$R = R_{\hat{r}\hat{r}} + R_{\hat{\theta}\hat{\theta}} + R_{\hat{\phi}\hat{\phi}} = 3[r^2(1-f^{-2})]_{,r}(2a^2r^3)^{-1} \equiv A$$

that is

$$r^2(1-f^{-2}) = Br^4 + C .$$

Since $f \to 1$ as $r \to 0$, we have $C = 0$. Thus

$$f^2 = (1-Br^2)^{-1} .$$

If $B \neq 0$, we can scale r by letting $r' = |B|^{\frac{1}{2}} r$. The metric then takes the prescribed form with radial coordinate r'.

Solution 19.5.

(a) Let

$$r = \begin{cases} \sin\chi & k = +1 \\ \chi & k = 0 \\ \sinh\chi & k = -1 \end{cases}$$

then

$$dr = \begin{cases} \cos\chi \;\; d\chi \\ d\chi \\ \cosh\chi \;\; d\chi \end{cases}$$

and

$$\frac{dr^2}{1-kr^2} = \begin{cases} d\chi^2 \\ d\chi^2 \\ d\chi^2 \end{cases}$$

$$\frac{dr^2}{1-kr^2} + r^2 d\Omega^2 = d\chi^2 + \Sigma^2(\chi)\, d\Omega^2 ,$$

where

$$\Sigma^2(\chi) = \begin{cases} \sin^2\chi\ , & k = +1 \\ \chi^2\ , & k = 0 \\ \sinh^2\chi\ , & k = -1\ . \end{cases}$$

(b) Let $dt = R(\eta)\,d\eta$ define a transformation from the variable t to the variable η. Then

$$\begin{aligned} ds^2 &= -dt^2 + R^2(t)(d\chi^2 + \Sigma^2 d\Omega^2) \\ &= R^2(\eta)(-d\eta^2 + d\chi^2 + \Sigma^2 d\Omega^2)\ . \end{aligned}$$

Solution 19.6. From the Robertson-Walker metric (in "trigonometric" form), the metric for the spacelike 3-surfaces is

$$d\sigma^2 = a^2\,[d\chi^2 + \sin^2\chi\,(d\theta^2 + \sin^2\theta\,d\phi^2)]\ .$$

By analogy with the metric for 2-spheres, it is easy to guess that this metric can be represented as a 3-sphere embedded in a Euclidean 4-space with cartesian coordinates W, X, Y, Z. In particular, if we define

$$\begin{aligned} W &= a\cos\chi \\ Z &= a\sin\chi\,\cos\theta \\ X &= a\sin\chi\,\sin\theta\,\cos\phi \\ Y &= a\sin\chi\,\sin\theta\,\sin\phi \end{aligned}$$

(which satisfy $W^2 + Z^2 + X^2 + Y^2 = A^2$), the metric becomes

$$d\sigma^2 = dW^2 + dZ^2 + dX^2 + dY^2\ .$$

In Euclidean 4-space one can find two planes (e.g. $X-Y$ and $W-Z$) which intersect only at the origin $(a=0)$. Thus, a rotation of the $X-Y$ plane changes all X and Y coordinates except the origin and leaves all $W-Z$ coordinates fixed; and vice versa for $W-Z$ rotations. Therefore, a combination of both rotations leaves no points fixed except the origin and (for nonzero a) takes all points of the 3-sphere into different points.

Solution 19.7. In some local proper frame suppose the bullet passes by with a proper velocity V. When it has moved a proper distance dr farther, it passes a cosmological observer whose velocity relative to the first proper frame is

$$\delta V = H\,dr = HV\,dt = \frac{\dot R}{R} V\,dt = V\,\frac{dR}{R}\ .$$

This observer sees it moving (according to the velocity addition formula) at velocity

$$V' = \frac{V-\delta V}{1-V\delta V} = V-(1-V^2)\delta V + \mathcal{O}(\delta V)^2$$
$$= V-(1-V^2)V\,\frac{dR}{R} + \mathcal{O}(\delta V)^2\ .$$

Thus, $dV/dR = -(1-V^2)V/R$ which is integrable, giving $\gamma V \equiv (1-V^2)^{-\frac{1}{2}}V = \text{constant}/R = \text{constant}\times(1+z)$. So V_1 and V_2 are related by $\gamma_2 V_2/\gamma_1 V_1 = (1+z)^{-1}$. For a particle with nonzero rest mass, this result says that the relativistic momentum redshifts away as one power of $(1+z)$ as the universe expands. For a photon, in the limit $V\to 1$, $\gamma V \to h\nu$ and we obtain the usual redshift factor. (Note that we have not taken into account the slowing down of the bullet due to the gravitational attraction of the cosmological matter at rest in the proper frame. The reader can verify that this is a higher order effect, with $dV \propto (dr)^2$, so it does not affect the differential equation for dV/dR.

Solution 19.8. Start with the η, χ form of the metric

$$ds^2 = R^2(\eta)[-d\eta^2 + d\chi^2 + \Sigma^2(\chi)d\Omega^2] \qquad (1)$$

(3rd form in Problem 19.5), where $\Sigma(\chi) = \sinh\chi$, χ, $\sin\chi$ for $k = -1$, 0, $+1$. Since conformal transformations preserve light cones, a natural first step is to go to null coordinates, e.g.

$$u = \frac{1}{2}(\eta + \chi)$$
$$v = \frac{1}{2}(\eta - \chi) \; . \qquad (2)$$

The metric now becomes

$$ds^2 = R^2(u+v)[-4\,du\,dv + \Sigma^2(u-v)\,d\Omega^2] \; . \qquad (3)$$

We now try a transformation which preserves the "nullness" of the coordinates, and also preserves the symmetry between u and v:

$$\begin{aligned} \alpha &= g(u) \qquad & u &= f(\alpha) \\ \beta &= g(v) \qquad & v &= f(\beta) \end{aligned} \qquad (4)$$

where g is the inverse function to f. With this transformation, Equation (3) becomes

$$\begin{aligned} ds^2 &= R^2(u+v)[-4f'(\alpha)f'(\beta)\,d\alpha\,d\beta + \Sigma^2(u-v)\,d\Omega^2] \\ &= R^2(u+v)f'(\alpha)f'(\beta)\left[-4\,d\alpha\,d\beta + \frac{\Sigma^2(u-v)}{f'(\alpha)f'(\beta)}\,d\Omega^2\right] . \end{aligned} \qquad (5)$$

If the metric is to be conformally flat, the term in brackets must have the flat space form $[-4\,d\alpha\,d\beta + (\alpha-\beta)^2\,d\Omega^2]$. Using $f'(\alpha) = du/d\alpha = (d\alpha/du)^{-1} = [g'(u)]^{-1}$, we get the condition

$$g'(u)\,g'(v)\,[\Sigma(u-v)]^2 = [g(u)-g(v)]^2 \; . \qquad (6)$$

For the case $k = 0$, we have $\Sigma(u-v) = u-v$, so $g(x) = x$ is a solution by inspection. For the other cases Equation (6) can be solved by first finding the solution for v close to u; that is, for $v = u + \epsilon$. Expanding Equation (6) in a Taylor series gives

$$\begin{aligned} &(g')^2[1+\epsilon g''/g' + \epsilon^2 g'''/(2g')\cdots]\,\epsilon^2(1-k\epsilon^2/6\cdots)^2 \\ &= \epsilon^2(g')^2[1+\epsilon g''/(2g') + \epsilon^2 g'''/(6g')\cdots]^2 \; , \end{aligned}$$

so that with $p = g'$, $q = p'$ and $k = +1$, we must solve

$$2q\,dq/dp - 4p = 3q^2/p\,.$$

This Bernoulli differential equation gives

$$q = p(Ap-4)^{\frac{1}{2}}\,, \qquad A = \text{constant}\,,$$

and integrating twice we get

$$u + B = \tan^{-1}(Ap/4-1)^{\frac{1}{2}}$$

$$g(u) = C\tan(u+B) + D\,.$$

We can now verify by substitution that this is in fact the general solution to Equation (6). Without loss of generality we can take $g = \tan u$. For $k = -1$, we can similarly find $g = \tanh u$. With these solutions, Equation (6) turns into the trigonometric identities

$$\sec u \sec v \sin(u-v) = \tan u - \tan v$$

$$\operatorname{sech} u \operatorname{sech} v \sinh(u-v) = \tanh u - \tanh v$$

and the metric of Equation (5) becomes

$$ds^2 = \frac{R^2[\tan^{-1}(\alpha) + \tan^{-1}(\beta)]}{(1+\alpha^2)(1+\beta^2)}\{-4\,d\alpha\,d\beta + (\alpha-\beta)^2\,d\Omega^2\}$$

or

$$ds^2 = \frac{R^2[\tanh^{-1}(\alpha) + \tanh^{-1}(\beta)]}{(1-\alpha^2)(1-\beta^2)}\{-4\,d\alpha\,d\beta + (\alpha-\beta)^2\,d\Omega^2\}$$

which is manifestly conformally flat. A conformally flat metric has zero Weyl tensor, so the Riemann tensor is only composed of the Ricci tensor and scalar curvature (see the introduction to Chapter 9).

$$R^{\alpha\beta}{}_{\gamma\delta} = 2\delta^{[\alpha}{}_{[\gamma}R^{\beta]}{}_{\delta]} - \tfrac{1}{3}\,\delta^{[\alpha}{}_{[\gamma}\delta^{\beta]}{}_{\delta]}R\,. \tag{8}$$

With this the Einstein equation

$$R^{\alpha}{}_{\beta} = 8\pi\left(T^{\alpha}{}_{\beta} - \frac{1}{2}\, g^{\alpha}{}_{\beta}\, T\right)$$

and the stress energy tensor

$$T^{\alpha}{}_{\beta} = (\rho + p)\, u^{\alpha} u_{\beta} + p g^{\alpha}{}_{\beta}$$

give $R^{\alpha\beta}{}_{\gamma\delta}$ in terms of $g_{\mu\nu}, u^{\mu}, \rho, p$.

Solution 19.9. If D is the physical size of an object, and δ is its angle subtended, $d_A \equiv D/\delta$. From the figure, and the Robertson-Walker metric, we see that $D = R(t) r_1 \delta$, so $d_A = r_1 R(t_1)$. If the object is moving transversely at a proper velocity V and with an apparent angular motion

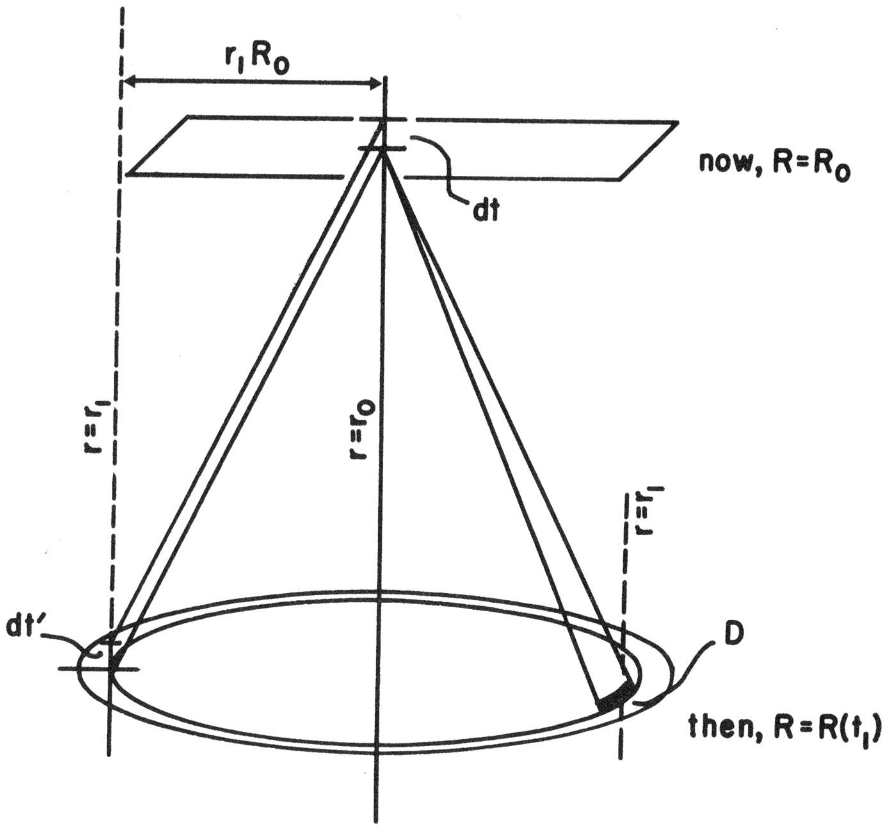

$d\delta/dt$, then $d_M \equiv V/(d\delta/dt)$. Let t' denote time measured at the emission of photons. Since $V = d(R(t_1)r_1\delta)/dt'$, and since $dt'/dt = R(t_1)/R_0$ because of the cosmological redshift, and noting that $R(t_1)$ can be considered constant since its change induces no transverse motion, we get $d_M = R_0r_1$. If the object has an intrinsic luminosity L, and we receive a flux ℓ, then $d_L \equiv (L/4\pi\ell)^{\frac{1}{2}}$. In a time dt' it emits an energy Ldt'. This energy is redshifted to the present by a factor $R(t_1)/R_0$, and is now distributed over a sphere of proper area $4\pi(r_1R_0)^2$ (see figure). Thus, $\ell = (Ldt'R/R_0)(4\pi r_1R_0)^{-2}/dt$ and $d_L = R_0^2 r_1/R(t_1)$. Using $R_0/R(t_1) = 1 + z$, we have now obtained $(1+z)^2 d_A = (1+z)d_M = d_L$.

Solution 19.10. From the solution to Problem 19.9 we have $d_L = R_0^2 r_1/R(t_1)$ where r_1 is the radial coordinate of the object and $R(t_1)$ is the scale factor at the time t_1 at which the light was emitted. Since $H_0 \equiv \dot{R}/R$ and $q_0 \equiv -\ddot{R}R/\dot{R}^2$, the first terms in the power series expansion of R are

$$R(t) = R_0\left(1 + H_0(t-t_0) - \frac{1}{2}q_0H_0^2(t-t_0)^2 + \cdots\right). \tag{1}$$

We can eliminate the factor $R(t_1)$ from d_L using $R_0/R(t_1) = 1+z$, but we still need to find expressions for R_0 and for r_1. Setting $ds^2 = 0$ to get the path of light rays in the Robertson-Walker metric gives

$$\int_{t_1}^{t_0} \frac{dt}{R(t)} = \int_0^{r_1} \frac{dr}{(1-kr^2)^{\frac{1}{2}}}. \tag{2}$$

Integrating Equation (2) to the lowest 2 orders in $(t_0 - t_1)$ and r_1 by means of Equation (1) gives

$$r_1 = \frac{1}{R_0}\left[(t_0-t_1) + \frac{1}{2}H_0(t_0-t_1)^2 + \cdots\right] \tag{3}$$

and inverting Equation (1) to get $t - t_0$ in terms of $1+z = R_0/R(t_1)$ gives

$$R_0r_1 = \frac{1}{H_0}\left[z - \frac{1}{2}(1+q_0)z^2 + \cdots\right]$$

so that

$$d_L = (1+z)R_0 r_1 = \frac{1}{H_0}\left[z + \frac{1}{2}(1-q_0)z^2 + \cdots\right]$$

or equivalently

$$\ell = \frac{L}{4\pi d_L^2} = \frac{L H_0^2}{4\pi z^2}\,[1 + (q_0 - 1)z + \cdots] .$$

Solution 19.11.

(a) Let $n(t_1)$ be the number of sources per unit volume at time t_1. The volume element is

$$|{}^{(3)}g|^{\frac{1}{2}} dr_1\, d\theta_1\, d\phi_1 = R^3(t_1)(1-kr_1^2)^{-\frac{1}{2}} r_1^2\, dr_1 \sin\theta_1\, d\theta_1\, d\phi_1 .$$

Thus the number of sources between r_1 and $r_1 + dr_1$ at time t_1 is

$$dN = 4\pi R^3(t_1)(1-kr_1^2)^{-\frac{1}{2}} r_1^2\, n(t_1)\, dr_1 .$$

The quantities r_1 and t_1 are related by the equation for null rays propagating in the Robertson-Walker metric (Equation (2) in the solution of Problem 19.10) i.e. $r_1 = r(t_1)$ where

$$dr_1 = (1-kr_1^2)^{\frac{1}{2}}\, dt_1/R(t_1) .$$

We have, therefore

$$dN = 4\pi R^2(t_1) r^2(t_1) n(t_1)\, |dt_1|$$

and

$$N(z) = \int_{t_z}^{t_0} 4\pi R^2(t_1) r^2(t_1) n(t_1)\, dt_1$$

where t_z, the cosmological time corresponding to redshift z, is defined implicitly by

$$\frac{R(t_z)}{R(t_0)} = \frac{1}{1+z} . \qquad (1)$$

Since the number density of sources is conserved, we have

$$n(t)R^3(t) = \text{constant} .$$

(This follows from $(nu^\mu)_{;\mu} = 0$) and hence

$$N(z) = 4\pi\, n(t_0)R^3(t_0)\int_{t_z}^{t_0} R^{-1}(t_1)\,r^2(t_1)\,dt_1 .$$

If z is small, $t_0 \approx t_z$, and we can use the expansions

$$R(t_1) = R(t_0)[1 - H_0(t_0 - t_1) + \cdots]$$

$$r(t_1) = \frac{t_0 - t_1}{R(t_0)}\left[1 + \frac{1}{2}H_0(t_0 - t_1) + \cdots\right] \tag{2}$$

(Solution 19.10, Equations (1) and (3)). Thus for small z

$$N(z) = 4\pi\, n(t_0)\int_{t_z}^{t_0} (t_0 - t_1)^2\,[1 + 2H_0(t_0 - t_1) + \cdots]\,dt_1$$

$$= \frac{4\pi}{3}\, n(t_0)(t_0 - t_z)^3\left[1 + \frac{3}{2}H_0(t_0 - t_z) + \cdots\right] .$$

To evaluate $t_0 - t_z$ to $\mathcal{O}(z^2)$ from Equation (1) we need the next term in Equation (2):

$$R(t_z) = R(t_0)\left[1 - H_0(t_0 - t_z) - \frac{1}{2}q_0H_0^2(t_0 - t_z)^2 + \cdots\right] .$$

This gives

$$H_0(t_0 - t_z) = z\left[1 - z\left(1 + \frac{q_0}{2}\right) + \cdots\right]$$

and so, finally,

$$N(z) = \frac{4\pi}{3}\frac{n(t_0)}{H_0^3}\, z^3\left[1 - \frac{3}{2}z(1 + q_0) + \cdots\right] .$$

Note that the field equations did not enter to this order in z.

(b) The flux received at the Earth is given by

$$S = \frac{LR^2(t_1)}{4\pi r_1^2 R^4(t_0)} \tag{3}$$

(cf. Problem 19.9, where S was called ℓ). Thus, as in part (a),

$$N(S) = \int_{t_s}^{t_0} 4\pi R^2(t_1) r^2(t_1) n(t_1) dt_1 \tag{4}$$

where, from Equation (3),

$$\frac{r^2(t_s)}{R^2(t_s)} = \frac{L}{4\pi SR^4(t_0)} . \tag{5}$$

When S is large, $t_0 - t_s$ is small and expanding the integral in Equation (4) as in part (a) gives

$$N(S) = \frac{4\pi}{3} n(t_0)(t_0 - t_s)^3 \left[1 + \frac{3}{2} H_0(t_0 - t_s) + \cdots\right] .$$

Inserting the expansions for $r(t_s)$ and $R(t_s)$ in Equation (5) gives

$$t_0 - t_s = \left(\frac{L}{4\pi S}\right)^{\frac{1}{2}} \left[1 - \frac{3}{2} H_0 \left(\frac{L}{4\pi S}\right)^{\frac{1}{2}} + \cdots\right]$$

and so

$$N(S) = \frac{4\pi}{3} n(t_0) \left(\frac{L}{4\pi S}\right)^{\frac{3}{2}} \left[1 - 3H_0 \left(\frac{L}{4\pi S}\right)^{\frac{1}{2}} + \cdots\right] .$$

Solution 19.12. Choose a new radial coordinate χ, defined by $d\chi^2 = dr^2/(1 - kr^2)$. Then for a radial photon $d\theta = d\phi = 0$ and the metric coefficients are independent of χ along the ray. Therefore χ is an ignorable coordinate, and p_χ = constant along the ray (see Problem 7.13). Raising this index with the metric gives

$$p^\chi = C/R^2(t)$$

where C is some constant. But $p^\chi = d\chi/d\lambda$, so

$$d\lambda = C^{-1}R^2(t)d\chi = C^{-1}\frac{R^2(t)dr}{(1-kr^2)^{\frac{1}{2}}} .$$

Note that since R is a function of t it is therefore a function of r along a past null ray.

Solution 19.13. In the comoving orthonormal frame we have $T^{\hat{0}}{}_{\hat{0}} = -\rho$, and $T^{\hat{r}}{}_{\hat{r}} = T^{\hat{\phi}}{}_{\hat{\phi}} = T^{\hat{\theta}}{}_{\hat{\theta}} = p$. The trace-reversed stress energy tensor $\overline{T}$ thus has components $\overline{T}^{\hat{0}}{}_{\hat{0}} = -\frac{1}{2}(\rho+3p)$, and $\overline{T}^{\hat{i}}{}_{\hat{i}} = \frac{1}{2}(\rho-p)$. We equate this to $1/8\pi G$ times the Ricci tensor. The Ricci tensor can be calculated e.g. from the results of Problem 9.20 (general spherically symmetric metric). The components are:

$$R^{\hat{0}}{}_{\hat{0}} = 3\ddot{R}/R$$

$$R^{\hat{i}}{}_{\hat{i}} = \frac{1}{R^2}(R\ddot{R} + 2\dot{R}^2 + 2k) .$$

The equations in the problem now follow immediately. If, equivalently, the Einstein tensor $G_{\mu\nu}$ (trace reverse Ricci tensor) had been equated to $8\pi T_{\mu\nu}$, two linear combinations of the equations would have been obtained.

Solution 19.14. Simply eliminating $\ddot{R}$ from the simultaneous equations in Problem 19.13 gives the first equation here. The second follows from this, from the identity $\frac{1}{2}\,d[(\dot{R})^2]/dR = \ddot{R}$, and from Equation (1) of Problem 19.13:

$$\frac{1}{2}\frac{d}{dR}\left(\frac{8\pi G}{3}\rho R^2\right) = \frac{1}{2}\frac{d}{dR}(\dot{R})^2 = \ddot{R} = -\frac{4}{3}\pi G(\rho+3p)R$$

$$\frac{d}{dR}(\rho R^2) = -(\rho+3p)R$$

$$\frac{d}{dR}(\rho R^3) = -3pR^2 .$$

Solution 19.15. Equation (1) follows immediately from the definition $H \equiv \dot{R}/R$ and the first-order Friedmann equation (Problem 19.14)

$$\dot{R}^2 + k = \frac{8\pi G}{3}\rho R^2 .$$

If we now take d/dR on this equation, use the identity $\frac{1}{2}d[(\dot{R})^2]/dR = \ddot{R}$, the other first order equation $d(\rho R^3)/dR = -3pR^2$ and the definition $q \equiv -\ddot{R}R/\dot{R}^2$, we get Equation (2). If $\rho >> p$ then the left side of Equation (2) is negligible compared with the right side, and Equation (3) results. Equation (4) is immediate upon substituting Equation (3) into Equation (1). If $p = 1/3\rho$, then eliminating ρ from Equations (1) and (2) gives Equation (5), while eliminating the k/R^2 term gives Equation (6).

Solution 19.16. Using $T^0_0 = -\rho$, and T^j_j (no sum) $= p$, and $p_{,j} = 0$ (homogeneity) the $\mu = j$ component is

$$\begin{aligned} 0 = T_{j\,;\nu}^{\nu} &= T_{j\,,\nu}^{\nu} - T_{\alpha}^{\nu}\Gamma^{\alpha}{}_{j\nu} + T_j^{\alpha}\Gamma^{\nu}{}_{\alpha\nu} \\ &= p_{,j} + \rho\Gamma^0{}_{j0} - p\Gamma^k{}_{jk} + p\Gamma^{\nu}{}_{j\nu} = (\rho+p)\Gamma^0{}_{j0} = 0 \end{aligned}$$

which is a tautology. The $\mu = 0$ component gives

$$\begin{aligned} 0 = T_{0\,;\nu}^{\nu} &= T_{0\,,\nu}^{\nu} - T_{\alpha}^{\nu}\Gamma^{\alpha}{}_{0\nu} + T_0^{\alpha}\Gamma^{\nu}{}_{\alpha\nu} \\ &= -\frac{d\rho}{dt} + \rho\Gamma^0{}_{00} - (\rho+p)(\log|g|^{\frac{1}{2}})_{,0} \end{aligned}$$

where $\Gamma^{\alpha}{}_{\alpha\beta} = (\log|g|^{\frac{1}{2}})_{,\beta}$ has been used (see Problem 7.7). Thus

$$d\rho/dt = -\frac{(\rho+p)}{R^3}\frac{d}{dt}(R^3)$$

and

$$d(\rho R^3) = -3pR^2\,dR$$

which we recognize as one of the first-order Friedmann equations (see Problem 19.14).

Solution 19.17. With $p = \rho = 0$, and $k = -1$, the first-order Friedmann equation

$$\dot{R}^2 + k = \frac{8\pi G}{3}\rho R^2$$

becomes $\dot{R} = 1$ which implies $R = t$ and thus the metric given is correct. Since the metric is spherically symmetric, the curvature radial coordinate is

$$r = t \sinh\chi .$$

It is not hard to guess a time coordinate

$$T = t \cosh\chi$$

and then to compute the transformed metric

$$ds^2 = -dT^2 + dr^2 + r^2 (d\theta^2 + \sin^2\theta\, d\phi^2)$$

which is empty Minkowski space.

Solution 19.18.

(a) When matter dominates, we ignore pressure so that mass-energy density decreases in proportion to the increase in volume of the universe, and

$$\rho = \rho_0 \left(\frac{R_0}{R}\right)^3 . \tag{1}$$

Define a new time coordinate ("development angle") by $d\eta = dt/R$. Then the Friedmann equation becomes

$$\left(\frac{\dot{R}}{R}\right)^2 = \left(\frac{dR/d\eta}{R^2}\right)^2 = \frac{8\pi G}{3}\rho_0 \left(\frac{R_0}{R}\right)^3 - \frac{k}{R^2} \tag{2}$$

or

$$R^{-\frac{1}{2}} \frac{dR}{d\eta} = 2 \frac{d}{d\eta} R^{\frac{1}{2}} = \left(\frac{8\pi G}{3}\rho_0 R_0^3 - kR\right)^{\frac{1}{2}}$$

which integrates to give

$$\frac{1}{2}\eta = \int_0^{R^{\frac{1}{2}}} \frac{dR^{\frac{1}{2}}}{\left(\frac{8\pi}{3}\rho_0 R_0^3 - kR\right)^{\frac{1}{2}}} = \begin{cases} \sin^{-1} \dfrac{R^{\frac{1}{2}}}{\left(\frac{8}{3}\pi G \rho_0 R_0^3\right)^{\frac{1}{2}}} & k=+1 \\ \dfrac{R^{\frac{1}{2}}}{\left(\frac{8\pi G}{3}\rho_0 R_0^3\right)^{\frac{1}{2}}} & k=0 \\ \sinh^{-1} \dfrac{R^{\frac{1}{2}}}{\left(\frac{8\pi G}{3}\rho_0 R_0^3\right)^{\frac{1}{2}}} & k=-1 \end{cases} \qquad (3)$$

From Problem 19.15 we know

$$q_0 = \frac{4\pi G}{3}\frac{\rho_0}{H_0^2} \qquad (4)$$

and

$$R_0^2 = \frac{k}{(2q_0-1)H_0^2} \qquad (k=\pm 1)\ . \qquad (5)$$

We see that since the left side is positive, $k = \operatorname{sign}(2q_0 - 1)$. Thus in Equation (3)

$$\frac{8\pi}{3}\rho_0 R_0^3 = \frac{2q_0}{H_0\,|(2q_0-1)|^{\frac{3}{2}}} \qquad (k=\pm 1)\ .$$

Inverting Equation (3) gives

$$R = \begin{cases} \dfrac{q_0}{H_0(2q_0-1)^{\frac{3}{2}}}\,(1-\cos\eta) & k=1 \\ \frac{1}{4} H_0^2 R_0^3 \eta^2 & k=0 \\ \dfrac{q_0}{H_0(1-2q_0)^{\frac{3}{2}}}\,(\cosh\eta - 1) & k=-1\ . \end{cases}$$

Now integrating $dt = R d\eta$ gives

$$t = \begin{cases} \dfrac{q_0}{H_0(2q_0-1)^{\frac{3}{2}}}(\eta - \sin\eta) & k = 1 \\ \frac{1}{12} H_0^2 R_0^3 \eta^3 & k = 0 \\ \dfrac{q_0}{H_0(1-2q_0)^{\frac{3}{2}}}(\sinh\eta - \eta) & k = -1 \;. \end{cases}$$

The fact that R_0 cannot be eliminated from the $k = 0$ case merely reflects the fact that the universe here has an arbitrary scaling and its geometry looks the same at all times. The value of R_0 will not enter the computation of any physically measurable quantity.

(b) When radiation dominates, the mass-energy in a given comoving volume is not constant. There is an extra decrease in density from the redshift of photons, so

$$\rho = \rho_0 \left(\frac{R_0}{R}\right)^4 .$$

The analog of Equation (2) is

$$\left(\frac{\dot{R}}{R}\right)^2 = \left(\frac{dR/d\eta}{R^2}\right)^2 = \frac{8\pi G}{3}\rho_0 \left(\frac{R_0}{R}\right)^4 - \frac{k}{R^2}$$

or

$$\frac{dR}{\left(\frac{8}{3}\pi G \rho_0 R_0^4 - kR^2\right)^{\frac{1}{2}}} = d\eta$$

which has solutions:

$$R = \left(\frac{8\pi}{3} G\rho_0 R_0^4\right)^{\frac{1}{2}} \times \begin{cases} \sin\eta & k = 1 \\ \eta & k = 0 \\ \sinh\eta & k = -1 \;. \end{cases} \tag{6}$$

Instead of Equation (4), we now get (Problem 19.15)

$$q_0 = \frac{8\pi G}{3} \frac{\rho_0}{H_0^2} ,$$

and instead of Equation (5), we have

$$R_0^2 = \frac{k}{(q_0 - 1) H_0^2} \qquad (k = \pm 1) \ .$$

Equation (6) therefore becomes

$$\frac{8\pi}{3} G\rho_0 R_0^4 = \begin{cases} \dfrac{q_0}{(q_0-1)^2 H_0^2} & k = \pm 1 \\ H_0^2 R_0^4 & k = 0 \end{cases}$$

and integrating for t gives

$$t = \begin{cases} \dfrac{1}{H_0}\left[\dfrac{q_0}{q_0 - 1}\right]^{\frac{1}{2}} (1 - \cos\eta) & k = 1 \\ \frac{1}{2} H_0 R_0^2 \eta^2 & k = 0 \\ \dfrac{1}{H_0}\left[\dfrac{q_0}{1 - q_0}\right]^{\frac{1}{2}} (\cosh\eta - 1) & k = -1 \ . \end{cases}$$

Solution 19.19. After the bullet has reached nonrelativistic velocity, we have (from Problem 19.7)

$$\frac{\text{constant}}{R} = \frac{dr_p}{dt} = \frac{dr_p}{dr}\frac{dr}{dR}\frac{dR}{dt} = \frac{dr}{dR}\dot{R}R \tag{1}$$

where r is the Robertson-Walker coordinate, related to proper separations ($r_p \equiv$ proper separation) by $dr_p = R\,dr$ (hence the last equality above). For a $k = 0$ Friedmann universe $R \propto t^{\frac{2}{3}}$ so $\dot{R} \propto R^{-\frac{1}{2}}$ and Equation (1) gives us $dr/dR \propto R^{-\frac{3}{2}}$ and hence

$$r = A + B/R^{\frac{1}{2}} \ .$$

As $t \to \infty$, the r_{bullet} approaches $r = A$. The proper distance between the $r = A$ observer and the bullet is $R\Delta r = BR^{\frac{1}{2}}$. Thus the *proper* distance becomes infinite even though the velocities (according to Equation (1)) approach each other.

If $k = -1$, then $\dot{R}$ becomes constant at large times and Equation (1) gives $dr/dR \propto R^{-2}$ so that

$$r = A + B/R .$$

As $t \to \infty$ the bullets coordinate r approaches A, but the proper distance to the $r = A$ observer $R\Delta r$ approaches B, a constant.

Solution 19.20. Measuring time in terms of development angle η defined by $d\eta = dt/R(t)$, the Friedmann metric for a radial $(d\theta = d\phi = 0)$ photon becomes

$$0 = ds^2 = R^2(\eta)(-d\eta^2 + d\chi^2)$$

where $d\chi^2 = dr^2/(1-r^2)$ is the "trigonometric" radial coordinate on the 3-sphere (see Problem 19.5). From Problem 19.18 we see that the universe lives an interval $\Delta\eta = 2\pi$ (time between two zeros of R), so in this time a photon can propagate $\Delta\chi = 2\pi$, i.e. exactly once around the universe.

Solution 19.21. For a $k = 0$ Friedmann universe we have

$$R(t)/R_0 = \left(\tfrac{3}{2} H_0 t\right)^{\frac{2}{3}} .$$

The energy emitted per source per unit proper time is L. This is redshifted by a factor $R(t)/R_0$, so the total energy per star which is *now* present is

$$\mathcal{E} = \int_0^{t_0} L \frac{R(t)}{R_0} dt = \frac{3}{5} L\left(\frac{3}{2} H_0\right)^{\frac{2}{3}} t_0^{5/3}$$

or $\mathcal{E} = 2L/5H_0$ since $t_0 = 2/3H_0$. The energy per volume, u, now is thus $\frac{2}{5} H_0^{-1} Ln$. Since the energy fills the universe isotropically, it corresponds to a flux

$$B = \frac{c}{4\pi} u = \frac{c}{10\pi} H_0^{-1} Ln .$$

[By comparison, a Newtonian universe where the stars "turned on" a time $t_0 = \left(\frac{3}{2} H_0\right)^{-1}$ ago would have

$$\mathcal{E} = \int_0^{t_0} L\,dt = \tfrac{2}{3} H_0^{-1} L$$

and

$$B = \tfrac{c}{6\pi} H_0^{-1} Ln.]$$

Solution 19.22. From the solution of Problem 19.18, if $q_0 > 0.5$ we must have a $k = +1$ cosmology which expands according to

$$R(t)/R_0 = \frac{q_0}{2q_0 - 1}(1 - \cos\eta) \ . \qquad (1)$$

Setting $R(t) = R_0$, so that $\eta = \eta_0$, gives a formula which relates η_0 to q_0 at the present epoch. The epoch is arbitrary, however, so it also relates η and q at any epoch. Putting $q = 0.5002$ gives $1 - \cos\eta = .0008$ at $z = 1500$. Substituting this back into Equation (1) and putting $R_0/R(t) = 1 + z = 1501$, we can solve for q_0, getting

$$q_0 = [2 - (1+z)(1 - \cos\eta)]^{-1} = 1.25 \ .$$

If $q_0 < 0.5$, the $k = -1$ analogy of Equation (1) is

$$R(t)/R_0 = \frac{q_0}{1 - 2q_0}(\cosh\eta - 1) \qquad (2)$$

so $q = .4998$ gives $\cosh\eta - 1 = .0008$ at $z = 1500$ and

$$q_0 = [2 + (1+z)(\cosh\eta - 1)]^{-1} = 0.312 \ .$$

Solution 19.23.

(a) At the present epoch $t = t_0$, the spacelike 3-surface has the metric

$$d\sigma^2 = R_0^2\left(\frac{dr^2}{1 - r^2} + r^2\,(d\theta^2 + \sin^2\theta\,d\phi^2)\right) .$$

Noting that $0 \le r \le 1$ covers half the 3-sphere, we have

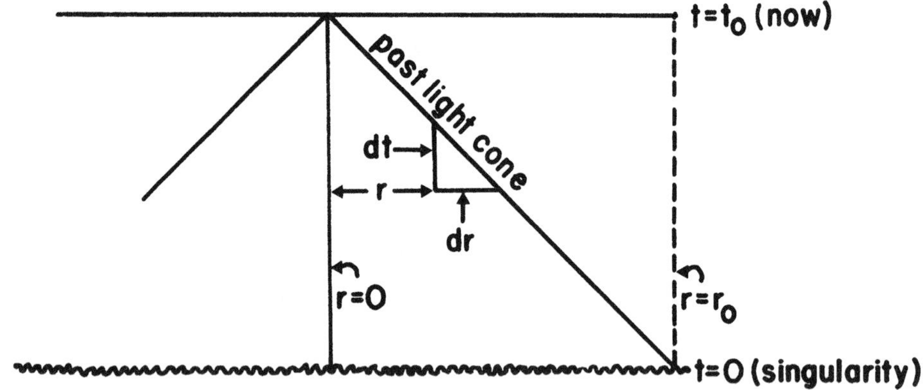

$$V = \int_{\text{3-sphere}} |{}^{(3)}g|^{\frac{1}{2}} d^3x = 2\int_0^1 dr \int_{-1}^1 d(\cos\theta) \int_0^{2\pi} d\phi \, \frac{R_0^3 r^2}{(1-r^2)^{\frac{1}{2}}} = 2\pi^2 R_0^3 .$$

For a matter dominated universe (see Problem 19.15)

$$R_0 = H_0^{-1} (2q_0 - 1)^{-\frac{1}{2}} \tag{1}$$

so

$$V = 2\pi^2 (2q_0 - 1)^{-\frac{3}{2}} H_0^{-3} . \tag{2}$$

(b) Our past light cone sweeps outward a proper radial distance $c\,dt$ in a time dt (see figure). From the metric, the proper area of a 2-sphere is $4\pi r^2 R^2$, where r and R are both functions of t along the light cone. Thus, the visible volume is ($c = 1$ now)

$$V = \int_0^{t_0} dt \; 4\pi r^2(t) R^2(t) . \tag{3}$$

The differential equation for $r(t)$ follows from its definition as a null cone:

$$-dt = \frac{R(t)}{(1-r^2)^{\frac{1}{2}}} \, dr , \tag{4}$$

and $r = 0$ at $t = t_0$. To solve Equation (4) and evaluate the integral in

Equation (3), we use the solution for $R(t)$ parameterized by "development angle" η (see the solution to Problem 19.18).

$$R = A(1-\cos\eta)\,, \qquad t = A(\eta - \sin\eta) \tag{5}$$

where

$$A = \frac{q_0}{(2q_0-1)^{\frac{3}{2}}\,H_0} \tag{6}$$

and the current value of η is

$$\eta_0 = \cos^{-1}\left(\frac{1-q_0}{q_0}\right) . \tag{7}$$

Equation (4) now becomes $d(\sin^{-1} r) = d\eta$ so that

$$r = \sin(\eta_0 - \eta) \tag{8}$$

and Equation (3) becomes

$$V = \int_0^{\eta_0} A(1-\cos\eta)\,d\eta\, 4\pi \sin^2(\eta_0-\eta)\, A^2(1-\cos\eta)^2 \ .$$

This straightforward integral gives

$$V = 4\pi A^3\left(\frac{61}{80}\sin 2\eta_0 + \frac{5}{4}\eta_0 - \frac{5}{2}\sin\eta_0 - \frac{3}{8}\eta_0\cos 2\eta_0 + \frac{1}{30}\sin 3\eta_0\right)$$

which can be reduced with Equations (6) and (7) to a function of q_0 and H_0.

(c) The farthest coordinate radius that we see is, according to Equation (8),

$$r_0 = \sin(\eta_0) = (2q_0-1)^{\frac{1}{2}}/q_0$$

where Equation (7) has been used. The volume out to this coordinate radius at the present epoch is (analogously to part (a))

$$\begin{aligned} V &= 4\pi\int_0^{r_0} \frac{R_0^3\, r^2\, dr}{(1-r^2)^{\frac{1}{2}}} = 2\pi R_0^3\,[\sin^{-1} r_0 - r_0(1-r_0^2)^{\frac{1}{2}}] \\ &= \frac{2\pi}{H_0^3(2q_0-1)^{\frac{3}{2}}}\left[\cos^{-1}\left(\frac{1-q_0}{q_0}\right) - \frac{(1-q_0)(2q_0-1)^{\frac{1}{2}}}{q_0^2}\right] . \end{aligned}$$

Solution 19.24.

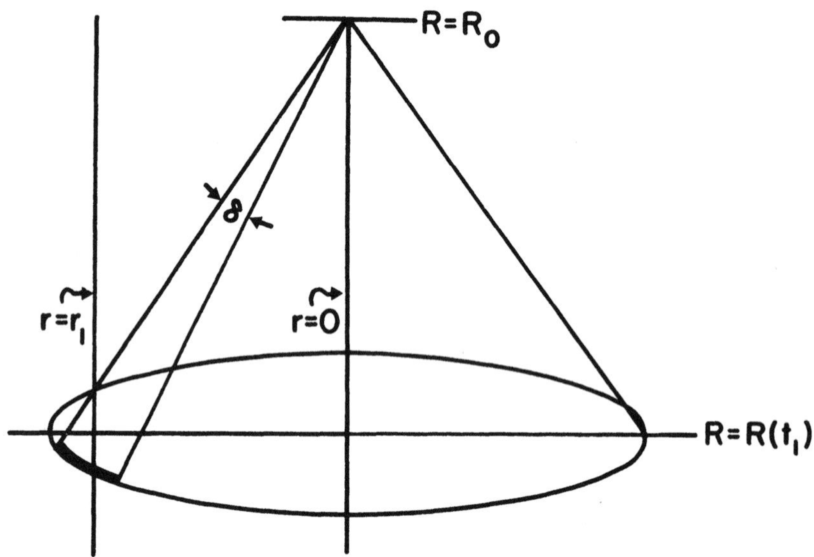

From the figure and the Robertson-Walker line element, $\ell = [r_1 R(t_1)]\delta$. We can express $R(t_1)$ as $R_0/(1+z)$; the hard part is now to compute r_1: Use the η,χ form of the metric from Problem 19.5. Then $r_1 = \Sigma(\chi_1)$. The path of a photon is $0 = R^2(d\chi^2 - d\eta^2)$, so $\chi_1 = \eta(t_0) - \eta(t_1)$, and using the formula for $\sin(A-B)$, $\sinh(A-B)$ we have:

$$r_1 = \Sigma(\eta_0)C(\eta_1) - \Sigma(\eta_1)C(\eta_0) . \tag{1}$$

[Here C means cos, 1, or cosh for $k = 1, 0, -1$.] Now we need the formula relating η of an epoch to its redshift and the current value of q (Problem 19.18):

$$1 - C(\eta) = \frac{2q_0 - 1}{q_0}\frac{1}{1+z}$$

$$C(\eta) = \frac{q_0 z - q_0 + 1}{q_0(1+z)} .$$

(This is true for all three values of k.) With this result we have

$$\Sigma(\eta) = \frac{|2q_0 - 1|^{\frac{1}{2}}(2q_0 z + 1)^{\frac{1}{2}}}{q_0(1+z)}$$

$$r_1 = \frac{|2q_0 - 1|^{\frac{1}{2}}}{q_0^2\,(1+z)}\,[1 - q_0 + q_0 z - (1-q_0)(1+2q_0 z)^{\frac{1}{2}}]\ .$$

So the apparent angle is

$$\delta = \ell/r_1 R(t_1) = \ell(1+z)/r_1 R_0\ ,$$

and when we use Equation (3) of Problem 9.15 to eliminate R_0 we get, finally,

$$\delta = \frac{\ell H_0 (1+z)^2 q_0^2}{[1 - q_0 + q_0 z - (1-q_0)(1+2q_0 z)^{\frac{1}{2}}]}\ .$$

Solution 19.25. We work with units which make $c = k = 1$. The density of matter (m_p = average baryon mass) is $\rho_{matt} = m_p n$. The density of radiation is $\rho_{rad} = KaT^4$, where a is the radiation constant = $(8\pi^5)/(15h^3)$ and K is 1 (considering photons only) or $1 + \frac{7}{4}\left(\frac{4}{11}\right)^{4/3} =$ 1.454 (considering photons and neutrinos – see Weinberg, p. 537). Equating these give

$$1 = \frac{\rho_{rad}}{\rho_{matt}} = \frac{KaT^4}{nm_p}\ . \tag{1}$$

The Saha equation for the fractional ionization x of hydrogen is

$$\frac{x^2}{1-x} = \frac{(2\pi m_e kT)^{\frac{3}{2}} \exp\left(-\frac{1}{2}\,\alpha^2 m_e/kT\right)}{nh^3}$$

(m_e = electron mass, $\alpha = 1/137$). If recombination is defined by the left side of this equation being about unity, Equation (1) gives us

$$\exp\left(\frac{1}{2}\,\alpha^2 m_e/kT\right) = \frac{15\,(2\pi m_e T)^{\frac{3}{2}}\, m_p}{8\pi^5\, KT^4}\ .$$

Obviously the exponential sets the scale of the solution, $T \sim \alpha^2 m_e$; guessing a coefficient, computing the right side, and iterating once gives more accurately: $T \approx \frac{1}{80}\alpha^2 m_e$ (cf. Problem 19.26). So, finally, when matter dominance and recombination occur at the same temperature T, we get

$$\sigma = \frac{4}{3}\frac{aT^3}{n} = \frac{4}{3}\frac{m_p}{TK}\left(\frac{KaT^4}{m_p n}\right)$$

$$= \frac{4}{3}\frac{m_p}{K\left(\frac{1}{80}\alpha^2 m_e\right)} = \frac{320}{3K}(137)^2(1836)$$

$$\approx 4 \times 10^9 .$$

Solution 19.26. The Saha equation for the equilibrium fractional ionization of hydrogen is (units with $k = c = 1$)

$$\frac{x^2}{1-x} = \frac{(2\pi m_e T)^{\frac{3}{2}} e^{-B/T}}{nh^3} \tag{1}$$

where B is the ionization energy $= \frac{1}{2}\alpha^2 m_e (\alpha = 1/137)$, and n is the number density of protons plus atoms. The value of n is related to σ, the entropy per baryon by

$$\sigma = \frac{4}{3}\frac{aT^3}{n}$$

so if we take $x = 1/2$ and use $a = 8\pi^5/15h^3$, Equation (1) becomes

$$\frac{2^{\frac{1}{2}} 4\pi^{7/2}}{45}\frac{1}{\sigma}\left(\frac{T}{m_e}\right)^{\frac{3}{2}} = \exp\left[-\frac{m_e/T}{2(137)^2}\right] .$$

This is easily solved by iteration. For example, iterating twice:

$$\frac{m_e}{T} = 2(137)^2\left(\log\frac{\sigma}{6.908} + \frac{3}{2}\log\frac{m_e}{T}\right)$$

$$T \approx \frac{m_e}{2(137)^2}\left[\log\frac{\sigma}{6.908} + \frac{3}{2}\left(10.53 + \log\log\frac{\sigma}{6.908}\right)\right]^{-1}$$

$$= \begin{cases} 4330^0\text{K for } \sigma = 10^8 \\ 4050^0\text{K for } \sigma = 10^9 . \end{cases}$$

Solution 19.27. From Problem 19.18, near the singularity in the radiation-dominated regime,

$$R = \left(\frac{8\pi}{3}\rho_0 R_0^4\right)^{\frac{1}{2}}\eta\,, \qquad t = \left(\frac{8\pi}{3}\rho_0 R_0^4\right)^{\frac{1}{2}}\frac{\eta^2}{2}\,. \tag{1}$$

(This equation holds for any value of k in the limit $\eta \to 0$.) Thus we have

$$t^2 = \frac{3R^4}{32\pi\rho_0 R_0^4} = \frac{3}{32\pi\rho} \tag{2}$$

where we have used $\rho \propto R^{-4}$ for the radiation-dominated era. The photon energy-density is

$$\rho_\gamma = \frac{8\pi}{h^3}\int_0^\infty \frac{q^3\,dq}{\exp(q/kT)-1} = aT^4$$

where a is the Stefan-Boltzmann constant $8\pi^5k^4/15h^3$. For highly relativistic electrons and positrons (which obey Fermi-Dirac Statistics)

$$\rho_{e+} = \rho_{e-} = \frac{8\pi}{h^3}\int_0^\infty \frac{q^3\,dq}{\exp(q/kT)+1} = \frac{7}{8}\,aT^4\,.$$

Thus the total mass-energy density is

$$\rho = \rho_\gamma + \rho_{e+} + \rho_{e-} = \frac{11}{4}\,aT^4$$

and so

$$T = \left(\frac{4}{11}\cdot\frac{3}{32\pi a}\right)^{\frac{1}{4}} t^{-\frac{1}{2}}\,.$$

If neutrinos are included and we assume they have zero chemical potential (cf. Weinberg, Section 15.6), then since they have only one spin state,

$$\rho_{\nu_e} = \rho_{\bar\nu_e} = \rho_{\nu_\mu} = \rho_{\bar\nu_\mu} = \frac{7}{16}\,aT^4$$

$$\rho = \frac{11}{4}\,aT^4 + \frac{7}{4}\,aT^4 = \frac{9}{2}\,aT^4$$

$$T = \left(\frac{2}{9}\cdot\frac{3}{32\pi a}\right)^{\frac{1}{4}} t^{-\frac{1}{2}}\,.$$

Solution 19.28. For relativistic particles, $E \approx p$, the equilibrium density at a temperature T is (see the solution to Problem 19.27):

$$\rho(T) = \int E\, n(p)\, dp = \begin{cases} aT^4 & \text{for photons} \\ \frac{7}{16}\, aT^4 & \text{for each neutrino species} \\ \frac{7}{8}\, aT^4 & \text{for each electron or muon species.} \end{cases}$$

Here a is Stefan's constant $= (8\pi^5 k^4)/(15h^3)$. For this problem, then, we have

$$\rho(T) = \left\{ \begin{array}{ll} \left(1+4\cdot\frac{7}{16}+4\cdot\frac{7}{8}\right) aT^4 & \text{at } R_1 \\ \left(1+4\cdot\frac{7}{16}+2\cdot\frac{7}{8}\right) aT^4 & \text{at } R_2 \end{array} \right\} \equiv KaT^4 ,$$

$$K_1 = 25/4 , \qquad K_2 = 18/4 .$$

We now use the fact that the expansion of the universe is isentropic. The entropy per volume of such a relativistic gas is calculated from the first law of thermodynamics to be

$$\frac{S}{V} = \frac{4}{3} KaT^3 .$$

Since V, the volume, varies as the expansion factor cubed, we have

$$\text{constant} = S \propto \frac{4}{3} Ka(RT)^3 ,$$

giving for isentropic expansion

$$K_1 R_1^3 T_1^3 = K_2 R_2^3 T_2^3$$

and

$$T_2 = T_1 \left(\frac{R_1}{R_2}\right)\left(\frac{25}{18}\right)^{\frac{1}{3}} .$$

Physically, this result says that in annihilating, the muons dump their energy (actually entropy) into the remaining particles, and increase their temperature by a factor $(25/18)^{\frac{1}{3}}$ over the expected first-power decrease of T with R.

Solution 19.29. The standard model produces helium and deuterium by (roughly speaking) (1) freezing out a nonequilibrium neutron/proton ratio when the weak interactions for $n \to p$ become slower then the timescale for the expansion to drop the temperature, (2) making deuterium by $n+p \to d+\gamma$ while most free neutrons have not yet decayed, and (3) using up almost all d by $d+d \to He^3$ or H^3 + etc., He^3 or $H^3+d \to He^4$ + etc.

We can thus answer the problem:

(i) There is no change in freezing out ratio, but higher baryon densities at any given temperature give more "cooking," therefore more complete transformation to He^4: more He^4, less d.

(ii) The $n \to p$ reactions is slower, so more frozen out neutrons; other rates not changed, so: more He^4, more d.

(iii) The reaction $n+\nu \leftrightarrow p+e^-$ is driven to the right by the degeneracy (Fermi) energy of the neutrinos; fewer neutrons give less He^4, less d.

(iv) This is similar to (iii) but for $\nu+p \leftrightarrow n+e^+$, giving more He^4, more d.

(v) The expansion time is related to density by

$$t = \left(\frac{3}{32\pi G\rho}\right)^{\frac{1}{2}} + \text{constant} .$$

Increased G thus means smaller times spent in each density stage. This gives a higher n/p ratio. The small change in G does not substantially affect the fact that the neutrons are almost all consumed to d, since the decay time is still long, so: more d, more He^4.

Solution 19.30. The cosmological constant is equivalent to a perfect fluid with effective density and pressure

$$\rho_{eff} = (8\pi)^{-1}\Lambda , \qquad p_{eff} = -(8\pi)^{-1}\Lambda .$$

The dynamical equations of Problem 19.14 become

$$\dot{R}^2 = \frac{1}{3}\Lambda R^2$$

with solution

$$R = R_0 e^{t/T_0}, \qquad T_0^{-1} = \frac{1}{3}\Lambda .$$

The metric is then

$$ds^2 = -dt^2 + R_0^2 e^{2t/T_0}(dr^2 + r^2 d\Omega^2) .$$

To transform this to a manifestly static form, introduce a "curvature coordinate"

$$r' = R_0 e^{t/T_0} r$$

so that the metric becomes

$$ds^2 = -dt^2 + (dr' - r' dt/T_0)^2 + r'^2 d\Omega^2 .$$

Now eliminate $g_{r't}$ by the transformation

$$t = t' + \frac{1}{2} T_0 \log\left(\frac{r'^2}{T_0^2} - 1\right)$$

and find

$$ds^2 = -\left(1 - \frac{r'^2}{T_0^2}\right) dt'^2 + \left(1 - \frac{r'^2}{T_0^2}\right)^{-1} dr'^2 + r'^2 d\Omega^2 . \tag{2}$$

The interpretation of Equation (1) as a nonstatic cosmological model is based on the assumption that the coordinates are comoving with the galaxies, the density of which is negligible in the dynamical equations.

Solution 19.31. The complete set of equations for isotropic, homogeneous universes is (see Problems 19.13 or 19.14)

$$\left(\frac{\dot{R}}{R}\right)^2 = \frac{-k}{R^2} + \frac{8\pi\rho}{3} \tag{1a}$$

$$2\frac{\ddot{R}}{R} = -\left(\frac{\dot{R}}{R}\right)^2 - \frac{k}{R^2} - 8\pi p \ . \tag{1b}$$

Since the universe contains only dust stress-energy, ρ_0, and "vacuum polarization" stress-energy Λ, we have

$$\rho = \rho_0 + \Lambda/8\pi \tag{2a}$$

$$p = -\Lambda/8\pi \ . \tag{2b}$$

Since all time derivatives of R higher than the second may be expressed as linear combinations of $\dot{R}$ and $\ddot{R}$ by Equations (1), the necessary and sufficient condition for a static solution is

$$\dot{R} = \ddot{R} = 0 \ . \tag{3}$$

With the conditions of Equations (3) and (2), Equations (1) become

$$k/R^2 = (8\pi\rho_0 + \Lambda)/3 \tag{4a}$$

$$k/R^2 = \Lambda \tag{4b}$$

which imply

$$\rho_0 = \Lambda/4\pi \tag{5a}$$

$$R = \Lambda^{-\frac{1}{2}} \ . \tag{5b}$$

Since $\rho_0 > 0$, Equation (4b) now implies $k = +1$. To investigate the stability of the "Einstein universe," first combine Equations (1) to give

$$\frac{2\ddot{R}}{R} = -\frac{8\pi}{3}\rho - 8\pi p = -\frac{8\pi}{3}\rho_0 + \frac{2}{3}\Lambda \tag{6}$$

and let

$$R = \Lambda^{-\frac{1}{2}} + \delta R \tag{7a}$$

$$\rho_0 = \frac{\Lambda}{4\pi} + \delta\rho_0 \tag{7b}$$

Equation (6), to order δR, now becomes

$$2\Lambda^{\frac{1}{2}}(\delta \ddot{R}) = -\frac{8\pi}{3}\delta\rho_0 . \tag{8}$$

Now, since the total amount of dust is conserved, the dust stress-energy obeys

$$\rho_0 R^3 = \text{constant}$$

$$\frac{\delta\rho_0}{\rho_0} = -3\frac{\delta R}{R} . \tag{9}$$

Equation (8) then gives

$$(\delta\ddot{R}) - \Lambda\delta R = 0 , \tag{10}$$

i.e. the perturbation δR grows exponentially in time. Hence, the universe is unstable.

Solution 19.32. The line element of the "Einstein universe" ($k=+1$) is (see Problem 19.5)

$$ds^2 = R^2[d\chi^2 + \sin^2\chi(d\theta^2 + \sin^2\theta\, d\phi^2)] \tag{1}$$

with

$$0 \leq \chi \leq \pi , \quad 0 \leq \theta \leq \pi , \quad 0 \leq \phi \leq 2\pi .$$

The volume V is therefore

$$V = \int_0^\pi 4\pi R^3 \sin^2\chi\, d\chi = 2\pi^2 R^3 . \tag{2}$$

From Problem 19.31, we have the result, for this universe, that

$$R = (4\pi\rho_0)^{-\frac{1}{2}} \tag{3}$$

where ρ_0 is the matter (dust) energy density. Substitution of Equation (3) into Equation (2) yields

$$V = 2\pi^2 \left(\frac{c^2}{4\pi G\rho_0}\right)^{\frac{3}{2}} \tag{4}$$

(where we have put in the factors of c and G).

Solution 19.33. Let Λ_E be the Einstein value of the cosmological constant so that the Einstein universe has $\rho_0 = \Lambda_E/4\pi$, and $R = \Lambda_E^{-\frac{1}{2}}$ (see Problem 9.31). All dust cosmologies, even with nonzero Λ, have $\rho R^3 =$ constant. Write this as

$$\rho R^3 = \frac{1+\epsilon}{4\pi\Lambda^{\frac{1}{2}}} \ . \tag{1}$$

This is equivalent to writing $\Lambda = \Lambda_E(1+\epsilon)^2$. The remaining dynamical equation of Problem 19.14 becomes

$$\dot{R}^2 = V(R) \equiv \frac{\Lambda R^2}{3} - 1 + \frac{2(1+\epsilon)}{3\Lambda^{\frac{1}{2}} R} \ . \tag{2}$$

For small R, $\dot{R}^2 \sim 1/R$, i.e. $R \sim t^{\frac{2}{3}}$. The expansion reaches a minimum when $\dot{R}$ is a minimum. Setting dV/dR equal to zero, we find that this occurs when

$$R = R_m = (1+\epsilon)^{\frac{1}{3}} \Lambda^{-\frac{1}{2}} \ . \tag{3}$$

When R is close to R_m,

$$\begin{aligned}\dot{R}^2 &\approx V(R_m) + \frac{1}{2}(R-R_m)^2 V''(R_m) + \cdots \\ &= (1+\epsilon)^{\frac{2}{3}} - 1 + [\Lambda^{\frac{1}{2}} R - (1+\epsilon)^{\frac{1}{3}}]^2 \ .\end{aligned}$$

From a table of integrals, we find

$$R = \frac{(1+\epsilon)^{\frac{1}{3}}}{\Lambda^{\frac{1}{2}}} [1 + \{1-(1+\epsilon)^{-\frac{2}{3}}\}^{\frac{1}{2}} \sinh \Lambda^{\frac{1}{2}}(t-t_m)] \tag{4}$$

where $R = R_m$ at $t = t_m$. For small ϵ, this becomes

$$R \approx R_m \left[1 + \frac{\epsilon}{3} \sinh \Lambda^{\frac{1}{2}}(t-t_m)\right] \ .$$

Thus R remains approximately equal to R_m until

$$\epsilon \sinh \Lambda^{\frac{1}{2}} (t - t_m) \approx 1$$

i.e. for a time

$$(t - t_m) \approx \Lambda^{-\frac{1}{2}} \log\left(\frac{1}{\epsilon}\right).$$

Hence if ϵ is made sufficiently small, the time can be made arbitrarily long. Eventually R continues to expand, and asymptotically

$$H^2 = \frac{\dot{R}^2}{R^2} \to \frac{1}{3}\Lambda$$

and we get the de Sitter universe of Problem 19.30.

If quasar formation occurred at $z = 2$, then $R = R_m$ at $z = 2$ i.e.

$$R_0 = R_m(1+z) = 3R_m = 3\Lambda^{-\frac{1}{2}} .$$

If this is so (with $\epsilon \ll 1$), then the density of matter now must be:

$$\rho_0 = (4\pi R_0^3 \Lambda^{\frac{1}{2}})^{-1} = (108\pi)^{-1} \Lambda$$

$$H_0 = \left(\frac{\dot{R}}{R}\right)_0 = \left(\frac{1}{3}\Lambda - \frac{1}{R_0^2} + \frac{2}{3\Lambda^{\frac{1}{2}} R_0^3}\right)^{\frac{1}{2}} = \frac{(20\Lambda)^{\frac{1}{2}}}{9} .$$

Taking $H_0 = 10^{-28}\text{cm}^{-1}$ gives $\Lambda = 4\times 10^{-56}\text{cm}^{-2}$, i.e. $\rho_0 = 1.2 \times 10^{-58}\text{cm}^{-2} = 1.6\times 10^{-30}\text{gm/cm}^3$.

Solution 19.34. The effective ρ_c density corresponding to Λ is

$$\rho_c = \frac{\Lambda}{8\pi} = \frac{10^{-57}\text{cm}^{-2}}{8\pi} \approx 4\times 10^{-31}\,\text{gm/cm}^3 .$$

Typical densities in the solar system are

$$\rho_{\text{body}} \sim 1\ \text{gm/cm}^3$$

$$\rho_{\text{solar system}} \sim 10^{33}\text{gm}/(10^{15}\text{cm})^3 \sim 10^{-12}\text{gm/cm}^3\ .$$

Thus, the effect of Λ in the solar system, in order of magnitude is

$$\frac{\rho_c}{\rho_{\text{solar system}}} \sim 10^{-19}\ .$$

Solution 19.35. The Friedmann equations (Problem 19.15) are

$$\ddot{R} = -\frac{4}{3}\pi G(\rho + 3p)R$$

$$R\ddot{R} + 2\dot{R} + 2k = 4\pi G(\rho - p)R^2\ .$$

For a static solution, all time derivatives must vanish, so we get two conditions, $\rho + 3p = 0$ and $\rho = 3k/8\pi GR^2$, which have no solution for a fluid with positive energy density and pressure (i.e. any known fluid).

Solutions 19.36 and 19.37. "Easy" cases: Since the metric is time invariant and reversible, $g_{0i} = 0$ and it has the form

$$ds^2 = -g_{00}(x^i)dt^2 + g_{ij}(x^k)dx^i dx^j\ .$$

If it is homogeneous (Problem 19.37), clocks must tick at the same rate everywhere, so $g_{00} =$ constant; if it is inhomogeneous but pressureless (Problem 19.36), it is possible to rescale the time coordinate locally to make $g_{00} =$ constant. (This was shown in Problem 16.25.) The standard formula for $\Gamma_{\mu\nu\sigma}$ now shows by inspection that the only nonvanishing components are Γ_{ijk}. (Any 0 component is either a time derivative of a component g_{0j} or a derivative of g_{00}.) Now R_{00} can be computed from the standard formula involving the Γ's; since every term contains a zero index on a Γ, it follows that $R_{00} = 0$ and hence

$$T^{\hat{0}}_{\hat{0}} - \frac{1}{2} T^{\hat{\alpha}}_{\hat{\alpha}} = 0$$

and hence $\rho + 3p = 0$. But this is impossible for a pressureless fluid, or for a physically reasonable perfect fluid (with positive p and ρ).

"Harder" cases: Let $\boldsymbol{\xi}$ be the time-symmetry Killing vector and $\mathbf{u}$ be the 4-velocity of the fluid. Since $\mathbf{u} \propto \boldsymbol{\xi}$ (Problem 13.9) and $\boldsymbol{\xi}$ is hypersurface-orthogonal, $\mathbf{u}$ is hypersurface-orthogonal and hence $\omega_{\alpha\beta} = 0$. In the Raychaudhuri equation (Problem 14.10)

$$\frac{d\theta}{d\tau} \propto \frac{\partial\theta}{\partial t} = 0 \quad ,$$

where $\boldsymbol{\xi} = \partial/\partial t$. Thus the Raychaudhuri equation becomes

$$0 = a^{\alpha}{}_{;\alpha} - 2\sigma^2 - \frac{1}{3}\theta^2 - 8\pi\left(T_{\alpha\beta} - \frac{1}{2} g_{\alpha\beta}T\right) u^{\alpha}u^{\beta} . \tag{1}$$

For $p = 0$ (Problem 19.36) we have $\mathbf{a} = 0$ from the Euler equation (Problem 14.3) and Equation (1) becomes

$$0 = -2\sigma^2 - \frac{1}{3}\theta^2 - 4\pi\rho$$

which has no solution for positive ρ.

When there is pressure, then in general $\mathbf{a}$ is nonzero; in fact,

$$\mathbf{a} = \nabla \log|\boldsymbol{\xi}\cdot\boldsymbol{\xi}|^{\frac{1}{2}} .$$

If we assume homogeneity (Problem 19.37), then $\mathbf{a} = 0$ and

$$0 = -2\sigma^2 - \frac{1}{3}\theta^2 - 4\pi(\rho + 3p) ,$$

which again has no solution for $\rho + 3p > 0$.

Solution 19.38.

(a) A galaxy moves with x^i = constant implying that $dx^i = 0$ and hence that $ds^2 = -d\tau^2$.

(b) For $d\tau = 0$, $ds^2 = g_{ij}\, dx^i dx^j$.

(c) If g_{0i} and g_{ij} are independent of x^i, then on constant τ hypersurfaces, nothing depends upon x^i; the space is homogeneous. On the other hand, a bad choice of coordinates and hence a seemingly distorted spacelike hypersurface may cause the metric functions of a homogeneous space to depend upon position.

(d) In Problem 14.9 we showed that if the spatial connecting vector between any two nearby geodesics is of constant length, then $\sigma_{\alpha\beta} = \theta = 0$ for the geodesic field. If g_{0i} and g_{ij} are independent of τ, then the connecting vector will have constant length.

(e) Since $u^0 = +1$, $u^i = 0$, we have

$$\begin{aligned} u_{i;j} &= -u^{\gamma}\Gamma_{\gamma ij} = -\Gamma_{0ij} = -\frac{1}{2}(g_{i0,j} + g_{0j,i} - g_{ij,0}) \\ &= \frac{1}{2} g_{ij,0} = \frac{1}{2} \dot{f} \bar{g}_{ij} . \end{aligned} \tag{1}$$

$$\begin{aligned} \theta \equiv u^{\alpha}{}_{;\alpha} &= |g|^{-\frac{1}{2}} (|g|^{\frac{1}{2}} u^{\alpha})_{,\alpha} = f^{-\frac{3}{2}} (f^{\frac{3}{2}})_{,0} \\ &= \frac{3}{2} \frac{\dot{f}}{f} . \end{aligned} \tag{2}$$

Now, from Equations (1), (2), and the formula (see Problem 5.18)

$$\sigma_{\alpha;\beta} = \frac{1}{2}(u_{\alpha;\mu} P^{\mu}{}_{\beta} + u_{\beta;\mu} P^{\mu}{}_{\alpha}) - \frac{1}{3} \theta P_{\alpha\beta} \tag{3}$$

we arrive at

$$\sigma_{ij} = \frac{1}{2}(u_{i;j} + u_{j;i}) - \frac{1}{3}\theta g_{ij} = \frac{1}{2}\dot{f}\bar{g}_{ij} - \frac{1}{2}\dot{f}\bar{g}_{ij} = 0 . \tag{4}$$

(f)
$$a^{\alpha} = u^{\alpha}{}_{;\beta} u^{\beta} = u^{\alpha}{}_{;0} = \Gamma^{\alpha}{}_{00}$$

$$a_{\alpha} = \Gamma_{\alpha 00} = \frac{1}{2}(g_{\alpha 0,0} + g_{\alpha 0,0} - g_{00,\alpha}) = g_{0\alpha,0} .$$

If $a_\alpha = 0$, then $g_{0\alpha,0} = 0$ and galaxies fall on geodesics if and only if $g_{0i,0} = 0$.

(g) From Problem 7.23 we showed that the necessary condition for a family of hypersurfaces to be orthogonal to $\mathbf{u}$ is $\omega_{\alpha\beta} = 0$. Since $\mathbf{u} = \partial/\partial\tau$ and the hypersurfaces of constant τ are spanned by $\partial/\partial x^i$, then $\omega_{\alpha\beta} \neq 0$ implies that $(\partial/\partial t)\cdot(\partial/\partial x^i) \neq 0$ and hence that $g_{0i} \neq 0$. If $\omega_{\alpha\beta,0} \neq 0$, then $g_{\alpha\beta,0} \neq 0$ and from part (f), this means that the galaxies do not fall on geodesics.

Solution 19.39. From Solution 14.9 we know that the spatial connecting vector, $\boldsymbol{\xi}$, between two nearby world lines obeys

$$\nabla_{\mathbf{u}}\boldsymbol{\xi} = \nabla_{\boldsymbol{\xi}}\mathbf{u} + (\boldsymbol{\xi}\cdot \mathbf{a})\mathbf{u} \ . \tag{1}$$

Putting $\boldsymbol{\xi} \equiv R\mathbf{n}$ in Equation (1) gives

$$Rn^\alpha{}_{;\beta}u^\beta + n^\alpha\dot{R} = Rn^\beta u^\alpha{}_{;\beta} + Ru^\alpha n^\gamma a_\gamma \tag{2}$$

where $\dot{R} \equiv dR/dt = u^\alpha R_{,\alpha}$. If we now dot n^α into Equation (2) and use the decomposition equation for $u_{\alpha;\beta}$ (see Problem 5.18)

$$u_{\alpha;\beta} = \omega_{\alpha\beta} + \sigma_{\alpha\beta} + \frac{1}{3}\theta\,(g_{\alpha\beta} + u_\alpha u_\beta) - a_\alpha u_\beta$$

we obtain

$$\dot{R} = Rn^\alpha n^\beta\left[\omega_{\alpha\beta} + \sigma_{\alpha\beta} + \frac{1}{3}\theta\,(g_{\alpha\beta} + u_\alpha u_\beta) - a_\alpha u_\beta\right] + Ru^\alpha n^\gamma n_\alpha a_\gamma \ . \tag{3}$$

Now the antisymmetry of $\omega_{\alpha\beta}$ and the fact that n^α is a spatial unit vector $(n^\alpha u_\alpha = 0)$ used in Equation (3) give us

$$\frac{\dot{R}}{R} = \sigma_{\alpha\beta}n^\alpha n^\beta + \frac{1}{3}\theta \ .$$

Averaging $\sigma_{\alpha\beta}n^{\alpha}n^{\beta}$ over directions produces a term proportional to the trace of $\sigma_{\alpha\beta}$. Since $\sigma_{\alpha\beta}$ is traceless, we are left with

$$\left\langle\frac{\dot{R}}{R}\right\rangle = \frac{1}{3}\theta .$$

Solution 19.40. Since $\mathbf{u} = \mathbf{e}_0$ we have

$$u_{\alpha;\beta} = u_{\alpha,\beta} - \Gamma^{\gamma}{}_{\alpha\beta}u_{\gamma} = 0 + \Gamma^{0}{}_{\alpha\beta}$$

so that

$$\theta = u_{\alpha;\beta}\,g^{\alpha\beta} = \Gamma^{0}{}_{\alpha\beta}\,g^{\alpha\beta} = -(-g)^{-\frac{1}{2}}\,[g^{0\alpha}(-g)^{\frac{1}{2}}]_{,\alpha} \tag{1}$$

(see part (f) of Problem 7.7). Hence for a Robertson-Walker cosmology

$$\theta = (-g)^{-\frac{1}{2}}\,(-g)^{\frac{1}{2}}_{,0} = 3\dot{R}/R .$$

The rotation is

$$\omega_{\alpha\beta} = P_{\beta}{}^{\gamma}u_{[\alpha;\gamma]} = 0$$

since $\Gamma^{0}{}_{\alpha\beta}$ is symmetric. For the shear we have

$$\sigma_{\alpha\beta} = P_{\beta}{}^{\gamma}u_{(\alpha;\gamma)} - \frac{1}{3}\,\theta P_{\alpha\beta}$$

$$\sigma_{0\alpha} = 0$$

$$\begin{aligned}\sigma_{ij} &= u_{(i,j)} - \frac{1}{3}\,\theta\, g_{ij}\\ &= -\frac{1}{2}\,(g_{0i,j} + g_{0j,i} - g_{ij,0}) - \frac{\dot{R}}{R}\,g_{ij}\\ &= \frac{1}{2}\,g_{ij,0} - \dot{R}/R\,g_{ij} = 0\end{aligned}$$

For the anisotropic metric, Equation (1) still holds, so

$$\theta = (-g)^{-\frac{1}{2}}\,(-g)^{\frac{1}{2}}_{,0} = \dot{a} + \dot{b} + \dot{c}$$

$$\omega_{\alpha\beta} = 0$$

$$\sigma_{0\alpha} = 0$$

$$\sigma_{ij} = \frac{1}{2}\,g_{ij,0} - \frac{1}{3}\,\theta\, g_{ij} = \frac{1}{3}\,A\, g_{ij}$$

where

$$A = \begin{cases} 2\dot{a} - \dot{b} - \dot{c} & \text{for} \quad i = j = x \\ 2\dot{b} - \dot{a} - \dot{c} & \text{for} \quad i = j = y \\ 2\dot{c} - \dot{a} - \dot{b} & \text{for} \quad i = j = z \ . \end{cases}$$

Solution 19.41. The quickest way to compute the Ricci tensor for this metric is to use the formulas of Problem 9.33. We have

$$K_{ij} = -\frac{1}{2}\dot{g}_{ij} \tag{1}$$

$$0 = R^t{}_t = K_{ij}K^{ij} - \dot{K} \tag{2}$$

$$0 = R^t{}_j = K^i{}_{j,i} - K_{,j} \tag{3}$$

$$0 = R^i{}_j = K^i{}_j K - \dot{K}^i{}_j \tag{4}$$

where a dot denotes d/dt. We have set ${}^{(3)}R_{ij}$ to zero in Equation (4) because the 3-geometry is flat, and we have replaced covariant derivatives by partial derivatives in Equation (3).

Contracting Equation (4) gives $\dot{K} = K^2$ and hence

$$K = -1/t \tag{5}$$

with a suitable choice of the constant of integration. Substituting Equation (5) back into Equation (4) gives

$$\dot{K}^i{}_j = -K^i{}_j/t$$

and hence

$$K^i{}_j = A^i{}_j/t \tag{6}$$

where $A^i{}_j$ is a constant matrix. We now choose the coordinates so that $K^i{}_j$ is diagonal at some time $t = t_0$. Then by Equation (6) $K^i{}_j$ remains diagonal and can be written in the form

$$K^i{}_j = -\delta^i{}_j P_j/t \tag{7}$$

where P_j = constant. For this equation to be consistent with Equation (5) we must have

$$\sum_i P_i = 1 \;. \tag{8}$$

Equation (3) is automatically satisfied, Equation (2) implies

$$\sum_i P_i^2 = 1 \tag{9}$$

and Equation (1) becomes

$$\dot{g}_{ij} = -2g_{im}\,K^m{}_j = 2g_{ij}\,P_j/t$$

and implies that

$$g_{ij} = \delta_{ij}\,t^{2P_i} \;.$$

Thus the final form of the metric ("Kasner metric") is

$$ds^2 = -dt^2 + t^{2P_1}dx^2 + t^{2P_2}dy^2 + t^{2P_3}dz^2 \tag{10}$$

subject to the constraints of Equations (8) and (9).

The volume of this universe is proportional to

$$[{}^{(3)}g_{ij}]^{\frac{1}{2}} = (t^{2\Sigma P_i})^{\frac{1}{2}} = t$$

which can be contrasted with the $t^{\frac{3}{2}}$ behavior of the radiation-dominated Friedmann model.

A convenient way of representing the constraints of Equations (8) and (9) is to draw a circle of radius 2/3 centered at $y = 1/3$ and to inscribe an equilateral triangle. The y-coordinates of the vertices of the triangle will satisfy Equations (8) and (9). From this we see that, except for the special case (1, 0, 0) which can easily be shown to represent flat space, two of the P's lie between 0 and 1 and one P lies between 0 and $-1/3$. Thus space must contract along one axis while expanding along the other two.

CHAPTER 20: SOLUTIONS

Solution 20.1. Note that there are no "real" (i.e. tidal) gravitational forces; the light goes in a straight line in any inertial frame, but in the accelerated frame of the tube it seems to bend.

Consider two inertial frames – one comoving with the tube at the instant the light enters it – the other comoving with the tube at the instant the light reaches the end. Since the length of the tube is seen in both frames as ℓ and since θ is always small, observers in either frame see it take ℓ/c for the light to cross the tube. At this time their relative velocity is $\beta = g\ell/c^2$. (An observer in the second frame sees the first frame moving upward with velocity β.) The direction angles of the photons path are related by (see Problem 1.8 and figure on p. 561):

$$\cos\psi = \sin\theta \approx \theta = \frac{\cos\psi' + \beta}{1+\beta\cos\psi'} = \beta .$$

Therefore

$$\theta = g\ell/c^2 , \qquad g/c^2 = 10^{-16}\,\mathrm{m}^{-1} ,$$

so, if the tube is 10 meters long $\theta = 10^{-15}$ radians, and the total displacement $\frac{1}{2}(g/c^2)\ell^2 \approx 10^{-14}$ meters. This is very small compared to the wave length of light; it is difficult to imagine interferometric (or any other) techniques capable of detecting this. When lasers are used for alignment (as in the Stanford linear accelerator) they are meant to correct for the curvature of the surface of the earth – a much larger effect, obviously, since a light ray at the earth's surface is not bent so strongly as to be pulled into a circular orbit.

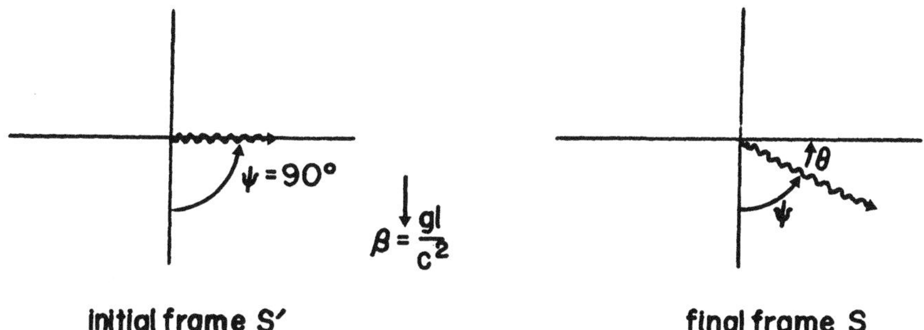

Solution 20.2.

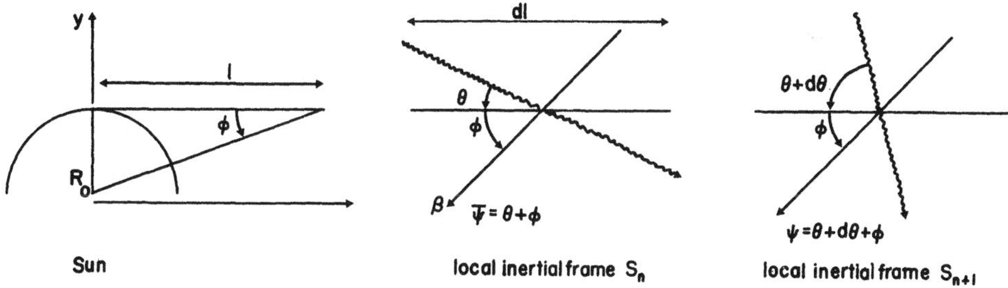

Consider a sequence of freely falling observers, each occupying a local inertial frame of length $d\ell$. Let the nth freely falling observer S_n be momentarily at rest relative to the sun when the photon hits the left side of his frame and let him see the photon move at an angle θ relative to its inertial direction (see figure). When the photon leaves his frame and enters S_{n+1}, which is now momentarily at rest, he is falling with speed $\beta = g d\ell = \frac{GM}{R^2}$ in the ϕ direction. This photon enters frame S_{n+1} moving in the direction $\theta + d\theta$: the transformation of photon angles (see Problem 1.8) is

$$\cos\psi = \frac{\cos\bar{\psi} - \beta}{1 - \beta\cos\bar{\psi}} = \cos\bar{\psi} - \beta\sin^2\bar{\psi} + \mathcal{O}(\beta^2)$$

so that

$$\cos\psi - \cos\bar{\psi} = -\beta\sin^2\bar{\psi}$$

and

$$d\theta = \beta \sin\psi \approx \beta \sin\phi = \left(\frac{GM}{R_0^2 + \ell^2}\right) \frac{R_0}{(R_0^2+\ell^2)^{\frac{1}{2}}}\, d\ell \,. \tag{1}$$

Thus the total deflection, δ is

$$\delta = \int d\theta = GMR_0 \int_{-\infty}^{\infty} \frac{d\ell}{(R_0^2+\ell^2)^{\frac{3}{2}}} = \frac{2GM}{R_0} \tag{2}$$

which is one half the observed value!

In performing the calculation we assumed that a sequence of rulers was placed along the line $y = R_0$ parallel to the line $y = 0$, with the $(n+1)^{th}$ ruler locally parallel to the nth ruler. It was the deflection with respect to these rulers that we calculated. But we did not correct for the fact that spacetime around the sun is curved (tidal forces!); in fact, a locally parallel sequence of rulers forms a curved line with respect to the $y = 0$ line and it is with respect to this curved line that we actually calculated the photon deflection. Thus the real deflection, as measured by a distant observer, must include both effects: the deflection of the photon with respect to a set of locally parallel rulers, and the bending of the line (with respect to $y = 0$) of the sequence of rulers.

Solution 20.3. We may consider purely equatorial motion with no loss of generality. In the gravitational field of the sun, the metric may then be approximated by

$$ds^2 \approx -\left(1-\frac{2M}{r}\right)dt^2 + \left(1+\frac{2M}{r}\right)(dr^2 + r^2\, d\phi^2)\,. \tag{1}$$

Null geodesics for this metric have the form

$$\frac{b}{r} = \sin\phi + \frac{2M}{r}(1-\cos\phi) \tag{2}$$

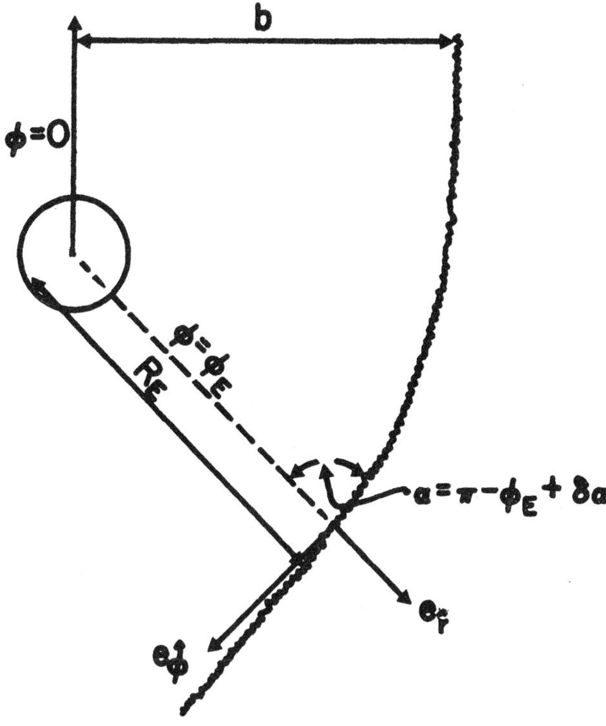

The angle α between the sun and the star (see figure), as measured by an astronomer on earth, satisfies

$$\tan\alpha = \tan(\pi - \phi_E + \delta\alpha) \approx -\tan\phi_E + \frac{\delta\alpha}{\cos^2\phi_E}$$

$$= \frac{u^{\hat{\phi}}}{u^{\hat{r}}} = \left[\frac{\left(1+\frac{M}{r}\right) r\, d\phi/d\lambda}{\left(1+\frac{M}{r}\right) dr/d\lambda}\right]_E = \left[r\,\frac{d\phi}{dr}\right]_E \tag{3}$$

where $u^{\hat{\phi}}$ and $u^{\hat{r}}$ are the orthonormal components of the photon's 4-velocity, and we have taken advantage of the isotropic form of the metric in Equation (1) to define angles. If Equation (2) is now used, Equation (3) becomes

$$\tan\phi_E - \frac{\delta\alpha}{\cos^2\phi_E} = \frac{\sin\phi_E \cdot \frac{2M}{b}(1-\cos\phi_E)}{\cos\phi_E + \frac{2M}{b}\sin\phi_E}$$

$$\approx \tan\phi_E - \frac{2M}{b}\frac{(1-\cos\phi_E)}{\cos^2\phi_E} \tag{4}$$

and hence

$$\delta\alpha \cong \frac{2M}{b}(1+\cos\alpha) \approx \frac{2M}{R_E}\left(\frac{1+\cos\alpha}{1-\cos\alpha}\right)^{\frac{1}{2}} . \tag{5}$$

It is $\delta\alpha$ which is the deflection angle measured by an astronomer at earth. [See also MTW, Section 40.3, and references given therein.]

Solution 20.4. Since we wish to compute only the lowest order deflection due to $\underset{\sim}{J}$, we can look for terms linear in $\underset{\sim}{J}$ and set $M = 0$ in the line element. The total first order deflection will then be the sum of the usual term linear in M and a term linear in $\underset{\sim}{J}$. Since the metric (for our purposes) is (see Problem 17.1),

$$ds^2 = -dt^2 - \frac{4J}{r}\sin^2\theta\, dt\, d\phi + dr^2 + r^2\, d\Omega^2 \tag{1}$$

($\underset{\sim}{J}$ points along z-axis) the equations of motion may be deduced from the variational principle (Problem 7.25)

$$\delta\int\left(-\dot{t}^2 - \frac{4J}{r}\sin^2\theta\,\dot{t}\dot{\phi} + \dot{r}^2 + r^2\dot{\theta}^2 + r^2\sin^2\theta\,\dot{\phi}^2\right)d\lambda = 0 \tag{2}$$

where $\dot{t} \equiv dt/d\lambda$ etc., and λ is affine parameter. The Euler equations for Equation (2), to first order in J, are

$$\frac{d}{d\lambda}\left(2\dot{t} + \frac{4J}{r}\sin^2\theta\,\dot{\phi}\right) = 0 \tag{3a}$$

$$\frac{d}{d\lambda}\left(2\dot{\phi}\, r^2\sin^2\theta - \frac{4J}{r}\sin^2\theta\,\dot{t}\right) = 0 , \tag{3b}$$

which imply that

$$P_0 \equiv \dot{t} + \frac{2J}{r}\sin^2\theta\,\dot{\phi} \tag{3c}$$

$$P_\phi \equiv \dot{\phi}\,r^2\sin^2\theta - \frac{2J}{r}\sin^2\theta\,P_0 \tag{3d}$$

are constants, and

$$\ddot{r} = r\dot{\theta}^2 + r\sin^2\theta\,\dot{\phi}^2 + \frac{4J}{r^2}\sin^2\theta\,\dot{\phi}\,P_0 \tag{3e}$$

$$\frac{d}{d\lambda}(r^2\dot{\theta}) = P_\phi^2\cos\theta/r^2\sin^3\theta\ . \tag{3f}$$

From Equation (3f), one may deduce a third constant of the motion (in addition to P_0 and P_ϕ):

$$P_\theta^2 + \frac{P_\phi^2}{\sin^2\theta} \equiv L^2 = \text{constant}\quad (P_\theta \equiv r^2\dot{\theta})\ . \tag{4}$$

We will now decompose the bending into three contributions, each defined by the relationship (see figure) of the ray trajectory (a straight line to lowest order) to $\underset{\sim}{J}$. (Since we are working only to linear order, the total bending must be the sum of such contributions.)

In both cases (a) and (b), the angular momentum of the light beam about the sun's center has zero projection onto $\underset{\sim}{J}$, in case (c) it has unit projection.

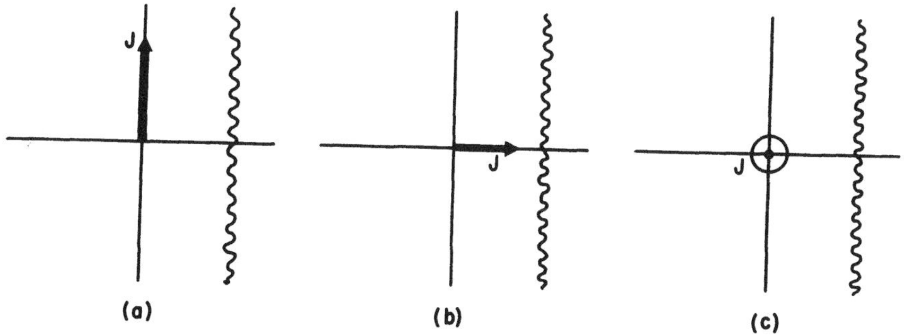

First we consider cases (a) and (b): In both cases (a) and (b) $P_\phi = 0$ (since P_ϕ is a constant of the motion). We wish to calculate $r(\theta)$. From Equation (3d) we have

$$\dot{\phi} = \frac{2JP_0}{r^3} = \mathcal{O}(J) \tag{5a}$$

and from Equation (4)

$$P_\theta = r^2\dot{\theta} = L = \text{constant} . \tag{5b}$$

Equations (3e) and (5a), to $\mathcal{O}(J)$, now give us

$$\ddot{r} = r\dot{\theta}^2 , \tag{5c}$$

but this result and Equation (5b) guarantee straight-line motion! Thus in cases (a) and (b) $\underset{\sim}{J}$ makes no contribution to the bending of light.

Now we consider case (c): In case (c), we may set $\theta = \pi/2$, and $\dot{\theta} = 0$ (Equation (3f) guarantees that $\ddot{\theta}$ will always then be 0) and solve our equations accordingly. Using

$$\frac{d}{d\lambda} = \frac{d\phi}{d\lambda}\frac{d}{d\phi} = \left(\frac{P_\phi}{r^2} + \frac{2J\,P_0}{r^3}\right)\frac{d}{d\phi} ,$$

and defining $u \equiv 1/r$, $\ell \equiv P_0/P_\phi$, we may put Equation (3e) in the form

$$u'' + u = -6u^2J\ell - 2J\ell u'^2 - 4J\ell uu'' \tag{6}$$

where $u' \equiv du/d\phi$. We will solve Equation (6) by perturbation techniques; we define

$$u = u_0 + u_1 \tag{7}$$

with $u_1 = \mathcal{O}(J)$. To lowest order, the equation is $u_0'' + u_0 = 0$ which implies

$$u_0 = \frac{1}{b}\cos\phi \tag{8}$$

where b is the impact parameter and where we have chosen our coordinate system such that the ray moves parallel to the y-axis. For the second order solution, u_1 must satisfy the equation

$$u_1'' + u_1 = -\frac{2J\ell}{b^2}(\sin^2\phi + \cos^2\phi) = -\frac{2J\ell}{b^2} \tag{9a}$$

or

$$u_1 = -\frac{2J\ell}{b^2} . \tag{9b}$$

Thus, from Equations (7), (8), (9b), the solution for u is

$$u \approx \frac{1}{b}\cos\phi - \frac{2J\ell}{b^2} . \tag{10}$$

The bending at each asymptote, α, is found by setting $u = 0$ $(r = \infty)$. At one asymptote $\phi \approx \pi/2 + \alpha$ so that $\cos\phi \approx -\alpha$ and

$$\alpha \approx -\frac{2J\ell}{b} . \tag{11}$$

There is an equal contribution at the other asymptote, so the total bending is

$$\delta\phi = 2\alpha = -\frac{4J}{b^2} \tag{12}$$

(where we have used the fact that $\ell = P_0/P_\phi \approx b^{-1}$).

From this result, and the results from cases (a) and (b), we may write the deflection of light by $\underset{\sim}{J}$ as

$$\delta\phi = -\frac{4\underset{\sim}{J}\cdot n}{b^2} \tag{13}$$

where n is a unit vector in the direction of the angular momentum of the ray about the Sun's center.

Note that the dimensionless small parameter in which we have been expanding is

$$\left(\frac{J}{b^2}\right)_{\text{sun}} = \frac{1.7\times 10^{48}\,\text{gm-cm}^2/\text{sec}}{(7\times 10^{10}\,\text{cm})^2} \approx 10^{-12} \approx 2\times 10^{-7}\,\text{arc sec} .$$

Solution 20.5. The index of refraction of a plasma is $n = (1-\nu_p^2/\nu^2)^{\frac{1}{2}}$ where ν_p is the plasma frequency $\nu_p^2 \equiv n_e e^2/\pi m_e$. The well-known equation describing the path of a ray through an inhomogeneous medium is [see, e.g. B. Rossi, *Optics* (Addison-Wesley, 1957), p. 54]

$$d(n\underset{\sim}{m})/d\ell = \underset{\sim}{\nabla} n \tag{1}$$

where $\underset{\sim}{m}$ is a unit vector tangent to the ray and ℓ is the distance along the ray. Since the deflection angle we are computing is small, we can correctly find the first-order change in $\underset{\sim}{m}$ by integrating Equation (1) along the unperturbed trajectory $y = b$, $-\infty < x < +\infty$. From the formula given for n_e we have

$$n \approx 1 - \frac{1}{2}\frac{\nu_p^2}{\nu^2} \approx 1 - .0101\left(\frac{r}{R_\odot}\right)^{-6.5}\left(\frac{\nu}{10^9\,\mathrm{Hz}}\right)^{-2}$$

or

$$\underset{\sim}{\nabla} n \approx \frac{.0657}{R_\odot}\left(\frac{r}{R_\odot}\right)^{-7.5}\left(\frac{\nu}{10^9\,\mathrm{Hz}}\right)^{-2}\hat{r}\,. \tag{2}$$

Thus, using $n = 1$ at $\pm\infty$ and $r = (x^2+y^2)^{\frac{1}{2}}$,

$$\underset{\sim}{m}|_{+\infty} - \underset{\sim}{m}|_{-\infty} \approx \frac{.0657}{R_\odot}\left(\frac{\nu}{10^9\,\mathrm{Hz}}\right)^{-2}\int_{-\infty}^{\infty}\left(\frac{r}{R_\odot}\right)^{-7.5}\left[\hat{x}\left(\frac{x}{r}\right) + \hat{y}\left(\frac{y}{r}\right)\right]dx\,. \tag{3}$$

The term in $\hat{x}$ vanishes by symmetry in $\pm x$, while the $\hat{y}$ term can be written as $2\int_{-\infty}^{\infty}$. Since $\underset{\sim}{m}$ is a unit vector, for small angles $(\underset{\sim}{m}|_\infty - \underset{\sim}{m}|_{-0})\cdot\hat{y}$ is the scattering angle we seek. Defining

$$\eta \equiv b/R_\odot\,, \qquad z \equiv r/b$$

and using

$$y = b\,,\quad x^2 = b^2(z^2-1)\,,\quad dx = bz(z^2-1)^{-\frac{1}{2}}\,dz\,,$$

we get

$$\theta_{\text{coronal}} \approx .131 \left(\frac{\nu}{10^9\,\text{Hz}}\right)^{-2} \eta^{-6.5} \int_1^\infty z^{-7.5}(z^2-1)^{-\frac{1}{2}}\,dz\,. \tag{4}$$

On putting $t = 1/z^2$, we recognize the integral in Equation (4) as a beta function, $\frac{1}{2}$ B(3.75, .5) = .158. Thus our final expression for the coronal deflection is

$$\theta_{\text{coronal}} \approx .021 \left(\frac{\nu}{10^9\,\text{Hz}}\right)^{-2} \eta^{-6.5}\,. \tag{5}$$

The general relativistic deflection, θ_{GR}, is

$$\theta_{GR} \approx \frac{8.5\times10^{-6}}{\eta} \tag{6}$$

(see e.g. Problem 15.6). The two deflections are equal when

$$\frac{8.5\times10^{-6}}{\eta} \approx .021 \left(\frac{\nu}{10^9\,\text{Hz}}\right)^{-2} \eta^{-6.5}$$

or

$$\eta \equiv \frac{b}{R_\odot} \approx 4.1 \left(\frac{\nu}{10^9\,\text{Hz}}\right)^{-.36}\,.$$

For smaller impact parameters b, the coronal deflection dominates.

Solution 20.6.

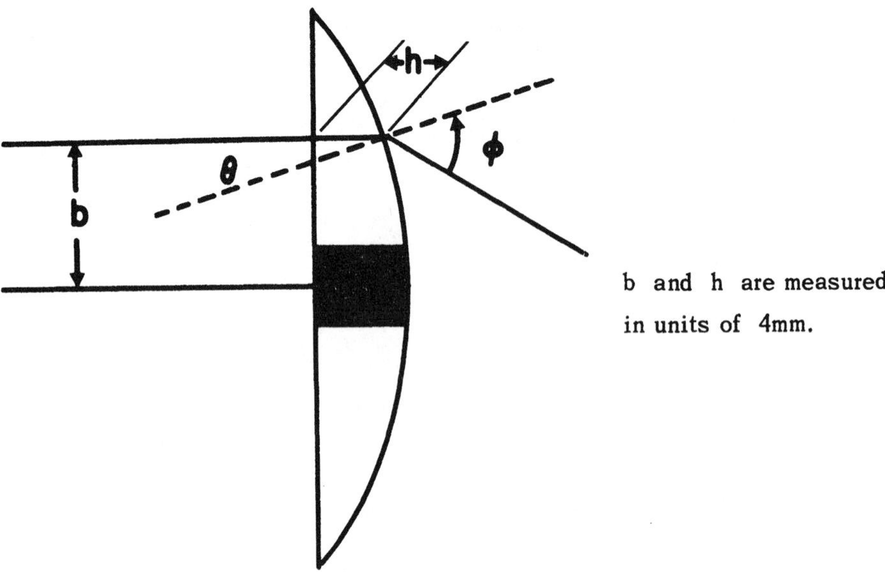

b and h are measured in units of 4mm.

The relation of ϕ and θ (see figure) is given by Snell's law

$$n = \phi/\theta \ . \tag{1}$$

The refraction required to simulate the sun's gravitational bending is

$$\delta \equiv \phi - \theta = 1.75''/b \ . \tag{2}$$

From Equations (1) and (2) θ must be

$$\theta = 1.75''/b(n-1) = 1.6 \times 10^{-5}/b \ . \tag{3}$$

From the geometry of the figure $\theta = -dh/db$, so Equation (3) gives

$$\frac{dh}{db} = -\frac{1.6 \times 10^{-5}}{b} \ ,$$

$$h = h_0 - 1.6 \times 10^{-5} \log(b) \ .$$

Solution 20.7. In the Schwarzschild metric, in curvature coordinates, the equation for noncircular orbits may be put into the form (see e.g. solution to Problem 15.9)

$$u'' + u = \frac{M}{\tilde{L}^2} + 3Mu^2 \equiv a + bu^2 \qquad (1)$$

where $u \equiv 1/r$, $u'' \equiv d^2u/d\phi^2$, and $\tilde{L} \equiv$ angular momentum per unit mass of the orbiting particle (= constant). The solution to the lowest order, the linear portion of Equation (1), is

$$u_0 = a + A\cos\phi \qquad (2)$$

where A is a constant. The second order solution to Equation (1), u_1, then satisfies the equation

$$u_1'' + u_1 \approx bu_0^2 = b(a^2 + 2aA\cos\phi + A^2\cos^2\phi)$$

$$= ba^2 + \frac{1}{2}bA^2 + 2abA\cos\phi + \frac{1}{2}bA^2\cos 2\phi\ . \qquad (3)$$

The inhomogeneous solution to Equation (3) is

$$u_1 = b\left(a^2 + \frac{1}{2}A^2\right) + abA\phi\sin\phi - (bA^2/6)\cos 2\phi \qquad (4)$$

and

$$u = \frac{1}{r} \approx u_0 + u_1\ . \qquad (5)$$

The second term in u_1 causes the nonperiodicity; we may put it into more useful form by use of the identity

$$\epsilon\phi\sin\phi = \cos(\phi - \epsilon\phi) - \cos\phi + \mathcal{O}(\epsilon^2)\,. \qquad (6)$$

Since ab is a small quantity according to Equation (1), Equation (6) may be used to put Equation (4) into the form

$$u_1 \approx b\left(a^2 + \frac{1}{2}A^2\right) + A\cos(\phi - ab\phi) - A\cos\phi - (bA^2/6)\cos 2\phi \quad (7)$$

and the perihelion shift, per revolution, is

$$\delta\phi = 2\pi ab = 6\pi M^2/\tilde{L}^2 . \tag{8}$$

We may express $\tilde{L}$ in terms of the semimajor axis a and eccentricity e of the orbit using standard classical mechanics

$$\tilde{L}^2 = Ma(1-e^2) . \tag{9}$$

Equation (8) then becomes

$$\delta\phi = \frac{6\pi M}{a(1-e^2)} , \tag{10}$$

which in the limit of small e agrees with the solution of Problem 15.7.

Solution 20.8.

(a) Yes. First note that particle mass is constant:

$$\frac{dm^2}{d\tau} = -\frac{dp\cdot p}{d\tau} = -2p\cdot\frac{dp}{d\tau} = 0 .$$

Since $p = mu$ and m is constant, a factor of m can be taken out of each term in the force equation leaving

$$du^\mu/d\tau = -(\eta^{\mu\nu}\Phi_{,\nu} + u^\alpha\Phi_{,\alpha}u^\mu)$$

and showing that particle motion is independent of mass.

(b) Yes. Near the earth the earth is flat. Take z to be in the vertical direction, then we have $\Phi = \Phi(z)$ and the "source equation" for Φ is $\partial^2\Phi/\partial z^2 = 0$. Above the earth's surface then, the potential is

$$\Phi = az + b .$$

The constant a is determined by dropping a massive particle from rest and setting its initial acceleration to g. With $u^0 = 1$, $u^j = 0$ in the force equation, we have

$$-mg = dp^z/d\tau = -m\Phi_{,z} = -ma$$

and hence $a = g$.

Now consider a photon moving vertically upward. It loses energy

$$dp^0 = -p^\alpha \Phi_{,\alpha} dx^0 = -p^z \Phi_{,z} dt$$

in time $dt = dz$. Since $p^z = p^0$ for a photon, the energy loss is then

$$dp^0 = -p^0 g\, dz .$$

For an earth-based experiment $g\Delta z << 1$ and

$$\Delta p^0/p^0 = \Delta\nu/\nu = -g\Delta z$$

in agreement with the Pound-Rebka experiment.

(c) No. For a photon $p_\beta dx^\beta = 0$ so the path of a photon is governed by

$$dp^\mu = -p^\alpha \Phi_{,\alpha} dx^\mu = -dx^\alpha \Phi_{,\alpha} p^\mu .$$

Since dp is proportional to p, the photon must move in a straight line; there is no bending.

[Note: Also in disagreement with observations, this theory predicts a *regression* of the perihelion of a planet's motion, of magnitude

$$\delta = 2\pi M_\odot^2/J^2$$

where δ = shift per orbit, J = angular momentum of planet per unit mass. For nearly circular orbits, this becomes

$$\delta = 2\pi M_\odot/r$$

which is $\approx$ 13″/century for Mercury's orbit.]

Solution 20.9. Any standardized clock can be calibrated against a clock at infinity (which measures time t). Then the ratio of the ticking rate of a clock at (r, θ), 4-velocity $\mathbf{u}$, to the rate of a clock at infinity is

$$\frac{d\tau}{dt} = \frac{1}{u^0(r,\theta)} . \tag{1}$$

In Solution 16.19, we showed that u^0 = constant on the surface of a rigidly rotating, equilibrium configuration of a perfect fluid. Thus

$$d\tau/dt = \text{constant}$$

and all clocks on the surface of the earth have the same ticking rate! (The doppler shift effect exactly cancels the redshift effect.)

CHAPTER 21: SOLUTIONS

Solution 21.1.

(i) One way of finding the formula for the derivative of a determinant is from the identity

$$\det A = e^{\mathrm{Tr}\{\log A\}}$$

valid for any nonsingular matrix A. (Here $\log A$ is the inverse function of e^A, where the exponential is defined as a power series.) Thus we have

$$\begin{aligned}\delta(\det A) &= e^{\mathrm{Tr}\{\log A\}}\,\delta(\mathrm{Tr}\{\log A\}) \\ &= (\det A)\,\mathrm{Tr}(A^{-1}\delta A)\end{aligned}$$

and for $g \equiv \det(g_{\alpha\beta})$

$$\delta g = g\, g^{\alpha\beta}\,\delta g_{\alpha\beta}$$

where we have used the symmetry of $g_{\alpha\beta}$. The desired result follows immediately:

$$\begin{aligned}\delta(-g)^{\frac{1}{2}} &= -\frac{1}{2}(-g)^{-\frac{1}{2}}\,\delta g \\ &= \frac{1}{2}(-g)^{\frac{1}{2}}\,g^{\alpha\beta}\,\delta g_{\alpha\beta}\ .\end{aligned}$$

(ii) From the definition of $g^{\alpha\beta}$,

$$g^{\alpha\beta}\,g_{\beta\gamma} = \delta^{\alpha}{}_{\gamma}\ ,$$

we have

$$\delta g^{\alpha\beta}\,g_{\beta\gamma} + g^{\alpha\beta}\,\delta g_{\beta\gamma} = 0$$

Multiplication by $g^{\gamma\delta}$ gives

$$\delta g^{\alpha\beta}\,\delta_{\beta}^{\delta} + g^{\gamma\delta}\,g^{\alpha\beta}\,\delta g_{\alpha\beta} = 0$$

and hence the answer:

$$\delta g^{\alpha\delta} = -g^{\gamma\delta}\,g^{\alpha\beta}\,\delta g_{\beta\gamma} \;.$$

Solution 21.2. The variation in L due to changes in Φ^A is given by

$$(-g)^{\frac{1}{2}}\delta L\,(\Phi^A, \Phi^A{}_{,\mu}, g_{\mu\nu}) = \left(\frac{\partial L}{\partial \Phi^A}\,\delta\Phi^A + \frac{\partial L}{\partial \Phi^A{}_{,\mu}}\,\delta\Phi^A{}_{,\mu}\right)(-g)^{\frac{1}{2}} \;. \qquad (1)$$

The operation of varying Φ^A commutes with partial differentiation

$$\delta(\Phi^A{}_{,\mu}) = (\delta\Phi^A)_{,\mu}$$

so Equation (1) becomes

$$(-g)^{\frac{1}{2}}\delta L = (-g)^{\frac{1}{2}}\,\frac{\partial L}{\partial \Phi^A}\,\delta\Phi^A + \left(\frac{\partial L(-g)^{\frac{1}{2}}}{\partial \Phi^A{}_{,\mu}}\,\delta\Phi^A\right)_{,\mu} - \frac{\partial}{\partial x^{\mu}}\left(\frac{\partial L(-g)^{\frac{1}{2}}}{\partial \Phi^A{}_{,\mu}}\right)\delta\Phi^A \,.$$

The second term is a perfect divergence, which gives a surface term in the integral δS, and hence is zero since $\delta\Phi^A$ is taken to vanish on the surface. Thus

$$0 = \frac{\delta L}{\delta \Phi^A} = \frac{\partial L}{\partial \Phi^A} - \frac{\partial}{\partial x^{\mu}}\left(\frac{\partial L(-g)^{\frac{1}{2}}}{\partial \Phi^A{}_{,\mu}}\right)\frac{1}{(-g)^{\frac{1}{2}}} \;.$$

Solution 21.3. Since $L(-g)^{\frac{1}{2}}\,d^4x$ is a scalar, it is unchanged under the infinitesimal coordinate transformation

$$\bar{x}^{\mu} = x^{\mu} + \xi^{\mu} \;.$$

By renaming the dummy variable of integration, $d^4\bar{x}$ can be changed back to d^4x. The net change in $L(-g)^{\frac{1}{2}}$ is therefore the Lie derivative (see Problem 8.17), so that

$$0 = \delta S = \int \mathcal{L}_{\boldsymbol{\xi}}(L(-g)^{\frac{1}{2}})d^4x$$

$$= \int \left[\frac{\delta L}{\delta \Phi^A}(\mathcal{L}_{\boldsymbol{\xi}}\Phi^A)(-g)^{\frac{1}{2}} + \frac{\delta(L(-g)^{\frac{1}{2}})}{\delta g_{\mu\nu}} \mathcal{L}_{\boldsymbol{\xi}} g_{\mu\nu} \right] d^4x \, . \tag{1}$$

The first term in the integrand vanishes by the equation of motion (Problem 21.2), thus

$$0 = \frac{1}{2} \int T^{\mu\nu} \mathcal{L}_{\boldsymbol{\xi}} g_{\mu\nu} (-g)^{\frac{1}{2}} d^4x \, .$$

Since

$$\mathcal{L}_{\boldsymbol{\xi}} g_{\mu\nu} = \xi_{\mu;\nu} + \xi_{\nu;\mu}$$

we get

$$0 = \int T^{\mu\nu}\xi_{\mu;\nu}(-g)^{\frac{1}{2}} d^4x$$

$$= \int (T^{\mu\nu}\xi_\mu)_{;\nu}(-g)^{\frac{1}{2}} d^4x - \int T^{\mu\nu}{}_{;\nu}\xi_\mu(-g)^{\frac{1}{2}} d^4x \, . \tag{2}$$

The first term is

$$\int (T^{\mu\nu}\xi_\mu)_{;\nu}(-g)^{\frac{1}{2}} d^4x = \int ((-g)^{\frac{1}{2}} T^{\mu\nu}\xi_\mu)_{,\nu} d^4x$$

and gives a surface integral which vanishes. Since ξ_μ is arbitrary, Equation (2) then gives

$$T^{\mu\nu}{}_{;\nu} = 0 \, .$$

Solution 21.4.

(a) Under the independent variations

$$g_{\mu\nu} \to g_{\mu\nu} + \delta g_{\mu\nu}$$

$$\Gamma^\alpha{}_{\mu\nu} \to \Gamma^\alpha{}_{\mu\nu} + \delta\Gamma^\alpha{}_{\mu\nu}$$

we have

$$\delta((-g)^{\frac{1}{2}} R) = (-g)^{\frac{1}{2}} g^{\mu\nu}\delta R_{\mu\nu} + R\delta(-g)^{\frac{1}{2}} + (-g)^{\frac{1}{2}} R_{\mu\nu}\delta g^{\mu\nu} \, . \tag{1}$$

Since

$$R_{\mu\nu} = \Gamma^{a}_{\mu\nu,a} - \Gamma^{a}_{\mu a,\nu} + \Gamma^{a}_{a\sigma}\Gamma^{\sigma}_{\mu\nu} - \Gamma^{a}_{\nu\sigma}\Gamma^{\sigma}_{\mu a}, \tag{2}$$

the first term on the right-hand side of Equation (1) becomes

$$\begin{aligned}(-g)^{\frac{1}{2}} g^{\mu\nu}\delta R_{\mu\nu} &= ((-g)^{\frac{1}{2}} g^{\mu\nu}\delta\Gamma^{a}_{\mu\nu})_{,a} - ((-g)^{\frac{1}{2}} g^{\mu\nu}\delta\Gamma^{a}_{\mu a})_{,\nu} \\ &- ((-g)^{\frac{1}{2}} g^{\mu\nu})_{,a}\delta\Gamma^{a}_{\mu\nu} + ((-g)^{\frac{1}{2}} g^{\mu\nu})_{,\nu}\delta\Gamma^{a}_{\mu a} \\ &+ (-g)^{\frac{1}{2}} g^{\mu\nu}(\delta\Gamma^{a}_{a\sigma}\Gamma^{\sigma}_{\mu\nu} + \Gamma^{a}_{a\sigma}\delta\Gamma^{\sigma}_{\mu\nu} - \delta\Gamma^{a}_{\nu\sigma}\Gamma^{\sigma}_{\mu a} - \Gamma^{a}_{\nu\sigma}\delta\Gamma^{\sigma}_{\mu a}).\end{aligned} \tag{3}$$

The first two terms on the right-hand side of Equation (3) are divergences, which therefore do not contribute to δS. (The integral of a divergence can be converted to a surface integral and $\delta\Gamma = 0$ on the surface of the region of integration.) By appropriate relabeling of dummy indices, Equation (3) can be written

$$(-g)^{\frac{1}{2}} g^{\mu\nu}\delta R_{\mu\nu} = (-g)^{\frac{1}{2}} (A^{\mu\nu}_{\sigma} + \delta^{\nu}_{\sigma} B^{\mu})\delta\Gamma^{\sigma}_{\mu\nu} \tag{4}$$

where

$$A^{\mu\nu}_{\sigma} \equiv g^{\mu\nu}\Gamma^{a}_{a\sigma} - (-g)^{-\frac{1}{2}} [(-g)^{\frac{1}{2}} g^{\mu\nu}]_{,\sigma} - g^{a\nu}\Gamma^{\mu}_{a\sigma} - g^{a\mu}\Gamma^{\nu}_{a\sigma} \tag{5}$$

$$B^{\mu} \equiv (-g)^{-\frac{1}{2}} [(-g)^{\frac{1}{2}} g^{\mu\beta}]_{,\beta} + g^{a\beta}\Gamma^{\mu}_{a\beta}\,. \tag{6}$$

By Problem 21.1 we have

$$\delta[(-g)^{\frac{1}{2}}] = -\frac{1}{2}(-g)^{\frac{1}{2}} g_{\mu\nu}\delta g^{\mu\nu} \tag{7}$$

so

$$\begin{aligned}0 = \delta S &= (16\pi)^{-1}\int (-g)^{\frac{1}{2}} d^4x\,(A^{\mu\nu}_{\sigma} + \delta^{\nu}_{\sigma}B^{\mu})\delta\Gamma^{\sigma}_{\mu\nu} \\ &+ (16\pi)^{-1}\int (-g)^{\frac{1}{2}} d^4x\left(R_{\mu\nu} - \frac{1}{2} g_{\mu\nu} R\right)\delta g^{\mu\nu} \\ &+ \int \frac{\delta(L_{\text{matter}}(-g)^{\frac{1}{2}})}{\delta g^{\mu\nu}}\delta g^{\mu\nu} d^4x\,.\end{aligned} \tag{8}$$

Since $\delta\Gamma^{\sigma}{}_{\mu\nu}$ is symmetric, the first term in Equation (8) gives

$$A^{\mu\nu}_{\sigma} + \frac{1}{2}\,\delta^{\nu}_{\sigma}B^{\mu} + \frac{1}{2}\,\delta^{\mu}_{\sigma}B^{\nu} = 0\,. \tag{9}$$

Contracting on σ and ν in Equation (9) yields

$$A^{\mu\sigma}_{\sigma} + 2B^{\mu} + \frac{1}{2}\,B^{\mu} = 0\,. \tag{10}$$

But contracting Equation (5) and comparing with Equation (6) shows

$$A^{\mu\sigma}_{\sigma} = -B^{\mu}\,. \tag{11}$$

Equations (10) and (11) imply $B^{\mu} = 0$, so by Equation (9)

$$A^{\mu\nu}_{\sigma} = 0\,. \tag{12}$$

If the explicit expression in Equation (5) is used, Equation (12) can be written

$$g^{\mu\nu}C_{\sigma} = g^{\mu\nu}{}_{,\sigma} + g^{\alpha\nu}\Gamma^{\mu}{}_{\alpha\sigma} + g^{\alpha\mu}\Gamma^{\nu}{}_{\alpha\sigma} \tag{13}$$

where

$$C_{\sigma} \equiv (-g)^{-\frac{1}{2}}\,(-g)^{\frac{1}{2}}_{,\sigma} - \Gamma^{\alpha}{}_{\alpha\sigma}\,. \tag{14}$$

When we multiply Equation (13) by $g_{\mu\nu}$ we find that

$$\begin{aligned} 4C_{\sigma} &= g^{\mu\nu}{}_{,\sigma}g_{\mu\nu} + 2\Gamma^{\alpha}{}_{\alpha\sigma} \\ &= -2(-g)^{-\frac{1}{2}}\,(-g)^{\frac{1}{2}}_{,\sigma} + 2\Gamma^{\alpha}{}_{\alpha\sigma} \\ &= -2C_{\sigma} \end{aligned}$$

(where Problem 21.1 has been used in the second line) and therefore that $C_{\sigma} = 0$. If we define

$$\Gamma_{\lambda\alpha\sigma} \equiv g_{\lambda\mu}\Gamma^{\mu}{}_{\alpha\sigma} \tag{15}$$

and if we now multiply Equation (13) by $g_{\lambda\mu}\,g_{\gamma\nu}$, we get

$$0 = -g_{\gamma\lambda,\sigma} + \Gamma_{\lambda\gamma\sigma} + \Gamma_{\gamma\lambda\sigma} \tag{16}$$

and thus

$$\frac{1}{2}(g_{\gamma\lambda,\sigma} + g_{\gamma\sigma,\lambda} - g_{\sigma\lambda,\gamma}) = \Gamma_{\gamma\lambda\sigma} \tag{17}$$

which is one of the required results.

We now return to Equation (8), and use Problem 21.3:

$$\begin{aligned}\frac{\delta(L_{\text{matter}}(-g)^{\frac{1}{2}})}{\delta g_{\mu\nu}}\delta g_{\mu\nu} &= \frac{\delta(L_{\text{matter}}(-g)^{\frac{1}{2}})}{\delta g^{\mu\nu}}\delta g^{\mu\nu} = \frac{1}{2}(-g)^{\frac{1}{2}} T^{\mu\nu}\delta g_{\mu\nu} \\ &= -\frac{1}{2}(-g)^{\frac{1}{2}} T_{\mu\nu}\delta g^{\mu\nu}.\end{aligned} \tag{18}$$

We conclude that

$$(16\pi)^{-1}\left(R_{\mu\nu} - \frac{1}{2} g_{\mu\nu} R\right) - \frac{1}{2} T_{\mu\nu} = 0$$

or

$$G_{\mu\nu} = 8\pi T_{\mu\nu}. \tag{19}$$

(b) Although $\Gamma^{\alpha}{}_{\mu\nu}$ (now taken to be the Christoffel symbols formed from $g_{\alpha\beta}$) is not a tensor, $\delta\Gamma^{\alpha}{}_{\mu\nu}$ is a tensor (Problem 8.26). Thus we can simplify the calculation of $\delta R_{\mu\nu}$ by working in a local inertial frame, where the Γ's vanish. In this frame, Equation (2) gives

$$\delta R_{\mu\nu} = \delta\Gamma^{\alpha}{}_{\mu\nu,\alpha} - \delta\Gamma^{\alpha}{}_{\mu\alpha,\nu} = \delta\Gamma^{\alpha}{}_{\mu\nu;\alpha} - \delta\Gamma^{\alpha}{}_{\mu\alpha;\nu}. \tag{20}$$

Since this is a tensor equation, it is valid in any coordinate system. The Γ's are Christoffel symbols, so $g_{\mu\nu;\alpha} = 0$ and

$$\begin{aligned}(-g)^{\frac{1}{2}} g^{\mu\nu}\delta R_{\mu\nu} &= (-g)^{\frac{1}{2}}[(g^{\mu\nu}\delta\Gamma^{\alpha}{}_{\mu\nu})_{;\alpha} - (\delta\Gamma^{\alpha}{}_{\mu\alpha} g^{\mu\nu})_{;\nu}] \\ &= ((-g)^{\frac{1}{2}} g^{\mu\nu}\delta\Gamma^{\alpha}{}_{\mu\nu})_{,\alpha} - (\delta\Gamma^{\alpha}{}_{\mu\alpha} g^{\mu\nu}(-g)^{\frac{1}{2}})_{,\nu}\end{aligned}$$

(where we have used Problem 7.7 (g) in the second line). Since this is a perfect divergence, it does not contribute to δS. The remaining terms in Equations (1) and (18) lead to Equation (19) as in part (a).

Solution 21.5. Taking the variation of the action

$$S = -(8\pi)^{-1} \int (\Phi_{;\alpha}\Phi^{;\alpha} + m^2\Phi^2)(-g)^{\frac{1}{2}}\, d^4x$$

gives us

$$0 = \frac{\delta L}{\delta \Phi} = \left[2m^2\Phi - 2(-g)^{-\frac{1}{2}}[(-g)^{\frac{1}{2}}\Phi^{,\alpha}]_{,\alpha}\right](-g)^{\frac{1}{2}}$$

or

$$\Box\Phi - m^2\Phi = 0 \; . \tag{1}$$

From Problem 21.3 the stress-energy is

$$\begin{aligned} T_{\mu\nu} &\equiv -2(-g)^{-\frac{1}{2}} \frac{\delta[(-g)^{\frac{1}{2}}L]}{\delta g^{\mu\nu}} \\ &= \frac{1}{4\pi}(-g)^{-\frac{1}{2}}\left[\Phi_{,\mu}\Phi_{,\nu}(-g)^{\frac{1}{2}} - \frac{1}{2}(\Phi_{,\alpha}\Phi^{,\alpha} + m^2\Phi^2)(-g)^{\frac{1}{2}}\, g_{\mu\nu}\right] \\ &= \frac{1}{4\pi}\left[\Phi_{,\mu}\Phi_{,\nu} - \frac{1}{2}\, g_{\mu\nu}(\Phi_{,\alpha}\Phi^{,\alpha} + m^2\Phi^2)\right] \end{aligned} \tag{2}$$

and hence its divergence is

$$T^{\mu\nu}{}_{;\nu} = \frac{1}{4\pi}(\Phi^{,\mu}{}_{;\nu}\Phi^{,\nu} + \Phi^{,\mu}\Box\Phi - \Phi_{,\alpha}{}^{;\mu}\Phi^{,\alpha} - m^2\Phi\Phi^{,\mu}) \; . \tag{3}$$

For a scalar field Φ, we also have the relation

$$\Phi^{,\mu}{}_{;\nu} = \Phi_{,\nu}{}^{;\mu} , \tag{4}$$

thus Equation (3) becomes

$$T^{\mu\nu}{}_{;\nu} = \frac{\Phi^{,\mu}}{4\pi}(\Box\Phi - m^2\Phi) \tag{5}$$

which, by Equation (1), vanishes.

Solution 21.6. Varying the action integral gives us

$$\frac{\delta(L[-g]^{\frac{1}{2}})}{\delta A_\mu} = -\frac{1}{4\pi}\frac{\delta\left(A_{[\mu;\nu]}A^{[\mu;\nu]}[-g]^{\frac{1}{2}}\right)}{\delta A_\mu} = -\frac{1}{4\pi}\frac{\delta\left(A_{[\mu,\nu]}A^{[\mu,\nu]}[-g]^{\frac{1}{2}}\right)}{\delta A_\mu}$$

$$= -\frac{1}{4\pi}\cdot 2\cdot 2\cdot \frac{\partial}{\partial x^\mu}\left(A^{[\mu,\nu]}[-g]^{\frac{1}{2}}\right) = -\frac{1}{2\pi}\frac{\partial}{\partial x^\mu}(F^{\nu\mu}[-g]^{\frac{1}{2}})$$

$$= -\frac{[-g]^{\frac{1}{2}}}{2\pi} F^{\nu\mu}{}_{;\mu} , \qquad (1)$$

where we have used Problem 7.7 (i).

The prescription for the stress-energy gives

$$T_{\alpha\beta} \equiv -2\frac{\delta L}{\delta g^{\alpha\beta}} + Lg_{\alpha\beta} = \frac{1}{8\pi}\frac{\delta(F_{\mu\nu}F_{\sigma\tau}g^{\mu\sigma}g^{\nu\tau})}{\delta g^{\alpha\beta}} - \frac{1}{16\pi}F_{\mu\nu}F^{\mu\nu}g_{\alpha\beta}$$

$$= \frac{1}{4\pi}F_\alpha{}^\nu F_{\beta\nu} - \frac{1}{16\pi}F_{\mu\nu}F^{\mu\nu}g_{\alpha\beta}. \qquad (2)$$

Consider now the Lagrangian density

$$L = \left(-\frac{1}{16\pi}F_{\mu\nu}F^{\mu\nu} - \frac{1}{4\pi}F^{\mu\nu}A_{\mu;\nu}\right)(-g)^{\frac{1}{2}} . \qquad (3)$$

Variation of L with respect to A_μ yields

$$F^{\mu\nu}{}_{;\nu} = 0 . \qquad (4)$$

Variation of L with respect to $F^{\mu\nu}$ yields

$$F^{\mu\nu} = 2A^{[\nu;\mu]} . \qquad (5)$$

Solution 21.7. Variation of the Lagrangian with respect to Φ yields

$$(-g)^{\frac{1}{2}}\left(R + \frac{\omega}{\Phi^2}\Phi_{,\alpha}\Phi^{,\alpha}\right) = -2\omega\frac{\partial}{\partial x^\alpha}\left(\frac{\Phi^{,\alpha}(-g)^{\frac{1}{2}}}{\Phi}\right)$$

and thus

$$2\omega\Phi^{-1}\Box\Phi - \omega\Phi^{-2}\Phi_{,\alpha}\Phi^{,\alpha} + R = 0 . \qquad (1)$$

(We have used here

$$\Box\Phi \equiv \Phi_{,\alpha}{}^{;\alpha} = \frac{1}{(-g)^{\frac{1}{2}}} \left(\Phi^{,\alpha}(-g)^{\frac{1}{2}}\right)_{,\alpha} .\Big)$$

Variation of L with respect to $g_{\alpha\beta}$ involves three terms:

(i) $$\frac{\delta(-\omega\Phi^{,\alpha}\Phi^{,\beta}g_{\alpha\beta}\Phi^{-1}(-g)^{\frac{1}{2}})}{\delta g_{\alpha\beta}} = -\frac{\omega}{\Phi}\left[\Phi^{,\alpha}\Phi^{,\beta}(-g)^{\frac{1}{2}} - \frac{1}{2}\Phi^{,\mu}\Phi_{,\mu}g^{\alpha\beta}(-g)^{\frac{1}{2}}\right] \tag{2a}$$

and

(ii) $$\delta(\Phi R(-g)^{\frac{1}{2}}) = (-g)^{\frac{1}{2}}\Phi\, G^{\alpha\beta}\delta g_{\alpha\beta} + \Phi(-g)^{\frac{1}{2}} g^{\alpha\beta}\delta R_{\alpha\beta} , \tag{2b}$$

$$(-g)^{\frac{1}{2}} g^{\alpha\beta}\delta R_{\alpha\beta} = \frac{\partial}{\partial x^{\gamma}}\left[(-g)^{\frac{1}{2}}(g^{\alpha\beta}\delta\Gamma^{\gamma}{}_{\alpha\beta} - g^{\alpha\gamma}\delta\Gamma^{\tau}{}_{\alpha\tau})\right] \tag{2c}$$

(see Problem 21.4). Also, we have from the expression for the Γ's as derivatives of the g's

$$\begin{aligned}\delta\Gamma^{\lambda}{}_{\mu\nu} &= -g^{\lambda\rho}\delta g_{\rho\sigma}\Gamma^{\sigma}{}_{\mu\nu} + \frac{1}{2}g^{\lambda\rho}[(\delta g_{\rho\mu})_{,\nu} + (\delta g_{\rho\nu})_{,\mu} - (\delta g_{\mu\nu})_{,\rho}] \\ &= \frac{1}{2}g^{\lambda\rho}[(\delta g_{\rho\mu})_{;\nu} + (\delta g_{\rho\nu})_{;\mu} - (\delta g_{\mu\nu})_{;\rho}] .\end{aligned} \tag{2d}$$

Using Equations (2c) and (2d), we can then write

$$\Phi(-g)^{\frac{1}{2}} g^{\alpha\beta}\delta R_{\alpha\beta} \to \delta g^{\alpha\beta}(-\Phi_{,\alpha;\beta} + g_{\alpha\beta}\Box\Phi)(-g)^{\frac{1}{2}} \tag{2e}$$

where $\to$ denotes the operation of dropping all exact divergences (which vanish when integrated over). Combining Equations (2b) and (2e) we have

$$\delta(\Phi R(-g)^{\frac{1}{2}}) = (-g)^{\frac{1}{2}}\delta g^{\alpha\beta}[-\Phi_{,\alpha;\beta} + g_{\alpha\beta}\Box\Phi + \Phi G_{\alpha\beta}] . \tag{2f}$$

The remaining term needed is

(iii) $$16\pi\frac{\delta(L_{\text{matter}}(-g)^{\frac{1}{2}})}{\delta g_{\alpha\beta}} = -8\pi(-g)^{\frac{1}{2}}\, T^{\alpha\beta} . \tag{2g}$$

From Equations (2a) and (2f), and (2g), we get the field equations obtained by varying $g^{\alpha\beta}$:

$$G_{\alpha\beta} + \Phi^{-1}(g_{\alpha\beta}\,\Box\Phi - \Phi_{,\alpha;\beta}) - \omega\,\Phi^{-2}\left(\Phi_{,\alpha}\Phi_{,\beta} - \frac{1}{2}\,g_{\alpha\beta}\Phi_{,\mu}\Phi^{,\mu}\right) - 8\pi\Phi^{-1}\,T_{\alpha\beta} = 0\,. \tag{3}$$

Solution 21.8. For simplicity choose coordinates (Gaussian normal coordinates; see Problem 8.25) in which the metric takes the form $ds^2 = dn^2 + {}^{(3)}g_{ij}\,dx^i dx^j$. The coordinates $x^i(i=1,2,3)$ span the surface layer which is taken to be at $n = 0$. From the initial value equations we have

$$G^i_{\ j} = {}^{(3)}G^i_{\ j} + (K^i_{\ j} - \delta^i_{\ j}\,\mathrm{Tr}K)_{,n} - (\mathrm{Tr}K)\,K^i_{\ j} + \frac{1}{2}\,\delta^i_{\ j}(\mathrm{Tr}K)^2 + \frac{1}{2}\,\delta^i_{\ j}\,\mathrm{Tr}(K^2) = 8\pi\,T^i_{\ j}\,.$$

We now integrate over n from $-\epsilon$ to $+\epsilon$, using the fact that n represents proper distance perpendicular to the 3-surface, and we let $\epsilon \to 0$. Since the intrinsic geometry of the 3-surface is well defined, the only term that survives this limiting integration is the term involving the derivatives of the extrinsic curvature, so for $[\mathbf{K}]$ the discontinuity in $\mathbf{K}$, we find

$$[K^i_{\ j}] - \delta^i_{\ j}\,\mathrm{Tr}\,[K] = 8\pi\,S^i_{\ j}$$

or

$$\mathrm{Tr}\,[K] = -4\pi\,S_i^{\ i} = -4\pi\,{}^{(3)}g_{ij}\,S^{ij}$$

and

$$[K^i_{\ j}] = 8\pi\left(S^i_{\ j} - \frac{1}{2}\,\delta^i_{\ j}\,S^k_{\ k}\right)\,.$$

Solution 21.9. From the initial value equations, we have

$$G^n_{\ i} = -\{K_i^{\ m}{}_{|m} - (\mathrm{Tr}K)_{,i}\} = -\{K_i^{\ m} - \delta_i^{\ m}\,\mathrm{Tr}(K)\}_{|m} = 8\pi\,T^n_{\ i}\,.$$

The discontinuity of this equation across the 3-surface is

$$\{[K_i^{\ m}] - \delta_i^{\ m}\,\mathrm{Tr}\,[K]\}_{|m} = -8\pi\,[T^n{}_i]\ .$$

By combining this with our result from Problem 21.8, we have then

$$S_i^{\ m}{}_{|m} = -\,[T^n{}_i]\ .$$

Solution 21.10. The surface stress energy of the thin shell of dust is $S^{\alpha\beta} = \sigma u^\alpha u^\beta$. From the solution to Problem 21.8, we have

$$[K^i_{\ j}] = 8\pi\left\{\sigma u^i u_j - \frac{1}{2}\,\delta^i_{\ j}\,(-\sigma)\right\}$$

or

$$[K_{ij}] = 8\pi\sigma\left\{u_i u_j + \frac{1}{2}\,{}^{(3)}g_{ij}\right\}. \tag{1}$$

Since there is no stress-energy outside the shell $[T^n{}_i] = 0$ and the equation of motion (see Problem 21.9) gives us

$$0 = S_i^{\ m}{}_{|m} = \sigma_{,m} u^m u_i + \sigma u^m{}_{|m} u_i + \sigma u_{i|m} u^m$$

$$= \frac{d\sigma}{d\tau}\, u_i + \sigma u_i u^m{}_{|m} + \sigma u_{i|m} u^m \tag{2}$$

By contracting this with $\mathbf{u}$ we verify the second relation

$$\frac{d\sigma}{d\tau} + \sigma u^m{}_{|m} = 0\ . \tag{3}$$

Comparison with Equation (2) shows that $u_{i|m} u^m = 0$. Thus

$$\mathbf{a} = \nabla_{\mathbf{u}}\mathbf{u} = u^i\,\nabla_i(u^j \mathbf{e}_j) = u^i(u^j{}_{|i}\mathbf{e}_j + u^j K_{ij}\,\mathbf{n}) = u^i u^j K_{ij}\,\mathbf{n}\ . \tag{4}$$

(We have used Equation (2) of Solution 9.32 to get the second-last equality.) The third relation now follows from Equations (1) and (4):

$$\mathbf{a}^+ - \mathbf{a}^- = u^i u^j\,[K_{ij}]\,\mathbf{n} = 4\pi\sigma\,\mathbf{n}\ .$$

To verify the fourth relation we need to show

$$(K^+_{ij} + K^-_{ij})\,u^i u^j = 0 .$$

We do this by using the initial value equation $[G^n{}_n] = 8\pi[T^n{}_n] = 0$. Since ${}^{(3)}R$ is continuous, this gives us

$$\begin{aligned}
0 &= [\mathrm{Tr}(K^2) - (\mathrm{Tr}K)^2] \\
&= K^{i+}_j K^{j+}_i - K^{i-}_j K^{j-}_i - (K^{i+}_i)^2 + (K^{i-}_i)^2 \\
&= (K^{i+}_j + K^{i-}_j)\{K^{j+}_i - K^{i-}_j - {}^{(3)}g^{ij}(K^{a+}_a - K^{a-}_a)\} \\
&= (K^{i+}_j + K^{i-}_j)\{[K^j_i] - \delta^j_i\,\mathrm{Tr}[K]\} ,
\end{aligned}$$

where we have used the easily verified fact that K^{i+}_j and K^{i-}_j commute. Since

$$[K^j{}_i] = 8\pi\sigma\left(u^j u_i + \frac{1}{2}\,{}^{(3)}\delta^j_i\right) .$$

it follows that

$$[K^i{}_i] = 4\pi\sigma$$

and that

$$[K^j{}_i] - \delta^j{}_i\,[K^i{}_i] = 8\pi\sigma\,(u^j u_i)$$

and therefore finally

$$(K^{i+}_j + K^{i-}_j)\,u^j u_i = 0 .$$

Solution 21.11. If τ is proper time measured by an observer at rest in the dust, then the motion of the shell is specified by $r(\tau)$, where r is the radial coordinate of the shell in either the interior or exterior metric. The metric of the shell can be specified by

$$ds^2 = -d\tau^2 + R^2(\tau)(d\theta^2 + \sin^2\theta\, d\phi^2) .$$

Here $4\pi R^2$ = the surface area of the shell at τ, and thus $R(\tau) = r(\tau)$.

From the second relation in Problem 21.10 we have that

$$0 = \frac{d\sigma}{d\tau} + \sigma u^i{}_{|i} = (\sigma u^i)_{|i} = (\sigma ({}^{(3)}g)^{\frac{1}{2}} u^i)_{,i} / ({}^{(3)}g)^{\frac{1}{2}} .$$

But $u^\tau = 1$ and ${}^{(3)}g = R^4(\tau)$ so that $(\sigma R^2)_{,\tau} = 0$. This implies that $4\pi R^2 \sigma = \mu$, which we identify with the rest mass of the shell, is a constant.

To find the equation of motion we use the junction condition from Problem 21.10

$$[K_{\theta\theta}] = 8\pi\sigma\left(u_\theta u_\theta + \frac{1}{2}\,{}^{(3)}g_{\theta\theta}\right) = 4\pi\sigma\,{}^{(3)}g_{\theta\theta} = \mu \ .$$

Evaluating $K_{\theta\theta}$ we have

$$K_{\theta\theta} = -n_{\theta;\theta} = n_\alpha \Gamma^\alpha{}_{\theta\theta} = -\frac{1}{2}\, n^r g_{\theta\theta,r} = -r\, n^r$$

and

$$[K_{\theta\theta}] = -r(n^{r+} - n^{r-}) = \mu$$

where n^{r+} and n^{r-} are the radial components of the normal evaluated in the exterior and the interior geometry.

By using $\mathbf{u}\cdot\mathbf{n} = 0$ and $\mathbf{n}\cdot\mathbf{n} = -\mathbf{u}\cdot\mathbf{u} = 1$, we can evaluate the components of $\mathbf{u}$ and $\mathbf{n}$. Exterior to the shell we have

$$1 = \left(1 - \frac{2M}{r}\right)(u^t)^2 - (u^r)^2 / \left(1 - \frac{2M}{r}\right)$$

$$0 = n_r u^r + n_t u^t$$

$$1 = -\left(1 - \frac{2M}{r}\right)^{-1} (n_t)^2 + \left(1 - \frac{2M}{r}\right)(n_r)^2 \ ,$$

and by eliminating u^t, n_t we find

$$n_r^+ = \left[\frac{1 + (u^r)^2 / \left(1 - \frac{2M}{r}\right)}{\left(1 - \frac{2M}{r}\right)}\right]^{\frac{1}{2}} .$$

On the shell $r = R(\tau)$ and $u^r = dR/d\tau \equiv \dot{R}$ so the contravariant component of **n**, exterior to the shell is

$$n^{r+} = \left(1 - \frac{2M}{R} + \dot{R}^2\right)^{\frac{1}{2}} .$$

The calculation for n^{r-} is identical, and we get the same answer as above with M set to zero. Thus we have

$$\mu = -R(n^{r+} - n^{r-}) = -R\left\{\left[1 - \frac{2M}{R} + \dot{R}^2\right]^{\frac{1}{2}} - [1 + \dot{R}^2]^{\frac{1}{2}}\right\}$$

which can be solved for M to give the equation of motion

$$M = \mu(1 + \dot{R}^2)^{\frac{1}{2}} - \mu^2/2R .$$

If $R = 0$ at $R = \infty$ (i.e. infall from rest at ∞), then $M = \mu$ and

$$\dot{R} = -\frac{\mu}{2R}\left(1 + \frac{4R}{\mu}\right)^{\frac{1}{2}} .$$

This can easily be integrated to give

$$\frac{4}{\mu}\tau = -\frac{1}{3}\left(1 + \frac{4R}{\mu}\right)^{\frac{3}{2}} + \left(1 + \frac{4R}{\mu}\right)^{\frac{1}{2}} - \frac{2}{3} ,$$

where we have taken $\tau = 0$ to correspond to $R = 0$.

Solution 21.12. To find an instantaneous metric, we need only to consider the metric and field equations on a spacelike Cauchy surface S, the surface of time symmetry. Six of the ten Einstein field equations determine the metric off S; these we do not need to worry about. The remaining four equations are initial-value equations and can be written in the form (see Equations (6) and (7) of Problem 9.33)

$$\frac{1}{2}\,{}^{(3)}R - \frac{1}{2}(K^i{}_i)^2 + \frac{1}{2}K_\ell{}^m K^\ell{}_m = 0 \tag{1a}$$

$$K_i{}^m{}_{|m} - K^m{}_{m|i} = 0 , \tag{1b}$$

where ${}^{(3)}R$ is the curvature scalar on S, $K^i_{\ m}$ is the extrinsic curvature, a slash represents covariant differentiation with respect to the three geometry ${}^{(3)}g_{ij}$ on S, and where we have used the vacuum initial value equations.

At a moment of time symmetry the (timelike) normal to S, from which K_{ij} is computed, goes into minus itself under time reversal. Thus $K_{ij} = -K_{ij}$ and hence the extrinsic curvature of S must vanish. For $K_{ij} = 0$ on S, Equation (1b) is trivially satisfied and Equation (1a) takes the form

$$ {}^{(3)}R = 0 \, . \tag{2} $$

As a solution to Equation (2), we begin with the *ansatz*

$$ g_{ij} = \Phi^4 \eta_{ij} \tag{3} $$

and see if there is a Φ which satisfies Equation (2):

$$ {}^{(3)}R = g^{ij}(\Gamma^m{}_{im,j} - \Gamma^m{}_{ij,m} + \Gamma^m{}_{\ell i}\Gamma^\ell{}_{jm} - \Gamma^\ell{}_{m\ell}\Gamma^m{}_{ij}) \, . \tag{4} $$

With the *ansatz* of Equation (3) we have, for the Christoffel symbols

$$ \Gamma^m{}_{\ell i} = 2\Phi^{-1}(\Phi_{,i}\delta^m{}_i + \Phi_{,\ell}\delta^m{}_i - \Phi^{,m}\eta_{\ell i}) \, , \tag{5a} $$

$$ g^{ij}\Gamma^m{}_{im,j} = -6\Phi^{-6}(\nabla\Phi)^2 + 6\Phi^{-5}\nabla^2\Phi \tag{5b} $$

$$ g^{ij}\Gamma^m{}_{ij,m} = 2\Phi^{-6}(\nabla\Phi)^2 - 2\Phi^{-5}\nabla^2\Phi \tag{5c} $$

$$ g^{ij}\Gamma^m{}_{\ell i}\Gamma^\ell{}_{jm} = -4\Phi^{-6}(\nabla\Phi)^2 \tag{5d} $$

$$ g^{ij}\Gamma^\ell{}_{m\ell}\Gamma^m{}_{ij} = -12\Phi^{-6}(\nabla\Phi)^2 \tag{5e} $$

where we have used the symbols

$$(\nabla\Phi)^2 \equiv \Phi_{,i}\Phi_{,j}\eta^{ij} \tag{5f}$$

$$\nabla^2\Phi \equiv \Phi_{,i,j}\eta^{ij} . \tag{5g}$$

Combining Equations (4)-(5), we see that Equation (2) reduces to

$$^{(3)}R = 0 = 8\Phi^{-5}\nabla^2\Phi , \tag{6}$$

which is satisfied by any solution of Laplace's equation. One such solution is

$$\Phi = 1 + \sum_i \frac{M_i}{2r_i} \tag{7}$$

leading to a metric

$$g_{ij} = \left(1 + \sum_i \frac{M_i}{2r_i}\right)^4 \eta_{ij} . \tag{8}$$

The lowest order term in M/r for the metric in Equation (8) is

$$g_{ij} \approx \left(1 + \sum_i \frac{2M_i}{r_i}\right) \eta_{ij}$$

which is the "Post-Newtonian" approximation to the spatial part of a metric generated by point masses M_i located at positions r_i. Thus the metric in Equation (8) is, in fact, the instantaneous spatial metric for point masses at arbitrary positions at a moment of time symmetry.

Solution 21.13. Begin by noticing that the square of U^{α}, which is a Lorentz invariant, is just the determinant of $U^{AA'}$. The analog of the Minkowski metric will then be a set of matrices which serve to raise and lower indices in constructing the determinant of $U^{AA'}$:

$$\delta^C_D \det(U^{AB}) = U^{FC}U^{EG}\varepsilon_{FE}\varepsilon_{DG}$$

where

$$\varepsilon_{FE} = \begin{pmatrix} 0 & 1 \\ -1 & 0 \end{pmatrix} \qquad \delta^C_D = \begin{pmatrix} 1 & 0 \\ 0 & 1 \end{pmatrix} .$$

Evidently, a pair of ε matrices plays the role of the Minkowski metric.

To find the analog of a Lorentz transformation, denote such a transformation by

$$U^{F'C'} = L^{F'C'}_{FC}\, U^{FC} .$$

Then, invariance of the dot product of two vectors $\mathbf{U}, \mathbf{V}$, requires

$$U^{AB}V^{CD}\varepsilon_{AC}\varepsilon_{BD} = L^{E'F'}_{EF}\, U^{EF} L^{G'H'}_{GH}\, V^{GH}\varepsilon_{E'G'}\varepsilon_{F'H'} .$$

Thus we must have

$$\varepsilon_{EG}\varepsilon_{FH} = L^{E'F'}_{EF}\, L^{G'A'}_{GH}\,\varepsilon_{E'G'}\varepsilon_{F'H'} .$$

This implies that

$$\delta^M_E\delta^K_H = L^{E'F'}_{EF}\, L^{G'A'}_{GH}\,\varepsilon_{E'G'}\varepsilon_{F'H'}\varepsilon^{GM}\varepsilon^{FK}$$

and hence that

$$\det(L^{E'F'}_{EF}) = 1 .$$

Thus the Lorentz transformation matrices must be unimodular (of unit determinant).

Solution 21.14.

(a) The result follows from the fact that the indices B, C and D range over only two values and hence cannot all be different.

(b) Writing out the result of part (a) and using the antisymmetry of ε_{AB} gives

$$\varepsilon_{AB}\varepsilon_{CD} + \varepsilon_{AC}\varepsilon_{DB} + \varepsilon_{AD}\varepsilon_{BC} = 0 .$$

Contracting now with ξ^{CD} gives (watch out for the signs!)

$$-\varepsilon_{AB}\xi^C_C + \xi_{AB} - \xi_{BA} = 0 .$$

Hence we have

$$\xi_{AB} = \xi_{(AB)} + \xi_{[AB]} = \xi_{(AB)} + \frac{1}{2}\varepsilon_{AB}\xi^C_C .$$

Solution 21.15. Since $\varepsilon_{ab}^{\ \ cd} = i(\delta_A^C \delta_B^D \delta_{B'}^{C'} \delta_{A'}^{D'} - \delta_B^C \delta_A^D \delta_{A'}^{C'} \delta_{B'}^{D'})$ (see e.g. F. A. E. Pirani, *Lectures on General Relativity*, p. 315) and $*T_{ab} = \frac{i}{2} \varepsilon_{ab}^{\ \ cd} T_{cd}$, the result follows immediately.

Solution 21.16. Since primed and unprimed indices commute, we have

$$\begin{aligned} T_{(ab)} &= \frac{1}{2}(T_{ABA'B'} + T_{BAB'A'}) \\ &= \frac{1}{4}(T_{ABA'B'} - T_{BAA'B'} + T_{BAB'A'} - T_{ABB'A'}) \\ &\quad + \frac{1}{2} T_{(ab)} + \frac{1}{4}(T_{BAA'B'} + T_{ABB'A'}) \ . \end{aligned}$$

With the result of Problem 21.14, this can be written as

$$\begin{aligned} T_{(ab)} &= 2T_{[AB][A'B']} + \frac{1}{2}(T_{BAA'B'} + T_{ABB'A'}) \\ &= \frac{1}{2} \varepsilon_{AB} \varepsilon_{A'B'} T_C{}^C{}_{C'}{}^{C'} + \frac{1}{2}(T_{BAA'B'} + T_{ABB'A'}) \ . \end{aligned}$$

Finally, using the result of Problem 21.15, we get

$$\begin{aligned} T_{BAA'B'} &= T_{(ab)} - \frac{1}{2} \varepsilon_{AB} \varepsilon_{A'B'} T_C{}^C{}_{C'}{}^{C'} + *T_{ab} \\ &= T_{(ab)} - \frac{1}{2} g_{ab} T_c^c + i *T_{ab} \ . \end{aligned}$$

INDEX

References are to problem numbers, except that integers (1.0, 2.0, etc.) refer to chapter introductions (Chapter 1, Chapter 2, etc.).